Electric Machines

STEADY-STATE THEORY AND DYNAMIC PERFORMANCE

Electric Machines

STEADY-STATE THEORY AND DYNAMIC PERFORMANCE

Mulukutla S. Sarma

NORTHEASTERN UNIVERSITY

CENGAGE
Learning™

Australia • Brazil • Japan • Korea • Mexico • Singapore • Spain • United Kingdom • United States

Electric Machines: Steady-State Theory and Dynamic Performance
Mulukutla S. Sarma

Sponsoring Editor: Bill Barter

Marketing Manager: Nathan Wilbur

Manufacturing Buyer: Andrew Christensen

For product information and technology assistance, contact us at
Cengage Learning Customer & Sales Support, 1-800-354-9706

For permission to use material from this text or product, submit all requests online at **www.cengage.com/permissions**
Further permissions questions can be e-mailed to
permissionrequest@cengage.com

Library of Congress Control Number: 93-28863

ISBN-13: 978-0-534-93843-7

ISBN-10: 0-534-93843-4

Cengage Learning
5191 Natorp Boulevard
Mason, OH 45040
USA

Cengage Learning is a leading provider of customized learning solutions with office locations around the globe, including Singapore, the United Kingdom, Australia, Mexico, Brazil, and Japan. Locate your local office at **www.cengage.com/global**

Cengage Learning products are represented in Canada by Nelson Education, Ltd.

To learn more about Cengage Learning, visit **www.cengage.com**

Purchase any of our products at your local college store or at our preferred online store **www.cengagebrain.com**

Printed in the United States of America
7 8 9 10

To My Children,
Satish, Santi, and Suresh
And My Wife,
Savitri

Contents

Appendix B   Special Machines   599

Answers to Odd-Numbered Problems   619

Index   639

Preface

I have always enjoyed engineering (teaching, research, and consultation), particularly because the subject area involves devices that move rather than static pieces of equipment such as electronic amplifiers. To some of our young would-be scientists and engineers who are inclined to believe that electrical machines is a difficult or dull subject and one in which the fundamental research has all been done, I have tried to convey some of the thrills of exploring the field of electromechanical energy conversion and, in most cases, succeeded in rapidly dispelling their misconceived ideas. The urgent need for updating of power engineering programs and production of highly competent engineers capable of working on the frontiers of fast-developing, competitive technology has been recognized.

Electrical machinery is one of the subject areas that has been most affected by changing curricula patterns. Whereas it once formed the core of the electrical engineering syllabus, with as many as three or four courses devoted to the subject, it must now constitute a more modest portion of the curriculum so that more time may be devoted to covering the vast range of developments in other fields. Nowadays there is barely a single course assigned for electrical machines, and in some rare instances a two-course sequence appears. Coupled with this significant reduction in time, however, is the increased scope of the subject to include steady-state performance as well as dynamic behavior of electrical machines, which is essential for analyzing the overall stability of a system of which the machine happens to be an integral part. The study of electric machines continues to be one of the most fundamental subjects in an engineering curriculum because it provides a basic physical and mathematical understanding of electromechanical energy conversion.

Objectives

As in the first edition, the focus is on electromechanical energy conversion, aimed at the typical undergraduate electrical engineering course of one or two semesters, providing a match with modern engineering curricula. The full text satisfies the needs of those whose training requires a deep understanding of various aspects of machine behavior for further study of power systems, control systems, robotics, power semiconductor-controlled drives, machine design, or general industrial applications. The main objective of this book is then to present the fundamentals of electromechanics, apply these to some of the basic configurations of electromechanical devices, and stimulate the reader for further investigation of new and complex situations with exciting possibilities for advanced development.

▄▄▄▄▄▄▄▄▄▄▄▄▄▄IIIIIIIII
Prerequisites and Background

In a typical undergraduate program, students will have taken courses in electric network theory including transient analysis, basic electromagnetic fields, and ordinary differential equations by the time a course on electric machines is offered. Besides a review of phasor diagrams, a brief chapter on three-phase circuits as well as some discussion on the magnetic aspect including magnetic circuits and permanent magnets are presented to aid the students in orienting themselves to the study of electromagnetic machines. Offering the much-needed background material right at the beginning of the text is considered by many users as a unique useful feature of this book.

▄▄▄▄▄▄▄▄▄▄▄▄▄▄IIIIIIII
Organization and Flexibility

The book is particularly developed to be student-oriented, comprehensive, and up-to-date, with necessary and sufficient detailed explanation at the level for which it is intended. The key word in the organization of the text is *flexibility.*

The book is divided into three parts in order to provide flexibility in meeting different circumstances, needs, and desires. A glance at the table of contents will show that Part 1 concerns itself with the basic concepts, including the magnetic aspect, magnetically-coupled systems and transformers, principles of voltage generation, and torque production, as well as the operating principles of elementary synchronous machines, direct-current machines, and induction machines. A chapter is also devoted to present the essentials of machine windings, as they form a unifying link between different machines. Part 2 of the text is a thorough treatment of steady-state theory and performance of basic rotating machinery: induction machines, synchronous machines, and direct-current machines. An introduction to the machine-design aspect of each of these devices is also included. Finally, Part 3 is devoted entirely to the treatment of the dynamic behavior of the electromechanical devices; A chapter is dedicated to the power semiconductor-controlled drives because of the special importance and development of power electronics in the present age.

The chapters need not be used in sequence with curricula in which prerequisite courses cover magnetic and/or three-phase circuits; the corresponding chapters in Part 1 may be skipped altogether. Without any loss of continuity in understanding the basic principles of electric machines, one need not cover the chapter on machine windings. In Parts 2 and 3, the sequence of presenting the induction machines, the synchronous machines, or the direct-current machines may be adapted to suit the circumstances, changing needs, and particular desires. The internal organization of each chapter also contributes to the flexibility.

▄▄▄▄▄▄II
Features

- The readability of the text and the level of presentation from the student's viewpoint—major strengths of the original edition—have been retained.
- Introductory chapters on phasor diagrams and magnetic aspect reinforce material previously learned by the students.
- Coverage on permanent magnets, multiwinding transformers, machine windings, and special machines, which are considered unique additions to the text in its original form, is kept in place.

- Most space is devoted to careful explanation of four main electromagnetic devices: the transformer, the induction motor, the synchronous machine, and the direct-current machine. A sound knowledge of these should permit the ready understanding of all other electric machines, which for the most part are modifications of these four types.

- The quantity of the subject matter, range of difficulty, coverage of topics, numerous illustrations and photographs, a large number of comprehensive worked examples, and a variety of end-of-chapter problems, which are viewed as major strengths of the first edition, are retained in the second edition.

- Similarities among main kinds of machines are brought out wherever possible. Similar lines of treatment are followed for each machine where this is practicable, in spite of the distinguishing features peculiar to each class of machines. Elementary concepts of the basic rotating machines have been presented under the chapter on "Principles of Electromechanical Energy Conversion," before detailed presentation on their steady-state theory and performance is taken up in Part 2 of the text.

- Fundamental physical concepts that underlie creative engineering—the most valuable as well as most permanent part of a student's background—are highlighted while giving due attention to mathematical techniques. To accomplish this in a shorter time than formerly available, much thought has gone into rationalizing the theory and conveying in a concise manner the essential details concerning the physical nature of machines. With a good grounding in basic concepts, a very wide range of engineering systems can be understood, analyzed, and devised.

- The theory has been developed from simple beginnings in such a manner that it can readily be extended to new and complex situations. The art of reducing a practical device to an appropriate mathematical model and recognizing its limitations has been adequately presented. Most of the models developed are in the form of equivalent circuits so that nonlinear parameters can be represented directly in relation to the controlling variables. Sufficient motivation is provided for the student to develop interest in the analytical procedures to be applied, and to realize that all models, being approximate representations of reality, should be no more complex than necessary for the application at hand.

- Consistent with modern practice, the International (SI) system of units has been used throughout the book. An appendix, "Units, Constants, and Conversion Factors for the SI System," is included for ready reference by the student.

- An appendix is devoted to some of the "Special Machines" that have features that distinguish them from the more conventional types.

- Broader aspects of electromechanical energy conversion are given due attention in order to present a proper perspective and motivate the reader to think beyond the devices that are treated in the text.

- Providing a match with the modern engineering curricula, steady-state performance as well as dynamic behavior of electric machines (which is essential for analyzing the overall stability of a system of which the machine happens to be an integral part) are presented in sufficient detail, giving the student a basic physical and mathematical understanding of electromechanical energy conversion.

- Since the essence of engineering is *design,* the end objective of each phase of preparatory study should be to increase the student's capability to design useful devices and systems to meet the needs of society. Toward that end, the student is motivated to go through the sequence of understanding physical processes, modeling, using analytical techniques, and finally, designing.

- Each vector and phasor quantity has been denoted by the use of a normal letter with a bar above the symbol rather than the common use of boldface type, since instructors and students cannot usually write in boldface.

So, What's New?

- A chapter on power semiconductor-controlled drives has been added, including a brief discussion of power electronic devices. This is in line with the modern development of the topics of drives and power electronics.

- An introduction to the machine-design aspect of transformers, induction machines, synchronous machines, and direct-current machines is added in the corresponding chapters to give the student a feel for the design aspect. Based on the presented material, the student could easily be motivated to pursue design projects on various electromagnetic devices. This unique addition of the design aspect and design-oriented problems should fill the need of the ABET-accreditation requirements on design methodologies.

- Machine applications have been added to the coverage of each of the electromagnetic devices, giving the student a sense of practical usage.

- New topics such as amorphous-steel core transformers, brushless dc motors, superconducting generators, stepper motors, and computer methods of fields analysis have been introduced.

- Dynamic analysis of dc machines with time-domain techniques has been added to complement the existing material presented on dynamic analysis with Laplace-transform-domain approach.

- The per-unit system and phase shift as applied to three-phase transformers have been included.

- A significant number of new problems have been added to the already substantial number at the end of each chapter. The problems draw on all the concepts of understanding, modeling, analysis, and design. Answers for the odd-numbered problems are included at the end of the text to reassure the reader of progress.

- A new solutions manual has been developed for the exclusive use of the instructors using the text for their courses. It is strongly suggested that the solution manual not be made available to students.

- A new laboratory manual based on the Faraday machines laboratory benches has been developed as a supplement to the text.

- The bibliography and references have been updated.

Pedagogy

The text is divided into three parts:

Part 1: Introduction   This part develops the basic background that is needed for the study of electromagnetic devices, and covers the principles of electromechanical energy conversion including elementary concepts of rotating machines.

Analysis by means of phasor diagrams is a very useful approach for the understanding of mathematical equivalent-circuit models of practical devices. As such, *A Review of Phasor Diagrams* is appropriate at the very beginning.

The modern interest in electronics has influenced circuit texts and courses to the detriment of the study of polyphase systems. Since this topic is fundamental to the study of commercial and industrial power applications that utilize polyphase devices, a chapter is devoted for the analysis of *Balanced Three-Phase Circuits.*

Electromagnetic devices utilize magnetic materials for shaping the magnetic fields that act as the medium for transferring and converting energy. The interaction between electric circuits and magnetic fields plays an important part in the operation of various types of equipment discussed in this book. That is why a chapter on *The Magnetic Aspect* including magnetic circuits and permanent magnets is presented.

The *Transformer,* although not an electromechanical energy conversion device, is an important auxiliary in the overall problem of energy conversion for the transformation of electrical power between different voltage-current levels. Concepts of transformer behavior and the analysis of mutually coupled circuits serve as a useful adjunct to the study of electromagnetic rotating machines. Besides single-phase and three-phase transformers, the topics of autotransformer and multiwinding transformers are also included along with some special topics, such as pulse transformers, under the chapter on transformers. The design aspect is also considered.

Machine windings are the means by which the theory is translated into practice, and they form a unifying link between different machines. Exciting as they are by their very nature, the last chapter in Part 1 is devoted for the essentials of these.

Part 2: Steady-State Theory and Performance   With the background in Part 1, the student is then directed in Part 2 to a thorough treatment of steady-state theory and performance of rotating machinery.

Induction Machines are presented first using the generalized-transformer approach. Besides polyphase machines, single-phase machines are also discussed in this chapter. The design aspect is also included.

Then *Synchronous Machines* are taken up including the effects of saliency and saturation. Steady-state stability and excitation systems are also discussed. The design aspect is also touched upon.

Finally, *Direct-Current Machines* are considered along with commutator action and armature reaction. The design aspect is also presented.

Part 3: Transients and Dynamics   This part is devoted entirely to the treatment of the dynamic behavior of electromechanical devices studied in Part 2.

Transients and Dynamics of AC Machines are first discussed based on mathematical-model descriptions of three-phase synchronous and induction machines.

Then the text goes into *Direct-Current Machine Dynamics* based on dynamic models and analysis. Methadynes and amplidynes are also introduced to the student in this chapter.

The last chapter in the text deals with *Power Semiconductor-Controlled Drives,* including an introduction to the power electronic devices. Then the solid-state control of dc, induction, and synchronous motors is discussed in sufficient detail.

Appendix A contains units, constants, and conversion factors for the SI system. Appendix B presents some special machines that have features which distinguish them from the more conventional types.

Chapter Introductions

Each chapter is introduced with a clear statement of its objectives, giving a sense of what to expect, and motivating the student with enough information to look forward to reading the chapter.

Illustrations

A large number of illustrations and photos have been included to support the subject matter and to give the student a practical feel for the devices and to motivate the student to pursue the topics further.

Examples

Numerous comprehensive examples have been worked out in detail in the text covering most of the theoretical points raised. An appropriate level of difficulty is chosen and sufficient stimulation is built in to go on to more challenging situations.

End-of-Chapter Problems

A good number of problems with proper graded-level of difficulty is included at the end of each chapter, allowing considerable flexibility to the instructors in assigning to their students.

Design-Oriented Problems

Based on the machine-design aspect introduced for the devices (transformers, induction machines, synchronous machines, and direct-current machines), design-oriented problems have been added to reinforce the suggested procedures. Beyond that, students could easily be motivated to pursue design projects on various electromagnetic devices. This unique addition of the design aspect and design-oriented problems/projects should nicely fill the need of the ABET-accreditation requirements on design methodologies.

Answers to Problems

Answers to odd-numbered problems are given at the end of the text (before the index) to aid the student in building his or her confidence in the solution procedure.

Supplements

A solution manual with complete detailed solutions for all the problems in the book is available to all instructors who adopt the text. Also, a laboratory manual based on the Faraday machines laboratory benches has been developed.

Acknowledgements

The material of this text is the result of a gradual development to meet the needs of the author's classes taught at universities in the United States and India over the past 30 years. It goes without saying that the author has incurred the indebtedness to many people during the planning and writing of the book. Assistance and encouragement were received from many sources who can be acknowledged only through the bibliography, as they are otherwise too numerous to mention. The profound influence of the earlier books on the subject (by authors such as A. E. Fitzgerald, L. W. Matsch, and D. C. White) and the developments made by various outstanding engineers are gratefully recognized. The following reviewers greatly improved this text through their thoughtful comments and useful suggestions.

Ali Abur
Texas A&M University

A. S. AlFuhaid
University of Florida

Alvin L. Day
Iowa State University of
  Science and Technology

A. A. El-Keib
University of Alabama

Jack Hanania
Northeastern University

Abdul Hye
University of Bridgeport

Ahmed Rubaai
Howard University

Phillip Schneider
University of Arkansas

T. Srinivasan
Wilkes University

Richard D. Stoy
Widener University

Elias G. Strangas
Michigan State University

I am indebted to my editor Peter C. Gordon as well as developing editor Sharon Adams and Lucy Paine Kezar of West Educational Publishing for their skillful guidance and helpful suggestions.

I would also like to thank my wife, Savitri, for her continued encouragement and support without which this project could not have been completed.

Mulukutla S. Sarma

PART

1

|||||||||||||||||||||||||||||||||||||||||||||||||||||||||||||||||||||||||||||||| ▬▬▬ ▬▬▬ ▬▬▬

Introduction

▬▬▬▬▬ |||

The theme of this book is electromechanical energy conversion. Electromechanical devices are involved in every industrial and manufacturing process of a technological society. The importance of these devices in almost every aspect of life does not need to be emphasized. An understanding of the principles of electromechanics is quite important for all those who desire to extend the usefulness of electrical technology in order to ameliorate the problems of energy, pollution, and poverty that presently face mankind.

The main purpose of this book is to present the fundamentals of electromechanics, apply these to some of the basic configurations of electromechanical devices, and stimulate the reader to further investigation of new and complex situations with exciting possibilities for advanced development. Most space is devoted to a careful explanation of four main electromagnetic devices: the transformer, the induction motor, the synchronous machine, and the direct-current machine. The book is up-to-date from the viewpoints of both subject matter and analytical procedures to meet the needs of those whose training requires a deep understanding of various aspects of machine behavior for further study of power systems, control systems, machine design, or general industrial applications.

The transformer, although not an electromechanical energy-conversion device, is an important auxiliary in the overall problem of energy conversion for the transformation of electrical power between different voltage-current levels. The concepts of the transformer behavior and the analysis of mutually coupled circuits serve as a useful adjunct to the study of electromagnetic rotating machines. The generators, converting mechanical energy to electrical energy, as well as the motors, converting electrical energy to mechanical energy, provide the power on which modern industrialized societies depend so much. Electromechanical energy conversion is effected by using the magnetic field as a coupling medium between a stationary member and a moving member. The great ease with which energy may be stored in magnetic fields accounts largely for the wide use of electromagnetic devices for the interconversion of electrical and mechanical energy.

Electromagnetic devices utilize magnetic materials for shaping the magnetic fields that act as the medium for transferring and converting energy. The interaction between the electric circuits and the magnetic fields plays an important part in the operation of various types of equipment discussed in this book. That is why a chapter on magnetic circuits, including those involving permanent magnets, is presented. Broader aspects of electromechanical energy conversion are given due attention in order to present a proper perspective and motivate the reader to think beyond the devices that are treated here.

Machine windings are the means by which the theory is translated into practice, and they form a unifying link between different machines; hence, a chapter is devoted for presenting the essentials of these.

The modern interest in electronics has influenced circuit texts and courses to the detriment of the study of polyphase systems. Since this topic is fundamental to the study of commercial and industrial power applications that utilize polyphase devices, a chapter is devoted for the analysis of balanced three-phase circuits.

Analysis by means of phasor diagrams is a very useful approach for the understanding of mathematical equivalent-circuit models of practical devices. As such a review of phasor diagrams is considered to be appropriate at the very beginning.

A REVIEW OF PHASOR DIAGRAMS

1.1 Phasors
1.2 Analysis with Phasor Diagrams

1.1 Phasors

Let us consider a complex function of time,

$$f(t) = re^{j\omega t} \tag{1.1.1}$$

whose magnitude, given by r, is a constant and whose angle, given by ωt radians at time t, changes with time. Such a function can be graphically interpreted in terms of a rotating *phasor* whose magnitude is given by r units and direction of rotation is counterclockwise (considered positive for positive ω and positive t). Further, if ω is a constant, the frequency of rotation becomes a constant equal to ω rad/s. $e^{-j\omega t}$ would imply clockwise (or negative) rotation.

By the use of Euler's identity, Equation 1.1.1 may be written as

$$f(t) = re^{j\omega t} = r(\cos \omega t + j \sin \omega t) \tag{1.1.2}$$

in which the real part (or the projection on the real axis) varies as the cosine of ωt, while the imaginary part (or the projection on the imaginary axis) varies as the sine of ωt. This concept is illustrated in Figure 1.1.1. The variation of the exponential function with time is sinusoidal in nature and corresponds to the case of the sinusoidal steady state. The sinusoidal-waveform assumption needs to be emphasized here.

The sum of rotating phasors of the same frequency ω is a rotating phasor of the same frequency. This property becomes a key to using rotating phasors for sinusoidal steady-state analysis. Also, it can be shown that if a rotating-phasor input signal is applied to a linear time-invariant network or system, the steady-state output is a rotating phasor of the same frequency.

Consider a general sinusoidal voltage of the form

$$v(t) = V_m \cos(\omega t + \theta) \tag{1.1.3}$$

Once the frequency ω is known, then v can be completely specified by its amplitude V_m and its phase θ. Based on the relationship

$$V_m \cos(\omega t + \theta) = \text{ Real part of } [V_m e^{j\theta} e^{\omega t}] \tag{1.1.4}$$

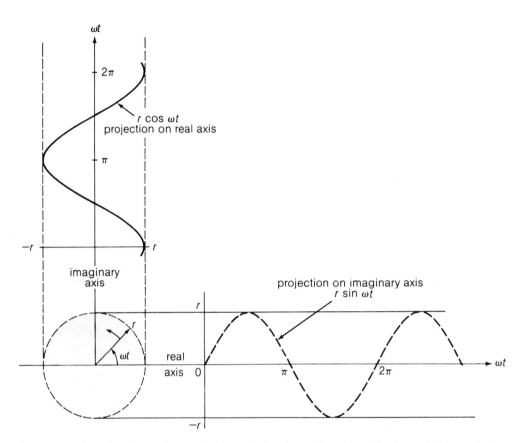

Figure 1.1.1 A rotating phasor whose projection on the imaginary axis varies as the sine and whose projection on the real axis varies as the cosine.

a related *phasor* can be defined such that

$$\bar{V} = V_m e^{j\theta} = V_m \angle \theta \tag{1.1.5}$$

and

$$v(t) = \text{Real part of } [\bar{V}e^{j\omega t}] \tag{1.1.6}$$

It should be pointed out that although we have chosen to represent sinusoids and their related phasors on the basis of cosine functions, we could represent them with sine functions just as easily. Useful equations in changing sine functions to cosines are

$$\sin(\omega t + \alpha) = \cos\left(\omega t + \alpha - \frac{\pi}{2}\right) \tag{1.1.7}$$

and

$$\cos(\omega t + \beta) = \sin\left(\omega t + \beta + \frac{\pi}{2}\right) \tag{1.1.8}$$

As an example in the time domain, let us consider an ac voltage of constant amplitude and constant frequency given by

$$v(t) = 100 \sqrt{2} \, \cos(\omega t + 30°) \text{ V} \qquad (1.1.9)$$

in which

$$\omega = (2 \times \pi \times 60) \text{ rad/s}$$

corresponding to a frequency f of 60 Hz. In the case of ac voltages and currents, it is more convenient to use the rms values than the amplitudes for the magnitudes of the phasors. Suppressing the explicit time variation, the phasor representation in the frequency domain becomes

$$\bar{V} = 100\angle 30° \text{ V} \qquad (1.1.10)$$

Starting from Equation 1.1.10, it is easy to go back to Equation 1.1.9 in the time domain by using Equation 1.1.6 and remembering that the amplitude is given by $\sqrt{2}$ times the rms (root-mean-square) value for sinusoidal variation. The phasor representation of a time-domain current such as

$$i(t) = 10\sqrt{2} \, \sin(\omega t + 30°) \text{ A} \qquad (1.1.11)$$

will be given by

$$\bar{I} = 10\angle -60° \text{ A} \qquad (1.1.12)$$

since by Equation 1.1.7

$$\sin(\omega t + 30°) = \cos(\omega t + 30° - 90°)$$
$$= \cos(\omega t - 60°) \qquad (1.1.13)$$

Note in Equation 1.1.12 that the rms value of the current has been chosen for the magnitude of the phasor.

Sinusoidal waveforms of the *same* wavelength or frequency can be represented by phasors that can be added or subtracted like coplanar vectors. Such waveforms may be functions of time, as in the case of ac voltages and currents of constant amplitude and constant frequency, or they can be space waves of flux density and magnetomotive force of constant amplitude and constant wavelength. Sinusoidal distributions of flux density and magnetomotive force in the air gaps of electric machines are treated as phasors. However, the magnitude of a space phasor is generally made equal to the amplitude of the wave, and the angles represent space angles rather than time angles.

For the three linear time-invariant passive elements, R (pure resistance), L (pure inductance), and C (pure capacitance), the relationships between voltage and current in the time and frequency domains are shown in Figure 1.1.2.

In general, the real solutions are time-domain functions, and their phasors are frequency-domain functions, functions of the frequency ω. Thus, in solving the time-domain problem, it is generally much easier to convert to phasors and solve the corresponding frequency-domain problem. Then, one can convert back to the time domain by finding the corresponding time function from its phasor representation. Figure 1.1.3 shows how to use phasors to solve such problems when all sources are sinusoidal and operating at the same frequency, and only the steady-state response is desired.

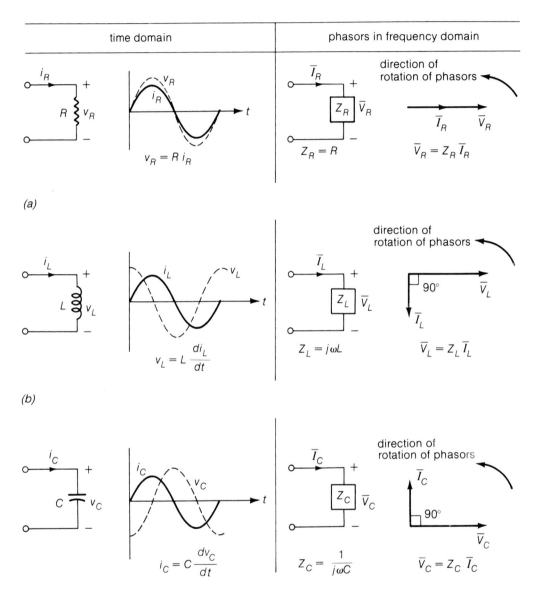

| time domain | phasors in frequency domain |

(a)

(b)

(c)

Figure 1.1.2 Voltage and current relationships in time and frequency domains for the elements R, L, and C. (a) The current is in phase with voltage in a purely resistive circuit (unity power factor). (b) The current lags the voltage by 90° in a pure inductor (zero power factor lagging). (c) The current leads the voltage by 90° in a pure capacitor (zero power factor leading).

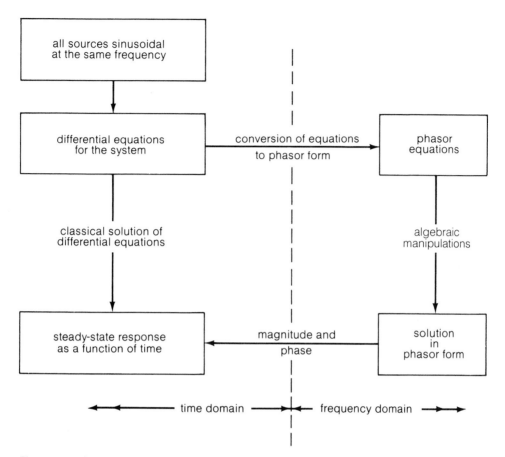

Figure 1.1.3 Sinusoidal steady-state analysis using phasors.

1.2 Analysis with Phasor Diagrams

We have seen that a phasor, being a complex number, can be represented in the complex plane in the conventional polar form as an arrow whose length corresponds to the phasor's magnitude and an angle (with respect to the positive real axis) representing the phasor's phase. In a *phasor diagram,* the various phasor quantities corresponding to a given network may be combined in such a way that at least one of Kirchoff's two laws is satisfied. The phasor method of analyzing circuits is generally credited to Charles Proteus Steinmetz (1865–1923), an electrical engineer for General Electric in the early twentieth century.

The phasor diagram may be drawn in the polar (or ray) form, in which all phasors originate at the origin; in the polygonal form, with each phasor located at the end of another; or in a combination of these depending on the convenience and the point to be made. Such diagrams provide insight into the geometry of network voltage and current relationships. They are particularly helpful for visualizing steady-state phenomena in analyzing networks with sinusoidal signals. Even in this age of computers, phasor diagrams are useful because of their visual impact for the viewer, since the interelationships of a circuit's variables can be seen at a glance.

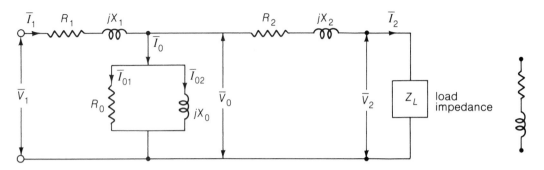

Figure 1.2.1 Series-parallel circuit.

In a phasor diagram, each sinusoidal voltage and current is represented by a phasor of length equal to the sinusoid's rms value and with an angular displacement from the positive real axis—the angle of the equivalent cosine function at t equal to zero. This use of the cosine and the rms value are arbitrary. The sine function could just as well be used, as could the amplitude rather than the rms value for the magnitude.

The *reference* of a phasor diagram is the line $\theta = 0$. It is not really necessary that any voltage or current phasor coincide with the reference, even though analysis is often more convenient when a given voltage or current phasor is used as the reference. Because a phasor diagram is a picture frozen at one instant in time that shows the relative locations of the various phasors and the whole diagram is assumed to be rotating at a constant frequency, different phasors may be referenced simply by rotating the diagram clockwise or counterclockwise.

For ease and convenience, it is sensible to choose the current phasor as the reference in the case of series circuits—circuits in which all the elements are connected in series and the common quantity for all the elements involved is the current. Similarly, for parallel circuits—circuits in which all the elements are connected in parallel and the common quantity for the elements is the voltage—a good choice for the reference is the voltage phasor. As for the series-parallel circuits, no rule applies to all situations.

We shall now consider the series-parallel circuit shown in Figure 1.2.1 and develop the corresponding phasor diagram. We shall assume the load to be inductive (a combination of resistance and inductance) and the corresponding load power-factor angle to be ϕ_2 lagging the voltage $\bar{V}_2$. That is, the current $\bar{I}_2$ lags the voltage $\bar{V}_2$ by an angle of ϕ_2, as shown in Figure 1.2.2(a). Phasor voltages and currents are indicated in Figure 1.2.1, and their magnitudes, unless otherwise stated, will be assumed to be their corresponding rms values, which would be the same as the meter readings if voltmeters and anmeters were to be connected appropriately.

For the sake of developing the phasor diagram, we shall assume that R_1 is approximately equal to R_2, X_1 is approximately equal to X_2, R_1 is less than X_1, R_0 is greater than R_1, and X_0 is greater than X_1.

$$R_1 \simeq R_2; \qquad X_1 \simeq X_2; \qquad R_1 < X_1; \qquad R_0 > R_1; \qquad X_0 > X_1 \qquad (1.2.1)$$

Given $\bar{V}_2$ and $\bar{I}_2$ at the load and the parameters of the circuit of Figure 1.2.1, we shall try to determine $\bar{V}_0$, $\bar{I}_0$, $\bar{I}_1$, and $\bar{V}_1$ by means of a phasor diagram. As seen in Figure 1.2.2, the voltage $\bar{V}_2$ is chosen as the *reference*. Starting with the reference voltage $\bar{V}_2$, $\bar{I}_2$ is drawn at an angle of ϕ_2 lagging the voltage $\bar{V}_2$ in Figure 1.2.2(a), because the load is assumed to be inductive. Looking at Figure 1.2.1 and using Kirchhoff's loop equation, the relationship to be satisfied by $\bar{V}_0$ and $\bar{V}_2$ is given by

$$\bar{V}_0 = \bar{V}_2 + \bar{I}_2 R_2 + j\bar{I}_2 X_2 \qquad (1.2.2)$$

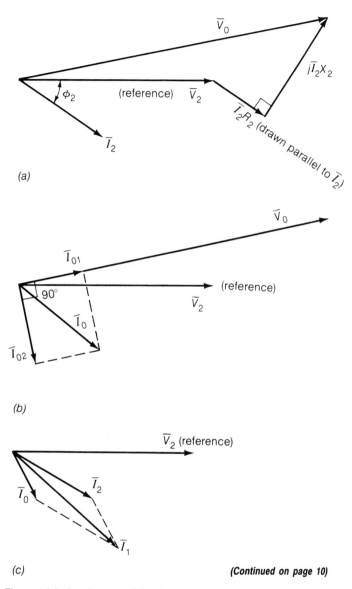

Figure 1.2.2 Development of the phasor diagram for the circuit shown in Figure 1.2.1 for an inductive load.

Figure 1.1.2(a) is drawn such that Equation 1.2.2 is satisfied. The voltage drop $\bar{I}_2 R_2$ is drawn parallel to $\bar{I}_2$, and $\bar{I}_2 R_2$ is added to $\bar{V}_2$ (by locating $\bar{I}_2 R_2$ at the end of the phasor $\bar{V}_2$). $j\bar{I}_2 X_2$ is drawn perpendicular to $\bar{I}_2$, and $j\bar{I}_2 X_2$ is added to $(\bar{V}_2 + \bar{I}_2 R_2)$, yielding $\bar{V}_0$.

Next, looking at the shunt portion of the series-parallel circuit of Figure 1.2.1 and knowing $\bar{V}_0, R_0$, and X_0, it is easy to find $\bar{I}_{01}$ and $\bar{I}_{02}$, which when added will yield $\bar{I}_0$.

$$\bar{I}_0 = \bar{I}_{01} + \bar{I}_{02} \tag{1.2.3}$$

This is what is shown in Figure 1.2.2(b). $\bar{I}_{01}$, the current in a pure resistor, is drawn in line with (i.e.,

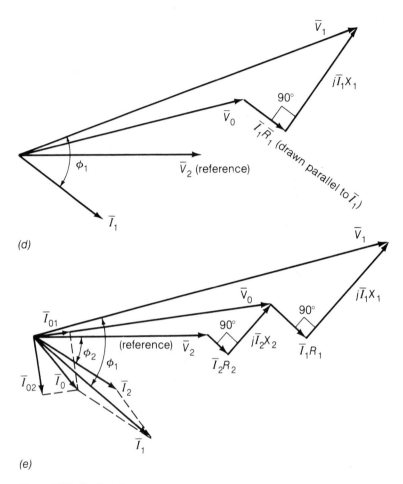

(d)

(e)

Figure 1.2.2 *Continued*

parallel to) $\bar{V}_0$; $\bar{I}_{02}$, the current in a pure inductor, is drawn perpendicular to $\bar{V}_0$, lagging $\bar{V}_0$ by 90°; $\bar{I}_{01}$ and $\bar{I}_{02}$ are then added, using the parallelogram law, to give $\bar{I}_0$.

The relationship that must be satisfied by the currents $\bar{I}_1, \bar{I}_0$, and $\bar{I}_2$ is given by

$$\bar{I}_1 = \bar{I}_2 + \bar{I}_0 \tag{1.2.4}$$

consistent with the notation shown in Figure 1.2.1. Figure 1.2.2(c) corresponds to Equation 1.2.4.

Then $\bar{V}_1$ can be found by solving the loop equation:

$$\bar{V}_1 = \bar{V}_0 + \bar{I}_1 R_1 + j\bar{I}_1 X_1 \tag{1.2.5}$$

This part is depicted in Figure 1.2.2(d). The power-factor angle ϕ_1 between $\bar{V}_1$ and $\bar{I}_1$ can also be found.

Thus, the entire phasor diagram is put together by superposition as shown in Figure 1.2.2(e). The mathematical steps associated with the steady-state solution of the network are omitted here, as they are covered in basic circuit-theory courses.

■■■■■■■IIII
BIBLIOGRAPHY

Budak, A. *Circuit Theory Fundamentals and Applications.* Englewood Cliffs, N.J.: Prentice-Hall, 1978.

Cruz, J. B., Jr., and M. E. Van Valkenburg. *Signals in Linear Circuits.* Boston: Houghton Mifflin, 1974.

Hayt, W. H., and J. E. Kemmerly. *Engineering Circuit Analysis,* 4th ed. New York: McGraw-Hill, 1986.

Johnson, D. E.; J. L. Hilburn; and J. R. Johnson. *Basic Electric Circuit Analysis.* Englewood Cliffs, N.J.: Prentice-Hall, 1978.

Neudorfer, P. O., and M. Hassul. *Introduction to Circuit Analysis.* Boston: Allyn and Bacon, 1990.

Smith, K. C. A., and R. E. Alley. *Electrical Circuits.* Cambridge, U.K: Cambridge University Press, 1992.

Van Valkenburg, M. E. *Network Analysis,* 3rd ed. Englewood Cliffs, N.J.: Prentice-Hall, 1974.

■■■■■■■■IIIIIIIIIII
PROBLEMS

1–1. Consider $\bar{A}$ to be an arbitrary phasor; relative to $\bar{A}$ sketch the following:

    a. $-j\bar{A}$ b. $2j\bar{A}$ c. $(1 - 2j)\bar{A}$ d. conjugate of $\bar{A}$, denoted by $\bar{A}*$

1–2. Using phasors, add the following:

    a. $10 \sin (377t + 30°)$ and $5 \cos (377t - 45°)$.

    b. $200 \sin (\omega t + \pi/6), 150 \cos (\omega t + \pi/6), 200 \sin (\omega t + 5\pi/6)$, and $-150 \cos (\omega t + 5\pi/6)$.

1–3. a. Evaluate these in rectangular form: *(i)* $8\angle 35° - 5e^{-j100°}$ *(ii)* $[4/(3 - j2)] + 1/(j3)$

    b. Evaluate these in polar form: *(i)* $2.5\angle -120° - j8$ *(ii)* $(4 + j3)e^{j(200t+50°)}$ at $t = 4$ ms

1–4. a. Find the resultant current in exponential form when a voltage $60e^{j(500t-50°)}$ V is applied to a parallel combination of 20 Ω and 0.1 mF.

    b. Find the resultant voltage in exponential form when a current $5e^{j(500t+90°)}$ A is applied to a series combination of 3 Ω and 2 mH.

1–5. a. Express each of these as a phasor:

      (i) $(6 \sin \omega t - 2 \cos \omega t)$ *(ii)* $[4 \cos (\omega t - 80°) - 3 \sin (\omega t + 30°)]$

    b. Find the instantaneous value at $t = 1$ ms of the voltage of frequency 60 Hz represented by each phasor:

      (i) $120\angle 0°$ V   *(ii)* $(80 + j75)$ V   *(iii)* $(50 + 80\angle 70°)$ V

      Note that the rms value is chosen for the magnitude of the phasor.

1–6. Given the frequency domain circuit with $\bar{V}_L = 66\angle 57°$ V and $\bar{V}_C = 50\angle -85°$ V, as shown in Figure P1–6, compute $\bar{V}_S$ and $\bar{I}_S$.

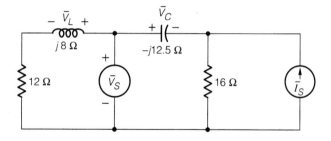

Figure P1–6

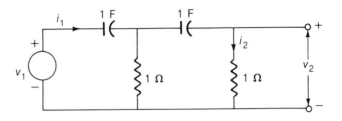

Figure P1-7

1-7. For the RC network shown in Figure P1–7, let $v_2(t) = \sqrt{2}\cos t$ volts.

    a. Draw a phasor diagram (approximately to scale, preferably on graph paper) showing all voltages and currents.

    b. Based on the results of the phasor diagram, find $v_1(t)$.

    c. What values would an rms meter record for the voltages $v_1(t)$ and $v_2(t)$?

1-8. a. Consider the frequency-domain circuit in Figure P1–8. Choose $\bar{V}_L = 1\angle 0°$ V and draw a phasor diagram with all phasor variables. Find the corresponding $\bar{V}_S$.

    b. Adjust and redraw the phasor diagram such that $\bar{V}_S$ becomes $1\angle 0°$. Find the corresponding $\bar{V}_L$.

1-9. A single-phase supply of $v(t) = 169.7\cos \omega t$ is applied across a load consisting of a resistance of 10 Ω in parallel with a capacitive reactance of 20 Ω. Find

    a. (i)   the current $i(t)$ supplied by the source,

        (ii) the source voltage and current in phasor form,

        (iii) the complex power supplied by the source, and

        (iv) the power factor of the supply (state if leading or lagging).

    b. Sketch the power triangle.

1-10. Consider the frequency-domain circuit in Figure P1–10. The supply voltage is 240 V and the frequency is 60 Hz.

    a. Calculate the impedance between terminals A and B, and sketch an impedance diagram for the circuit.

    b. Find the currents in each branch of the circuit and then find the voltage of node D with respect to B.

    c. Verify the node voltage of part (b) by using nodal analysis.

    d. Sketch a phasor diagram for the complete circuit that shows all phasor variables.

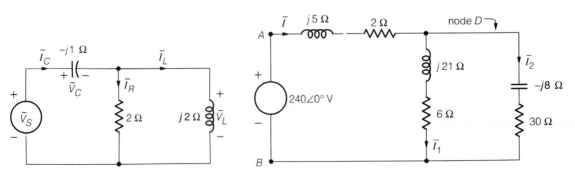

Figure P1-8

Figure P1-10

1–11. The circuit shown in Figure P1–11 is supplied with a constant voltage of variable frequency. Obtain an expression for the frequency (in terms of the circuit parameters) at which the voltage across AB is in phase with the source voltage.

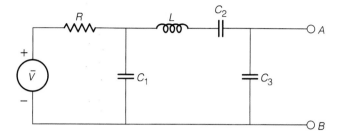

Figure P1–11

1–12. Consider the circuit in Figure P1–12; $R = 20\ \Omega$; $L_1 = 2$ H; $C_1 = 0.01$ F; $C_2 = 5$ mF.

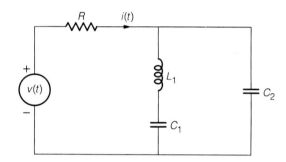

Figure P1.12

   a. Calculate $z(j\omega)$ seen by the voltage source $v(t) = 80\cos 10t$. Find $i(t)$.

   b. Sketch the phasor diagram for the circuit.

   c. Find the resonance frequencies at which the reactance $X(j\omega)$ becomes (i) zero (ii) infinite.

1–13. For the network shown in Figure P1–13, at what frequency is (a) $R_{in} = 200\ \Omega$; (b) $X_{in} = -200\ \Omega$; (c) $G_{in} = 1/200$ S; and (d) $B_{in} = 1/240$ S.

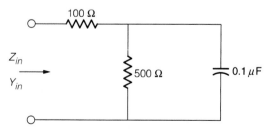

Figure P1–13

1–14. Consider the series-parallel circuit of Figure 1.2.1.

   a. Draw the corresponding phasor diagrams [similar to Figure 1.2.2(e)] for these two cases:

    i.   Let the load be purely resistive, in which case $\dot{I}_2$ will be in phase with $\bar{V}_2$, with the load power factor being unity.

    ii.   Let the load be capacitive (a combination of resistance and capacitance), for which case $\dot{I}_2$ will be leading $\bar{V}_2$ by an angle ϕ_2.

  b. It appears from Figure 1.2.2 that the magnitude of $\bar{V}_1$ is greater than that of $\bar{V}_2$. Would this be true for all types of loads under all conditions? If not, try to identify case in which it is not and sketch the corresponding phasor diagram.

1-15.   Consider the circuit shown in Figure P1–15. Sketch the corresponding phasor diagram, and find the corresponding V_2 for these cases:

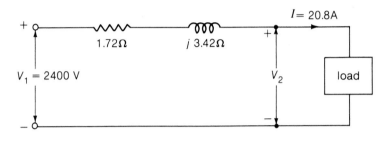

Figure P1–15

  a. A load power factor of 0.8 lagging

  b. A load power factor of 0.8 leading

  c. Unity load power factor

2

Three-Phase Circuits

2.1 Three-Phase Source Voltages and Phase Sequence
2.2 Balanced Three-Phase Loads
2.3 Power Measurement

2.1 Three-Phase Source Voltages and Phase Sequence

Due to its definite economic and operating advantages, the three-phase system is by far the most common polyphase system used for generation, transmission, and heavy power utilization of ac electric energy. An ideal three-phase source generates three sinusoidal voltages of equal amplitudes displaced from each other by an angle of 120° in time phase. The voltages generated by the giant synchronous generators in power stations are practically sinusoidal with a frequency of 60 Hz in the United States—and 50 Hz in the United Kingdom and many other countries. Even though voltages and currents are sinusoidal, the power delivered to a balanced load is constant for a three-phase system, whereas it is sinusoidal and pulsating for a single-phase system. Three-phase power transmission is the simplest polyphase method that offers the advantages of using the ac mode, constant power flow, and high power-transfer capability.

The elementary three-phase, two-pole generator shown in Figure 2.1.1 has three identical stator coils of one or more turns (aa', bb', and cc') that are displaced by 120° in space from each other. The rotor carries a field winding excited by the dc supply through brushes and slip rings. When the rotor is driven counterclockwise at a constant speed, voltages are generated in the three phases in accordance with Faraday's law. Each stator coil constitutes one phase in this single generator. If the field structure is so designed that the flux is distributed sinusoidally over the poles, the flux linking any phase will vary sinusoidally with time, and sinusoidal voltages will be induced in the three phases. These three induced voltage waves will be displaced by 120 electrical degrees in time, because the stator phases are displaced by 120° in space. Figure 2.1.2 shows the wave forms and the corresponding phasors of the three voltages. The time origin and the reference axis are chosen on the basis of analytical convenience. In a balanced system, all three phase voltages are equal in magnitude but differ in phase by 120°. The sequence of voltages in Figure 2.1.2b corresponding to that of Figure 2.1.2a is known as the *positive sequence* (*a-b-c*), while that of 2.1.2c is called *the negative sequence* (*a-c-b*). Notice that in positive sequence, $\bar{E}_{bb'}$ lags $\bar{E}_{aa'}$ by 120° and $\bar{E}_{cc'}$ lags $\bar{E}_{bb'}$ by 120°, whereas in negative sequence, $\bar{E}_{cc'}$ lags $\bar{E}_{aa'}$ by 120° and $\bar{E}_{bb'}$ lags $\bar{E}_{cc'}$ by 120°.

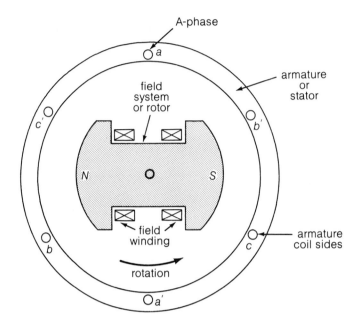

Figure 2.1.1 Elementary three-phase, 2-pole ac generator.

Stator phase windings can be connected in either *wye* (known also as "star" and symbolized as *Y*) or *delta* (known also as "mesh" and symbolized as Δ), as shown schematically in Figure 2.1.3. The stator phase windings of almost all alternating-current generators (otherwise known as *alternators*) are connected in wye. When either all three primed terminals or all three unprimed terminals are connected to form the neutral of the wye, a wye connection results. If a neutral conductor is brought out, the system is known as a four-wire, three-phase system; otherwise, it is a three-wire, three-phase system. A delta connection is effected for the armature of the generator by connecting terminal a' to b, terminal b' to c, and terminal c' to a. The generator terminals A, B, and C (and sometimes N for a wye connection) are brought out as shown in Figure 2.1.3. In the Δ-connection, no neutral exists, so only a three-wire, three-phase system can be formed. Note that a phase is one of the three branch circuits making up a three-phase circuit. In a wye connection, a phase consists of those circuit elements that connect one line and neutral; in a delta circuit, a phase consists of those circuit elements that connect two lines.

From the nature of the connections shown in Figure 2.1.3, it can be seen that the line-to-line voltage (V_{L-L} or V_L) equals the phase voltage V_{ph} for the Δ-connection, and the line current equals the phase current for the Y-connection. A balanced wye-connected, three-phase source and its phasor diagram are shown in Figure 2.1.4. It can be seen in this figure that the line-to-line voltage is $\sqrt{3}$ times the phase voltage (or the line-to-neutral voltage). The student should be able to reason along similar lines and conclude that for a balanced delta-connnected, three-phase source, the line current will be $\sqrt{3}$ times the phase current.

The notation for subscripting is such that V_{AB} is the potential at point A with respect to point B, I_{AB} is a current with positive flow from point A to point B, and I_A, I_B, and I_C are line currents with

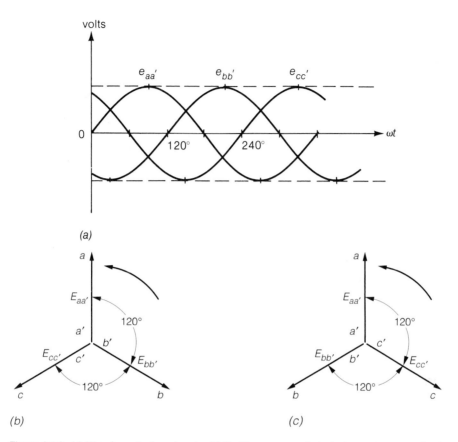

Figure 2.1.2 (a) Waveforms in time domain. (b) Positive-sequence (*a-b-c*) phasors corresponding to Figure 2.1.2a. (c) Negative-sequence (*a-c-b*) phasors.

positive flow from source to the load. These are shown in Figure 2.1.5 (page 20). The notation used is rather arbitrary; in some textbooks a different notation for the voltage is adopted such that the order of subscripts indicates the direction in which the voltage rise is taken. The student should be careful not to get confused and to be consistent with any chosen conventions. The rms values are usually chosen as phasor magnitudes for convenience. It is customary to use the symbol E for generated emf and V for terminal voltage. Sometimes the two are equal; other times not. If we should neglect the existence of the generator winding impedance, the generated emf will be equal to the terminal voltage of the generator. Although a set of three-phase voltages is generated in one three-phase alternator, for analytical purposes it is modeled as three identical, interconnected, single-phase sources.

The one-line equivalent circuit of the balanced wye-connected, three-phase source is shown in Figure 2.1.4c. The line-to-neutral (otherwise known as *phase*) voltage is used; it can serve as a reference with a phase angle of zero for convenience. This procedure yields the equivalent single-phase circuit, in which all quantities correspond to those of one phase of the three-phase circuit. Except for the 120° phase displacements in the currents and voltages, the conditions in the other two phases are the same, and

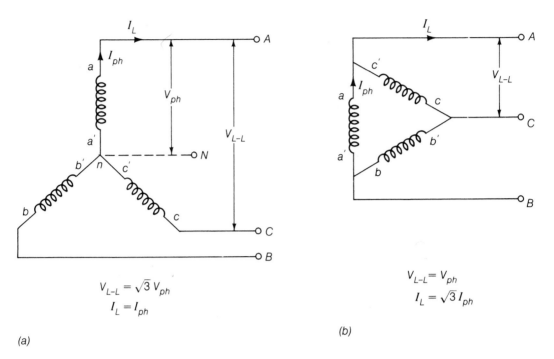

Figure 2.1.3   Schematic representation of generator windings: (a) Balanced wye connection. (b) Balanced delta connection.

there is no need to investigate them individually. Line currents in the three-phase system are the same as in the single-phase circuit, and total three-phase real power, reactive power, and volt-amperes are three times the corresponding quantities in the single-phase circuit. Line-to-line voltages, in magnitude, can be obtained by multiplying voltages in the single-phase circuit by $\sqrt{3}$.

When a system of sources is so large that its voltage and frequency remain constant regardless of the power delivered or absorbed, it is known as an *infinite bus*. The voltage and frequency of an infinite bus are unaffected by external distrubances, and it is treated as an ideal voltage source.

Phase Sequence

It is a standard practice in the United States to designate the phases *A-B-C* such that under balanced conditions the voltage and current in *A*-phase lead in time the voltage and current in *B*-phase by 120° and in *C*-phase by 240°. This is known as *positive phase sequence A-B-C*. The phase sequence should be observed either from the waveforms in the time domain shown in Figure 2.1.2a or from the phasor diagram shown in Figure 2.1.2b or 2.1.4b, rather than from the space or schematic diagrams such as Figures 2.1.3 and 2.1.4a. If the rotation of the generator in Figure 2.1.1 is reversed, or if any two of the three leads from the armature (not counting the neutral) to the generator terminals are reversed, the phase sequence becomes *A-C-B* (or *C-B-A* or *B-A-C*), which is known as *negative phase sequence*.

Only balanced three-phase sources are considered in this chapter. Selection of one voltage as the reference with a phase angle of zero determines the phase angles of all the other voltages in the system for a given phase sequence.

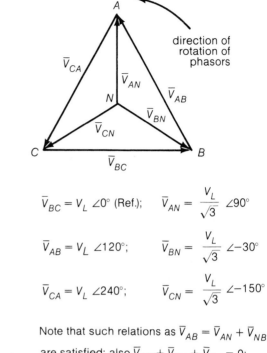

direction of rotation of phasors

$\overline{V}_{BC} = V_L \angle 0°$ (Ref.); $\overline{V}_{AN} = \dfrac{V_L}{\sqrt{3}} \angle 90°$

$\overline{V}_{AB} = V_L \angle 120°$; $\overline{V}_{BN} = \dfrac{V_L}{\sqrt{3}} \angle -30°$

$\overline{V}_{CA} = V_L \angle 240°$; $\overline{V}_{CN} = \dfrac{V_L}{\sqrt{3}} \angle -150°$

Note that such relations as $\overline{V}_{AB} = \overline{V}_{AN} + \overline{V}_{NB}$ are satisfied; also $\overline{V}_{AN} + \overline{V}_{BN} + \overline{V}_{CN} = 0$;

$\overline{V}_{AB} + \overline{V}_{BC} + \overline{V}_{CA} = 0$.

(a)

(b)

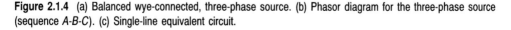

(c)

Figure 2.1.4 (a) Balanced wye-connected, three-phase source. (b) Phasor diagram for the three-phase source (sequence A-B-C). (c) Single-line equivalent circuit.

As indicated before, the reference phasor is arbitrarily chosen for convenience. In Figure 2.1.4b, $\overline{V}_{BC}$ is the reference phasor, and with the counterclockwise rotation (assumed positive) of all the phasors at the same frequency, the sequence is *A-B-C*. Unless otherwise stated, the positive phase sequence is assumed.

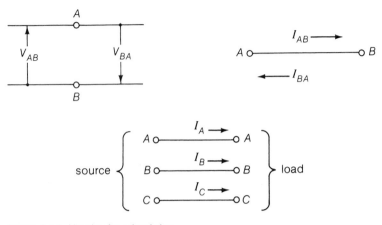

Figure 2.1.5 Notation for subscripting.

2.2 Balanced Three-Phase Loads

Three-phase loads also can be connected in either wye ("star" or Y) or delta ("mesh" or Δ). If the load impedances in the three phases are the same in both magnitude and phase angle, the load is said to be balanced.

For analyzing network problems, it is useful to know the transformations for converting a delta-connected network to an equivalent wye-connected network and vice versa. The relationships for interconversion of wye and delta networks are given in Figure 2.2.1. These can be obtained by imposing the condition of equivalence; that is, that the impedance between any two terminals of one network equals the impedance between the corresponding terminals of the other network. (Students are encouraged to investigate and discover the details of this condition on their own.) For the balanced case, each Y-impedance is one third of each Δ-impedance; conversely, each Δ-impedance is three times each Y-impedance.

Balanced Wye-Connected Load

Let us consider a three-phase, four-wire, 208-volt supply system connected to a balanced wye-connected load with impedance of $10\angle20°$ ohms, as shown in Figure 2.2.2a. We will solve for the line currents and draw the corresponding phasor diagram.

Conventionally, 208 volts is assumed as the rms value of the line-to-line voltage of the supply system, and the phase sequence is assumed to be positive (A-B-C) unless stated otherwise. The magnitude of the line-to-neutral (or phase) voltages is given by $208/\sqrt{3}$ or 120 volts. Selecting the line currents returning through the neutral conductor as shown in Figure 2.2.2(a), we have

$$\bar{I}_A = \frac{\bar{V}_{AN}}{Z} = \frac{(208/\sqrt{3})\angle90°}{10\angle20°} = 12\angle70° \tag{2.2.1}$$

$$\bar{I}_B = \frac{\bar{V}_{BN}}{Z} = \frac{(208/\sqrt{3})\angle-30°}{10\angle20°} = 12\angle-50° \tag{2.2.2}$$

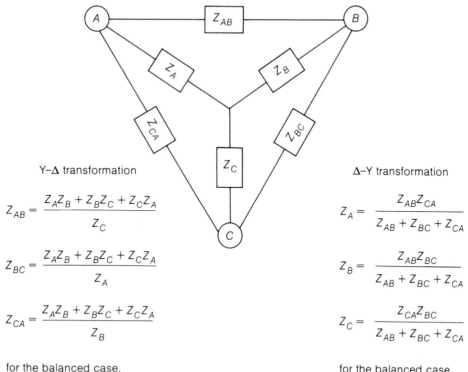

Y–Δ transformation

$$Z_{AB} = \frac{Z_A Z_B + Z_B Z_C + Z_C Z_A}{Z_C}$$

$$Z_{BC} = \frac{Z_A Z_B + Z_B Z_C + Z_C Z_A}{Z_A}$$

$$Z_{CA} = \frac{Z_A Z_B + Z_B Z_C + Z_C Z_A}{Z_B}$$

for the balanced case,

$$Z_{AB} = Z_{BC} = Z_{CA} = Z_\Delta = 3Z_Y$$

Δ–Y transformation

$$Z_A = \frac{Z_{AB} Z_{CA}}{Z_{AB} + Z_{BC} + Z_{CA}}$$

$$Z_B = \frac{Z_{AB} Z_{BC}}{Z_{AB} + Z_{BC} + Z_{CA}}$$

$$Z_C = \frac{Z_{CA} Z_{BC}}{Z_{AB} + Z_{BC} + Z_{CA}}$$

for the balanced case,

$$Z_A = Z_B = Z_C = Z_Y = \frac{1}{3} Z_\Delta$$

Figure 2.2.1 Wye–delta and delta–wye transformations.

$$\bar{I}_C = \frac{\bar{V}_{CN}}{Z} = \frac{(208/\sqrt{3})\angle -150°}{10\angle +20°} = 12\angle -170° \qquad (2.2.3)$$

Note that $\bar{V}_{BC}$ has been arbitrarily chosen as the reference phasor as in Figure 2.1.4b. Assuming the direction of the neutral current toward the load as positive, we obtain

$$\bar{I}_N = -(\bar{I}_A + \bar{I}_B + \bar{I}_C) = -(12\angle 70° + 12\angle -50° + 12\angle 170°) = 0 \qquad (2.2.4)$$

That is to say, the system neutral and the star point of the wye-connected load are at the same potential, even if they are not connected electrically. It makes no difference whether they are interconnected or not. *Thus for a* balanced *wye-connected load, the neutral current is always zero; the line currents and phase currents are equal (in magnitude); and the line-to-line voltage (in magnitude) is $\sqrt{3}$ times the phase voltage.* The phasor diagram is Figure 2.2.2b. From this it can be observed that the balanced line (or phase) currents lag the corresponding line-to neutral voltages by the impedance angle (20° in this case). The load power factor, given by cos 20°, is said to be *lagging*, as the impedance angle is positive, and the phase current lags the corresponding phase voltage by that angle.

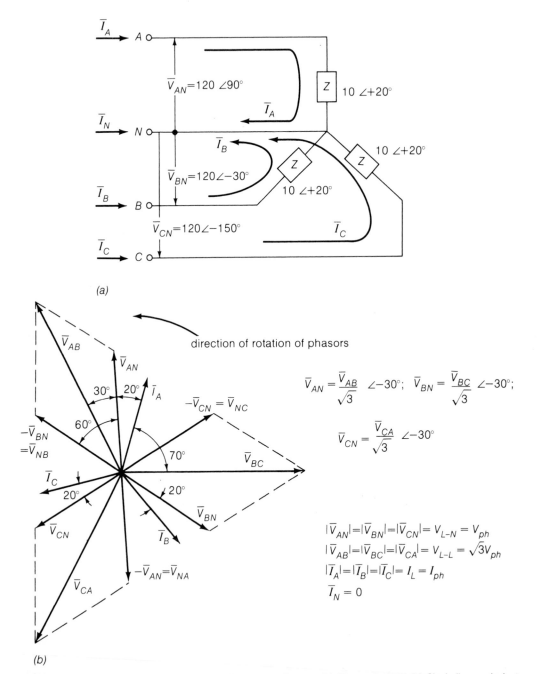

Figure 2.2.2 Balanced wye-connected load: (a) Connection diagram. (b) Phasor diagram. (c) Single-line equivalent circuit.

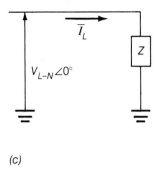

(c)

Figure 2.2.2 *continued*

The problem can be solved in a simpler way by using the single-line equivalent circuit shown in Figure 2.2.2c

$$\bar{I}_L = \frac{\bar{V}_{L-N}}{Z} = \frac{(208/\sqrt{3})\angle 0°}{10\angle 20°} = 12\angle -20° \tag{2.2.5}$$

in which $\bar{V}_{L-N}$ is chosen as the reference for convenience. The magnitude of the line current and the power-factor angle are known; the negative sign associated with the angle indicates that the power factor is lagging. By knowing that the line (or phase) currents $\bar{I}_A$, $\bar{I}_B$, and $\bar{I}_C$ lag their respective voltages $\bar{V}_{AN}$, $\bar{V}_{BN}$, and $\bar{V}_{CN}$ by 20°, the phase angles of various voltages and currents, if desired, can be obtained with respect to any chosen reference such as $\bar{V}_{BC}$.

Balanced Delta-Connected Load

Next let us consider the case of a balanced delta-connected load with impedance of $5\angle 45°$ ohms supplied by a three-phase, three-wire, 100-volt system, as shown in Figure 2.2.3a. We will determine the line currents and draw the corresponding phasor diagram.

With the assumed positive phase sequence (A-B-C) and with $\bar{V}_{BC}$ as the reference phasor, the line-to-line voltages $\bar{V}_{AB}$, $\bar{V}_{BC}$, and $\bar{V}_{CA}$ are shown in Figure 2.2.3. The rms value of the line-to-line voltages is 100 volts for this example. Choosing the positive directions of the line and phase currents as in Figure 2.2.3a, we have

$$\bar{I}_{AB} = \frac{\bar{V}_{AB}}{Z} = \frac{100\angle 120°}{5\angle 45°} = 20\angle 75° \tag{2.2.6}$$

$$\bar{I}_{BC} = \frac{\bar{V}_{BC}}{Z} = \frac{100\angle 0°}{5\angle 45°} = 20\angle -45° \tag{2.2.7}$$

$$\bar{I}_{CA} = \frac{\bar{V}_{CA}}{Z} = \frac{100\angle 240°}{5\angle 45°} = 20\angle 195° \tag{2.2.8}$$

By applying Kirchhoff's current law at each vertex of the delta-connected load, we obtain

$$\bar{I}_A = \bar{I}_{AB} + \bar{I}_{AC} = 20\angle 75° - 20\angle 195° = 34.64\angle 45° \tag{2.2.9}$$

$$\bar{I}_B = \bar{I}_{BA} + \bar{I}_{BC} = -20\angle 75° + 20\angle -45° = 34.64\angle -75° \tag{2.2.10}$$

$$\bar{I}_C = \bar{I}_{CA} + \bar{I}_{CB} = 20\angle 195° - 20\angle -45° = 34.64\angle 165° \tag{2.2.11}$$

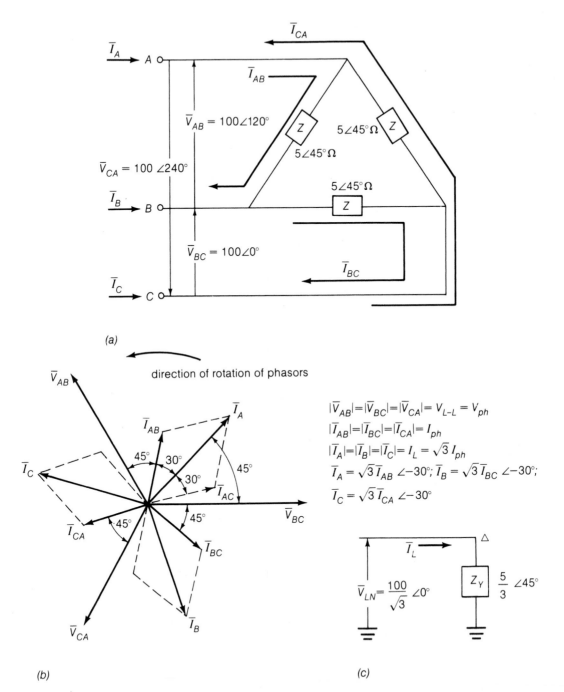

Figure 2.2.3 Balanced delta-connected load: (a) Connection diagram. (b) Phasor diagram. (c) Single-line equivalent circuit.

The phasor diagram showing the line-line voltages, phase currents, and line currents appears in Figure 2.2.3b. The load power factor is lagging, given by $\cos 45°$.

*For a **balanced** delta-connected load, the phase voltages and the line-to-line voltages are equal (in magnitude); and the line current (in magnitude) is $\sqrt{3}$ times the phase current.*

The example could also be solved by the one-line equivalent method, for which the delta-connected load is replaced by its equivalent wye-connected load. The single-line equivalent circuit is shown in Figure 2.2.3c. The details are left as an excercise for the student.

Power in Balanced Three-Phase Circuits

The total power delivered by a three-phase source or consumed by a three-phase load is found simply by adding the powers in the three phases. In a balanced circuit, however, this is the same as multiplying the average power in any one phase by 3, since the average power is the same for all phases. Thus we have

$$P = 3V_{ph}I_{ph}\cos\phi \tag{2.2.12}$$

where V_{ph} and I_{ph} are the magnitudes of any phase voltage and phase current, $\cos\phi$ is the load power factor, and ϕ is the power-factor angle between the phase voltage, $\bar{V}_{ph}$, and the phase current, $\bar{I}_{ph}$, corresponding to any phase. In view of the relationships between the line and phase quantities for balanced wye- or delta-connected loads, Equation 2.2.12 can be rewritten in terms of the line-to-line voltage and the line current for either wye-connected or delta-connected balanced loading as

$$P = \sqrt{3}V_L I_L \cos\phi \tag{2.2.13}$$

V_L and I_L are the magnitudes of the line-to-line voltage and the line current; ϕ is still the load power-factor angle as in Equation 2.2.12, being the angle between the phase voltage and the corresponding phase current.

In a balanced three-phase system, the sum of the three individually pulsating phase powers is a constant, nonpulsating total power of magnitude three times the average real power in each phase. That is, in spite of the sinusoidal nature of the voltages and currents, the total instantaneous power delivered into the three-phase load is a constant equal to the total average power. The real power P is expressed in watts when the voltage and current are expressed in volts and amperes, respectively. You may recall that the instantaneous power in single-phase ac circuits absorbed by a pure inductor or capacitor is a double-frequency sinusoid with zero average value. The instantaneous power absorbed by a pure resistor has a nonzero average value plus a double-frequency term with zero average value. The instantaneous reactive power is alternately positive and negative, indicating the reversible flow of energy to and from the reactive component of the load. Its amplitude, or maximum value, is known as the *reactive power.*

The total reactive power Q (expressed as reactive volt-amperes, VAR) for wye-connected and delta-connected balanced loadings is given by

$$Q = 3V_{ph}I_{ph}\sin\phi \tag{2.2.14}$$

which can be expressed in terms of the line values as

$$Q = \sqrt{3}V_L I_L \sin\phi \tag{2.2.15}$$

The total volt-amperes (in magnitude) for three-phase balanced loadings is given by

$$|S| = \sqrt{P^2 + Q^2} = 3V_{ph}I_{ph} = \sqrt{3}V_L I_L \tag{2.2.16}$$

and the complex power, S, is expressed as

$$S = P + jQ \tag{2.2.17}$$

In speaking of a three-phase system, unless specified otherwise, balanced conditions are assumed; and the terms *voltage, current,* and *power,* unless identified otherwise, conventionally imply the line-to-line voltage (rms value), the line current (rms value), and the total power of all three phases. In general, the ratio of the real or average power P to the apparent power or the magnitude of the complex power $|S|$ is the power factor, which happens to be $\cos \phi$ in the sinusoidal case.

IIIIIIIIIIII▬▬▬▬▬▬▬▬▬▬▬▬▬▬▬

EXAMPLE 2.2.1

A wye-connected generator is to be designed to supply a 20-kV three-phase line.

a. Find the terminal line-to-neutral voltage of each phase winding.

b. Determine the output line-to-line voltage if the windings of the generator are delta connected.

c. Let the 20-kV generator supply a line current of 10 A at a lagging power factor of 0.8. Compute the kilovolt-amperes (kVA), kilowatts (kW), and reactive kilovolt-amperes (kVAR) supplied by the alternator.

Solution

a. $V_{L-N} = V_{ph} = 20/\sqrt{3} = 11.547$ kV

b. $V_L = V_{ph} = 11.547$ kV

c. kVA $= \sqrt{3}(20)10 = 346.4$
   kW $= (346.4)(0.8) = 277.12$
   kVAR $= (346.4)(0.6) = 207.84$

▬▬▬▬▬▬▬▬▬▬▬▬▬▬▬**IIIIIII**

2.3 Measurement of Power

A wattmeter is an instrument with a potential coil and a current coil so arranged that its deflection is proportional to $VI \cos \theta$, where V is the voltage (rms value) applied across the potential coil, I is the current (rms value) passing through the current coil, and θ is the angle between $\bar{V}$ and $\bar{I}$. By inserting a wattmeter to measure the average real power in each phase (with its current coil in series with one phase of the load and its potential coil across the phase of the load), the total real power in a three-phase system can be determined by computing the sum of the wattmeter readings. In practice this may not be possible, however, due to the nonaccessibility of either the neutral of the wye connection or the individual phases of the delta connection. Hence it is more desirable to develop a method for measuring the total real power drawn by a three-phase load when one has access to only three line terminals.

The three-phase power can be measured by three wattmeters having current coils in each line and potential coils connected across the given line and any common junction. Since this common junction is completely arbitrary, it may be placed on any one of the three lines. When this is done the wattmeter connected in that line will indicate zero power, because its potential coil has no voltage across it. Hence

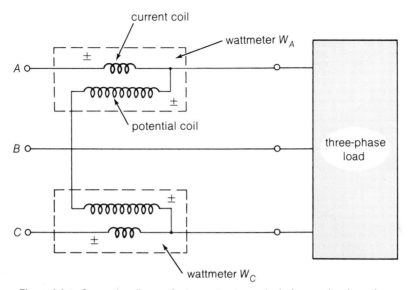

Figure 2.3.1 Connection diagram for two-wattmeter method of measuring three-phase power.

that wattmeter can be dispensed with, and three-phase power can be measured with only two wattmeters that have a common potential junction on any of the three lines in which there is no current coil. This is known as the two-wattmeter method of measuring three-phase power. In general, m-phase power can be measured by means of $m - 1$ wattmeters; the method is valid for both balanced and unbalanced circuits in which either the load or the source is unbalanced.

Figure 2.3.1 is a connection diagram for the two-wattmeter method of measuring three-phase power. The total real power delivered to the load is given by the *algebraic sum* of the two wattmeter readings:

$$P = W_A + W_C \tag{2.3.1}$$

The significance of the algebraic sum is explained in the paragraphs that follow. Two wattmeters can be connected with their current coils in any two lines, and their potential coils are connected to the third line, as shown in Figure 2.3.1. The wattmeter readings are given by

$$W_A = V_{AB} \cdot I_A \cdot \cos \theta_A \tag{2.3.2}$$

where θ_A is the angle between the phasors $\overline{V}_{AB}$ and $\overline{I}_A$, and

$$W_C = V_{CB} \cdot I_C \cdot \cos \theta_C \tag{2.3.3}$$

where θ_C is the angle between the phasors $\overline{V}_{CB}$ and $\overline{I}_C$.

Applying the two-wattmeter method to *balanced* loads yields interesting results. Considering either wye-connected or delta-connected balanced loads and referring to the corresponding phasor diagrams for the phase sequence *A-B-C* (Figures 2.2.2 and 2.2.3), it can be seen that the angle between $\overline{V}_{AB}$ and $\overline{I}_A$ is $(30° + \phi)$ and that the angle between $\overline{V}_{CB}$ and $\overline{I}_C$ is $(30° - \phi)$, where ϕ is the load power-factor angle or the angle associated with the load impedance. Thus we have

$$W_A = V_L I_L \cos (30° + \phi) \tag{2.3.4}$$

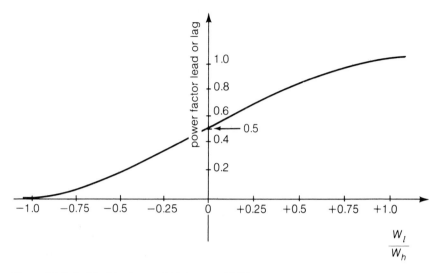

Figure 2.3.2 A plot of load power factor versus W_l/W_h.

and

$$W_C = V_L I_L \cos(30° - \phi) \tag{2.3.5}$$

where V_L and L_L are the magnitudes of the line-to-line voltage and line current, respectively. Simple manipulations yield

$$W_A + W_C = \sqrt{3} V_L I_L \cos\phi \tag{2.3.6}$$

and

$$W_C - W_A = V_L I_L \sin\phi \tag{2.3.7}$$

from which

$$\tan\phi = \sqrt{3}\, \frac{W_C - W_A}{W_C + W_A} \tag{2.3.8}$$

When the load power factor is unity, corresponding to a purely resistive load, both wattmeters will indicate the same wattage. In fact, both of them should read positive. If one wattmeter has a below-zero indication in the laboratory, an upscale deflection can be obtained by simply reversing the leads of either the current coil or the potential coil of the wattmeter. The sum of the wattmeter readings then gives the total power absorbed by the load.

At zero factor—corresponding to a purely reactive load—both wattmeters will have the same wattage indication but opposite signs; their algebraic sum will yield zero power absorbed, as it should. The transition from negative value to positive value occurs when the load power factor is 0.5 (i.e., ϕ equals 60°); at this power factor, one wattmeter reads zero, while the other one reads the total real power delivered to the load.

For power factors (leading or lagging) greater than 0.5, both wattmeters read positive, and the sum of the two readings is the total power. For a power factor less than 0.5 (leading or lagging), the smaller wattmeter reading should be given a negative sign, and the total real power absorbed by the load (which must be positive) is the difference between the two wattmeter readings.

Figure 2.3.2 shows a plot of the load power factor versus the ratio (W_l/W_h), where W_l and W_h are the lower and higher wattmeter readings, respectively.

Another method is sometimes useful in a laboratory environment for determining whether the total power is the sum or the difference of the two readings, and it is described here. To start with, make sure that both wattmeters have an upscale deflection. To perform the test, remove the lead of the potential coil of the lower-reading wattmeter from the common line that has no current coil, and touch the lead to the line that has the current coil of the higher-reading wattmeter. If the pointer of the lower-reading wattmeter deflects upward, add the two wattmeter readings; if the pointer deflects in the below-zero direction, subtract the wattage reading of the lower-reading wattmeter from that of the higher-reading wattmeter.

Given the two wattmeter readings from applying the two-wattmeter method to a three-phase *balanced* load, it is possible to find the tangent of the phase-impedance angle. It is $\sqrt{3}$ times the ratio of the difference between the two readings and their sum, based on Equation 2.3.8. If one knows the system sequence and the lines in which the current coils of the wattmeters are located, the angles's sign can be determined by using these expressions:

For sequence *A-B-C*,

$$\tan \phi = \sqrt{3} \, \frac{W_C - W_A}{W_C + W_A} = \sqrt{3} \, \frac{W_A - W_B}{W_A + W_B} = \sqrt{3} \, \frac{W_B - W_C}{W_B + W_C} \tag{2.3.9}$$

and for sequence *C-B-A*,

$$\tan \phi = \sqrt{3} \, \frac{W_A - W_C}{W_A + W_C} = \sqrt{3} \, \frac{W_B - W_A}{W_B + W_A} = \sqrt{3} \, \frac{W_C - W_B}{W_C + W_B} \tag{2.3.10}$$

The two-wattmeter method we have explained for measuring three-phase power uses single-phase wattmeters. It should be noted that three-phase wattmeters also are available that, when connected appropriately, indicate the total real power absorbed. The total reactive power associated with the three-phase *balanced* load is given by

$$Q = \sqrt{3} \, V_L I_L \sin \phi = \sqrt{3}(W_C - W_A) \tag{2.3.11}$$

based on the two wattmeter readings of the two-wattmeter method.

With the generator action of the source assumed, $+P$ for the real power indicates that the source is supplying real power to the load; $+Q$ for the reactive power indicates that the source is delivering inductive VAR and that the current lags the voltage (i.e., the power factor is lagging), and $-Q$ for the reactive power indicates that the source is delivering capacitive VAR or absorbing inductive VAR and that the current leads the voltage (i.e., the power factor is leading).

▐▌▌▌▌▌▌▌▌▌▌█████████████████████

EXAMPLE 2.3.1

Considering Figure 2.3.1, let balanced positive-sequence three-phase voltages with $\overline{V}_{AB} = 100\sqrt{3}\angle 0°$ V (rms) be applied to terminals *A*, *B*, and *C*. The three-phase wye-connected balanced load consists of per-phase impedance of $(10 + j10)$ Ω. Determine the wattmeter readings, W_A and W_C. Then find the total 3-phase power delivered to the load.

Solution

$$\overline{V}_{AB} = 100\sqrt{3}\angle 0° \text{ V}; \qquad \overline{V}_{BC} = 100\sqrt{3}\angle -120°; \qquad \overline{V}_{CA} = 100\sqrt{3}\angle 120°$$

$$\overline{V}_{AN} = 100\angle -30° \text{ V}; \qquad \overline{V}_{BN} = 100\angle -150°; \qquad \overline{V}_{CN} = 100\angle 90°$$

$$\bar{I}_A = \frac{\bar{V}_{AN}}{Z} = \frac{100\angle -30°}{10\sqrt{2}\angle 45°} = 5\sqrt{2}\angle -75° \text{ A (rms)}$$

$$\bar{I}_C = \frac{\bar{V}_{CN}}{Z} = \frac{100\angle 90°}{10\sqrt{2}\angle 45°} = 5\sqrt{2}\angle 45° \text{ A (rms)}$$

$$WA = V_{AB}I_A \cos\theta_A = (100\sqrt{3})(5\sqrt{2})\cos 75° = 317 \text{ W}$$

$$W_C = V_{CB}I_C \cos\theta_C = (100\sqrt{3})(5\sqrt{2})\cos 15° = 1183 \text{ W}$$

Total power delivered to the load = 317 + 1183 = 1500 W, which checks with $\sqrt{3}V_L I_L \cos\phi = \sqrt{3}(100\sqrt{3})(5\sqrt{2})\cos 45° = 1500$ W.

■■■■■IIII
BIBLIOGRAPHY

Cruz, J. B., Jr., and M. E. Van Valkenburg, *Signals in Linear Circuits.* Boston: Houghton Mifflin, 1974.

Gingrich, H. W. *Electrical Machinery, Transformers, and Control.* Englewood Cliffs, N.J.: Prentice-Hall, 1979.

Hayt, W. H., Jr., and J. E. Kemmerly, *Engineering Circuit Analysis.* 4th ed. New York: McGraw-Hill, 1986.

Johnson, D. E.; J. L. Hilburn; and J. R. Johnson, *Basic Electric Circuit Analysis.* Englewood Cliffs, N.J.: Prentice-Hall, 1978.

Smith, K. C. A., and R. E. Alley, *Electrical Circuits.* Cambridge, UK: Cambridge University Press, 1992.

■■■■■■■■■■■IIIIIIIIIIII
PROBLEMS

2-1. The line-to-line voltage of a balanced wye-connected three-phase source is given as 100 V. Choose V_{AB} as the reference.

   a. For the phase sequence *A-B-C*, sketch the phasor diagram of voltages and find the expressions for the phase voltages.

   b. Repeat part a for the phase sequence *C-B-A*.

2-2. A three-phase, three-wire, 208-volt system is connected to a balanced three-phase load. The line currents $\bar{I}_A$, $\bar{I}_B$, and $\bar{I}_C$ are given to be in phase with the line-to-line voltages $\bar{V}_{BC}$, $\bar{V}_{CA}$, and $\bar{V}_{AB}$, respectively. If the line current is measured as 10 A. What is the per-phase impedance of the load

   a. if the load is wye-connected?   b. if the load is delta-connected?

2-3. Consider a three-phase 25-kVA, 440-V, 60-Hz alternator operating at full load (i.e., delivering its rated kVA) under balanced steady-state conditions. Find the magnitudes of the alternator line current and the phase current

   a. if the generator windings are wye-connected.

   b. if the generator windings are delta-connected.

2-4. Let the alternator of Problem 2–3 supply a line current of 20 A at a power factor of 0.8 lagging. Determine

   a. the kVA supplied by the machine.

    b. the real power kW delivered by the generator.

    c. the reactive power kVAR delivered by the alternator.

2–5. A balanced wye-connected load with a per-phase impedance of $(4 + j3)$ Ω is supplied by a 173-V, 60-Hz, three-phase source.

    a. Find the line current, the power factor, and the total volt-amperes, real power, and reactive power absorbed by the load.

    b. With $\bar{V}_{AB}$ as the reference, sketch a phasor diagram that shows all the voltages and currents.

    c. What would the meter read if the star point of the load is connected to the system neutral through an ammeter?

2–6. A balanced delta-connected load with a per-phase impedance of $(12 + j9)$ Ω is supplied by a 173-V, 60-Hz, three-phase source.

    a. Determine the line current, the power factor, and the total volt-amperes, real power, and reactive power absorbed by the load.

    b. Compare the results of part a with those obtained for part a of Problem 2–5. Explain why they are the same.

    c. With $\bar{V}_{AB}$ as the reference, sketch a phasor diagram showing all voltages and currents.

2–7. A 60-Hz, 440-V, three-phase system feeds two balanced wye-connected loads in parallel. One load has a per-phase impedance of $(8 + j3)$ Ω, and the other, $(4 - j1)$ Ω. Compute the real power in kilowatts delivered to

    a. the inductive load.   b. the capacitive load.

2–8. Balanced wye-connected loads drawing 10 kW at 0.8 pf (power factor) lagging and 15 kW at 0.9 pf leading are connected in parallel and supplied by a 60-Hz, 300-V, three-phase system. Find the line current delivered by the source.

2–9. A balanced delta-connected load with a per-phase impedance of $(30 + j10)$ Ω is connected in parallel with a balanced wye-connected load with a per-phase impedance of $(40 - j10)$ Ω. This load combination is connected to a balanced three-phase supply through three conductors, each of which has a resistance of 0.4 Ω. The line-to-line voltage at the terminals of the load combination is measured to be 2,000 V. Determine the power in kilowatts

    a. delivered to the delta-connected load.

    b. delivered to the wye-connected load.

    c. lost in the line conductors.

2–10. Three identical impedances of $30\angle30°$ Ω are connected in delta to a three-phase, 173-V system by conductors that have impedances of $(0.8 + j0.6)$ Ω each. Compute the magnitude of the line-to-line voltage at the terminals of the load.

2–11. Repeat Problem 2–10 for a delta-connected set of capacitors with a per-phase reactance of $(-j60)$ Ω connected in parallel with load, and compare the result with that obtained for Problem 2–10.

2–12. Two balanced, 3-phase, wye-connected loads are in parallel across a balanced, 3-phase supply. Load 1 draws a current of 20 A at 0.9 power-factor leading, and load 2 draws a current of 30 A at 0.8 power-factor lagging. For the combined load, compute

    a. the current taken from the supply and the supply power factor.

    b. the total real power in kilowatts supplied by the source if the supply voltage is 400 V.

2–13. Two balanced, 3-phase, wye-connected loads are in parallel across a balanced, 3-phase supply. Load 1 draws 15 kVA at 0.8 power-factor lagging, and load 2 draws 20 kVA at 0.6 power-factor leading. Determine

    a. the total real power supplied in kW.

   b.   the total kVA supplied.

   c.   the overall power factor of the combined load.

2–14.   Two balanced, wye-connected, 3-phase loads are in parallel across a balanced, 3-phase, 60-Hz, 208-V supply. The first load takes 12 kW at 0.6 power-factor lagging, and the second load takes 15 kVA at 0.8 power-factor lagging.

   a.   Draw the power triangle for each load and for the combination.

   b.   What is the total kVA supplied?

   c.   Find the supply power factor and specify whether it is lagging or leading.

   d.   Compute the magnitude of the current drawn by each load and drawn from the supply.

   e.   If wye-connected capacitors are placed in parallel with the two loads in order to improve the supply power factor, determine the value of capacitance in farads needed in each phase to bring the overall power factor to unity.

2–15.   A balanced 3-phase load draws 100 kW at 0.8 power-factor lagging. In order to improve the supply power factor to 0.95 leading, a synchronous motor drawing 50 kW is connected in parallel with the load. Compute the kVAR, the kVA, and the power factor of the motor (specify whether it is lagging or leading).

2–16.   a.   A balanced wye-connected load with per-phase impedance of $(20 + j10)$ Ω is connected to a balanced 415-V three-phase supply through three conductors, each of which has a series impedance of $(2 + j4)$ Ω. Find the line current, the voltage across the load, the power delivered to the load, and the power lost in the line conductors.

   b.   Repeat the problem of part a for three per-phase load impedances connected in delta.

2–17.   A balanced electrical industrial-plant load of 9.8 MW with 0.8 lagging power factor is supplied by a 3-phase, 60-Hz system having a maximum rating (load-carrying capacity) of 660 A at 11 kV (line-to-line voltage).

   a.   Determine the apparent power and reactive power drawn by the load.

   b.   An additional load of 1.5 MW and 0.7 MVAR lagging is to be installed in the plant. Compute the minimum rating in MVA of the power-factor correction capacitor that must be installed so that the rating of the line will not be exceeded. Then find the system power factor.

   c.   If the capacitor consists of three equal sections and is connected in delta across the supply lines, what capacitance will be required in each section?

2–18.   Derive the relationships given in Figure 2.2.1 for the wye–delta and delta–wye transformations.

2–19.   Two three-phase generators supply a common balanced, three-phase load of 30 kW at 0.8 power-factor lagging. The per-phase impedance of the lines connecting generator G_1 to the load is $(1.4 + j1.6)$ Ω, and that of the lines connecting generator G_2 to the load is $(0.8 + j1)$ Ω. If generator G_1 supplies 15 kW at 0.8 pf lagging while operating at a terminal voltage of 800 V (line-to-line), find the voltage at the load terminals, the terminal voltage of generator G_2, and the real-power and reactive-power output of generator G_2.

2–20.   Determine the wattmeter readings when the two-wattmeter method is applied to Problem 2–5, and check the total power obtained.

2–21.   When the two-wattmeter method for measuring three-phase power is applied to a particular balanced load, readings of 1,200 W and 400 W are obtained (with no reversals). Determine the delta-connected load impedances if the system voltage is 440 V. With only this information, is it possible to determine if the load impedance is capacitive or inductive?

2–22.   The two-wattmeter method for measuring three-phase power is applied to a balanced wye-connected load as shown in Figure 2.3.1, and the readings are $W_C = 836$ W and $W_A = 224$ W. If the system voltage is 100 V, what is the per-phase impedance of the load? Is it possible to specify the capacitive or inductive nature of the impedance?

2-23. Two wattmeters are used as shown in Figure 2.3.1 to measure the power absorbed by a balanced delta-connected load. Determine the total power in kilowatts, the power factor, and the per-phase impedance of the load if the supply-system voltage is 120 V and the wattmeter readings are

a. $W_A = -500$ W   and $W_C = 1,300$ W.

b. $W_A = 1,300$ W   and $W_C = -500$ W.

2-24. Refer to Problem 2–16b. With A-phase voltage as the reference phasor, with positive-phase sequence supply system, two single-phase wattmeters are used to measure the power delivered to the load. The wattmeter current coils are arranged as in Figure 2.3.1 to carry currents I_A and I_C, and the potential coils share a common connection at node B. Determine the readings of wattmeters W_A and W_C, and verify that the sum of these readings is the total power in the load.

3

The Magnetic Aspect

In engineering practice solutions to problems generally must be in a convenient form and easy to interpret and visualize. This accounts for the widely accepted use of the *circuit concept,* as distinct from the *field concept.* Although all phenomena occur in space and time—and these are field phenomena—it is possible to approximate them by time relations only, summarizing the actual physical quantities by means of a few parameters, preferably of constant value. However, it is sometimes necessary to re-examine the basis of the circuit concept by studying the electromagnetic fields involved. For producing the smallest, most economical, most efficient device, accurate knowledge of the fields present in the device, expecially the magnetic field, is desirable. It is necessary to analyze the fields and modify the design for optimization. Thus the *field theory,* as opposed to the *circuit theory,* is increasingly useful for solving engineering problems. A significant part of designing an electromechanical device centers on the design of the magnetic system. The purposes of this chapter are to review the basic concepts of field theory and to present some concepts of magnetism that are valuable in designing and analyzing electromagnetic devices.

3.1 Review of Electromagnetic Field Theory

The fundamental laws governing electromagnetic phenomena at any point in space can be expressed by the Maxwell equations. Assuming that the dielectric properties of a material are not significant compared with the conduction properties, as is true in low-frequency problems, the dielectric effects or the displacement currents can be neglected, and the material regions can be considered as void of the volume charge density. Then the relevant equations can be written in differential and integral forms as

$$\nabla \cdot \bar{B} = 0; \qquad \oiint_S \bar{B} \cdot d\bar{S} = 0 \qquad\qquad (3.1.1; \ 3.1.2)$$

$$\nabla \times \bar{H} = \bar{J}; \qquad \oint_c \bar{H} \cdot d\bar{l} = I \qquad\qquad (3.1.3; \ 3.1.4)$$

$$\nabla \times \bar{E} = -\frac{\partial \bar{B}}{\partial t}; \qquad \oint_c \bar{E} \cdot d\bar{l} = -\iint_S \frac{\partial \bar{B}}{\partial t} \cdot d\bar{S} \qquad\qquad (3.1.5; \ 3.1.6)$$

In the standard international (SI) system of units, $\bar{B}$ is the magnetic flux density in webers per square meter (Wb/m^2) or tesla (T); $\bar{H}$ is the magnetic field intensity in amperes (or ampere-turns) per meter (A/m or At/m); $\bar{J}$ is the current density in amperes per square meter (A/m^2); I is the current in amperes (A); $\bar{E}$ is the electric field intensity in volts per meter (V/m); and t is the time in seconds (s). Other systems of units such as the centimeter-gram-second (CGS) system and the English system have been widely used, and the relationships among the systems are given in Appendix A (Table A.4).

Equation 3.1.2, known as Gauss's law for the magnetic field, states that the total magnetic flux passing through a closed surface is zero. Equation 3.1.4, Ampere's law, states that the line integral of the magnetic field intensity around a closed line contour is equal to the current enclosed by the contour. The current-continuity relation, known as Kirchhoff's current law, is a consequence of Equation 3.1.3 and is given by

$$\nabla \cdot \bar{J} = 0; \qquad \oint_S \bar{J} \cdot d\bar{S} = 0 \qquad\qquad \text{(3.1.7; 3.1.8)}$$

Equation 3.1.6 is Faraday's law, according to which an electric field is produced in space when the magnetic field is allowed to vary with time. The line integral is taken along the closed contour around the surface through which the magnetic flux density passes. As for the direction, if the line-contour integral is taken in the counterclockwise (positive) direction along the fingers of the right hand oriented in the direction of the closed path, the thumb then indicates the positive direction for the right sides of Equations 3.1.4 and 3.1.6. A flux density $\bar{B}$ in the direction of $d\bar{S}$, *decreasing* with time, results in a positive direction for $\bar{E}$ along the closed path.

The scalar form of Faraday's law is given by

$$e = -\frac{d\phi}{dt} \quad \text{or} \quad -\frac{d\lambda}{dt} \qquad\qquad \text{(3.1.9)}$$

where

$$\phi = \iint_S \bar{B} \cdot d\bar{S} \qquad\qquad \text{(3.1.10)}$$

and

$$\lambda = N\phi \qquad\qquad \text{(3.1.11)}$$

where ϕ is the magnetic flux in webers (Wb) passing across an arbitrary surface, and λ is the flux linkage in weber-turns given by the product of a number of turns (N) and the flux (ϕ) linking N.

It can be seen from Equation 3.1.9 that a time-varying flux linking a stationary coil yields a time-varying voltage. Also, a conductor or a coil moving through a magnetic field will have an induced voltage, known as the *motional emf*, given by

$$\text{Motional emf} = \oint_c (\bar{U} \times \bar{B}) \cdot d\bar{l} \qquad\qquad \text{(3.1.12)}$$

where $\bar{U}$ is the velocity vector of the conductor. The voltage induced in a conductor of length l moving perpendicular to the direction of a non-time-varying, uniform magnetic field of flux density $\bar{B}$ with a constant velocity $\bar{U}$ can be expressed as

$$\text{Motional emf} = BlU \qquad\qquad \text{(3.1.13)}$$

This is often called the *cutting-of-flux* equation, in which $\bar{B}$, $\bar{l}$, and $\bar{U}$ are mutually perpendicular. The direction for the motional emf can be worked out either from the right side of Equation 3.1.12 or from the simple *right-hand-rule* in cases where Equation 3.1.13 is applicable. If the thumb, first, and second fingers of the right hand are extended so that they are mutually perpendicular to each other, and if the thumb represents the direction of $\bar{U}$ and the first finger the direction of $\bar{B}$, the second finger will then represent the direction of the emf along $\bar{l}$. Equation 3.1.12 is quite useful in analyzing rotating machines.

Thus we see that the voltage effect can be caused by a time-varying flux with stationary coils, as seen later to be the case in a transformer. The same effect would result from the relative motion of a coil and a constant-amplitude spatial flux-density wave, as seen later to be the case in a rotating machine. Rotation, in effect, introduces the time element and transforms a space distribution of flux density into a time variation of voltage.

In addition to Ampere's law and Faraday's law, another significant field relationship is the Lorentz force equation,

$$d\bar{F} = I \, d\bar{l} \times \bar{B} \tag{3.1.14}$$

where I is the current flowing through a differential conductor of length $d\bar{l}$. Only the effect of the magnetic field is considered here, while that of the electric field, if any, is neglected. Applying Equation 3.1.14 to a simple situation in which a current-carrying conductor is located in a uniform magnetic field and the direction of the conductor is perpendicular to the direction of the magnetic field, the force is given by

$$F = BIl \tag{3.1.15}$$

which is often used in machine analysis. The direction of the force is orthogonal (perpendicular) to the direction of both the conductor and the magnetic field.

Another law used for analytical purposes is the Biot–Savart law, given by

$$d\bar{H} = \frac{I \, d\bar{l} \times \bar{a}_R}{4\pi R^2}; \qquad \bar{H} = \oint \frac{I \, d\bar{l} \times \bar{a}_R}{4\pi R^2} \tag{3.1.16; 3.1.17}$$

from which the magnetic field intensity $d\bar{H}$ can be evaluated at a point in space due to a differential current-carrying element of length $d\bar{l}$ carrying a current I at a distance R in the direction from the current element given by the unit vector $\bar{a}_R$.

By manipulating Maxwell's field equations one can come up with partial-differential equations of Laplace's or Poisson's type in terms of the scalar potential or the vector potential. Then their solution can be determined either analytically or numerically on a digital computer, subject to the boundary conditions to be satisfied. Such methods of finding the flux distribution are quite involved; in view of the size and the scope of this book, they are not considered here.

3.2 Magnetic Materials

For material media, the magnetic flux density, $\bar{B}$, and the field intensity, $\bar{H}$, are related through the relationship

$$\bar{B} = \mu \bar{H} \tag{3.2.1}$$

where μ stands for the permeability of the material. The free-space permeability, μ_0, is a constant given by the following in the SI system of units:

$$\mu_0 = 4\pi \times 10^{-7} \text{ henry/meter} = 1.257 \times 10^{-6} \text{ H/m} \tag{3.2.2}$$

For a linear (i.e., with a straight-line relationship between B and H), homogeneous (i.e., having uniform quality), isotropic (i.e., with the same properties in any direction) material, the permeability is a constant given by the slope of the linear B-H characteristic, and it is related to the free-space permeability as

$$\mu = \mu_r \mu_0 \tag{3.2.3}$$

where μ_r is the relative permeability, a dimensionless constant. If the B-H characteristic is nonlinear, as with a number of common materials, then the permeability varies as a function of the magnetic induction.

Some classifications of materials and their distinguishing characteristics are given here.

- *paramagnetic:* materials whose relative permeability is slightly greater than unity.
- *diamagnetic:* materials whose relative permeability is slightly less than unity.
- *nonmagnetic:* materials whose relative permeability is essentially equal to 1.0.
- *superconducting:* materials whose relative permeability is essentially zero and that exhibit "perfect diamagnetism" at temperatures near absolute zero. No magnetic field can be established.
- *ferromagnetic:* materials that exhibit a high degree of magnetizability; generally subdivided into hard and soft materials. *Soft* ferromagnetic materials include most of the soft steels and many four-component alloys of iron, nickel, cobalt, and a rare-earth element. *Hard* ferromagnetic materials include the permanent-magnet materials such as the alnicos, the alloys of cobalt with a rare-earth element such as samarium, the chromium steels, the copper-nickel alloys, and other metal alloys.
- *ferrimagnetic:* materials that are ferrites (composed of iron oxides); subdivided into hard and soft materials. *Hard* ferrimagnetic materials include permanently magnetic ferrites, usually barium or strontium ferrites. *Soft* ferrimagnetic materials include nickel-zinc and manganese-zinc ferrites.
- *super-paramagnetic:* magnetic materials composed of powdered iron particles or other magnetic material suspended in a nonferrous matrix such as epoxy or other plastic. Of the powdered materials, molybdenum-nickel-iron powder, known as *permalloy,* is one of the best known.
- *ferro-fluidic:* magnetic fluids consisting of three components: a carrier fluid, iron oxide (Fe_3O_4) particles suspended in the fluid, and a stabilizer.
- *amorphous:* magnetic materials with a new-type spin ordering not classified by those of crystalline sytems because of their topologically disordered nature. Typical examples of amorphous ferromagnetic materials are transition metal–metalloid alloys such as $Fe_{80-x} B_{20-x}$ and $(Fe_x Ni_{1-x})_{80}B_{10}P_{10}$. Transition metal–rare-earth gadolinium (Gd) alloy systems such as $GdCO_3$ are examples of amorphous ferrimagnets. Canted or noncollinear ferrimagnets are the transition metal–heavy rare-earth alloy systems such as amorphous $DyCO_3$, where Dy stands for dysprosium. Most amorphous materials are obtained by quenching the corresponding liquids at relatively high cooling rates; a few are obtained by deposition techniques in a vacuum.

The permeability ratio between good and poor magnetic materials over a typical working range of operation is of the order of 10^4, while the electrical conductivity ratio between a good electrical conductor (such as copper) and a good insulator (such as polystyrene) is of the order of 10^{24}. Thus, no material that can be described as a good magnetic insulator exists except for the superconductors.

For proper use of magnetic materials in various electromagnetic systems, a number of magnetic properties (in addition to mechanical properties) must be considered: (1) the magnetization characteristic giving the magnetic field intensity (H) at various levels of the magnetic flux density (B), (2) the saturation flux density, (3) the hysteresis characteristics, (4) the permeability at various levels of the flux density,

Table 3.2.1 Typical Characteristics of Some Soft Magnetic Materials

Trade Name	Saturation Flux Density (T)	Maximum Relative Permeability ($\times 10^3$)	Electrical Resistivity (ohm-meter $\times 10^{-6}$)	Curie Temperature (°C)
Carpenter				
Silicon core iron	2–2.1	4–5	0.25-0.60	800
Electrical iron	2.15	2.2–5.5	0.10	760
430F solenoid quality	1.47	1.1–1.6	0.60	671
High-permeability 49	1.6	30–120	0.48	450
Hy mu 80	0.78	70–75	0.58	460
Hiperco 27	2.36	2.8	0.19	925
Silectron	1.97	10–20	0.50	732
2 V Permendur	2.30	8	0.40	932
Monimax	1.45	40–100	0.65	398
Deltamax	1.60	100–200	0.45	499
4-79 Molybdenum permalloy	0.80	100–400	0.55	454
Ferrite	0.22–0.45	0.16–10	$(0.1 \times 10^6) - (10 \times 10^6)$	135–500

(5) the variation of permeability with temperature, (6) the electrical resistivity, (7) the Curie temperature, and (8) the loss coefficients. The *Curie temperature,* or the *Curie point,* is the critical temperature beyond which a ferromagnetic material loses its essential magnetic properties. Typical characteristics of some of the soft magnetic materials are given in Table 3.2.1.

Relative permeabilities of a number of materials of various classifications are given in Table 3.2.2. Ferromagnetic materials obviously have very high relative permeability. Ferrites with low electrical conductivity are employed in inductor cores for ac applications.

Ferromagnetic materials are quite easily magnetized, since they have high relative permeability. The magnetization characteristic (or the *B-H* curve) is a result of the domain changes within the magnetic material. A typical *B-H* characteristic is shown in Figure 3.2.1. Applying weak magnetic fields causes the domains to undergo boundary displacement. An increase in magnetic-field intensity produces a sudden orientation of the domains in the direction of the applied field. Further increases result in slower orientation of the domains, and when the *B-H* curve is practically flat, the material is said to be *saturated.* The small increase in flux density beyond the *saturation flux density* is exhibited by the linear relationship $B = \mu_0 H$.

Figure 3.2.1 shows three regions of a typical nonlinear magnetization curve, which is single-valued. The ratio of B to H at any point on the curve is known as *amplitude permeability.* The permeability in region I is said to be the *initial permeability;* in region II, the *B-H* curve for many materials is relatively straight, so that linear theory can be applied if a magnetic device is operated only in this region.

Nonlinearity of the *B-H* relationship of magnetic materials is a principal impediment to accurate analysis of electromechanical systems. Digital computations may be more convenient for analyzing

Table 3.2.2 Classifications and Relative Permeabilites of Selected Materials

Material	Classification	Relative Permeability, μ_r
Bismuth	diamagnetic	0.99983
Silver	diamagnetic	0.99998
Lead	diamagnetic	0.999983
Copper	diamagnetic	0.999991
Water	diamagnetic	0.999991
Vacuum	nonmagnetic	1
Air	paramagnetic	1.0000004
Aluminum	paramagnetic	1.00002
Palladium	paramagnetic	1.0008
2-81 Permalloy powder (2 Mo, 81 Ni, Iron)	super-paramagnetic	130
Cobalt	ferromagnetic	250
Nickel	ferromagnetic	600
Ferroxcube 3 (Mn-Zn-ferrite powder)	ferrimagnetic	1,500
Ferrites	ferrimagnetic	160~10,000
Mild steel (0.2 C)	ferromagnatic	2,000
Iron (0.2 impurity)	ferromagnatic	5,000
Silicon iron (4 Si)	ferromagnatic	7,000
78 Permalloy (78.5 Ni)	ferromagnatic	100,000
Mumetal (75 Ni, 5 Cu, 2 Cr)	ferromagnatic	100,000
Purified iron (0.05 impurity)	ferromagnatic	200,000
Superalloy (5 Mo, 79 Ni)	ferromagnatic	1,000,000

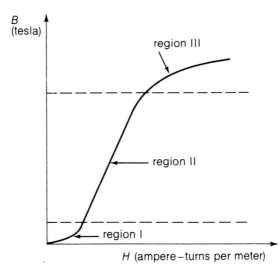

Figure 3.2.1 Typical magnetization characteristic showing three regions of domain behavior.

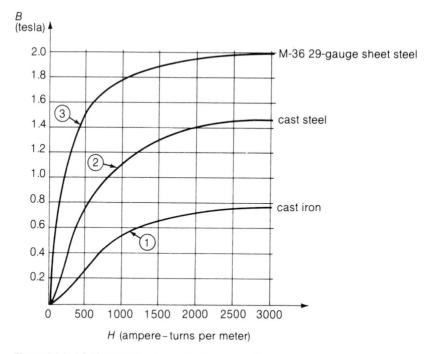

Figure 3.2.2 DC Magnetization curves for three magnetic materials.

some relations between B and H. Typical expressions (out of the many proposed) are given here:

$$B = \frac{aH}{1 + bH} \tag{3.2.4}$$

$$B = \frac{a_0 + a_1 H + a_2 H^2 + \cdots}{1 + b_1 H + b_2 H^2 + \cdots} \tag{3.2.5}$$

$$H = [k_1 \exp(k_2 B^2) + k_3] B \tag{3.2.6}$$

where a, b, a_0, a_1, a_2, ..., b_1, b_2, ..., k_1, k_2, and k_3 are constants to be determined from the experimental points on the measured B-H curve for a given material.

Alternatively, by determining a sufficiently large number of points on the B-H curve and storing the data in the computer, either linear interpolation or some other curve-fitting technique could be used to obtain the intermediate values of the magnetization characteristic.

The magnetic permeability of ferromagnetic materials varies with the magnetic field intensity, starting at a relatively low value, increasing to a maximum, and then falling off with increasing saturation. Typical dc magnetization curves for three magnetic materials are given in Figure 3.2.2. The characteristic can be obtained in two ways: the virgin B-H curve, obtained from a totally demagnetized sample; or the normal B-H curve, obtained by joining the tips of hysteresis loops of increasing magnitude. The slight differences between the two methods are not of sufficient importance to discuss here.

When the magnetization is carried through a complete cycle, the permeability for increasing flux density generally differs from that for decreasing flux density at the same value of the magnetic field

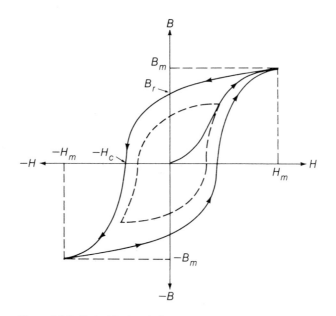

Figure 3.2.3 Typical hysteresis loop.

intensity, as exhibited by the typical *hysteresis loop,* shown in Figure 3.2.3. As seen from the figure, when H is zero there is a residual value of the flux density, B_r, which depends on the material, its crystal structure, and the value of B_m. To bring the flux density down to zero from B_r requires a demagnetizing coercive force whose magnitude is given by H_c. As H is varied unidirectionally from $+H_m$ to $-H_m$ and back again unidirectionally to $+H_m$, a complete hysteresis loop is traced out. The normal single-valued *B-H* curve follows the locus of (B_m, H_m) in the first quadrant for a series of steady-state hysteresis loops, taken up to various values of the peak flux density, as shown in Figure 3.2.3.

The area of the loop is a function of B_m and is found experimentally to vary as B_m^n up to moderate values of the flux density; the index n ranges from 1.5 to 2.5. The loop area is related to the energy required to reverse the magnetic domain walls as the magnetizing force is reversed. This is an irreversible energy and results in an energy loss known as the *hysteresis loss*. The area of the loop varies with temperature and the frequency of the reversal of H for a given material.

Soft or easily magnetized materials have relatively thin hysteresis loops with smaller enclosed area, whereas hard magnetic materials exhibit hysteresis loops with greater enclosed area. Square-loop ferrites exhibit nearly square hysteresis loops at low frequencies with relatively low values of the coercive force and high resistivities. Such ferrite materials are found to be suitable for high-speed memory storage and switching elements.

The most common amorphous metallic alloys contain ferromagnetic metals (Fe, Co, Ni, or rare earths). Most amorphous ferromagnets can be described as magnetically soft. Recent advances in material-science technology have resulted in the development of various amorphous metals with high resistivity and much lower core loss at high frequency because of small strip thickness. With higher permeability over a wide frequency range, these amorphous metals exhibit square, regular, or even flat hysteresis loops, depending on the thermal treatment. Applications of them include magnetic cores in switched-mode power supplies, high-frequency transformers, sensors, and distribution transformers, because of their superior magnetic characteristics and high stability.

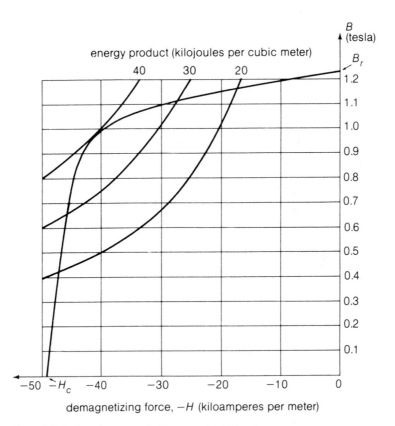

Figure 3.2.4 Normal demagnetization curve for Alnico 5.

Amorphous metals have been developed in strip thicknesses of 16 to 24 μm, with saturation flux density of 0.53 to 1.5 T, coercive force of 0.32 to 4.2 A/m, remanence ratio (B_r/B_m) of 50% to 95%, core loss of 300 to 2000 kw/m^3 at 0.2 T and 100 kHz, Curie temperature of 180° to 420° C, and crystallization temperature of 446° to 540° C. Depending on the application, materials with high remanence ratio/low core loss, high permeability/low core loss, or high flux density will be employed. *Finemet*, an Fe-based, soft magnetic alloy composed of ultrafine crystal grains obtained through a super-quick cooling process, has recently been brought out. Its excellent soft magnetic properites are comparable to those of Co-based amorphous alloys—low magnetic distortion, high permeability, high saturation flux density of 1.35 T, and Curie temperature of 570° C.

The second quadrant of the hysteresis loop, known as the *demagnetization curve,* is used in analyzing devices that contain *permanent magnets*. The characteristic for Alnico 5 is shown in Figure 3.2.4. The *coercive force, H_c,* is the magnitude of the demagnetizing force corresponding to zero flux density and gives a measure of the material's resistance to demagnetization. The magnitude of B on the vertical axis (for $H = 0$) is known as the *residual flux density* or the *residual induction B_r.* Also shown in Figure 3.2.4 is the *energy product,* the product of the flux density (B) and the magnitude of the demagnetizing force (H) for a point on the demagnetization curve. This energy product becomes a maximum at a point on the demagnetization curve. The maximum energy product is often used to compare permanent-magnet materials on the basis of the amount of energy available to an external magnetic circuit.

The units, symbols, constants, and conversion factors for the SI system are given in Appendix A.

Iron Losses

Iron core losses are usually divided into two components: the *hysteresis loss* and the *eddy-current loss*. Time-varying fluxes produce core losses in ferromagnetic materials. No such losses occur in iron cores carrying flux that does not vary with time.

The area of the hysteresis loop represents the energy loss during one cycle in a unit cube of the core material. The hysteresis loss per cycle in a core of volume (Vol) possessing a uniform flux density B throughout its volume is given by

$$P_h \text{ per cycle} = \text{Vol} \oint H \, dB \qquad (3.2.7)$$

where the closed line integral represents the area of the hysteresis loop. The hysteresis loss per second is commonly expressed empirically by

$$P_h = k_h \cdot \text{Vol} \cdot f \cdot B_m^n \qquad (3.2.8)$$

since a cyclic variation of the flux at f Hz results in f hysteresis loops per second; k_h is a proportionality constant dependent on the characteristics of iron and the units used; the exponent n, as stated earlier, ranges from 1.5 to 2.5 (a value of 2.0 is often used for estimating purposes); and B_m is the maximum value of the flux density.

Since iron is a conductor, a changing flux induces voltages and currents, called *eddy currents,* that circulate within the iron mass. These eddy currents produce losses, heating, and demagnetization. If the current paths are assumed to be wholly resistive and the redistribution of flux due to demagnetization is neglected so that the flux density can be regarded as uniform, a simple expression for the eddy-current loss can be obtained. Neglecting saturation, the eddy-current loss in a sinusoidally excited material is usually expressed by

$$P_e = k_e \cdot \text{Vol} \cdot f^2 \cdot \tau^2 \cdot B_m^2 \qquad (3.2.9)$$

where k_e is a proportionality constant dependent on the characteristics of iron and the units used; Vol is the volume of iron; f is the frequency; τ is the lamination thickness; and B_m is the maximum value of the flux density.

To reduce eddy-current loss, ferromagnetic materials that carry pulsating flux are *laminated* (built up from thin sheets electrically insulated from one another). To reduce the eddy currents, the thin sheets must be oriented in a direction parallel to the flow of magnetic flux. The eddy-current loss is inversely proportional to the electrical resistivity of the ferromagnetic material and directly proportional to the square of the frequency, the square of the lamination thickness, and the square of the maximum flux density, as shown by Equation 3.2.9.

A common thickness of lamination for continuous operation at 60 Hz is 0.35 mm. Lamination thicknesses typically vary from 0.3 to 5 mm in electromagnetic devices meant for power applications and from 0.01 to 0.5 mm in devices for electronics applications. Photo 3.2.1 shows typical stator- and rotor-core laminations.

Laminating a magnetic material generally results in an increase in the overall volume, with the increase depending upon the method used to bond the laminations together. The *stacking factor* is defined as the ratio of the volume occupied by the magnetic material to the overall (or gross) volume of the magnetic part. As the lamination thickness increases, the stacking factor approaches unity.

Equations 3.2.8 and 3.2.9 are to be regarded as guides to the relative effects of the various factors; the actual core loss should be determined from experimental data. Most manufacturers of magnetic

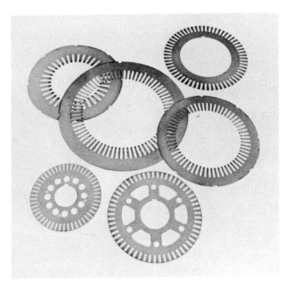

Photo 3.2.1 Typical stator- and rotor-core laminations for induction motors. (Photo courtesy of Westinghouse Electric Corporation.)

materials provide core-loss data for most of their products corresponding to the condition of sinusoidal excitation. However, such data are not generally adequate when the electromagnetic devices are applied in circuits in which the voltages and currents have nonstandard waveshapes, such as those frequently caused by the switching action of semiconductors. In general, it is a good idea to measure the core loss with the device excited by a source whose waveform is as close as possible to that under which the device will be operated in actual use.

In producing an alternating flux in minimum section with minimal loss, the important considerations are the permeability and the core loss. Magnetic sheet materials are alloy steels in which the chief alloy constituent is silicon, which raises the permeability at low flux densities and reduces the hysteresis and eddy-current losses. The silicon content, however, affects the mechanical properties; it increases the tensile strength, while it impairs the ductility. For example, whereas a 5-percent silicon steel cannot be punched or sheared easily, adding nickel to it makes *cold-rolling* possible, which offers some important advantages. Where core loss is not a primary consideration, a low silicon content is used. A high-grade, low-loss steel is utilized if the losses are to be minimized and the cooling is critical. The directional properties of *cold-rolled, grain-oriented* sheet and strip materials are highly advantageous in some cases, because the magnetic properties in the rolling direction are far superior to those on any other axis. Such materials can sustain higher flux densities than the high-resistance steel of 4–5 percent silicon content.

EXAMPLE 3.2.1

a. Estimate the hysteresis loss at 60 Hz for a toroidal (doughnut-shaped) core of 300-mm mean diameter and a square cross section of 50 mm × 50 mm. The symmetrical hysteresis loop for M-36 electric sheet steel (of which the torus is made) is given in Figure 3.2.5.

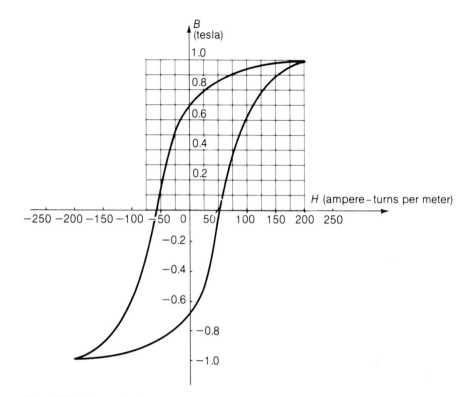

Figure 3.2.5 Symmetrical hysteresis loop for M-36 electric sheet steel.

b. Now suppose that all the linear dimensions of the core are doubled. How will the hysteresis loss differ?

c. Next, suppose that the torus (which was originally laminated for reducing the eddy-current losses) is redesigned so that it has half the original lamination thickness. Assume the stacking factor to be unity in both cases. What would be the effect of such a change in design on the hysteresis loss?

d. Suppose that the torodial core of part a is to be used on 50-Hz supply. Estimate the change in hysteresis loss if no other conditions of operation are changed.

Solution

a. Refer to Figure 3.2.5. The area of each square represents

$$(0.1 \text{ T}) \times (25 \text{ A/m}) = 2.5 \frac{\text{Wb}}{\text{m}^2} \times \frac{\text{A}}{\text{m}}$$

$$= 2.5 \frac{\text{V} \times \text{s} \times \text{A}}{\text{m}^3}$$

$$= 2.5 \text{ J/m}^3$$

By counting, the number of squares in the upper half of the loop is found to be 43; the area of the hysteresis loop is given by

$$2 \times 43 \times 2.5 = 215 \text{ J/m}^3$$

The toroidal volume is

$$\pi \times 0.3 \times 0.05^2 = 2.36 \times 10^{-3} \text{ m}^3$$

Hysteresis loss in the torus is then given by

$$2.36 \times 10^{-3} \times 60 \times 215 = 30.44 \text{ W}$$

b. When all the linear dimensions of the core are doubled, its volume will be eight times the previous volume. Hence the new core hysteresis loss will be

$$8 \times 30.44 = 243.52 \text{ W}$$

c. The lamination thickness has no bearing on the hysteresis loss; it changes only the eddy-current loss. Hence, the hysteresis loss remains unchanged.

d. Since the hysteresis loss is directly proportional to the frequency, the loss on the 50 Hz will be

$$30.44 \times \frac{50}{60} = 25.37 \text{ W}$$

■■■■■■||||||
3.3 Magnetic Circuits

The concept of a magnetic circuit is useful in estimating the flux produced by coils wound on ferromagnetic materials. Magnetic circuit analysis follows the lumped-parameter approach of simple dc electrical circuit analysis and is applicable to systems excited by dc sources and, by means of an incremental approach, to systems with low-frequency ac excitation. Calculations of excitation are usually based on *Ampere's law,* given by

$$\oint \bar{H} \cdot d\bar{l} = \text{current enclosed} \quad \text{(or ampere-turns enclosed)} \tag{3.3.1}$$

which is strictly valid for steady (or time-invariant) currents and is approximately valid when the time variation of current is so slow that electromagnetic radiation is negligible. The direction of the current enclosed by the contour and the reference direction of the contour are related by the *right-hand rule.*

Equation 3.3.1 can be written in a scalar form as

$$\oint (\cos \theta) H \, dl = \text{current enclosed} \tag{3.3.2}$$

in which θ is the angle between the vectors $\bar{H}$ and $d\bar{l}$. If the closed contour is considered to be made up of n small incremental lengths $dl_k (k = 1 \text{ to } n)$, the integral may be replaced with a summation:

$$\sum_{k=1}^{n} \cos \theta_k H_k dl_k = \text{current enclosed} \tag{3.3.3}$$

where H_k is assumed to be a constant over the corresponding dl_k. Further, if the corresponding H_k and dl_k are in the same direction, $\cos \theta_k$ becomes unity, simplifying the calculations.

The *magnetomotive force (mmf)*, or the *magnetic potential difference* that tends to produce a magnetic field, has the unit of amperes, or ampere-turns in the SI system of units. (Note that the number of turns is a dimensionless constant.) The two sources of mmf in magnetic circuits are the electric current and the permanent magnet. The current source is commonly a coil of N turns carrying a current known as the *exciting current.*

Analogous to Ohm's law for dc electric circuits, we have this relation for magnetic circuits:

$$\phi = \frac{\mathbf{F}}{\mathbf{R}} \tag{3.3.4}$$

ϕ is the magnetic flux, $\mathbf{F}$ is the mmf, and $\mathbf{R}$ is the *reluctance* given by

$$\mathbf{R} = \frac{l}{\mu A} \tag{3.3.5}$$

for uniform permeability μ and the cross-sectional area A; l stands for the corresponding portion of the length of the magnetic circuit. The reciprocal of the reluctance is known as the *permeance* with the SI unit of henry. The permeance can also be expressed as the ratio of two quantities; the first is the flux through any cross section of a tubular portion of a magnetic circuit bounded by lines of force and by two equipotential surfaces; the second is the magnetic potential difference between the surfaces, taken within the portion under consideration.

The analogy between the electric and magnetic circuits holds in a number of aspects, and it is summarized in Table 3.3.1. One should not infer, however, that electric and magnetic circuits are analogous in all respects. For example, there are no magnetic insulators similar to the electrical insulators known to exist for electric circuits. Also, energy must be continuously supplied when a direct current is established and maintained in an electric circuit; a similar situation does not prevail in the case of a magnetic circuit in which a flux is established and maintained constant. The reluctance of a magnetic circuit is somewhat analogous to the resistance in an electric circuit with the exception that the reluctance is not an energy-loss component.

The magnetic circuit calculations can be carried out in any one of the systems of units that have been in practice from time to time. Refer to Table A.4 of Appendix A for the conversion factors for the various systems of units.

In the case of ferromagnetic systems containing air gaps, a useful approximation for making quick estimates is to consider the ferromagnetic material to have infinite permeability. This is somewhat justified, since the permeability of ferromagnetic materials used in electromagnetic devices is of the order of a few thousand times that of air. The relative permeability of iron is considered so high that practically all the ampere-turns of the winding are consumed in the air gaps alone.

Two other terms commonly associated with the analysis of magnetic circuits are *leakage* and *fringing.* That portion of the flux that does not pass through the air gap or the useful part of the magnetic circuit is generally known as *leakage flux.* When used in conjunction with a coupled circuit with two or more windings, it refers to that part of the flux that links one coil but not the other. Leakage, being a characteristic of all practical magnetic circuits, can never be completely eliminated. The flux lines can be better directed along a chosen path by using magnetic shielding—thin sheets of high-permeability material—at dc or very low ac excitation frequencies. At higher frequencies of excitation, electrical shielding—aluminum foil or some other conducting material—is employed to reduce the leakage flux by dissipating its energy in the form of induced currents in the shield.

Whenever there is an air gap in a ferromagnetic system, there is a tendency for the flux to bulge outward along the edges of the air gap, instead of passing straight through the air gap parallel to the

Table 3.3.1 Analogy between Electric and Magnetic Circuits

	Electric Circuit	Unit	Magnetic Circuit	Unit
Driving force	emf (V)	V	mmf ($\mathbf{F}$)	At
Response	current (I)	A	flux (ϕ)	Wb
Impedance	resistance (R)	Ω	reluctance ($\mathbf{R}$)	1/H
Equivalent circuit	$V = IR$		$\mathbf{F} = \phi\mathbf{R}$	
Field intensity relationship	$\oint \bar{E} \cdot d\bar{l} = V$	V	$\oint \bar{H} \cdot d\bar{l} = I$	A
Potential difference	$V = IR$	V	$\mathbf{F} = \phi\mathbf{R}$	At
Other relations	$J = \dfrac{I}{A} = \dfrac{V}{AR} = \dfrac{El}{A(\rho l/A)}$ $= \dfrac{E}{\rho} = \sigma E$ or $E = \rho J = J/\sigma$, where J is the current density, ρ is the resistivity, and σ is the conductivity	A/m^2 $\Omega\cdot$m $1/(\Omega\cdot$m)	$B = \dfrac{\phi}{A} = \dfrac{\mathbf{F}}{A\mathbf{R}} = \dfrac{Hl}{A(l/\mu A)}$ $= \mu H = H/\nu$ or $H = B/\mu = \nu B$, where B is the flux density, μ is the permeability, and ν is the reluctivity	T or Wb/m^2 H/m m/H
Admittance	conductance $G = 1/R$	S	permeance $\mathbf{P} = 1/\mathbf{R}$	H

edges, as shown in Figure 3.3.1. This is referred to as *fringing*. Fringing occurs because the reluctances of different paths available near the air gap are quite comparable to each other, and the flux lines spread out just as the current divides into parallel branches in electric circuits. The effect of fringing is usually taken into account in magnetic circuit calculations by recognizing that the effective area of the air gap is greater than the actual area. Note that the relative effect of fringing increases with the length of the air gap.

Ampere's law, applied to the analysis of a magnetic circuit, states that the algebraic sum of the magnetic potentials around any closed path is zero. This is analogous to the Kirchhoff's voltage relationship in electrical circuits. Series and parallel magnetic circuits can be analyzed by means of their corresponding electric-circuit analogs. A few examples with numerical values will illustrate these statements.

Calculating the mmf, or the excitation ampere-turns for simple electromagnetic structures, is rather straightforward for a given value of the flux or flux density. However, it is not so simple to determine the flux or flux density when the mmf is given, because of the nonlinear characteristic of the ferromagnetic material.

Graphical methods or successive approximation by iterative methods (best suited for the digital computer) can be used. The more complicated general field boundary-value problems in two or three

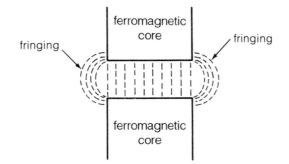

Figure 3.3.1 Fringing of magnetic flux at an air gap.

dimensions (with or without time as a variable) are solved by analytical (if possible), graphical, experimental, or analogue techniques, or more recently by numerical methods with iterative relaxation procedures on the digital computer.

With the advent of digital computers, numerical solutions such as finite-difference schemes, image methods, finite elements (variational formulations), and integral equations have gained currency. In the finite-element method, the field region is subdivided into elements—(subregions). The unknown quantities, such as scalar or vector potentials, are represented by suitable interpolation functions that contain the values of the potential at the respective nodes of each element. Minimizing the energy functional by using such interpolation functions generates an algebraic system of equations, as in the finite-difference methods,[1] and the potential values at the nodes can be determined by direct or iterative methods. The first general nonlinear variational formulation of electromagnetic field problems in electrical machines is due to Silvester[2] and Chari.[3] The finite-element method[4] is now accepted widely by electrical engineers for field analysis. Another approach recently introduced for solving boundary-value problems is the boundary integral equation formulation on which the boundary-element method[5,6] is based. Figure 3.3.2a shows a typical grid system for numerical analysis by the finite-difference method, and part b of the figure depicts the flux distribution in the cross section of a salient-pole synchronous machine. Figure 3.3.3a shows a typical cross section of half a pole pitch of a 2-pole turbogenerator, subdivided into finite-element triangles; part b of the figure represents the no-load flux distribution typically obtained by numerical analysis with the finite-element method.

Several software packages are currently available for two- and three-dimensional analyses of electromagnetic field problems related to electrical machinery and other devices. Finding field solutions for optimized designs would be excellent projects for students to pursue under faculty guidance. The

[1] M. S. Sarma, "End-Winding Leakage of High-Speed Alternators by Three-Dimensional Field Determination," *IEEE Transactions on Power Apparatus and Systems,* vol. PAS-90, no. 2 (March/April 1971), pp. 465.

[2] P. P. Silvester and R. L. Ferrari, *Finite Elements for Electrical Engineers.* (Cambridge, U.K.: Cambridge University Press, 1990).

[3] M. V. K. Chari, and P. P. Silvester, eds., *Finite Elements in Electrical and Magnetic Field Problems* (New York: John Wiley & Sons, 1980).

[4] D. S. Burnett, *Finite Element Analysis* (Reading, MASS.: Addison-Wesley Publishing Company, 1988).

[5] M. H. Lean, and D. S. Bloomberg, "Nonlinear Boundary Element Method for Two-Dimensional Fields," *Journal of Applied Physics.* 55, no. 6 (March 1984).

[6] Y. B. Yildir, et al., "Three-Dimensional Analysis of Magnetic Fields using the Boundary Element Method," Joint EEIC/ICWA Conference, Boston, MASS., Oct. 1991.

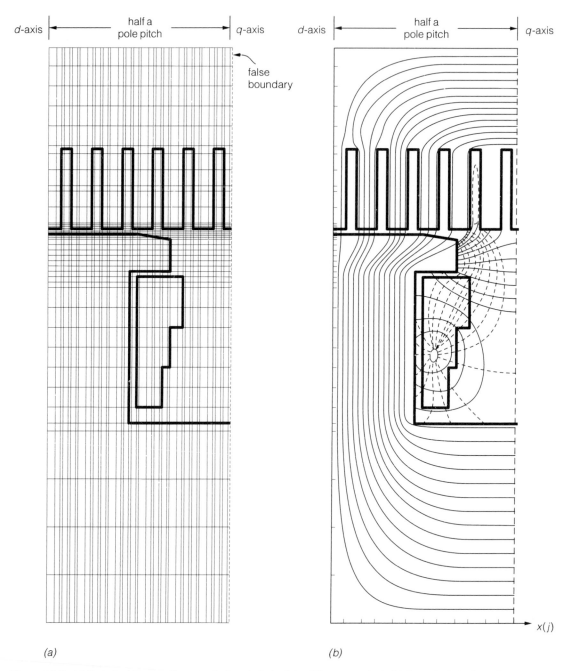

(a)   *(b)*

Figure 3.3.2 (a) Typical grid system for numerical analysis by finite-difference method. (Half a polepitch of a salient-pole synchronous machine is shown here.) (b) Flux distribution in the cross section of a salient-pole synchronous machine, shown in part (a).

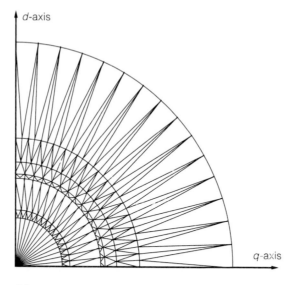

(a)

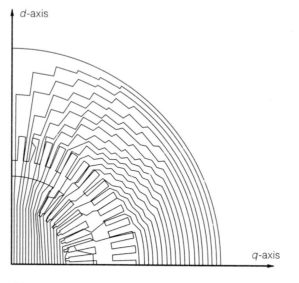

(b)

Figure 3.3.3 (a) Typical cross section of half a pole pitch of a 2-pole turbo-generator, subdivided into finite-element triangles. (b) No-load flux distribution in the cross section part (a) obtained by numerical analysis with the finite-element method.

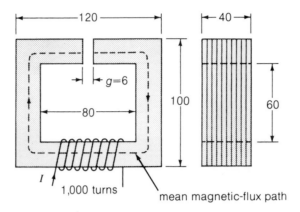

Figure 3.3.4 Laminated electromagnet with an air gap for Example 3.3.1 (all dimensions in mm).

references cited in the footnotes should give the student a very good start. Although such exotic methods of field analysis are not presented in this book, a few examples using the simple magnetic-circuit approach with simple structures are considered next.

EXAMPLE 3.3.1

The magnetic circuit shown in Figure 3.3.4 consists of a laminated core of USS Transformer 72, 29-gauge steel with an air gap g of 6 mm and an exciting winding of 1,000 turns. The stacking factor for the core is given to be 0.94. Neglect leakage but correct for fringing by adding the length of the air gap to each of the other two dimensions of the cross section. The dc magnetization characteristic for USS Transformer 72, 29-gauge steel is given in Figure 3.3.5.

Determine the current in the exciting winding that produces a core flux of 1×10^{-3} Wb.

Solution

Since the leakage is neglected, the magnetic flux will be confined to the paths through the iron and the air gap in series; the mean path along the direction of the flux is shown in Figure 3.3.4. The equivalent magnetic circuit and its electric-circuit analog are given in Figure 3.3.6. Notice that the polarity of the mmf **F** in the equivalent magnetic circuit must be consistent with the direction of the net flux in the core produced by the exciting winding as per the right-hand rule in Figure 3.3.4.

Net cross-sectional area of the core $= (20 \times 10^{-3})(40 \times 10^{-3})0.94 = 0.752 \times 10^{-3}$ m^2

Length of the mean path in iron $= 2(100 + 80) - 6 = 354$ mm $= 0.354$ m

Uniform flux density in iron, $B_i = \dfrac{\phi}{A_i} = \dfrac{1 \times 10^{-3}}{0.752 \times 10^{-3}} = 1.33$ T

The corresponding H_i for iron from the dc magnetic characteristic is 560 At/m.

The mmf for the iron portion of the magnetic circuit is given by

$$\mathbf{F}_i = H_i l_i = 560 \times 0.354 = 198 \text{ At}$$

The cross-sectional area of the air gap corrected for fringing is

$$A_g = (20 + 6)10^{-3}(40 + 6)10^{-3} = 26 \times 46 \times 10^{-6} = 1.196 \times 10^{-3} \text{ m}^2$$

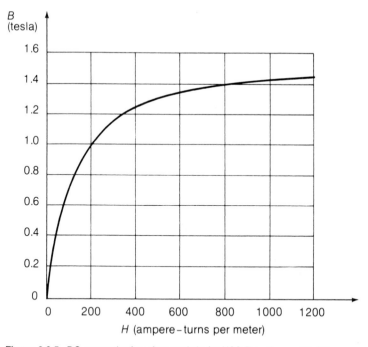

Figure 3.3.5 DC magnetization characteristic for USS Transformer 72, 29-gauge steel.

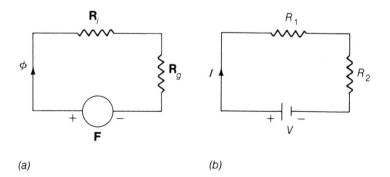

(a)   (b)

Figure 3.3.6 (a) Equivalent magnetic circuit for Example 3.3.1. (b) Electric-circuit analog for Example 3.3.1.

Note that the stacking factor does not apply to the air-gap region; it applies only to the laminated iron. The air-gap flux density is then

$$B_g = \frac{\phi}{A_g} = \frac{1 \times 10^{-3}}{1.196 \times 10^{-3}} = 0.836 \text{ T}$$

$$H_g = \frac{B_g}{\mu_0} = \frac{0.836}{1.257 \times 10^{-6}} = 0.665 \times 10^6 \text{ At/m}$$

The mmf for the air gap is given by

$$\mathbf{F}_g = H_g \cdot g = (0.665 \times 10^6)(6 \times 10^{-3}) = 3{,}990 \text{ At}$$

The total mmf for the iron portion and the air gap in series is

$$\mathbf{F}_t = \mathbf{F}_i + \mathbf{F}_g = 198 + 3{,}990 = 4{,}188 \text{ At}$$

and the corresponding current in the exciting winding is

$$\mathbf{I} = \frac{\mathbf{F}_t}{N} = \frac{4{,}188}{1{,}000} = 4.2 \text{ A}$$

Note that the mmf $\mathbf{F}_i$ for the iron portion is less than 5 percent of the total ampere-turns $\mathbf{F}_t$. A good first approximation for the ampere-turns could have been obtained by considering just the air gap.

EXAMPLE 3.3.2

Given the total mmf in Example 3.3.1, suppose that you are asked to find the resultant flux in the core. Describe a graphical method of solution.

Solution
The total mmf is given by

$$\mathbf{F}_t = \mathbf{F}_i + \mathbf{F}_g \qquad \text{or} \qquad \mathbf{F}_i = \mathbf{F}_t - \mathbf{F}_g = \mathbf{F}_t - \mathbf{R}_g \phi$$

where $\mathbf{R}_g$ is the reluctance of the air gap. Based on the numerical values of Example 3.3.1, we have

$$\mathbf{F}_i = 0.354 H_i; \qquad \phi = B_i A_i = (0.752 \times 10^{-3}) B_i$$

From the magnetization characteristic and by using the above relations, plot the flux ϕ as a function of $\mathbf{F}_i$, as shown in Figure 3.3.7.

$$\mathbf{R}_g = \frac{l_g}{\mu_0 A_g}$$

Using numerical values for l_g and A_g from Example 3.3.1, in SI system,

$$\mathbf{R}_g = \frac{6 \times 10^{-3}}{(1.257 \times 10^{-6})(1.196 \times 10^{-3})} = 3.991 \times 10^6$$

Next, plot the flux ϕ as a function of $(\mathbf{F}_t - \mathbf{R}_g \phi)$, knowing that

$$\mathbf{F}_t = 4{,}188 \qquad \text{and} \qquad \mathbf{R}_g = 3.991 \times 10^6$$

Note that the graph is a straight line as shown in Figure 3.3.7, with the intercept on the flux-axis given by

$$\frac{\mathbf{F}_t}{\mathbf{R}_g} = \frac{4{,}188}{3.991 \times 10^6} = 1.049 \times 10^{-3} \text{ Wb}$$

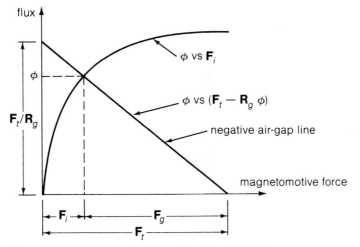

Figure 3.3.7 Graphical method for solving Example 3.3.2.

and the intercept on the mmf axis given by the total mmf

$$\mathbf{F}_t = 4{,}188 \text{ At}$$

The resultant core flux in the magnetic circuit is then determined by the intersection of the two plots, given by 1×10^{-3} Wb. The mmf for the iron portion ($\mathbf{F}_i$) and for the air gap ($\mathbf{F}_g$) can be found as shown in Figure 3.3.7.

The student is encouraged to follow this procedure with other values of total mmf in order to gain confidence. The straight-line graph of ϕ versus ($\mathbf{F}_t - \mathbf{R}_g\phi$) is sometimes referred to as the *negative air-gap line*. For the same magnetic structure and other values of total mmf, the resultant flux is obtained by simply shifting the air-gap line parallel to itself so that it intersects the selected value of mmf on the horizontal mmf axis. On the other hand, to find the resultant flux for a given value of mmf and varying values of air-gap length, you need to adjust only the slope of the negative air-gap line corresponding to the reluctance of the air gap.

This procedure is generally applicable only to fairly simple magnetic structures. More involved calculations may be needed for analyzing complex systems.

IIIIIIIIIII ▬▬▬▬▬▬▬▬▬▬▬▬▬▬▬▬
EXAMPLE 3.3.3

Figure 3.3.8 shows an electromagnet with two air gaps in parallel. Note that the magnetic circuit has two parallel paths linking the winding. Neglecting leakage, the iron's reluctance, and fringing at the air gaps, calculate the flux and flux density in each leg of the magnetic circuit when the current in the 1,000-turn exciting winding is 0.25 ampere. Assume the stacking factor to be unity.

Solution

The mean paths of the flux are shown in Figure 3.3.8. The equivalent magnetic circuit and its electric-circuit analog are given in Figure 3.3.9. With two air gaps in parallel and with iron's negligible reluctance, the entire mmf is expended across each of the air gaps. We shall convert to the SI system of units right from the beginning and solve the problem.

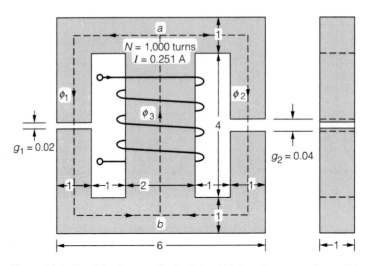

Figure 3.3.8 Parallel-path magnetic circuit in which two air gaps are in parallel (all dimensions in inches).

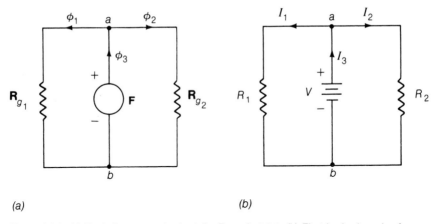

(a)   (b)

Figure 3.3.9 (a) Equivalent magnetic circuit for Example 3.3.3. (b) Electric circuit analog for Example 3.3.3.

The cross-sectional area of the air gaps is

$$A_{g1} = A_{g2} = 1 \text{ in.}^2 \times \frac{1 \text{ m}^2}{39.37 \text{ in.}^2} = 0.645 \times 10^{-3} \text{ m}^2$$

The air-gap lengths are

$$g_1 = 0.020 \text{ in.} \times \frac{1 \text{ m}}{39.37 \text{ in.}} = 0.508 \times 10^{-3} \text{ m}$$

$$g_2 = 0.040 \text{ in.} \times \frac{1 \text{ m}}{39.37 \text{ in.}} = 1.016 \times 10^{-3} \text{ m}$$

The reluctances of the air-gaps are given by

$$\mathbf{R}_{g1} = \frac{g_1}{\mu_0 A_{g1}} = \frac{0.508 \times 10^{-3}}{(1.257 \times 10^{-6})(0.645 \times 10^{-3})} = 0.626 \times 10^6 \text{ At/Wb}$$

$$\mathbf{R}_{g2} = \frac{g_2}{\mu_0 A_{g2}} = \frac{1.016 \times 10^{-3}}{(1.257 \times 10^{-6})(0.645 \times 10^{-3})} = 1.252 \times 10^6 \text{ At/Wb}$$

The fluxes in the legs of the magnetic circuit are then calculated as

$$\phi_1 = \frac{F}{\mathbf{R}_{g1}} = \frac{1,000 \times 0.25}{0.626 \times 10^6} = 0.4 \times 10^{-3} \text{ Wb}$$

$$\phi_2 = \frac{F}{\mathbf{R}_{g2}} = \frac{1,000 \times 0.25}{1.252 \times 10^6} = 0.2 \times 10^{-3} \text{ Wb}$$

$$\phi_3 = \phi_1 + \phi_2 = 0.6 \times 10^{-3} \text{ Wb}$$

The cross-sectional areas of the legs of the magnetic circuit are

$$A_{i1} = A_{i2} = 0.645 \times 10^{-3} \text{ m}^2 \qquad \text{and} \qquad A_{i3} = 1.29 \times 10^{-3} \text{ m}^2$$

Then the corresponding flux densities are given by

$$B_{i1} = \frac{\phi_1}{A_{i1}} = \frac{0.4 \times 10^{-3}}{0.645 \times 10^{-3}} = 0.619 \text{ Wb/m}^2$$

$$B_{i2} = \frac{\phi_2}{A_{i2}} = \frac{0.2 \times 10^{-3}}{0.645 \times 10^{-3}} = 0.31 \text{ Wb/m}^2$$

$$B_{i3} = \frac{\phi_3}{A_{i3}} = \frac{0.6 \times 10^{-3}}{1.29 \times 10^{-3}} = 0.464 \text{ Wb/m}^2$$

Note that the total flux ϕ_3 in the center leg can also be found from the total permeance and the mmf as follows:

$$\mathbf{P}_{g1} = \frac{1}{\mathbf{R}_{g1}} = \frac{1}{0.626 \times 10^6} = 1.6 \times 10^{-6} \text{ H}$$

$$\mathbf{P}_{g2} = \frac{1}{\mathbf{R}_{g2}} = \frac{1}{1.252 \times 10^6} = 0.8 \times 10^{-6} \text{ H}$$

With two air gaps in parallel, their combined permeance is the sum of their individual permeances:

$$\mathbf{P}_t = \mathbf{P}_{g1} + \mathbf{P}_{g2} = 2.4 \times 10^{-6} \text{ H}$$

Then the total flux ϕ_3 is given by

$$\phi_3 = \mathbf{F P}_t = 1,000 \times 0.25(2.4 \times 10^{-6}) = 0.6 \times 10^{-3} \text{ Wb}$$

Note that the reluctances of *n* components in *series* have a resultant total reluctance of

$$\mathbf{R}_t = \mathbf{R}_1 + \mathbf{R}_2 + \cdots + \mathbf{R}_n$$

just as do resistances in series. The permeances of *n* components in *parallel* have a total permeance given by

$$\mathbf{P}_t = \mathbf{P}_1 + \mathbf{P}_2 + \cdots + \mathbf{P}_n$$

The various parts of a magnetic structure may be made of different magnetic materials and may have different cross sections. Analogous to electric circuits, the magnetic circuits may be arranged in series, parallel, or series-parallel configuration.

EXAMPLE 3.3.4

The magnetic circuit of Example 3.3.3 is solved under the assumption that the permeability of iron is infinite and hence the mmf of the iron parts of the magnetic path is zero. Now suppose that the core material is USS Transformer 72, 29-gauge steel, for which the dc magnetization characteristic is given in Figure 3.3.5. Assuming the stacking factor to be unity, compute the flux densities in the three legs of the magnetic structure. Neglect leakage and fringing.

Solution
The problem can be solved graphically by following the procedure of Example 3.3.2. This is left as a desirable excercise for the student. The successive-approximation method is illustrated here, taking the solution of Example 3.3.3 as the first approximation. The mean magnetic-path lengths between points *a* and *b* are

$$l_1 \simeq 10 \text{ in.} = 0.254 \text{ m}$$

$$l_2 \simeq 10 \text{ in.} = 0.254 \text{ m}$$

$$l_3 \simeq 5 \ \text{ in.} = 0.127 \text{ m}$$

The values of the magnetizing force, from the dc magnetization characteristic for the *first* approximation, are

$$\text{for} \quad B_1 = 0.619 \text{ Wb/m}^2, \qquad H_1 = 80 \text{ At/m}$$

$$\text{for} \quad B_2 = 0.31 \ \text{ Wb/m}^2, \qquad H_2 = 42 \text{ At/m}$$

$$\text{for} \quad B_3 = 0.464 \text{ Wb/m}^2, \qquad H_3 = 60 \text{ At/m}$$

The mmfs for the paths are then given by

$$\mathbf{F}_1 = H_1 l_1 = 80 \times 0.254 = 20 \text{ At}$$

$$\mathbf{F}_2 = H_2 l_2 = 42 \times 0.254 = 11 \text{ At}$$

$$\mathbf{F}_3 = H_3 l_3 = 60 \times 0.127 = 8 \text{ At}$$

For the *second* approximation of the flux densities, we reduce the winding ampere-turns by the mmfs of the core and apply the result to the air gaps.

$$B_1 = \frac{\mu_0}{g_1}(\mathbf{F}_t - \mathbf{F}_1 - \mathbf{F}_3) = \frac{1.257 \times 10^{-6}}{0.508 \times 10^{-3}}(250 - 20 - 8) = 0.549 \text{ Wb/m}^2$$

$$B_2 = \frac{\mu_0}{g_2}(\mathbf{F}_t - \mathbf{F}_2 - \mathbf{F}_3) = \frac{1.257 \times 10^{-6}}{1.016 \times 10^{-3}}(250 - 11 - 8) = 0.286 \text{ Wb/m}^2$$

The flux ϕ_3 in the center leg of the magnetic structure is

$$\phi_3 = B_1 A_1 + B_2 A_2$$

and

$$B_3 = \frac{\phi_3}{A_3} = \frac{B_1 A_1 + B_2 A_2}{A_3} = \frac{B_1 + B_2}{2} = 0.418 \text{ Wb/m}^2$$

The corresponding new values of the magnetizing force are read from the dc magnetization characteristic as

$$H_1 = 70 \text{ At/m}$$
$$H_2 = 40 \text{ At/m}$$
$$H_3 = 50 \text{ At/m}$$

Using this second set of H-values to calculate the *third*-round approximation to the B-values gives

$$B_1 = 0.549 \frac{250 - 24}{250 - 28} = 0.559 \text{ Wb/m}^2$$

$$B_2 = 0.286 \frac{250 - 16}{250 - 19} = 0.29 \text{ Wb/m}^2$$

$$B_3 = \frac{B_1 + B_2}{2} = 0.425 \text{ Wb/m}^2$$

The values of B converge rapidly to the final values. Note that these values are within 10 percent of the values calculated in Example 3.3.3 with negligible core-material reluctance.

■■■■■■■■■IIIII
3.4 Energy Storage

We have mentioned earlier that the greater ease of storing energy in magnetic fields largely accounts for the common use of electromagnetic devices for electromechanical energy conversion. The volume energy density stored in a magnetic field is given by

$$w_m = \frac{1}{2} \bar{B} \cdot \bar{H} \tag{3.4.1}$$

The potential energy stored in a magnetic field is expressed by the volume integral

$$W_m = \frac{1}{2} \int_{\text{Vol}} \bar{B} \cdot \bar{H} \, dv \tag{3.4.2}$$

This magnetic energy is distributed in the space that is occupied by the field. For a magnetic medium with no losses and a constant permeability, the magnetic energy density is given by

$$w_m = \frac{1}{2} BH = \frac{1}{2} \frac{B^2}{\mu} = \frac{1}{2} \frac{B^2}{\mu_r \mu_0} \tag{3.4.3}$$

expressed in SI units of joules per cubic meter. Equation 3.4.3 shows that the greater the value of relative permeability μ_r, the less energy is stored in the magnetic material for a given value of the flux density B. It is easy to see why the amount of energy stored in an air gap is several times that stored in a much greater volume of iron.

In a linear region such as that of an air gap, the energy stored can be expressed as

$$W_{m\,air} = \frac{1}{2} \left(\frac{B_g^2}{\mu_0} \right) \text{Vol} = \frac{1}{2} \mathbf{F}\phi = \frac{1}{2} \mathbf{R}\phi^2 = \frac{1}{2} i\lambda \tag{3.4.4}$$

This equation expresses the amount of energy stored upon increasing the flux denstiy from zero to B_g. With nonlinearity and hysteresis neglected, the reluctance and permeance are constant with ϕ and $\mathbf{F}$. The flux and mmf are directly proportional, as in air, for the entire magnetic circuit. In practical electromagnetic devices built with air gaps, the magnetic nonlinearity and the core losses often may be neglected, and a linear analysis may be justified.

Going back to Example 3.3.1, the energy stored in the air gap can be calculated as

$$\begin{aligned} W_{m\,air} &= \frac{1}{2} \frac{B_g^2}{\mu_0} \text{Vol} \\ &= \frac{1}{2} \frac{(0.836)^2}{1.257 \times 10^{-6}} \times (1.196 \times 10^{-3}) \times (6 \times 10^{-3}) \\ &= 1.995 \text{ J} \end{aligned}$$

or it can be alternatively calculated as

$$W_{m\,air} = \frac{1}{2} \mathbf{F}_g \phi = \frac{1}{2}(3,990)(1 \times 10^{-3}) = 1.995 \text{ J}$$

The energy stored in the magnetic material is given by

$$W_{m\,iron} = \frac{1}{2} \mathbf{F}_i \phi = \frac{1}{2}(198)(1 \times 10^{-3}) = 0.1 \text{ J}$$

The energy stored in the air gap in this case is about 20 times that stored in the iron, even though the volume of the air gap (even after accounting for fringing) is only about 1/37 that of the iron. It is obvious that most of the energy is required to establish the flux in the air gap.

In a singly excited system, a change in flux density from a value of zero initial flux density to B requires an energy input to the field occupying a given volume:

$$W_m = \text{Vol} \int_0^B H \, dB \tag{3.4.5}$$

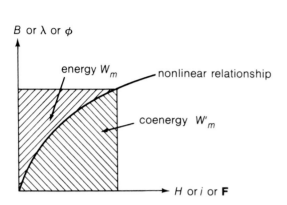

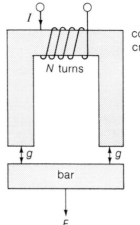

core or yoke of uniform
cross-sectional area A

N turns

bar

F

Figure 3.4.1 Graphical interpretation of energy and coenergy in a singly excited, nonlinear system.

Figure 3.4.2 Magnetic circuit for weight lifting.

which can also be expressed by

$$W_m = \int_0^\lambda i(\lambda)\, d\lambda = \int_0^\phi \mathbf{F}(\phi)\, d\phi \tag{3.4.6}$$

Note that the current is a function of the flux linkages and that the mmf is a function of the flux; their relations depend on the geometry of the coil, the magnetic circuit, and the magnetic properties of the core material. Equations 3.4.5 and 3.4.6 may be interpreted graphically as the area labeled as *energy* in Figure 3.4.1. The other area, labeled as *coenergy* in the figure, can be expressed as

$$W_m' = \text{Vol} \int_0^H B\, dH = \int_0^i \lambda(i)\, di = \int_0^{\mathbf{F}} \phi(\mathbf{F})\, d\mathbf{F} \tag{3.4.7}$$

For a linear system in which B and H, λ and i, or ϕ and $\mathbf{F}$ are proportional, it is easy to see that the energy and coenergy are numerically equal. For a nonlinear system, on the other hand, the energy and coenergy differ as shown in Figure 3.4.1; but the sum of the energy and coenergy for a singly excited system is given by

$$W_m' + W_m = \text{Vol} \cdot BH = \lambda i = \phi \mathbf{F} \tag{3.4.8}$$

A magnetic crane used for lifting weights can be modeled and analyzed as a simple magnetic circuit. Let us consider a magnetic circuit configuration consisting of two distinct pieces of the same magnetic material with two air gaps, as shown in Figure 3.4.2. We shall calculate the weight the crane can lift for a given set of parameters, while making reasonable approximations.

The magnetic energy stored in an incremental volume of space is given by

$$dW_m = \frac{1}{2}(\bar{B} \cdot \bar{H})\, dv \tag{3.4.9}$$

In view of the symmetry, working with one pole of the magnetic circuit, the incremental change in volume is

$$dv = A\, dg \tag{3.4.10}$$

where A is the area of crosssection perpendicular to the plane of paper. Since the directions of $\bar{B}$ and $\bar{H}$ coincide, neglecting leakage and fringing, we have

$$dW_m = \frac{1}{2}\, BH\, dv = \frac{1}{2}\, BHA\, dg = \frac{1}{2}\left(\frac{B^2}{\mu_0}\right) A\, dg \tag{3.4.11}$$

The definition of work gives us

$$dW = \bar{F} \cdot d\bar{l} = F\, dg \tag{3.4.12}$$

While a magnetic pull is exerted upon the bar, an energy dW equal to the magnetic energy dW_m stored in the magnetic field is expended. Thus

$$dW_m = dW \tag{3.4.13}$$

or

$$\frac{1}{2}\left(\frac{B^2}{\mu_0}\right) A\, dg = F\, dg \tag{3.4.14}$$

which yields the pulling force *per pole* on the bar to be

$$F = \frac{1}{2}\left(\frac{B^2}{\mu_0}\right) A \tag{3.4.15}$$

The total pulling force (in newtons in SI system of units) on the bar is then given by

$$F_{total} = 2\left(\frac{1}{2}\right)\left(\frac{B^2}{\mu_0}\right) A = \left(\frac{B^2}{\mu_0}\right) A \tag{3.4.16}$$

3.5 Inductance

An electric circuit in which the current links the magnetic flux is said to have *inductance*. A single loop or an inductor carrying a current i is linked by its own flux, as shown in Figure 3.5.1. If the medium in the flux path has a linear magnetic characteristic (i.e., constant permeability), then the relationship between the flux linkages λ and the current i is linear, and the slope of the linear $\lambda - i$ characteristic gives the *self-inductance,* defined as flux linkage per ampere:

$$L = \frac{\lambda}{i} = \frac{N\phi}{i} = \frac{N^2}{\mathbf{R}} = N^2 \mathbf{P} \tag{3.5.1}$$

This equation illustrates that the inductance is a function of geometry and permeability, and that in a linear system, it is independent of voltage, current, and frequency. The energy stored in the magnetic field of an inductor in a linear medium is given by

$$W_m = \frac{1}{2}\, i\lambda = \frac{Li^2}{2} \tag{3.5.2}$$

When more than one loop or circuit is present, the flux produced by the current in one loop may link

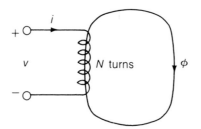

Figure 3.5.1 A single inductive loop.

another loop, thereby inducing a current in that loop. Such loops are said to be *mutually coupled,* and a *mutual inductance* exists between such loops. The mutual inductance between two circuits is defined as the flux linkage produced in one circuit by a current of one ampere in the other circuit. Let us now consider the pair of mutually coupled inductors illustrated in Figure 3.5.2. The self-inductances L_{11} and L_{22} of inductors 1 and 2, respectively, are given by

$$L_{11} = \frac{\lambda_{11}}{i_1} \tag{3.5.3}$$

and

$$L_{22} = \frac{\lambda_{22}}{i_2} \tag{3.5.4}$$

where λ_{11} is the flux linkage of inductor 1 produced by its own current i_1, and λ_{22} is the flux linkage of inductor 2 produced by its own current i_2. The mutual inductances L_{12} and L_{21} are given by

$$L_{12} = \frac{\lambda_{12}}{i_2} \tag{3.5.5}$$

and

$$L_{21} = \frac{\lambda_{21}}{i_1} \tag{3.5.6}$$

where λ_{12} is the flux linkage of inductor 1 produced by the current i_2 in inductor 2, and λ_{21} is the flux linkage of inductor 2 produced by the current i_1 in inductor 1.

If a current of i_1 flows in inductor 1 and no current flows in inductor 2, as shown in Figure 3.5.2a the equivalent fluxes are given by

$$\phi_{11} = \frac{\lambda_{11}}{N_1} \tag{3.5.7}$$

and

$$\phi_{21} = \frac{\lambda_{21}}{N_2} \tag{3.5.8}$$

where N_1 and N_2 are the number of turns of inductors 1 and 2, respectively. The portion of inductor 1's flux that does not link any turn of inductor 2 is known as inductor 1's equivalent *leakage flux:*

$$\phi_{l1} = \phi_{11} - \phi_{21} \tag{3.5.9}$$

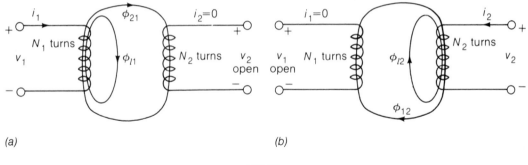

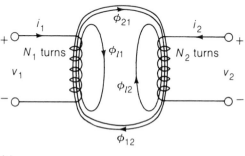

Figure 3.5.2 Mutually coupled inductors.

Similar reasoning with respect to Figure 3.5.2b yields

$$\phi_{l2} = \phi_{22} - \phi_{12} \tag{3.5.10}$$

The situation when both coils are carrying currents simultaneously is shown in Figure 3.5.2c. The *coefficient of coupling* is given by

$$k = \sqrt{k_1 k_2}, \qquad \text{where} \quad k_1 = \phi_{21}/\phi_{11} \text{ and } k_2 = \phi_{12}/\phi_{22}$$

When k approaches unity, the two inductors are said to be "tightly coupled"; when k is much less than unity, they are said to be "loosely coupled." Although the coefficient of coupling cannot exceed unity, it may be as high as 0.998 for iron-core transformers and smaller than 0.5 for air-core transformers.

   When there are only two inductively coupled circuits, the symbol M is frequently used to represent the mutual inductance. It can be shown that the mutual inductance between two electric circuits coupled by a homogeneous medium of constant permeability is reciprocal:

$$M = L_{12} = L_{21} = k\sqrt{L_{11}L_{22}} \tag{3.5.12}$$

The energy considerations that lead to such a conclusion are taken up in a problem at the end of the chapter.

   Let us next consider the energy stored in a pair of mutually coupled inductors:

$$W_m = \frac{i_1 \lambda_1}{2} + \frac{i_2 \lambda_2}{2} \tag{3.5.13}$$

where λ_1 and λ_2 are the total flux linkages of inductors 1 and 2, respectively. Equation 3.5.13 can be rewritten as

$$W_m = \frac{i_1}{2}(\lambda_{11} + \lambda_{12}) + \frac{i_2}{2}(\lambda_{22} + \lambda_{21}) = \frac{1}{2}L_{11}i_1^2 + \frac{1}{2}L_{12}i_1 i_2 + \frac{1}{2}L_{22}i_2^2 + \frac{1}{2}L_{21}i_1 i_2$$

or

$$W_m = \frac{1}{2}L_{11}i_1^2 + Mi_1 i_2 + \frac{1}{2}L_{22}i_2^2 \tag{3.5.14}$$

Equation 3.5.14 is valid whether the inductances are constant or variable, so long as the magnetic field is confined to a uniform medium of constant permeability.

When there are n coupled circuits, the energy stored in the magnetic field can be expressed as

$$W_m = \sum_{j=1}^{n} \sum_{k=1}^{n} \frac{1}{2}L_{jk}i_j i_k \tag{3.5.15}$$

Going back to the pair of mutually coupled inductors shown in Figure 3.5.2, the flux-linkage relations and the voltage equations for circuits 1 and 2 are given by the following, while neglecting the resistance associated with the coils:

$$\lambda_1 = \lambda_{11} + \lambda_{12} = L_{11}i_1 + L_{12}i_2 = L_{11}i_1 + Mi_2 \tag{3.5.16}$$

$$\lambda_2 = \lambda_{21} + \lambda_{22} = L_{21}i_1 + L_{22}i_2 = Mi_1 + L_{22}i_2 \tag{3.5.17}$$

$$v_1 = p\lambda_1 = \frac{d\lambda_1}{dt} = L_{11}\frac{di_1}{dt} + M\frac{di_2}{dt} \tag{3.5.18}$$

$$v_2 = p\lambda_2 = \frac{d\lambda_2}{dt} = M\frac{di_1}{dt} + L_{22}\frac{di_2}{dt} \tag{3.5.19}$$

where p is the derivative operator d/dt.

For the terminal voltage and current assignments shown in Figure 3.5.2, the coil windings are such that the fluxes produced by currents i_1 and i_2 are additive. In such a case the algebraic sign on the mutual voltage term is positive, as in Equations 3.5.18 and 3.5.19.

In order to avoid drawing detailed sketches of windings showing how each coil is wound, a *dot convention* is used. The pair of mutually coupled inductors of Figure 3.5.2 are represented by this system in Figure 3.5.3. The notation is such that a current i entering a dotted (undotted) terminal in one coil induces a voltage $M[di/dt]$ with a positive polarity at the dotted (undotted) terminal of the other coil. For example, if currents i_1 and i_2 are entering (or leaving) the dotted terminals, the convention is that the fluxes produced by i_1 and i_2 will aid each other, and the mutual and self-inductance terms for each terminal pair will have the same sign. Otherwise, they will have opposite signs.

Although just a pair of mutually coupled inductors are considered here for the sake of simplicity, complicated magnetic coupling situations do occur in practice. For example, Figure 3.5.4 shows the coupling between coils 1 and 2, 1 and 3, and 2 and 3.

In nonlinear models, the fluxes and currents are not related by linear equations. For example, with nonlinear current-controlled, coupled coils, the flux-linkage relationships have the general form

$$\lambda_1 = f_1(i_1, i_2) \tag{3.5.20}$$

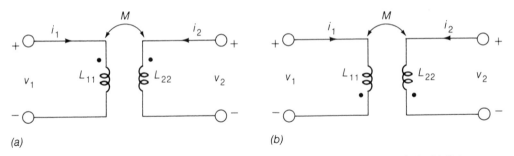

Figure 3.5.3 Dot notation for a pair of mutually coupled inductors. (a) Dots on upper terminals. (b) Dots on lower terminals.

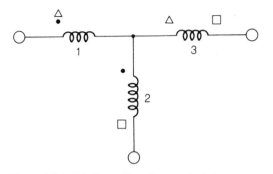

Figure 3.5.4 Polarity markings for complicated magnetic coupling situations.

and

$$\lambda_2 = f_2(i_1, i_2) \tag{3.5.21}$$

For the time-varying case, the functions f_1 and f_2 will also depend explicitly on time t.

▰▰▰▰▰|||||||
3.6 Permanent Magnets

A permanent magnet is a body that is capable of maintaining a magnetic field at other than cryogenic (extreme low) temperatures with no expenditure of power. The basic purpose of any magnet is to store energy or to convert energy from one form to another. This is generally achieved by setting up a magnetic field at some region in space, commonly called the *air gap*, or by storing the energy within the magnetic material. A magnetic field can be established either by an electrical excitation coil energized from a source or by a permanent magnet. Permanent magnets are magnetic circuit components used widely in a variety of industrial equipment and devices. They are utilized in such simple devices as latches for equipment doors and in more complex applications such as field magnets for electrical rotating machines and magnetic focusing of electron beams.

Various classes of permanent-magnet materials are available: ductile metallic magnets, typified by Cunife; ceramic magnets such as Indox; brittle metallic magnets such as Alnico (cast and sintered); and rare-earth–cobalt magnets such as samarium–cobalt. The size, shape, and other physical properties of the permanent magnet must be considered along with design constraints in selecting the most appropriate magnetic material for a specific application. In a practical design, the choice will be based on cost factors, availability, structural requirements for mechanical design, space limitations, and the environment in which the magnet will be situated, as well as the magnetic and electrical performance specifications. Since most permanent magnets are not machinable, they may have to be used in the devices in the form available from the manufacturer.

The shape of a permanent magnet influences the amount of flux produced. Equal volumes of magnetic material produce various amounts of flux, depending on their shape. In a circuit excited by a permanent magnet, the operating conditions of the permanent magnet are determined to a great extent by the external magnetic circuit. Since there are no magnetic insulators, it is rather difficult to direct the flux to those parts of the magnetic circuit where it can be usefully applied. Considering that the flux produced by a permanent magnet consists of useful flux and leakage flux, a fundamental problem in magnetic circuit design is to minimize the unavoidable flux leakage so as to improve the efficiency of the magnetic circuit.

The operating point and subsequent performance of a permanent magnet are functions of where the magnet is physically located in the magnetic circuit and whether it is magnetized before or after installation. In a number of applications, the magnet must be put through a stabilizing process prior to its use. In addition to the intrinsic magnetic properties of the material, other factors that influence the stability of a magnet include external magnetic-field effects, temperature effects, structural stability, mechanical stress, and radiation.

Effective magnet specification requires knowledge of available magnet materials and their magnetic characteristics, magnet test procedures, and various factors that affect the flux produced by a magnet. Too-tight specifications may require extra manufacturing, testing, and quality-control operations, all of which increase costs. On the other hand, a loose specification could allow shipment of magnets that would not work in the particular application. In addition to magnetic properties, specifications on surface finishes and physical dimensions should be evaluated for each application.

Demagnetization curves are an essential tool in selecting the magnetic material that best meets the designer's requirements. Figure 3.2.4 shows the normal second-quadrant demagnetization curve for Alnico-5 permanent-magnet material.

Designers of permanent magnets use another important graph, in which the left half carries the second-quadrant normal demagnetization curve, and the right half has a curve called the (BH) or energy-product curve. These two curves are in mutual correspondence, as shown in Figure 3.6.1. A straight line drawn from the origin to the point with coordinates B_r and $-H_c$ will intersect the demagnetization curve at the point of maximum energy. Ideally, one wants to design the permanent magnet to operate at this optimum point (where the available external energy is at its maximum) by using the corresponding magnitudes of H_d and B_d. Although this is usually not achieved in practice, the majority of the magnet material in a good design will operate in the vicinity of the maximum energy point on the knee of the *B-H* curve.

The solution to a magnet design problem is achieved by first designing a magnet that provides a specified flux across a specified air gap. Then the magnet is proportioned to have minimum volume. In practice, various magnetic materials will be considered in arriving at an efficient design for the space allocated to the permanent magnet. Thus a wide variety of potential solutions may be obtained prior to final selection of the material to be used.

For a specific material, if the magnet is allowed to operate at an arbitrarily chosen point $(B_d, -H_d)$, the first step in the design process is to determine the length of the magnet l_m. Figure 3.6.2 shows a typical configuration consisting of a permanent magnet, two L-shaped soft-iron pole pieces, and an air

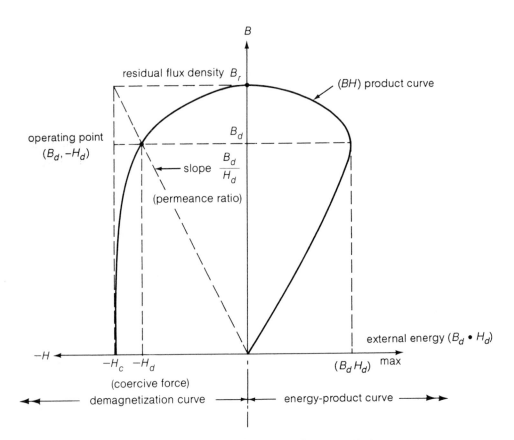

residual flux density B_r

(BH) product curve

operating point $(B_d, -H_d)$

B_d

slope $\dfrac{B_d}{H_d}$

(permeance ratio)

external energy $(B_d \bullet H_d)$

$-H$

$-H_c$ $-H_d$

$(B_d H_d)^{\max}$

(coercive force)

demagnetization curve · energy-product curve

Figure 3.6.1 Second-quadrant normal demagnetization curve and energy-product curve.

gap. Neglecting the reluctance of the soft-iron pieces and applying Ampere's law, one obtains

$$H_d l_m = H_g l_g \tag{3.6.1}$$

where

H_d is the magnitude of the demagnetizing force of the magnet,

l_m is the length of the magnet,

H_g is the intensity of the air-gap magnetic field, and

l_g is the length of the air gap parallel to the line of flux.

Equation 3.6.1 is true only if there is no leakage or fringing and if the reluctance drop in soft-iron pole pieces can be neglected. In order to account for these in practical systems, one may introduce a *reluctance factor, f,* defined as the ratio of the total mmf produced by the permanent magnet to the mmf in the air gap. Then it follows that

$$H_d l_m = f H_g l_g = f B_g l_g / \mu_0$$

and

$$l_m = \frac{f B_g l_g}{\mu_0 H_d} \tag{3.6.2}$$

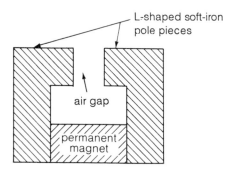

Figure 3.6.2 Typical configuration consisting of a permanent magnet, soft-iron pole pieces, and an air gap.

where all the quantities are in SI units. The reluctance factor ranges from 1.1 to 1.5 in different magnetic circuit configurations; normally a value of about 1.25 is quite adequate. A large reluctance factor may be caused by magnetic saturation in some portion of the magnetic circuit or by disproportionate gap geometry.

The cross-sectional area of the magnet can be calculated from the flux required in the air gap by introducing a *leakage coefficient, F;*

$$B_d A_m = B_g A_g F$$

or

$$A_m = \frac{F B_g A_g}{B_d} \tag{3.6.3}$$

where

B_d is the flux density in the magnet,

A_m is the cross-sectional area of the magnet (orthogonal to the direction of magnetization),

B_g is the flux density of the air gap,

A_g is the cross-sectional area of the gap (orthogonal to the line of flux), and

F is the leakage factor, the ratio of the flux produced by the magnet to the flux in the air gap.

It is difficult to evaluate the leakage factor F, which may range from 1.2 to more than 60. It may be kept small by placing the magnet material as close as possible to the air-gap region. Large steel surfaces between the magnet and the gap tend to increase the leakage factor considerably. The leakage factor for a particular type of magnetic system may be determined by field computation, by estimation on the basis of design experience with similar systems, or by making measurements on a model. In a number of magnetic-circuit problems, the leakage factor can be estimated by calculating the permeance of various flux paths about the gap and magnet. The ratio of the sum of all the permeances to the gap permeance gives the leakage factor. In problems where it is critical, one may have to resort to detailed flux distribution maps, which can now be obtained by numerical methods on digital computers or by other experimental methods such as the electrolytic tank.

The magnet volume can then be obtained from Equations 3.6.2 and 3.6.3 as

$$V_m = A_m l_m = \frac{f F B_g^2 V_g}{\mu_0 B_d H_d} \tag{3.6.4}$$

where $V_g = A_g l_g$, the volume of the air gap, and all quantities are expressed in SI units.

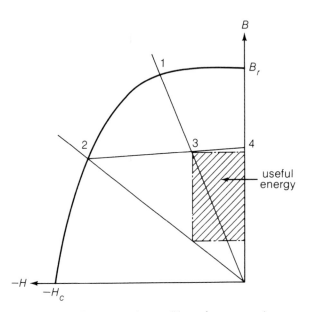

Figure 3.6.3 Different operating conditions of a permanent magnet.

The *permeance coefficient*, P_c, is defined by the slope (B_d/H_d) shown in Figure 3.6.1 and is obtained from Equations 3.6.2 and 3.6.3 as

$$P_c = \frac{B_d}{H_d} = \frac{F}{f} \cdot \frac{l_m}{A_m} \cdot \frac{A_g}{l_g} \cdot \mu_0 \tag{3.6.5}$$

The line joining the origin and the operating point ($B_d, -H_d$) on the normal demagnetization curve is known as the *load line*. The slope of the load line depends only on the physical dimensions of the magnetic circuit and is independent of the properties of the magnetic material. Even though different sections of a magent may practically operate on different permeance coefficients, generally a specified load line refers to the central or neutral section of the magnet.

Some magnets are used in open-circuited condition without steel pole pieces, in which case the permeance is strictly a function of the magnet geometry. When assembled in a magnetic circuit where soft pole pieces direct the flux path, one needs to consider the permeance of the complete magnetic circuit. Let the permeance coefficient line for the assembly shown in Figure 3.6.2 cause the magnet to operate at a point 1 as shown in Figure 3.6.3 when the magnet is magnetized in place. When the magnet is removed or open-circuited, its operating point will drop to point 2 (as shown in Figure 3.6.3), which is determined by the open-circuited permeance coefficient of the magnet.

If the magnet should be replaced in the assembly, the operating point will shift from point 2 to some point 3 rather than to point 1, as shown in Figure 3.6.3. It moves up a *recoil line* until it intersects the assembly permeance coefficient line at point 3. This recoil line is approximately parallel to the slope of the demagnetization curve at point B_r. From point 1 to point 3, there has been a loss of flux. Where such a point 3 will be located for a particular permanent magnet depends greatly on the shape of the demagnetization curve and the position of the operating point on the curve. The hatched rectangular

Table 3.6.1 Characteristics of Selected Permanently Magnetic Materials (typical values)

Material (Trade name)		Residual Flux Density B_r (T)	Coercive Force H_c (kA/m)	Maximum Energy product (kJ/m³)	Average Recoil Permeability (H/m × 10⁻⁶)
Cast alnico	5	1.28	51	44	2.1
	5–7	1.34	58	60	1.9
	6	1.05	62	31	5.0
Sintered alnico	5	1.09	49	31	2.0
	6	0.94	64	23	5.0
Cunife		0.55	42	11	1.7
Indox	1	0.22	145	8	1.15
	2	0.29	193	14	1.15
	3	0.335	187	21	1.1
	4	0.255	183	12	1.1
Rare-earth–cobalt	18	0.87	637	143	1.05
Incor	16	0.81	629	127	1.05
36% Cobalt steel		1.04	18	8	10

area in the figure represents the total useful energy. It can be shown graphically that this is maximal when point 3 is halfway between points 2 and 4.

The slope of the recoil line, called the *recoil permeability* is an intrinsic property of the magnetic material and is one of the most important considerations in designing variable-gap applications. Changes in magnetic circuit reluctance occur in many applications, such as holding magnets, electromechanical energy converters, and magnetic drives. Magnetic materials used in such applications should be those whose demagnetization curves most closely approximate straight lines. Alternatively, one may design so that the lowest operating point is above the knee of the demagnetization curve.

In many Indox and samarium–cobalt magnet assemblies, the loss in flux may be essentially zero, because these magnets have approximately straight demagnetization curves. On the other hand, in loudspeakers using Alnico-5 magnets, the loss may be as much as 70 percent. To avoid the loss due to reluctance changes and for efficient magnet usage, in-circuit magnetization is used in many applications. To protect magnets against demagnetizing influences in shipping and storage, *keepers* (magnetic conductors) are used to complete the magnetic circuit of the permanent magnet.

Table 3.6.1 gives some characteristics of some common permanent-magnet materials. Figure 3.2.4 and 3.6.4 show the normal demagnetization curves for Alnico-5 and samarium–cobalt, respectively.

The most striking aspects of the rare-earth magnet materials, developed during the late 1960s, are their high coercive force (see Figure 3.6.1) and potential maximum energy level, while the most gratifying aspect is their predictable performance. They constitute a major step forward in the art of making magnets, much as the alnicos did in the 1930s and the ferrites in the 1950s. The cost of rare-earth magnets is relatively high, but is decreasing. In order to make the best use of a high-quality and relatively expensive magnetic material such as $SmCo_5$, one should strive for optimal designs. For accurate analysis of difficult applications and increased magnet efficiency, computer-aided design, which

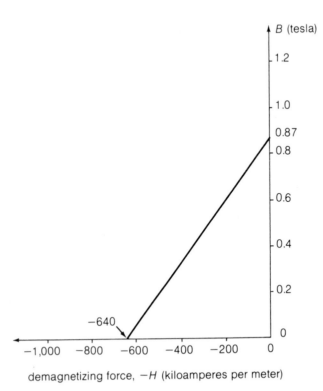

Figure 3.6.4 Normal demagnetization curve for samarium–cobalt permanent-magnet material

can consider the effects of a large number of variables, has become an important tool for the magnet designer.

Let us now take up a simple illustrative example of permanent-magnet design.

EXAMPLE 3.6.1

The problem is to determine the length and area of a permanent magnet that can establish a specified flux density through an air gap of known dimensions. For the sake of simplicity, let us neglect leakage and fringing and allow the operating point on the normal demagnetization curve of Alnico-5 (Figure 3.2.4) to correspond to the point (B_d = 0.96 T and $-H_d$ = −42 kA/m) at which the product ($B_d H_d$) is maximized. It is further given that

$$B_g = 0.5 \text{ T}; \qquad A_g = 2 \text{ cm}^2; \qquad l_g = 0.5 \text{ cm}$$

Solution
From Equation 3.6.2 we have

$$l_m = \frac{B_g l_g}{\mu_0 H_d} = \frac{0.5(0.5 \times 10^{-2})}{(4\pi \times 10^{-7})(42 \times 10^3)} = 4.74 \times 10^{-2} \text{ m} = 4.74 \text{ cm}$$

and from Equation 3.6.3 we get

$$A_m = \frac{B_g A_g}{B_d} = \frac{0.5(2 \times 10^{-4})}{0.96} = 1.04 \times 10^{-4} \text{ m}^2 = 1.04 \text{ cm}^2$$

■■■■■■■ IIII
BIBLIOGRAPHY

Buchanan, R. C. *Ceramic materials for Electronics.* New York: Marcel Dekker, Inc., 1987.

Durney, C. H., and C. C. Johnson. *Introduction to Modern Electromagnetics.* New York: McGraw-Hill, 1969.

Indiana General Corporation. *manual No. 9: Permanent magnet Design Manual.* Valparaiso, Ind., 1968.

Magnetic materials Producers Association. *Standard No. 0100-78: Standard Specifications for Permanent magnet materials.* Chicago, 1978.

Parker, J. J., and R. J. Studders. *Permanent magnets and Their Applications.* New York: John Wiley & Sons, 1962.

Perry, M. P. *Low-Frequency Electromagnetic Design.* New York: Marcel Dekker, Inc., 1987

Roters, H. C. *Electromagnetic Devices.* New York: John Wiley & Sons, 1941.

Strnat, K. J., ed. "Proceedings of the Second International Workshop on Rare-Earth Cobalt Permanent magnets and Their Applications." University of Dayton, Dayton, Ohio, 1976.

Thomas & Skinner, Inc. *Bulletin No. M303: Permanent magnet Design.* Indianapolis, 1967.

Veinott, C. G., Engineering Math. Routines (Curve-fits and Simul. Equations), Sarasota, Florida, 1991. (Published by Author)

Woodson, H. H., and J. R. Melcher. *Electromechanical Dynamics, Part 1: Discrete Systems.* New York: John Wiley & Sons, 1968.

■■■■■■■■■■■■■ IIIIIIIIIII
PROBLEMS

3–1. Write the general maxwell equations that govern the electromagnetic phenomena at any point in space, and specify the assumptions that lead to Equations 3.1.1, 3.1.3, and 3.1.5. What is the unmodified current-continuity relation?

3–2. Show that the magnetic field associated with a long, straight, current-carrying wire is given by

$$B_\phi = \frac{\mu_0 I}{2\pi r}$$

where subscript ϕ denotes the ϕ-component in the circular cylindrical coordinate system, μ_0 is the free-space permeability, I is the current carried by the wire, and r is the radius from the current-carrying wire. Solve the problem by applying (a) Ampere's circuital law and (b) the Biot–Savart law.

3–3. A magnetic force exists between two adjacent, parallel current-carrying wires. Let I_1 and I_2 be the currents carried by the wires and r the separation between them. Use the result of Problem 3–2 to find the force between the wires. Discuss the nature of the force when the wires carry currents in the same direction, and in opposite directions.

3–4.   A rectangular loop is placed in the field of an infinitely long straight conductor carrying a current of I amperes as shown in Figure P3–4. Find an expression for the total flux linking the loop, assuming a medium of permeability μ.

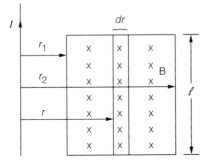

Figure P3–4

3–5.   Consider the arrangement shown in Figure P3–5. A conductor-bar of length L is free to move along a pair of conducting rails. The bar is driven by an external force at a constant velocity of U meters per second. A constant uniform magnetic field $\bar{B}$ is present pointing into the paper of the book page. Neglect the resistance of the bar and rails, as well as the friction between the bar and the rails.

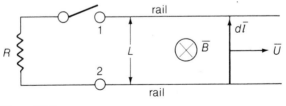

Figure P3–5

a. Determine the expression for the motional voltage across terminals 1 and 2. Is terminal 1 positive with respect to terminal 2?

b. If an electrical load resistance R is connected across the terminals, what is the current and power dissipated in the load resistance? Show the direction of current on the figure.

c. Find the magnetic field force exerted on the moving bar, and the mechanical power required to move the bar. How is the principle of energy conservation satisfied?

d. Since the moving bar is not accelerating, the net force on the bar must be equal to zero. How can you justify this?

3–6.   A toroidal coil is wound on a plastic ring of rectangular cross section with inside diameter 30 cm, outside diameter 40 cm, and height 10 cm. The coil has 200 turns of round copper wire of 3-mm diameter.

a. Calculate the magnetic flux density at the mean diameter of the coil when the coil current is 50 A.

b. Assuming the answer of part a to be the uniform flux density within the coil, compute the flux linkages of the coil.

c. Evaluate the percentage error involved in assuming uniform flux density in the coil.

d. Find the resistance of the wire, given that the volume resistivity of copper is 17.2×10^{-9} ohm-meter.

3-7. Assuming an ideal core ($\mu_i \rightarrow \infty$), calculate the flux density in the air gap of the magnetic circuit shown in Figure P3–7.

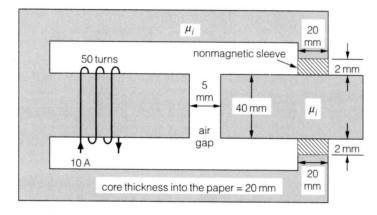

Figure P3-7

3-8. A toroid is made up of three ferromagnetic materials whose segments are arranged in series. One material is nickel-iron alloy having a mean arc length of 0.3 m. The second material is medium silicon steel with a mean arc length of 0.2 m. The third material is cast steel with a mean arc length of 0.1 m. Each of the materials has a cross-sectional area of 0.001 m^2. The excitation coil has 100 turns. Figure P3–8 shows the magnetization curves for the materials. The leakage may be neglected.

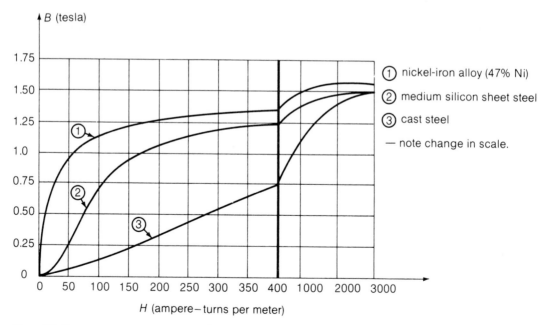

Figure P3-8

a. Determine the excitation current to establish a magnetic flux of 0.5×10^{-3} Wb.

b. Calculate the relative permeability and the reluctance of each segment.

3-9. For the toroid of Problem 3–8, compute the magnetic flux established by an applied mmf of 50 ampere-turns, while neglecting the leakage. Use a cut-and-try procedure consisting of successive approximations.

3-10 A composite magnetic circuit of varying cross section is shown in part a of Figure P3-10; the iron portion has the B-H characteristic of part b of the figure. Given: $N = 100$ turns; $\ell_1 = 4\ell_2 = 40$ cm; $A_1 = 2A_2 = 10 \text{ cm}^2$; $\ell_g = 2$ mm; leakage flux, $\phi_l = 0.01$ mWb. Calculate I required to establish an air-gap flux density of 0.6 T.

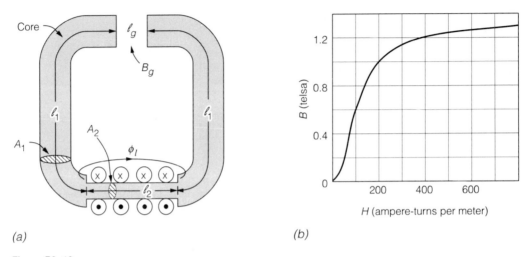

(a)   (b)

Figure P3–10

3-11. Determine the mmf to establish an air-gap flux of 0.7×10^{-3} Wb in the magnetic circuit shown in Figure P3–11, which consists of two different ferromagnetic materials in series with an air gap. Neglect leakage and fringing. The permeability of the materials at a working flux density of 1.4 tesla are $\mu = 1,400\ \mu_0$ for the rolled steel and $\mu = 620\ \mu_0$ for the cast steel. Assume a uniform cross-sectional area of 5 cm² throughout.

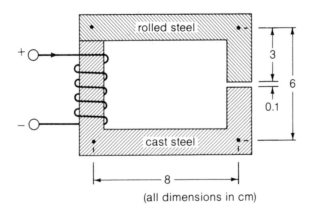

(all dimensions in cm)

Figure P3–11

3-12. The configuration of a magnetic circuit is given in Figure P3–12. Assume the permeability of the ferromagnetic material to be $\mu = 1,000\ \mu_0$. Neglect leakage, but correct for fringing by increasing each linear dimension of the cross-sectional area by the length of the air gap. The magnetic material has a square cross-sectional area of 4 cm^2. Find the air-gap flux, the air-gap flux density, and the magnetic field intensity in the air gap.

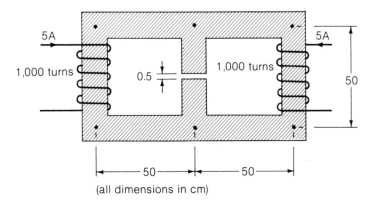

(all dimensions in cm)

Figure P3–12

3–13.   Compute the excitation current needed for a 1,000-turn coil in order to establish an air-gap flux of 0.4 × 10^{-3} Wb for the magnetic circuit shown in Figure P3–13. Neglect leakage and fringing; assume that the magnetic material (cast steel) has a thickness of 5 cm. Refer to Figure 3.2.2 for the dc magnetization curve of cast steel.

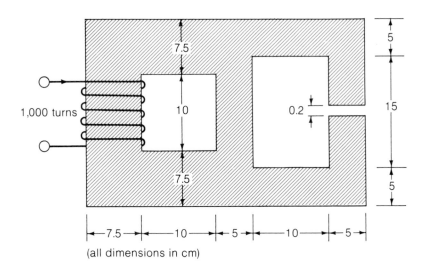

(all dimensions in cm)

Figure P3–13

3–14.   Consider the magnetic circuit shown in Figure P3–14. Assume the cross-sectional area to be the same throughout and the relative permeability of the magnetic material to be 1,000. The excitation coil has 500 turns. Determine what current is needed in the coil to make the flux density in the center limb 1.00 T.

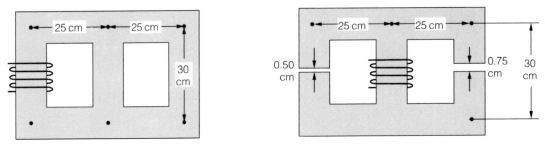

Figure P3-14

Figure P3-15

3-15. In the magnetic circuit shown in Figure P3-15 the center leg has the same cross-sectional area as each of the outer legs. The coil has 400 turns. The permeability of iron may be assumed to be infinite. If the air-gap flux density in the left leg is to be 1.2 T, find

a. the flux density in the air gap of the right leg.

b. the flux density in the center leg.

c. the current needed in the coil.

3-16. Figure P3-16 shows the cross section of a rectangular iron core with two airgaps g_1 and g_2. The ferromagnetic iron can be assumed to have infinite permeability. The coil has 500 turns.

a. Gaps g_1 and g_2 are each equal to 0.1 cm, and a current of 1.83 A flows through the winding. Compute the flux densities in the two air gaps, B_{g1} in the center gap g_1 and B_{g2} in the end gap g_2.

b. Let gap g_2 be now closed by inserting an iron piece of the correct size and infinite permeability, so that only the center gap g_1 = 0.1 cm remains. If a flux density of 1.25 T is needed in gap g_1, find the current that is needed in the winding.

3-17. In the magnetic circuit shown in Figure P3-17 the coil of 500 turns carries a current of 4 A. The air gaps are g_1 = g_2 = 0.25 cm; and g_3 = 0.4 cm. The cross-sectional areas are related such that A_1 = A_2 = $0.5A_3$. The permeability of iron may be assumed to be infinite. Determine the flux densities B_1, B_2, and B_3 in the gaps g_1, g_2 and g_3, respectively.

3-18. Consider the magnetic circuit in Figure P3-18, in which all parts have the same cross section. The coil has 200 turns and carries a current of 5 A. The air gaps are g_1 = 0.4 cm and g_2 = 0.5 cm. Assuming the core has infinite permeability, compute the flux density in tesla in

a. gap g_1.   b. gap g_2.   c. the left limb.

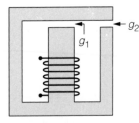

Figure P3-16

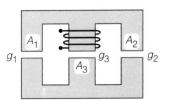

Figure P3-17

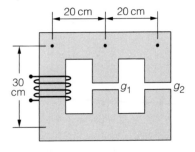

Figure P3-18

3-19. A toroid of rectangular cross section is shown in Figure P3–19. The mean diameter is large in comparison with the core thickness in the radial direction, so that the core flux density is uniform. Derive an expression for the inductance of the toroid, and evaluate it if $r_1 = 80$ mm, $r_2 = 100$ mm, and $N = 200$ turns. The core relative permeability is 900.

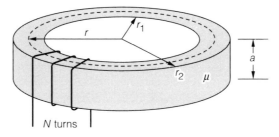

Figure P3–19

3-20. An inductor made of a highly permeable material has N turns. The dimensions of the core and coil are as shown in Figure P3–20. Calculate the input power to the coil to establish a given flux density B in the air gap. The coil's winding space-factor is k_s and its conductivity is σ.

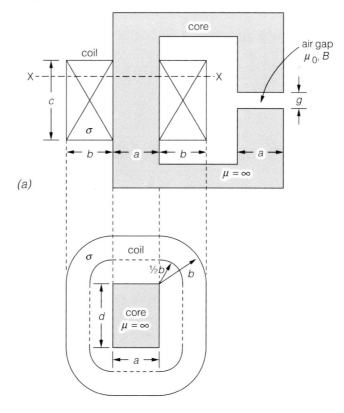

(a)

(b)   Cross section at XX

Figure P3–20

3-21. The inductor of Problem 3–20 is made of magnet wire. If the dimensions in Figure P3–20 are

$$a = b = c = d = 25 \text{ mm} \quad \text{and} \quad g = 2 \text{ mm}$$

and the core flux density is 0.8 T, what are the input power and the number of turns? Assume that $k_s = 0.8$, $\sigma = 5.78 \times 10^7$ S/m, and the coil current $= 1$ A.

3-22. For the inductor of Problem 3-21, find the area of the conductor cross section. What is the time constant of the coil, and at what voltage may it be operated?

3-23. The λ-i characteristic of a magnetic circuit, as shown in Figure P3–23, consists of two straight-line segments. Compute the energy W_m and coenergy W'_m for the magnetic circuit at point a and at point b.

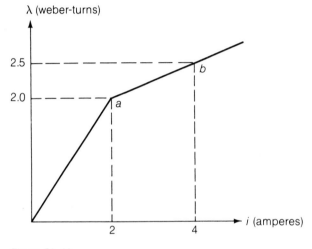

Figure P3–23

3-24. The self-inductances of two coupled coils are L_{11} and L_{22}, and the mutual inductance between them is M. Show that the effective inductance of the two coils in series is given by

$$L_{\text{series}} = L_{11} + L_{22} \pm 2M$$

and that the effective inductance of the two coils in parallel is given by

$$L_{\text{parallel}} = \frac{L_{11}L_{22} - M^2}{L_{11} \mp 2M + L_{22}}$$

Specify the conditions corresponding to different signs of the term $(2M)$.

3-25. Referring to the circuit of Figure 3.5.3, let

$$L_{11} = L_{22} = 0.1 \text{ H} \quad \text{and} \quad M = 10 \text{ mH}$$

Determine v_1 and v_2 if

a. $i_1 = 10$ mA and $i_2 = 0$.

b. $i_1 = 0$ and $i_2 = (10 \sin 100t)$ mA.

c. $i_1 = (0.1 \cos t)$ A and $i_2 = [0.3 \sin (t + 30°)]$ A.

Also find the energy stored in each of the above cases at $t = 0$.

3-26. Consider a pair of coupled coils as shown in Figure 3.5.3 with currents, voltages, and polarity dots as indicated. Show that the mutual inductance is $L_{12} = L_{21} = M$ by following these steps:

a. Starting at time t_0 with $i_1(t_0) = i_2(t_0) = 0$, maintain $i_2 = 0$ and increase i_1 until at time $t_1, i_1(t_1) = I_1$ and $i_2(t_1) = 0$. Determine the energy accumulated during this time. Now maintaining $i_1 = I_1$, increase i_2 until at time $t_2, i_2(t_2) = I_2$. Determine how much energy is accumulated and the total energy stored at time t_2.

b. Repeat the process in the reverse order, allowing the currents to reach their final values. Compare the expressions obtained for the total energy stored, and obtain the desired result, $L_{12} = L_{21} = M$.

3-27. For the coupled coils shown in Figure P3–27, a dot has been arbitrarily assigned to a terminal as indicated. Following the dot convention presented in this chapter, place the other dot in the remaining coil and justify your placement with an explanation. Comment on whether the polarities are consistent with Lenz's law, which states that an induced electric current flows in a direction such that the current opposes the change that induced it.

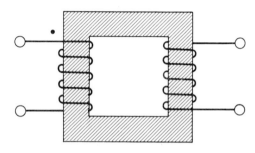

Figure P3–27

3-28. For the configurations of the coupled coils shown in Figure P3–28, obtain the voltage equations for v_1 and v_2.

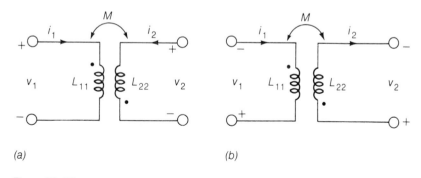

(a)   (b)

Figure P3–28

3-29. A long, straight cylindrical conductor of radius R meters and length l meters has a relative permeability of unity, and it carries a current of I amperes distributed uniformly over its cross section. Determine the following in terms of I, R, and l:

a. The total magnetic flux within the conductor.

b. The magnetic energy density within the conductor as a function of the distance ρ from the center.

c. The energy stored in the magnetic field within the conductor.

d. The corresponding self-inductance due to the magnetic field within the conductor.

3-30.   The λ–i characteristic of a nonlinear singly-excited magnetic circuit is given by

$$\lambda = \frac{2.8i}{0.12 + i}$$

Find the energy and coenergy for the magnetic circuit corresponding to a current of 0.10 amperes.

3-31.   Refer to Example 3.6.1. Suppose that the dimensions of the Alnico-5 magnet and the air gap are known and that you need to calculate the air-gap flux density. Neglect leakage and fringing as before, and let

$$l_m = 4.74 \text{ cm}; \qquad A_m = 1.04 \text{ cm}^2; \qquad A_g = 2 \text{ cm}^2; \qquad l_g = 0.5 \text{ cm}$$

3-32.   An Alnico-5 permanent magnet has the shape shown in Figure P3–32, with a mean diameter of 15 cm and a cross-sectional area of 5 cm^2. The demagnetization curve is shown in Figure 3.2.4 for Alnico-5. Neglect fringing and determine the length of the air gap across which this magnet can produce a flux density of 1.05 tesla.

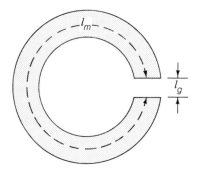

Figure P3-32

3-33.   A permanent magnet is required to produce a flux density of 0.6 tesla in an air gap of length 0.4 cm and cross-sectional area of 20 cm^2. A magnet of the shape shown in Figure 3.6.2 is to be used.

   a. Neglecting the reluctance in the L-shaped soft-iron portions of the magnetic circuit, and ignoring fringing and leakage flux, determine the minimum volume of Alnico-5 material that is required. Specify the dimensions of the magnet.

   b. Repeat part a assuming a leakage factor F of 2.5 and a reluctance factor f of 1.2.

   c. Suppose that the air gap in the magnet system is closed by a keeper of ferromagnetic material whose permeability is assumed to be infinite. If the recoil permeability of the magnetic material is 2×10^{-6} H/m, find the flux density in the magnet.

   d. Repeat part a using Indox V material (for which $B_d = 0.2$ tesla and $H_d = 140$ kA/m, corresponding to the maximum energy product of 28 kJ/m^3). Compare the shape and volume of this magnet with that made of Alnico-5.

4

Transformers

The transformer is basically a static device in which two or more stationary electric circuits are coupled magnetically, the windings being linked by a common time-varying magnetic flux. Its analysis involves many of the principles essential to the study of rotating electrical machinery. Even though the static transformer is not an energy-conversion device and involves only the interchange of electrical energy between two or more electric systems, it is an extremely important component in many energy-conversion systems. Thus the transformer deserves serious study.

The transformer performs many useful functions in several fields of electrical engineering. In the widespread ac power systems the transformer provides the much-needed capability to change the voltage and current levels rather easily. Differing requirements of generation, transmission, and utilization of electric power dictate different optimum values for the voltage and current combinations. The transformer is utilized in stepping up the generator voltage to an appropriate transmission voltage for power transfer, and in stepping down the transmission voltage at various levels for proper distribution and power utilization. In communication systems ranging in frequency from audio to radio levels, the transformer performs such functions as input transformer, interstage transformer, output transformer, impedance-matching device for improved power transfer, insulation device between electric circuits, and isolation device for blocking the dc signal while maintaining the ac continuity between circuits.

Transformers are used in circuits of various voltage levels, from the microvolts of electronic circuits to the very high power-system transmission-voltage levels of 765 kV. Transformer applications are seen throughout the frequency spectrum utilized in electrical circuits, from near dc to several hundreds of megahertz, with both continuous sinusoidal and pulse waveforms.

All that is really necessary for transformer action to occur is for two coils to be so positioned that some of the flux produced by a current in one coil links some of the turns of the other coil. Some air-core transformers employed in communications equipment are no more elaborate than this. However, transformers used in power-system networks are usually much more elaborately constructed in order to minimize energy loss, to produce a large flux in the ferromagnetic core by a current in any one coil, and to ensure that as much flux as possible links as many turns of the other coils on the core as possible. Thus, transformers come in various sizes, from very small ones weighing only a few ounces to very large ones weighing hundreds of tons.

The ratings of transfomers cover a very wide range. Whereas transformers applied for electronic circuits and systems usually have ratings of 300 volt-amperes or less, power-system transformers that transmit and distribute electric power have the highest volt-ampere ratings (a few kVA to several MVA) as well as the highest continuous voltage ratings. Also, transformers (potential transformers and current transformers) with very small volt-ampere ratings are used in instruments for sensing voltages or currents.

Transformers may be classified by their frequency range: power transformers, which usually operate at a fixed frequency; audio and ultrahigh frequency transformers; wide-band and narrow-band frequency transformers; and pulse transformers. Transformers employed in supplying power to electronic systems are generally known as "power transformers." In power-system applications, however, the term *power transformers* denotes those that are used to transmit power in ratings larger than those associated with distribution transformers, usually more than 500 kVA at the voltage levels of 67 kV and above.

Conventional transformers have two windings, but others—(known as *autotransformers*)—have only one winding, and still others—*multiwindings*—have more than two windings. Transformers used in polyphase circuits are known as *polyphase transformers*. In the most popular three-phase system, the most common connections are the wye ("star" or Y) and the delta ("mesh" or Δ) connections.

A two-winding transformer essentially consists of two windings interlinked by a mutual mangetic field. The winding that is excited or energized by connecting it to an input source is usually referred to as the *primary* winding; the other, to which the electrical load is connected and from which the output energy is taken, is known as the *secondary* winding. Depending on the voltage level at which the winding is operated, the windings are classified as *HV (high-voltage)* or *LV (low-voltage)* windings. The terminology of *step-up* or *step-down* transformer is also used if the main purpose of the transformer is to raise or lower the voltage level. In a step-up transformer, the primary winding is a low-voltage winding and the secondary winding is a high-voltage winding. The opposite is true for a step-down transformer.

4.1 Constructional Features of Transformers

An elementary model of a transformer is shown in Figure 4.1.1. Although it is a convenient model for analyzing and illustrating the principles involved, practically, it would be a very unacceptable design— mainly because of the large leakage flux. In order to minimize the leakage flux and improve the magnetic coupling between the two windings, a variety of core and winding configurations have been adopted. Basically these can be classified as the *core type* and the *shell type*.

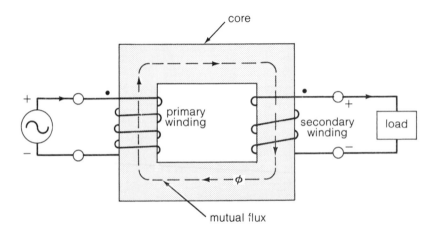

Figure 4.1.1 Elementary model of a transformer.

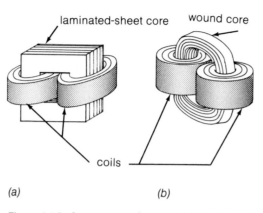

Figure 4.1.2   Core-type transformers. (a) With a core of stacked laminated sheets. (b) With a wound core.

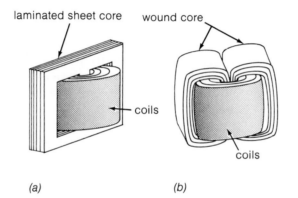

Figure 4.1.3   Shell-type transformers. (a) With a core of stacked laminated sheets. (b) With wound cores.

In order to confine most of the flux to a definite path linking the windings, the core is usually made of some ferromagnetic material. For reducing the losses caused by eddy currents in the core, the core is commonly comprised of a stack of thin laminations.

The winding and core arrangements for core-type transformer are shown in Figure 4.1.2; those for shell-type transformers are shown in Figure 4.1.3. In a core-type transformer, the magnetic circuit takes the form of a single ring encircled by two or more groups of primary and secondary windings distributed around the periphery of the ring. In a shell-type transformer, the primary and secondary windings take the form of a common ring that is encircled by two or more rings of magnetic material distributed around its periphery.

The core may consist of stacks of steel laminations comprised of flat punchings of various shapes, or it may be wound of a long, continuous strip of steel in the direction in which the steel was rolled during manufacture. Magnetizing the wound cores in the direction of rolling results in a lower core loss and requires a lower exciting current. Grain-oriented steel materials are also used in large transformers in the forms of flat strips stacked so that as much of the flux path as possible coincides with the direction of rolling.

Silicon-steel laminations 0.35 mm thick are generally used for transformers that operate at frequencies below a few hundred hertz. Desirable properties of the silicon steel include its low cost, its low core loss, and its high permeability at flux-density levels of 1 to 1.4 tesla. While power-transformer cores are generally constructed from soft magnetic materials in the form of punched laminations or wound tapes of appropriate thickness (depending upon the operating frequency), cores of high-frequency electronics transformers are often constructed of soft ferrites. Some communication transformer cores are also made of compressed powdered ferromagnetic alloys such as permalloy.

In the early 1980s, the development of amorphous steel for use as transformer core material had progressed sufficiently to allow its use in distribution transformers. The use of amorphous material in transformer cores reduces core losses to one third of those that occur with conventional silicon steel cores. Another feature of the amorphous material is that the manufacturing process consists of just casting liquid metal onto a moving substrate and cooling the metal at a rate of more than one million degrees per second. Thus, amorphous metals are potentially less expensive to produce than silicon steels, which involve numerous manufacturing steps. The Electric Power Research Institute developed an *amorphous steel core distribution transformer* technology that reduces core losses by 60%–70%. In a utility field trial, the first 1,000 25-kVA distribution transformers using amorphous steel cores displayed excellent performance, with core losses as low as 13 watts.[1]

In order to reduce the leakage, the windings are subdivided into sections placed as closely together as possible. In the core-type construction, shown in Figure 4.1.4a, each winding consists of two sections—one section placed on each of the two legs of the core—and the primary and secondary windings are arranged to be concentric coils. In the shell-type construction, as shown in Figure 4.1.4b, the windings usually consist of a number of thin "pancake" coils assembled in a stack, with primary and secondary coils interleaved.

The characteristic features of the core-type transformer are a long mean length of the magnetic circuit and a short mean length of the windings, while those of the shell-type are a short mean length of the magnetic circuit and a long mean length of the windings. Thus, for a given output and performance, the core type will have a smaller area of core and a larger number of turns than the corresponding shell-type. Except for certain extreme ratings, the choice between core- and shell-type constructions is largely a matter of manufacturing facilities and of individual preference. Generally, the steel-to-copper weight ratio is greater in the shell type transformer.

The constructional details of transformers vary greatly, depending upon the specific applications, the winding voltage and current ratings, and the operating frequencies. Whereas many transformers used for electronic circuits and systems are quite simple in their construction, transformers used in electric power systems are usually much more complicated. The main constructional elements of a power-system transformer include:

- cores comprising limbs, yokes, and clamping devices;
- primary, secondary, and sometimes also tertiary phase windings, coil formers, spacers, and conductor insulation;
- interwinding and winding/earth insulation and bracing;
- tanks, coolers, dryers, conservators, and other auxiliaries; and
- terminals and bushings, connections, and tapping switches.

See Photos 4.1.1, 4.1.2, and 4.1.3.

---

[1] Electric Power Research Institute, *EPRI report TR-100295*, vols. 1 and 2, April 1992.

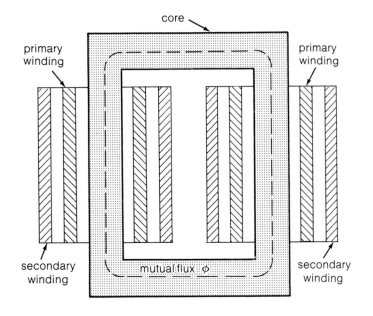

(a)

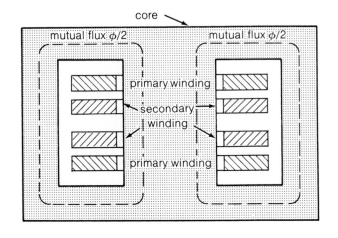

(b)

Figure 4.1.4 Transformer winding arrangements. (a) Core-type. (b) Shell-type.

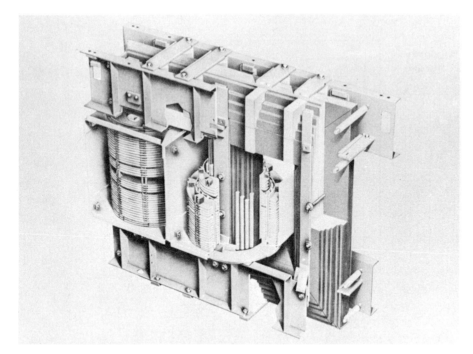

Photo 4.1.1 Rigidily braced core and coil assemblies of a typical medium-power core-type transformer. (Photo courtesy of Westinghouse Electric Corporation.)

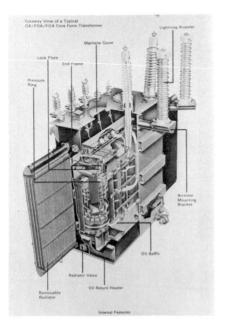

Photo 4.1.2 Cutaway view of a typical self-cooled, forced oil, forced-air cooled medium-power core-type transformer. (Photo courtesy of Westinghouse Electric Corporation.)

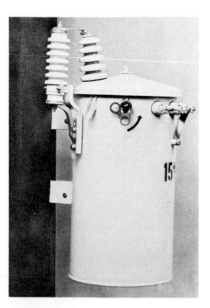

Photo 4.1.3 Typical single-phase, oil-filled, pole-mounted, overhead distribution transformer with surge arrester. (Photo courtesy of Westinghouse Electric Corporation.)

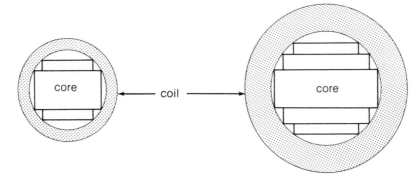

Figure 4.1.5 Stepped transformer-core cross sections.

In small transformers, cores with limbs of square cross section may be used; either square or circular coils are fitted on them. In large transformers, however, either of these arrangements is wasteful of copper, thereby increasing the copper loss. Stepped cores of cross sections such as those shown in Figure 4.1.5 are then employed whenever economically justified in order to embrace the maximum cross-sectional area of steel with the minimum length of copper conductor.

In order to avoid overheating and consequent breakdown of insulation, the heat produced in a transformer must be removed. Some transformers are air-cooled; the heat is removed by radiation and convection either immediately from the core and coil surfaces or from the surrounding protective enclosure. The so called *dry transformers* have simple natural air cooling or sometimes sealed into cabinets filled with nitrogen. The great majority of industrial power-system transformers are immersed in oil for cooling, insulation, and mechanical protection. The transformer oil removes the heat by convection and conveys it to the walls of the tank (and the radiators), from where it is transferred to the atmosphere. Conservator tanks may be required to take up the cycle expansion and contraction of the transformer oil without allowing it to come in contact with the ambient air, from which it may absorb moisture. In very large power-system transformers, radiators and forced oil circulation in the tank as well as forced air cooling are employed. Recent studies have shown that it is technically feasible to build large power transformers with sheet-wound coils that are insulated by compressed gas (such as sulfur hexafluoride, SF_6) and cooled by forced circulation of a liquid.[2]

The windings for transformers applied for electronic systems are usually made of "magnet wire" designated by an American Wire Gauge (AWG) number—increasing gauge numbers correspond to decreasing conductor cross section—and by a letter indicating insulation class—*A, B, C, F,* and *H*—corresponding to safe operating temperatures of 105° C to 180° C. The windings of power-system transformers, however, generally are conductors with heavier insulation and are assembled with much greater mechanical support. Often, the windings are preformed and the transformer core is built around them by stacking the laminations. The main constructional features include:

a. Provision of insulation strength adequate to withstand the voltages occurring in service or in tests, both to ground and between coils, turns, and phases as well as full-wave and chopped-impulse voltages imposed to prove capability against switching and lightning surges.

b. Design that limits losses in order to obtain better efficiency.

---

[2] "Evaluation of Advanced Technologies for Power Transformers," General Electric Company Report prepared for the U.S. Dept. of Energy, DOE/RA/2134-01, June 1980.

c. Provision of cooling means to meet the specified limits on the windings' temperature rise.

d. Provision of adequate leakage reactance.

e. Provision of adequate mechanical strength to withstand short-circuit forces.

4.2 Transformer Theory

The most important aspects of transformer action can be illustrated by idealizing the transformer. Figure 4.2.1 shows schematically a transformer having two windings with N_1 and N_2 turns, respectively, on a common magnetic circuit. Note that this is an elaboration of Figure 4.1.1. The *ideal transformer* is one that has no losses (associated with iron or copper) and no leakage fluxes; it has a core of infinite magnetic permeability and of infinite electrical resistivity. That is to say, all the flux is confined to the core linking both windings; the winding resistances are negligible; and the core losses are negligible. Only a negligible mmf is required to establish the flux; and the electric fields produced by the windings are negligible. Although these properties are never actually achieved in practice, they are closely approached.

When a time-varying voltage v_1 is applied to the primary winding (assumed to have zero resistance), a core flux ϕ is established and a counter emf e_1 is developed such that

$$v_1 = e_1 = \frac{d\lambda_1}{dt} = N_1 \frac{d\phi}{dt} \tag{4.2.1}$$

where $\lambda_1 (= N_1\phi)$ is the flux linkage with the primary N_1-turn winding. The polarity of the primary induced voltage e_1 with respect to that of the applied voltage v_1 is as shown in Figure 4.2.1, satisifying Lenz's law, which states that an induced electric current produces flux in a direction that opposes the change of flux linkage that induced it.

Since there is no leakage flux with ideal transformer, the core flux ϕ also links all N_2 turns of the secondary winding and produces an induced emf e_2.

$$e_2 = \frac{d\lambda_2}{dt} = N_2 \frac{d\phi}{dt} \tag{4.2.2}$$

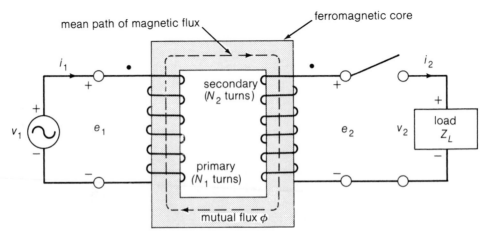

Figure 4.2.1 Ideal transformer.

When the flux is increasing, the secondary induced voltage e_2 has the polarity shown in Figure 4.2.1 by the use of dots. If a passive external load circuit is connected to the secondary-winding terminals, then the terminal voltage v_2 will cause a current i_2 to flow as shown in Figure 4.2.1. Since the resistance of the secondary winding is assumed to be zero, the secondary terminal voltage will be equal to the secondary induced voltage e_2.

$$v_2 = e_2 = \frac{d\lambda_2}{dt} = N_2 \frac{d\phi}{dt} \tag{4.2.3}$$

From Equations 4.2.1 and 4.2.3 it follows that

$$\frac{v_1}{v_2} = \frac{e_1}{e_2} = \frac{N_1}{N_2} = a \tag{4.2.4}$$

where a is the turns ratio. Thus, in an ideal transformer, voltages are transformed in the direct ratio of the turns.

The applications of Ampere's law around the closed contour of the mean path of magnetic flux (shown in Figure 4.2.1) yields

$$N_1 i_1 - N_2 i_2 = 0 \tag{4.2.5}$$

since the core is assumed to have infinite magnetic permeability. Consequently the magnetic field intensity is zero everywhere in the core, even though the magnetic flux and the magnetic flux density are finite; that is, the net mmf acting on the core of an ideal transformer at any instant is zero. Equation 4.2.5 implies that

$$\frac{i_1}{i_2} = \frac{N_2}{N_1} = \frac{1}{a} \tag{4.2.6}$$

Thus, for an ideal transformer, currents are transformed in the inverse ratio of the turns.

The instantaneous power input equals the instantaneous power output in an ideal transformer:

$$v_1 i_1 = v_2 i_2 \tag{4.2.7}$$

Let us next consider the case in which the waveforms of the applied voltage and flux are sinusoidal. If the flux as a function of time is given by

$$\phi = \phi_{max} \sin \omega t \tag{4.2.8}$$

where ϕ_{max} is the maximum value of the flux, and ω is $2\pi f$, f being the frequency, then the induced voltage e_1 is given by

$$e_1 = N_1 \frac{d\phi}{dt} = \omega N_1 \phi_{max} \cos \omega t \tag{4.2.9}$$

The induced emf leads the flux by 90° for the positive directions shown in Figure 4.2.1. The rms value of the induced emf is given by

$$E_1 = \frac{2\pi}{\sqrt{2}} f N_1 \phi_{max} = 4.44 f N_1 \phi_{max} \tag{4.2.10}$$

If the resistance drop in the winding is neglected, the counter emf equals the applied voltage. Thus

$$V_1 = E_1 = 4.44 f N_1 \phi_{max}$$

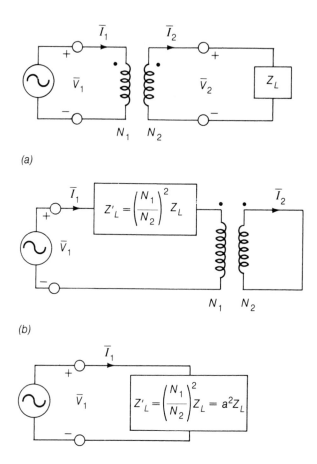

(a)

(b)

(c)

Figure 4.2.2 Equivalent circuits viewed from source terminals when the transformer is ideal.

or

$$\phi_{max} = \frac{V_1}{4.44 f N_1}$$

(4.2.11)

where V_1 is the rms value of the applied voltage. The flux is thus determined solely by the applied voltage, the frequency of the applied voltage, and the number of turns in the winding.

For the case of a sinusoidal voltage applied to the primary winding and a load impedance connected to the secondary winding, the phasor notation can conveniently be used. Equivalent circuits viewed from the source terminals, when the transformer is ideal, are shown in Figure 4.2.2, in which the rms or effective magnitudes of the variables are used. The ends of the two windings at which the dots are placed become positive in potential simultaneously with respect to the other ends of the windings. The voltages $\bar{V}_1$ and $\bar{V}_2$ are in phase, and so also the currents $\bar{I}_1$ and $\bar{I}_2$. The direction of the currents are such that their net mmf is zero. In phasor form, the voltage and current relations are given by

$$\bar{V}_1 = \frac{N_1}{N_2}\bar{V}_2 = a\bar{V}_2 \tag{4.2.12}$$

$$\bar{I}_1 = \frac{N_2}{N_1}\bar{I}_2 = \frac{1}{a}\bar{I}_2 \tag{4.2.13}$$

and

$$\bar{V}_2 = \bar{I}_2 Z_L \tag{4.2.14}$$

from which it follows that

$$\frac{\bar{V}_1}{\bar{I}_1} = \left(\frac{N_1}{N_2}\right)^2 \frac{\bar{V}_2}{\bar{I}_2} = \left(\frac{N_1}{N_2}\right)^2 Z_L = a^2 Z_L \tag{4.2.15}$$

where Z_L is the complex impedance of the load, as shown in Figure 4.2.2. The consequence of Equation 4.2.15 is that an impedance Z_L in the secondary circuit can be replaced by an equivalent impedance Z'_L in the primary circuit insofar as the effect at the source terminals is concerned.

$$Z'_L = \left(\frac{N_1}{N_2}\right)^2 Z_L = a^2 Z_L \tag{4.2.16}$$

Thus the circuits shown in Figure 4.2.2 are indistinguishable when viewed from the source terminals. Z'_L is the load impedance *referred to the primary side*. Transferring an impedance from one side of the transformer to the other in this manner is called *referring the impedance*. With the use of Equations 4.2.12 and 4.2.13, voltages and currents also may be *referred to* one side or the other.

Based on our knowledge of circuit analysis, the equivalent circuit of Figure 4.2.2c suggests a way of obtaining the maximum power transfer from a source of internal impedance Z_S to a load impedance Z_L through *impedance matching*. This is done by choosing the turns ratio such that

$$Z'_L = \left(\frac{N_1}{N_2}\right)^2 Z_L = a^2 Z_L = Z_S^* \tag{4.2.17}$$

where Z_S^* is the complex conjugate of Z_S.

▌▌▌▌▌▌▌▌▌▌▌▌
EXAMPLE 4.2.1

A 60-Hz, 100-kVA, 2,400/240-V transformer (considered to be ideal) is used as a step-down transformer from a transmission line to a distribution system.

a. Determine the turns ratio.

b. What secondary load impedance will cause the transformer to be fully loaded, and what is the corresponding primary current?

c. Find the value of the load impedance that is referred to the primary side of the transformer.

Solution

a. The turns ratio is the same as the voltage ratio specified in the rating of the transformer. The primary is the high-voltage winding and the secondary is the low-voltage winding for a step-down transformer.

b. When the transformer is fully loaded, it should deliver 100 kVA based on the given rating. So

$$V_2 I_2 = 100 \times 10^3 \text{ VA} \quad \text{or} \quad I_2 = \frac{100 \times 10^3}{240} = 416.67 \text{ A}$$

The corresponding primary current may be obtained as

$$I_1 = \frac{I_2}{a} = \frac{416.67}{10} = 41.67 \text{ A} \quad \text{or} \quad V_1 I_1 = 100 \times 10^3$$

from which

$$I_1 = \frac{100 \times 10^3}{2,400} = 41.67 \text{ A}$$

Note that the values given for the currents and voltages are the rms values.
The secondary load impedance, when the transformer is fully loaded, is given by

$$|Z_L| = \frac{V_2}{I_2} = \frac{240}{416.67} = 0.576 \ \Omega$$

In practice, the load impedance is usually a combination of the resistance and the inductive reactance with a lagging power factor. In some cases, however, the impendence corresponds to unity power factor (i.e., purely resistive load) or to the leading power factor (i.e., capacitive load).

c. $|Z_L'| = \left(\dfrac{N_1}{N_2}\right)^2 |Z_L| = a^2 |Z_L| = 10^2 \times 0.576 = 57.6 \ \Omega$

Transformer on No-Load

When there is no current in the secondary as shown in Figure 4.2.3a, the transformer is said to be on *no-load*. Even when the transformer is on no-load, a current known as the *exciting current* flows in the primary (see Figure 4.2.3) because of the core losses and the finite permeability of any practical (realistic) transformer core. The exciting current can be considered as having two components, the *core-loss current* and the *magnetizing current*.

The core-loss current is in phase with the induced primary voltage and is given by

$$I_c = \frac{P_c}{E_1} \tag{4.2.18}$$

where P_c is the core loss given by the sum of the hystersis and eddy-current losses. It is manifested as heat generated in the core.

In linear magnetic circuits, the *B-H* characteristic is a straight line, and the permeability has a finite constant value. In such cases the magnetizing current is proportional to the flux and is in phase with the flux, thereby lagging the primary induced voltage by 90 degrees.

The case of a transformer on no-load is represented by Figure 4.2.3a, and the corresponding phasor diagram is part b of the figure. The no-load current is given by

$$I_0 = \sqrt{I_c^2 + I_m^2} \tag{4.2.19}$$

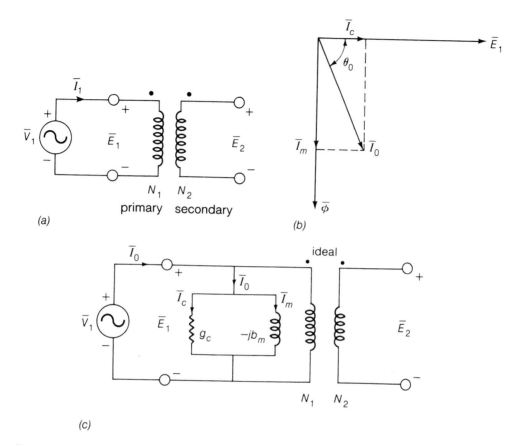

Figure 4.2.3 Transformer on no-load. (a) Transformer with no-load current. (b) Phasor diagram. (c) Equivalent circuit taking the no-load exciting current into account.

and the no-load power factor is given by

$$\cos \theta_0 = \frac{I_c}{I_0} \tag{4.2.20}$$

Taking the exciting current into account, the equivalent circuit can be constructed as in Figure 4.2.3c.

In the case of ferromagnetic circuits mostly used in transformers, the magnetic properties are nonlinear; hence the waveform of the exciting current differs that of the flux. Based on the hysteresis loop or the flux-mmf loop of the core material, it is possible to obtain graphically the waveform of the exciting current as a function of time. This graphical procedure is shown in Figure 4.2.4. The waveforms of the flux (Equation 4.2.8) and the voltage (Equation 4.2.9) as well as the flux-mmf loop of the ferromagnetic core material are drawn as shown. Values of mmf corresponding to various values of flux are then found. For example, at time t_1 the instantaneous value of the flux is ϕ_1 while the flux is increasing; the corresponding value of the mmf read from the increasing-flux portion of the flux-mmf loop is $\mathbf{F_1}$, based on which a value of the exciting current i_0' is plotted at time t_1. At time t_2, although the flux has the same instantaneous value ϕ_1, the flux is decreasing; the corresponding values of the

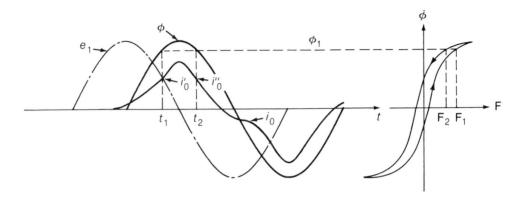

Figure 4.2.4 Waveform of no-load current of a transformer.

mmf and the exciting current are F_2 and i_0''. In this manner the complete curve of the exciting current i_0 as a function of time can be plotted as shown in Figure 4.2.4.

As seen from Figure 4.2.4 the waveform of the exciting current is not sinusoidal. However, it is symmetrical, because $i_0(t + \pi) = -i_0(t)$. The waveform can therefore be represented by a series of odd harmonics including the fundamental. Analyzing by the Fourier-series method, it can be seen that the predominant harmonic is the third. It is usually of the order of 40 percent of the exciting current for typical power transformers, while the exciting current itself is about 5 percent of the full-load rated current. The nonsinusoidal nature of the exciting current waveform is not usually accounted for except in problems that deal directly with the effect of harmonics. The exciting current is then represented by its equivalent sine wave, which has the same effective value and frequency and causes the same average power as the actual wave. With such a representation it becomes possible to draw a phasor diagram (such as Figure 4.2.3b), even for the nonlinear ferromagnetic circuits. The phasor $\bar{I}_o$ now represents the equivalent sinusoidal exciting current.

The equivalent circuit of Figure 4.2.3c shows the exciting admittance added to the ideal transformer circuit as

$$y_0 = g_c - jb_m = \frac{\bar{I}_0}{\bar{E}_1} \tag{4.2.21}$$

where the conductance $g_c = I_c/E_1$ and the susceptance $b_m = I_m/E_1$. The circuit element g_c absorbs a power corresponding to that dissipated in core losses such that

$$P_c = I_c E_1 = g_c E_1^2 \tag{4.2.22}$$

EXAMPLE 4.2.2

A 200-turn winding on a laminated magnetic-steel core is excited by a 60-Hz sinusoidal source of 200 V (rms).

a. Neglecting the resistance of the winding and any leakage flux, find the maximum flux density in the core if the uniform cross-sectional area of the core is 5 cm $\times$ 5 cm.

b. Corresponding to the maximum flux density of 1.5 Wb/m^2, the core loss and the exciting rms volt-amperes for the core are given to be $P_c = 50$ W and $(VI)_{rms} = 500$ VA. Determine the exciting current, the core-loss component, the magnetizing component, and the corresponding power-factor angle.

c. Find the exciting admittance for part b and the corresponding equivalent circuit elements g_c and b_m.

Solution

a. Since the resistance drop in the winding is neglected, the induced emf is the same as the applied voltage.

$$E_1 = V_1 = 200 \text{ V}$$

From Equation 4.2.11 one obtains

$$\phi_{max} = \frac{200}{4.44 \times 60 \times 200} = 0.00375 \text{ Wb}$$

The corresponding flux density is given by

$$B_{max} = \frac{0.00375}{0.05 \times 0.05} = 1.5 \text{ Wb/m}^2$$

b. Exciting current, $I_0 = \dfrac{500}{200} = 2.5$ A

Core-loss component, $I_c = \dfrac{50}{200} = 0.25$ A

Magnetizing component, $I_m = \sqrt{I_0^2 - I_c^2} = \sqrt{(2.5)^2 - (0.25)^2} = 2.49$ A

Power factor, $\cos \theta_0 = \dfrac{P_c}{VI} = \dfrac{I_c}{I_0} = \dfrac{0.25}{2.5} = 0.1$

and the corresponding power-factor angle θ_0 is 84.3°. Note that θ_0 is nearly 90°; I_0 and I_m are nearly the same in magnitude.

c. The exciting admittance is given by

$$y_0 = \frac{\bar{I}_0}{\bar{E}_1} = \frac{2.5\angle-84.3°}{200\angle0°}$$
$$= 0.0125\angle-84.3° = (0.00125 - j0.01245) \text{ S}$$

g_c and b_m may also be found as

$$g_c = \frac{I_c}{E_1} = \frac{0.25}{200} = 0.00125 \text{ S}$$

$$b_m = \frac{I_m}{E_1} = \frac{2.49}{200} = 0.01245 \text{ S}$$

The equivalent circuit is shown in Figure 4.2.5.

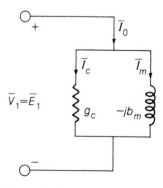

Figure 4.2.5

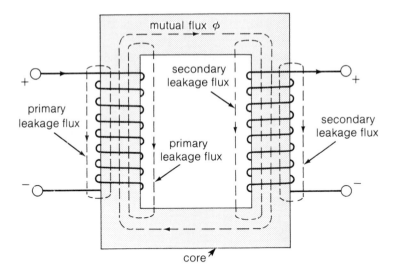

Figure 4.2.6   Mutual flux and leakage fluxes in an elementary core-type transformer.

Transformer Leakage Impedance and Equivalent Circuit

As a departure from the properties assumed for an ideal transformer, we have considered the exciting no-load current. Next we will consider modeling the effects of winding resistances and magnetic leakage. The winding capacitances also may have to be included in order to analyze high-frequency transformer problems, transient conditions encountered in pulse transformers, and effects of voltage surges caused by lightning or switching on power-system transformers. However, we will neglect the capacitances of the windings for now.

In order to account for magnetic leakage, the total flux linking the primary winding can be considered to be made up of two components: the resultant mutual flux ϕ linking both windings, and the primary leakage flux ϕ_{l1} linking only the primary. These components of an elementary core-type transformer are shown in Figure 4.2.6. ϕ_{l2} is the secondary leakage flux linking only the secondary winding. Under load conditions, the windings develop sizable magnetomotive forces that act on spaces external to the core, thereby giving rise to leakage fluxes. The actual paths taken by the leakage fluxes in practical transformers are more complex than those shown in Figure 4.2.6 but the essential features remain the same. The resultant mutual flux linking both windings is produced by the combined effect of the primary and secondary currents.

The primary mmf must not only counteract the demagnetizing effect of the secondary mmf; it must also have sufficient additional mmf to create the resultant mutual flux. Thus, the primary current can be considered to be made up of two components: a load component that is the same as the secondary current referred to the primary current as in an ideal transformer, and the exciting no-load component $\bar{I}_0$ discussed earlier.

In order to account for leakage fluxes, a leakage inductance is assigned to each winding on the basis of its own leakage-flux linkages per unit current. The corresponding *leakage reactances* (X_{l1} and X_{l2} are obtained by multiplying the leakage inductances by $(2\pi f)$, to be introduced as additional external inductive reactances in the equivalent circuit. External resistances (R_1 and R_2) are added to represent the resistance of each winding. Assuming that the ideal transformer carries only the resultant mutual flux, the equivalent circuit—accounting for the exciting current, the winding resistances, and the magnetic leakage fluxes—is shown in Figure 4.2.7a.

$\bar{I}'_2$ is the secondary current referred to the primary as shown in Figure 4.2.7a. The primary current $\bar{I}_1$ is the phasor summation of the load component $\bar{I}'_2$ and the exciting no-load component $\bar{I}_0$.

$$\bar{I}_1 = \bar{I}'_2 + \bar{I}_0$$

$$= \left(\frac{N_2}{N_1}\right)\bar{I}_2 + \bar{I}_0$$

$$= \frac{\bar{I}_2}{a} + \bar{I}_0 \tag{4.2.23}$$

The primary induced voltage $\bar{E}_1$ and the primary impressed voltage $\bar{V}_1$ will differ by an impedance drop as in this phasor relationship:

$$\bar{V}_1 = \bar{E}_1 + \bar{I}_1(R_1 + jX_{l1})$$

$$= \bar{E}_1 + \bar{I}_1 Z_{l1} \tag{4.2.24}$$

Similarly, on the secondary side we have

$$\bar{E}_2 = \bar{V}_2 + \bar{I}_2(R_2 + jX_{l2})$$

$$= \bar{V}_2 + \bar{I}_2 Z_{l2} \tag{4.2.25}$$

The induced voltages $\bar{E}_1$ and $\bar{E}_2$ produced by the resultant mutual flux are related by the turns ratio:

$$\frac{\bar{E}_1}{\bar{E}_2} = \frac{N_1}{N_2} = a \tag{4.2.26}$$

The actual transformer is thus equivalent to an ideal transformer plus external impedances, as shown in Figure 4.2.7a. It is convenient to refer all quantities to either the primary or secondary, in which case the ideal transformer would be moved to the right or left, respectively, of the equivalent circuit, as shown in parts b and c of Figure 4.2.7. R'_2 and X'_{l2} are the secondary resistance and leakage reactance referred to the primary, while R'_1 and X'_{l1} are the primary resistance and leakage reactance referred to the secondary. In Figure 4.2.7b the load can also be referred to the primary, and the ideal transformer can be eliminated from the equivalent circuit as was done in part c of the figure. When all quantities are referred to the secondary, the ideal transformer may also be omitted from the diagram and remembered mentally; the exciting shunt branch conductance and susceptance should also be referred appropriately as shown in Figure 4.2.7c. The phasor diagram for the equivalent circuit can be drawn as explained in Chapter 1.

The commonly used equivalent circuit developed in Figure 4.2.7 and shown in Figure 4.2.8a is often called as the *T-circuit* of a transformer. Several modifications and simplifications of this basic complete equivalent circuit (with winding capacitances neglected here, of course) are used in practice, depending upon the requirements of the problem being solved and the degree of accuracy desired. Approximately equivalent circuits commonly used for the constant-frequency power-system transformer

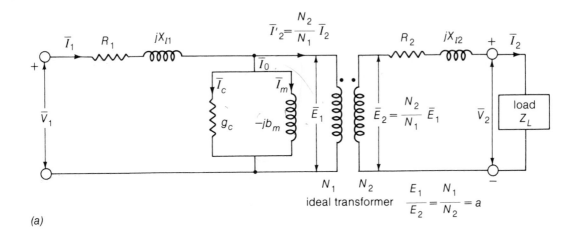

(a)

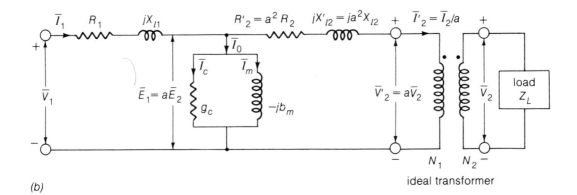

(b)

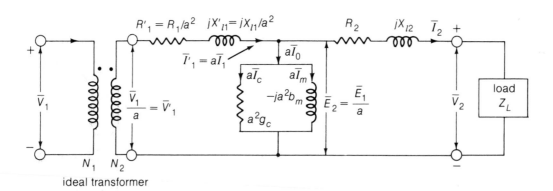

(c)

Figure 4.2.7 Transformer equivalent circuits. (a) Acounting for the exciting current, winding resistances, and magnetic leakage fluxes. (b) Referred to primary. (c) Referred to secondary.

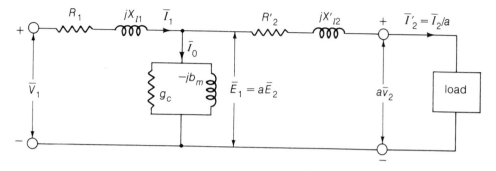

(a)

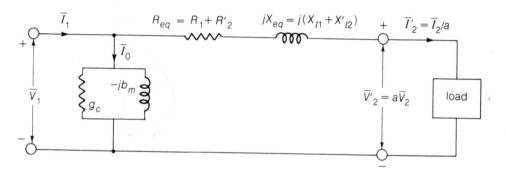

(b)

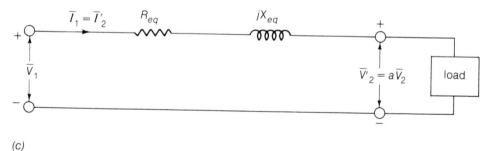

(c)

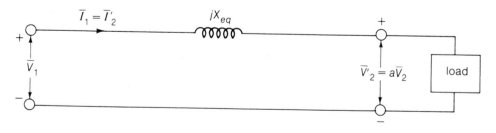

(d)

Figure 4.2.8 Other equivalent circuits of a transformer.

analysis are shown in parts b, c, and d of Figure 4.2.8. By moving the exciting admittance from the middle of the T-circuit to either the left (as shown in part b) or the right, computational labor can be greatly reduced with little error. The resistances R_1 and R_2' can then be combined in series to form an *equivalent resistance* R_{eq}. Similarly, the leakage reactances X_{l1} and X_{l2}' can be combined in series to form an *equivalent* reactance X_{eq}, as shown in Figure 4.2.8b. Further simplification is gained by neglecting the exciting current altogether, as in Figure 4.2.8c, which represents the transformer as an equivalent series impedance. When R_{eq} is small compared to X_{eq}, as in the case of large power system transformers, R_{eq} may frequently be neglected. The transformer is then modeled by its equivalent reactance X_{eq} only, as shown in Figure 4.2.8d.

Although the equivalent circuits in Figure 4.2.8d are referred to the primary, the student should have no difficulty drawing those referred to the secondary. Phasor diagrams for all parts of Figure 4.2.8 appear in Figure 4.2.9—following the step-by-step procedure of Section 1.2—corresponding to a lagging load power factor $(\cos \theta_L)$. For the cases of unity power-factor loads and leading power-factor loads, the diagrams would need to be appropriately modified.

EXAMPLE 4.2.3

A single-phase, 50-kVA, 2,400:240-volt, 60-Hz distribution transformer has these parameters:

   Resistance of the 2,400-volt winding $R_1 = 0.75\ \Omega$

   Resistance of the 240-volt winding $R_2 = 0.0075\ \Omega$

   Leakage reactance of the 2,400-volt winding $X_{l1} = 1\ \Omega$

   Leakage reactance of the 240-volt winding $X_{l2} = 0.01\ \Omega$

   Exciting admittance on the 240-volt side $y_0 = (0.003 - j0.02)\ S$

a. Draw the equivalent circuit referred to the high-voltage side and referred to the low-voltage side; label the impedances numerically.

b. The transformer is used as a step-down transformer at the load end of a feeder whose impedance is $(0.5 + j2.0)$ ohms. Determine the voltage V_s at the sending end of the feeder when the transformer delivers rated load at rated secondary voltage and 0.8 lagging power factor. Neglect the exciting current of the transformer.

Solution

a. (1) The equivalent circuit referred to the high-voltage side is shown in Figure 4.2.10 (page 104). The quantities, referred to the high-voltage side from the low-voltage side, are calculated as

$$R_2' = a^2 R_2 = \left(\frac{2,400}{240}\right)^2 (0.0075) = 0.75\ \Omega$$

$$X_{l2}' = a^2 X_{l2} = \left(\frac{2,400}{240}\right)^2 (0.01) = 1.0\ \Omega$$

The exciting branch conductance and susceptance referred to the high-voltage side are given by

$$\frac{1}{a^2}(0.003) \quad \text{or} \quad \frac{1}{100} \times 0.003 = 0.03 \times 10^{-3}\ S$$

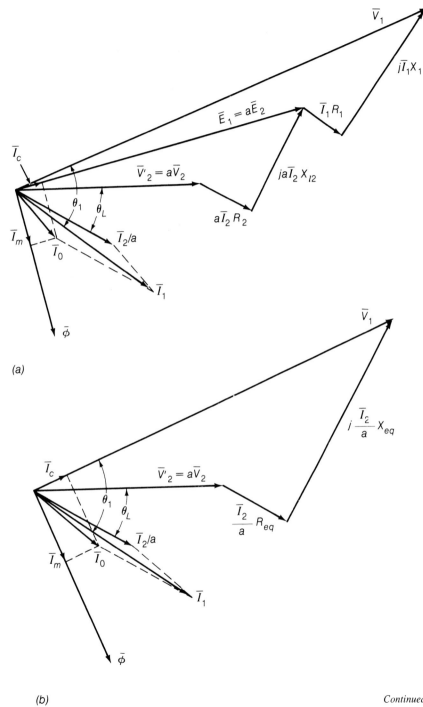

(a)

(b)

Continued

Figure 4.2.9 Phasor diagrams of a transformer on load.

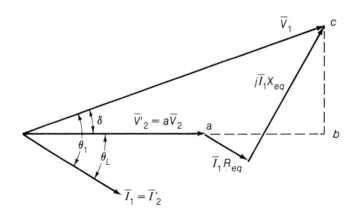

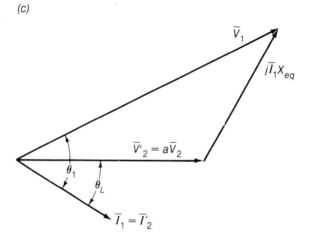

(c)

(d)

Figure 4.2.9 *Continued*

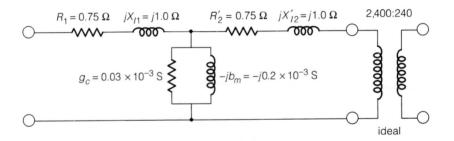

Figure 4.2.10

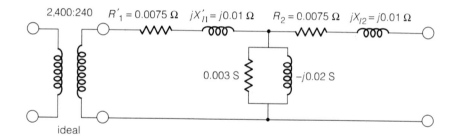

Figure 4.2.11

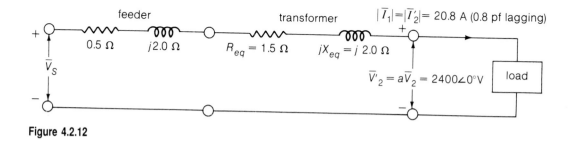

Figure 4.2.12

and

$$\frac{1}{a^2}(0.02) \quad \text{or} \quad \frac{1}{100} \times 0.02 = 0.2 \times 10^{-3} \text{ S}$$

a. (2) The equivalent circuit referred to the low-voltage side is given in Figure 4.2.11. Note the following points: The voltages specified on the nameplate of a transformer yield directly the turns ratio while neglecting the small leakage-impedance voltage drops under load. The actual voltages at the terminals of either the high-voltage or low-voltage side depend on the particular conditions of the problem; they are generally near the specified voltages. The turns ratio in this problem is 10 to 1.

Impedances are referred by either multiplying or dividing by 100, which is the square of the turns ratio in this problem. The value of an impedance referred to the high-voltage side is larger than its value referred to the low-voltage side, while the value of an admittance referred to the high-voltage side is smaller than its value referred to the low-voltage side. Since admittance is the reciprocal of impedance, the reciprocal of the referring factor for impedance must be used when referring admittance from one side to the other.

The ideal transformer can usually be omitted from the diagram, but be certain to keep it in mind.

b. Figure 4.2.12 shows the equivalent circuit of the transformer (along with the feeder impedance) referred to the high-voltage (primary) side, neglecting the exciting current of the transformer. From the given conditions of the problem, the voltage at the load terminals referred to the primary or the high-voltage side is 2,400 V. Further, the load current corresponding to the rated load condition is given by

$$I = \frac{50 \times 10^3}{2,400} = 20.8 \text{ A}$$

which is also referred to the high-voltage side.

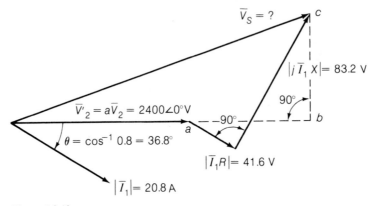

Figure 4.2.13

The feeder impedance can be combined with the transformer impedance in series:

$$Z = (2.0 + j4.0) \ \Omega = R + jX$$

The phasor diagram in Figure 4.2.13 corresponds to the lagging 0.8 power factor.

$$
\begin{aligned}
\bar{V}_s &= \bar{V}'_2 + \bar{I}_1(R + jX) \\
&= 2{,}400\angle 0° + 20.8\angle{-36.8°}(2.0 + j4.0) \\
&= 2{,}400 + j0 + (16.64 - j12.48)(2 + j4) \\
&= 2{,}400 + j0 + 83.2 + j41.6 \\
&= 2{,}483.2 + j41.6 = 2{,}483.5\angle 0.96°
\end{aligned}
$$

The voltage at the sending end of the feeder is 2,483.5 volts, and the power factor at the sending end is $\cos(36.8 + 0.96)°$ or 0.79. Alternatively, noting that

$$ab = I_1 R \cos\theta + I_1 X \sin\theta; \qquad bc = I_1 X \cos\theta - I_1 R \sin\theta$$

V_s could be solved for.

Per-Unit Values

A per-unit system, essentially a system of dimensionless parameters, is used extensively for computational convenience and for readily comparing the performance of a set of transformers or a set of rotating machines. Per-unit quantities simplify analysis of complex power systems involving transformers of different ratios. When expressed in a per-unit system related to their rating, the parameters of transformers and machines lie in a reasonably narrow numerical range. The per-unit system is also useful in simulating complex power-system problems on analog and digital computers for the purpose of analyzing transient and dynamic behavior.

The process of normalization is carried out by dividing one physical parameter by another of the same dimension. The denominator is referred to as the *base* quantity. Base quantities are chosen because of the ways they characterize particular features of the physical system. The magnitudes of some base quantities may be chosen freely and quite arbitrarily; those of others cannot, because they follow laws that govern the physical nature of the system. Bases should be selected that minimize computational effort, and that make evaluation and understanding of the main characteristics as simple and direct as possible. Normally, for any device, the principal per-unit variables assume unit magnitude under rated conditions.

The per-unit (pu) quantity is defined as

$$\text{Quantity in pu} = \frac{\text{actual quantity}}{\text{base-value quantity}} \tag{4.2.27}$$

Base values of apparent power $(VA)_{\text{base}}$ and voltage V_{base} are usually chosen first. The base current I_{base} for a single-phase system is then calculated as

$$I_{\text{base}} = \frac{(VA)_{\text{base}}}{V_{\text{base}}} \tag{4.2.28}$$

Other base quantities are then established as follows for a single-phase system:

$$P_{\text{base}} = Q_{\text{base}} = |S|_{\text{base}} = (VA)_{\text{base}} \tag{4.2.29}$$

$$R_{\text{base}} = X_{\text{base}} = |Z|_{\text{base}} = \frac{V_{\text{base}}}{I_{\text{base}}} \tag{4.2.30}$$

$$G_{\text{base}} = B_{\text{base}} = |Y|_{\text{base}} = \frac{I_{\text{base}}}{V_{\text{base}}} \tag{4.2.31}$$

The per-unit impedance is generally expressed by

$$|Z|_{\text{pu}} = \frac{\text{actual impedance in ohms}}{\text{base impedance in ohms}} = |Z|_{\text{ohms}} \frac{(VA)_{\text{base}}}{(V_{\text{base}})^2} \tag{4.2.32}$$

When only one electrical device, such as a transformer, is involved, its own rating is generally taken for the volt-ampere base. In order that the per-unit impedance of a transformer may have the *same value* whether referred to the high-voltage or the low-voltage side, the rated (or nominal) voltages of the respective sides of the transformer are chosen as the base voltages. Thus, for a transformer, the values of V_{base} are different on the two sides and are in the same ratio as the turns of the transformer. The advantage of such a choice of base quantities is that the equivalent circuit in per-unit quantities will be the same whether referred to the high-voltage or low-voltage side. The next example illustrates this with numerical values.

In studies of power systems involving several devices, an arbitrary choice of the volt-ampere base is usually made, and the same base is used for the overall system. In such a case, per-unit values may have to be changed from one volt-ampere base to another, while the voltage base may or may not be the same. These relations may then be applied:

$$(P, Q, VA)_{\text{pu (base 2)}} = (P, Q, VA)_{\text{pu (base 1)}} \frac{(VA)_{\text{base 1}}}{(VA)_{\text{base 2}}} \tag{4.2.33}$$

$$(R, X, |Z|)_{\text{pu (base 2)}} = (R, X, |Z|)_{\text{pu (base 1)}} \frac{(VA)_{\text{base 2}}}{(VA)_{\text{base 1}}} \left(\frac{V_{\text{base 1}}}{V_{\text{base 2}}}\right)^2 \tag{4.2.34}$$

$$(G, B, |Y|)_{\text{pu (base 2)}} = (G, B, |Y|)_{\text{pu (base 1)}} \frac{(VA)_{\text{base 1}}}{(VA)_{\text{base 2}}} \left(\frac{V_{\text{base 2}}}{V_{\text{base 1}}}\right)^2 \tag{4.2.35}$$

|||||||||||||

EXAMPLE 4.2.4

Consider the transformer of Example 4.2.3. Draw its equivalent circuit in per-unit quantities referred to the high-voltage side and to the low-voltage side.

Solution

The base quantities for high-voltage and low-voltage sides are given as

	High-Voltage Side	**Low-Voltage Side**
$(VA)_{base}$	50,000 VA	50,000 VA
V_{base}	2,400 V	240 V
I_{base}	$\dfrac{50{,}000}{2{,}400} = 20.8$ A	$\dfrac{50{,}000}{240} = 208$ A
$\|Z\|_{base}$	$\dfrac{2{,}400}{20.8} = 115.4\ \Omega$	$\dfrac{240}{208} = 1.154\ \Omega$
$\|Y\|_{base}$	$\dfrac{20.8}{2{,}400} = 0.0087$ S	$\dfrac{208}{240} = 0.87$ S

Figure 4.2.14 shows the equivalent circuit in per-unit quantities. It comes out to be exactly the same, whether referred to the high-voltage or the low-voltage side, since

$$R_{pu(1)} = R_{pu(2)} = \frac{0.75}{115.4} = \frac{0.0075}{1.154} = 0.0065 \text{ pu}$$

$$X_{pu(1)} = X_{pu(2)} = \frac{1.0}{115.4} = \frac{0.01}{1.154} = 0.0087 \text{ pu}$$

$$g_c = \frac{0.3 \times 10^4}{0.0087} = \frac{0.003}{0.87} = 0.00345 \text{ pu}$$

$$b_m = \frac{2 \times 10^4}{0.0087} = \frac{0.2}{0.87} = 0.023 \text{ pu}$$

in which the actual impedance or admittance values are taken from the solution of Example 4.2.3. The base system is so chosen that the per-unit values come out to be the same when referred to either side. Note that the per-unit value is a dimensionless quantity.

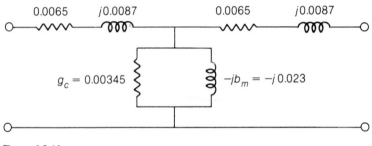

Figure 4.2.14

IIIIIIIIIIII
EXAMPLE 4.2.5

In the single-phase electric system shown in Figure 4.2.15, two transformers are used to interconnect three circuits A, B, and C. Their ratings are

> Transformer A–B: 15 MVA, 13.2:132 kV, equivalent leakage reactance 0.1 pu.
>
> Transformer B–C: 15 MVA, 66:132 kV, equivalent leakage reactance 0.08 pu

Choosing the bases in circuit B as 15 MVA and 132 kV, determine the per-unit value of the 400-Ω resistive load in circuit C referred to circuits C, B, and A. Draw the equivalent circuit diagram with per-unit quantities, neglecting the exciting currents of the transformers, the transformers resistances, and the line impedances.

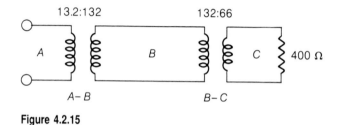

Figure 4.2.15

Solution

Since the base voltage for circuit B is chosen as 132 kV, the base voltage for Circuits A and C will be 13.2 kV and 66 kV, respectively, based on the turns ratios of the transformers.

$$\text{Base impedance of circuit C} = \frac{V_{base}}{I_{base}} = \frac{(V_{base})^2}{(VA)_{base}} = \frac{(66 \times 10^3)^2}{15 \times 10^6} = 290.4 \ \Omega$$

$$\text{Per-unit impedance of the load in Circuit C} = \frac{400}{290.4} = 1.38 \ \text{pu}$$

$$\text{Base impedance of circuit B} = \frac{(132 \times 10^3)^2}{15 \times 10^6} = 1.161.6 \ \Omega$$

$$\text{Impedance of the load referred to circuit B} = 400 \left(\frac{132}{66}\right)^2 = 400 \times 4 = 1,600 \ \Omega$$

$$\text{Per-unit impedance of the load referred to B} = \frac{1,600}{1,161.6} = 1.38 \ \text{pu}$$

$$\text{Base impedance of circuit A} = \frac{(13.2 \times 10^3)^2}{15 \times 10^6} = 11.616 \ \Omega$$

$$\text{Impedance of the load referred to circuit A} = 400 \left(\frac{132}{66}\right)^2 \left(\frac{13.2}{132}\right)^2 = 400 \times 4 \times 0.01 = 16 \ \Omega$$

$$\text{Impedance of the load referred to circuit A} = \frac{16}{11.616} = 1.38 \ \text{pu}$$

Note that the per-unit impedance of the load referred to any part of the system is the same, because the voltage bases in different parts of the system are based on the turns ratios of the transformers. Figure 4.2.16 shows the desired equivalent circuit diagram with per-unit quantities.

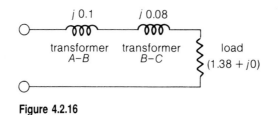

Figure 4.2.16

So far we have considered per-unit values related to a single-phase system. In dealing with balanced three-phase sytems, the three-phase volt-ampere base and the line-to-line voltage base are usually chosen first. The base line current is then calculated as

$$I_{\text{L base}} = \frac{(VA)_{3\phi\,\text{base}}}{\sqrt{3}V_{\text{L base}}} = \frac{3(VA)_{\text{base per phase}}}{\sqrt{3}V_{\text{L base}}} \tag{4.2.36}$$

Other base quantities are then established from the relations that hold for a balanced three-phase system:

$$P_{3\phi\,\text{base}} = Q_{3\phi\,\text{base}} = (VA)_{3\phi\,\text{base}} = 3(VA)_{\text{base per phase}} \tag{4.2.37}$$

$$V_{\text{L base}} = \sqrt{3}V_{\text{base line-to-neutral}} \tag{4.2.38}$$

$$I_{\text{base per phase of }\Delta} = \frac{1}{\sqrt{3}}I_{\text{base per phase of Y}} \tag{4.2.39}$$

From Equations 4.2.38 and 4.2.39 it follows that

$$|Z|_{\text{base per phase of }\Delta} = 3\,|Z|_{\text{base per phase of Y}} \tag{4.2.40}$$

Equations 4.2.29, 4.2.30, and 4.2.31 still apply to the base values per phase. The base impedance per phase of Y is given by

$$|Z|_{\text{base per phase of Y}} = \frac{V_{\text{base line-to-neutral}}}{I_{\text{L base}}} = \frac{V_{\text{L base}}/\sqrt{3}}{I_{\text{L base}}} \tag{4.2.41}$$

Multiplying the numerator and denominator by $(\sqrt{3}V_{\text{L base}})$, one obtains

$$|Z|_{\text{base per phase of Y}} = \frac{(V_{\text{L base}})^2}{(VA)_{3\phi\,\text{base}}} \tag{4.2.42}$$

The per-unit impedance per phase of Y is then given by

$$|Z|_{\text{pu per phase of Y}} = |Z|_{\text{ohms line-to-neutral}}\frac{(VA)_{3\phi\,\text{base}}}{(V_{\text{L base}})^2} \tag{4.2.43}$$

Balanced three-phase problems can be solved in per unit as if they were single-phase problems with a single line and a neutral return.

If, for example, for a balanced three-phase system

$$(VA)_{3\phi\,\text{base}} = 10{,}000 \text{ VA} \qquad \text{and} \qquad V_{L\,\text{base}} = 1{,}000 \text{ V}$$

then it follows that

$$(VA)_{1\phi\,\text{base}} = \frac{10{,}000}{3} \text{ VA} \qquad \text{and} \qquad V_{\text{base line-to-neutral}} = \frac{1{,}000}{\sqrt{3}} \text{ V}$$

Now to express an actual line-to-line voltage of 900 V or the corresponding line-to-neutral voltage of $[900/\sqrt{3}]$ V, we have

$$\text{Per-unit voltage} = \frac{900}{1{,}000} = \frac{900/\sqrt{3}}{1{,}000/\sqrt{3}} = 0.9 \text{ pu}$$

If the power per phase is 2,000 W, the total three-phase power is 6,000 W. The power in per unit is expressed as

$$\text{Per-unit power} = \frac{6{,}000}{10{,}000} = \frac{2{,}000}{10{,}000/3} = 0.6 \text{ pu}$$

▪▪▪▪▪▪▪▪▪ ‖‖‖‖
4.3 Transformer Tests

Most transformers are supplied by their manufacturers with polarity markings such as dot markings. Accrding to the dot-notation system that has been explained, if the primary current is flowing into the dotted terminal of one winding, the secondary load current will be flowing out of the dotted terminal of the other winding. Another polarity-notation convention is used for power-system transformers. For single-phase transformer windings, the terminals on the high-voltage side are identified as H_1 and H_2, and those on the low-voltage side are labeled as X_1 and X_2. The terminals with subscript 1 (or 2) in this convention are equivalent to the dotted terminals in the dot-notation system. Figure 4.3.1a shows the notation for a single-phase, two-winding transformer. If there is doubt about the transformer polarity, it

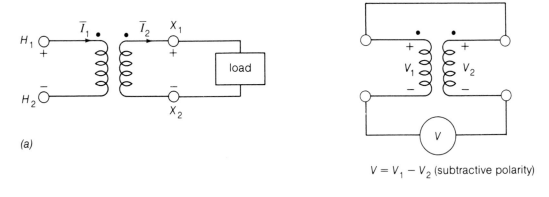

(a)

(b)

$V = V_1 - V_2$ (subtractive polarity)

Figure 4.3.1   (a) Polarity markings of a single-phase, two-winding transformer. (b) Polarity test.

can easily be checked by the polarity test shown in part b of the figure and the dot marks can then be made according to the convention. In the polarity test, terminals in the closest physical proximity—one from each winding—are connected as shown. By applying the rated voltage (say V_1) to one winding, the voltage V across the two remaining terminals is measured. If the measured voltage is larger than the input voltage, the polarity is said to be additive; if smaller, the polarity is subtractive, as shown in Figure 4.3.1b.

The parameters of the equivalent circuit of a transformer are usually determined by conducting two tests: the open-circuit (or no-load) test, and the short-circuit test.

Open-Circuit Test

The open-circuit test is conducted in order to determine the exciting admittance of the transformer equivalent circuit, the no-load loss, the no-load exciting current, and the no-load power factor. While one winding is open-circuited, a voltage (usually rated voltage at rated frequency) is applied to the other winding, and measurements of voltage, current and real power are made with voltmeter, ammeter, and wattmeter, respectively, as shown in Figure 4.3.2a. Usually the high-voltage winding is open-circuited, and the test is conducted by placing the instruments on the low-voltage side.

Let V be the applied voltage in volts as read by the voltmeter V, I_0 the exciting no-load current in amperes as read by the ammeter A, and P_{oc} the real power in watts as read by the wattmeter W. The voltage drop in the leakage impedance of the winding (which is excited in the open-circuit test) caused by the usually small exciting current is normally neglected. The approximate equivalent circuit is then as shown in Figure 4.3.2b. The exciting admittance in siemens and its conductance and susceptance components are then given by

$$|y_0| = \frac{I_0}{V} \tag{4.3.1}$$

$$g_c = \frac{P_{oc}}{V^2} \tag{4.3.2}$$

$$b_m = \sqrt{|y_0|^2 - g_c^2} \tag{4.4.3}$$

The values so obtained are referred to the side on which the instruments are connected and the winding is excited in the open-circuit test. Neglecting the copper loss caused by the exciting no-load current I_0, the power input P_{oc} will be equal to the core loss P_c of the transformer. The no-load power factor is obtained as

$$\cos \theta_0 = \frac{P_{oc}}{VI_0} \tag{4.3.4}$$

The no-load phasor diagram can then be easily drawn by showing the exciting current $\bar{I}_0$ lagging the applied voltage $\bar{V}$ by an angle of θ_0.

As a part of the open-circuit test, the voltage at the terminals of the open-circuited winding is sometimes measured to check the turns ratio of the transformer. Since the no-load power factor is generally very low (in the range of 0.05 to 0.2), it becomes very difficult to obtain accurate power measurements; electronic multiplier wattmeters are generally used for no-load transformer power measurements. Because of the small quantities involved in most communication transformers, ac bridges or other suitable devices may be used instead of voltmeters, ammeters, and wattmeters. Care must be exercised during the test procedure: the high-voltage winding terminals must be guarded to ensure safety for test personnel.

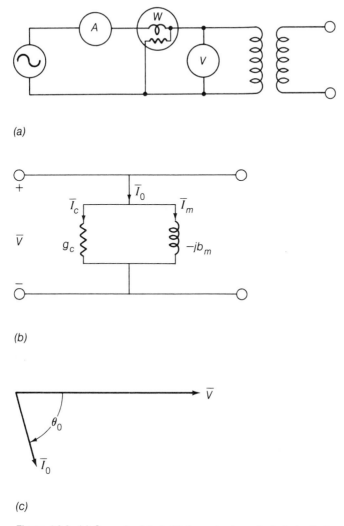

(a)

(b)

(c)

Figure 4.3.2 (a) Open-circuit test. (b) Approximate equivalent circuit of a transformer on open-circuit. (c) No-load phasor diagram.

Short-Circuit Test

The short-circuit test is conducted by short-circuiting one winding (usually the low-voltage winding) and applying a *reduced* voltage at rated frequency such that the rated current results. Figure 4.3.3a is a connection diagram for performing the short-circuit test on power-system transformers. Let V_{sc}, I_{sc}, and P_{sc}, respectively, be the impressed voltage, the input short-circuit current, and the power input as measured by the indicating instruments.

Neglecting the exciting current, which is usually small compared to the rated (i.e., full-load) current, the approximate equivalent circuit of a transformer under short-circuit condition is shown in Figure 4.3.3b. The short-circuit impedance Z_{sc}, which is then equal to the transformer's equivalent series

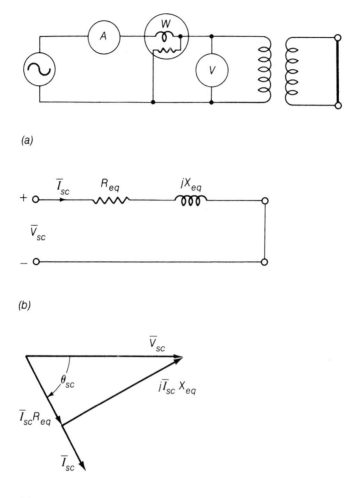

(a)

(b)

(c)

Figure 4.3.3 (a) Short-circuit test. (b) Approximate equivalent circuit of a transformer on short-circuit. (c) Phasor diagram under short-circuit conditions.

impedance, is given by

$$|Z_{eq}| = |Z_{sc}| = \frac{V_{sc}}{I_{sc}} \tag{4.3.5}$$

Neglecting the core loss at the low value of V_{sc}, the equivalent series resistance and leakage reactance are given by

$$R_{eq} = R_{sc} = \frac{P_{sc}}{I_{sc}^2} \tag{4.3.6}$$

$$X_{eq} = X_{sc} = \sqrt{|Z_{eq}|^2 - R_{eq}^2} \tag{4.3.7}$$

The values so obtained are referred to the side on which the voltage is applied and the measuring instruments are connected. The short-circuit power factor is obtained as

$$\cos \theta_{sc} = \frac{P_{sc}}{V_{sc}I_{sc}} \tag{4.3.8}$$

The phasor diagram under short-circuit conditions is Figure 4.3.3c.

EXAMPLE 4.3.1

These data were obtained when open-circuit and short-circuit tests were performed on a single-phase, 50-kVA, 2,400:240-V, 60-Hz distribution transformer:

	Voltage (volts)	Current (amps)	Power (watts)
With high-voltage winding open-circuited	240	4.85	173
With low-voltage winding short-circuited	52	20.8	650

Determine the equivalent circuit of the transformer referred to its high-voltage side and to its low-voltage side.

Solution

The exciting admittance *referred to the low-voltage side* is obtained from the open-circuit test data as

$$|y|_{LV} = \frac{4.85}{240} = 0.0202 \text{ S}$$

$$g_{LV} = \frac{173}{(240)^2} = 0.003 \text{ S}$$

and

$$b_{LV} = \sqrt{(0.0202)^2 - (0.003)^2} = 0.02 \text{ S}$$

These values can be referred to the high-voltage side by dividing by the square of the turns ratio.

The equivalent series impedance of the transformer referred to its high-voltage side is obtained from the short-circuit test data:

$$|Z_{eq}|_{HV} = \frac{52}{20.8} = 2.5 \text{ }\Omega$$

$$R_{eq\,HV} = \frac{650}{(20.8)^2} = 1.5 \text{ }\Omega$$

$$X_{eq\,HV} = \sqrt{(2.5)^2 - (1.5)^2} = 2.0 \text{ }\Omega$$

These values can be referred to the low-voltage side by dividing by the square of the turns ratio. Approximate equivalent circuits with the exciting shunt branch moved to the low-voltage or high-voltage terminals can now be drawn. See Figure 4.3.4.

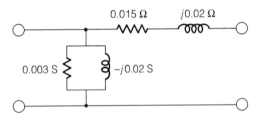

values referred to the low-voltage side

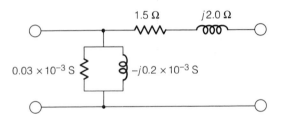

values referred to the high-voltage side

Figure 4.3.4

If an equivalent T-circuit is desired, the values of individual primary and secondary resistances and leakage reactances, *referred to the same side,* are usually assumed to be equal:

$$R_1 = R_2' = \frac{R_{eq\,HV}}{2} = 0.75 \ \Omega$$

$$X_{l1} = X_{l2}' = \frac{X_{eq\,HV}}{2} = 1.0 \ \Omega$$

$$R_1' = R_2 = \frac{R_{eq\,LV}}{2} = \frac{0.015}{2} = 0.0075 \ \Omega$$

$$X_{l1}' = X_{l2} = \frac{X_{eq\,LV}}{2} = \frac{0.02}{2} = 0.01 \ \Omega$$

The subscripts 1 and 2 refer to the high-voltage and low-voltage sides, respectively. The T-equivalent circuits referred to the high-voltage and low-voltage sides are then the same as those shown for Example 4.2.3.

There are several other important tests for electronics transformers. For the details of them, refer to the IEEE *Standards* listed in the chapter bibliography.

4.4 Transformer Performance

The losses in a transformer consist of (*a*) the core loss, (*b*) the copper loss, and (*c*) the stray loss. The core loss is practically independent of the load for a given voltage and frequency of operation. It is taken to be approximately equal to the no-load loss P_{oc} as read by the wattmeter in the no-load (open-circuit) test. The copper loss or the I^2R loss is due to the resistance of the windings. The stray loss is largely due to eddy currents induced by the leakage flux in the tank and other structural parts. The sum of the copper loss and the stray loss is known as the *load loss,* being equal to I^2R_{eq} as determined from the short-circuit test by the wattmeter reading P_{sc}. The term *copper loss* is often used instead of *load loss* when including the stray loss. Copper loss is a function of the load or the load current for constant voltage operation. In fact, it varies as the square of the load current.

The efficiency of any power-transmitting device that includes a transformer is given by

$$\text{Per-unit efficiency } \eta = \frac{\text{power output}}{\text{power input}} = \frac{\text{input} - \text{losses}}{\text{Input}} = 1 - \frac{\text{losses}}{\text{input}} \tag{4.4.1}$$

The efficiency of power-system transformers at rated load is quite high; it may be as high as 90 to 99 percent or 0.9 to 0.99 per unit. The higher the rating of a transformer, the greater is its efficiency.

Most power-system transformers used to transmit power operate normally near their rated capacities and are switched out of circuit when their output is not required. In such cases it makes sense to design the transformer for its maximum efficiency at or near rated output. On the other hand, for a distribution transformer that is in the system 24 hours a day but operates well below its rated output for a good number of those hours, it is desirable to design for maximum efficiency at or near the average output.

The ratio of core loss to copper loss at rated output varies in transformers used for different types of service. It can be shown that the maximum efficiency of a transformer operating at a constant output voltage and power factor occurs at that load for which the copper loss equals the core loss. On the basis of the approximate equivalent circuit of Figure 4.2.8b, it follows that

$$\eta = \frac{V_2' I_2' \cos \theta}{V_2' I_2' \cos \theta + P_c + (I_2')^2 R_{eq}} \tag{4.4.2}$$

where the core loss $P_c (= V_1^2 g_c)$ is a constant, the copper loss $(I_2')^2 R_{eq}$ varies with the square of the load current, and $\cos \theta$ is the load power factor. Maximum efficiency occurs when $1/\eta$ is minimal. For constant output voltage and power factor, differentiating with respect to I_2' and setting the derivative equal to zero, one obtains the condition for maximum efficiency to be

$$(I_2')^2 R_{eq} = P_c \tag{4.4.3}$$

which states that the core loss should be equal to the copper loss for maximum operating efficiency at a given load.

For distribution transformers, an all-day (or energy) efficiency becomes important. It is given by

$$\eta_{AD} = \frac{\text{energy output over 24 hours}}{\text{energy input over 24 hours}} \tag{4.4.4}$$

This can be determined once the load cycle of the transformer is known.

Although the equivalent circuits can be used to calculate the losses for a given output of the transformer, it is usually more convenient to use the open-circuit and short-circuit test data directly.

EXAMPLE 4.4.1

Compute the efficiency of the transformer of Example 4.3.1 corresponding to (a) full load, 0.8 power factor lagging, and (b) one-half load, 0.6 power factor lagging.

Solution

a. Corresponding to full load, 0.8 power factor lagging,

$$\text{Output} = 50,000 \times 0.8 = 40,000 \text{ W}$$

The load loss (or the copper loss) at rated (full) load equals the real power measured in the short-circuit test at *rated current*:

$$\text{Copper loss} = I_{HV}^2 R_{eq\,HV} = P_{sc} = 650 \text{ W}$$

where subscript *HV* refers to the high-voltage side.

$$\text{Core loss} = P_{oc} = 173 \text{ W}$$

$$\text{Total losses at full load} = 650 + 173 = 823 \text{ W}$$

$$\text{Input} = 40{,}000 + 823 = 40{,}823 \text{ W}$$

The full-load efficiency at 0.8 power factor is given by

$$\eta = 1 - \frac{\text{losses}}{\text{input}} = 1 - \frac{823}{40{,}823} = 1 - 0.02 = 0.98$$

b. Corresponding to one-half rated load, 0.6 power factor lagging,

$$\text{Output} = \frac{1}{2} \times 50{,}000 \times 0.6 = 15{,}000 \text{ W}; \quad \text{Copper loss} = \frac{1}{4} \times 650 = 162.5 \text{ W}$$

Note that the current at one-half rated load is half of the full-load current, and that the copper loss is one-quarter of that at rated current value.

$$\text{Core loss} = P_{oc} = 173 \text{ W}$$

which is considered to be unaffected by the load, as long as the secondary terminal voltage is at its rated value.

$$\text{Total losses at one-half rated load} = 162.5 + 173 = 335.5 \text{ W}$$

$$\text{Input} = 15{,}000 + 335.5 = 15{,}335.5 \text{ W}$$

Efficiency at one-half rated load and 0.6 power factor is given by

$$\eta = 1 - \frac{335.5}{15{,}335.5} = 1 - 0.022 = 0.978$$

EXAMPLE 4.4.2

The distribution transformer of Example 4.3.1 is supplying a load at 240 V and 0.8 power factor lagging.

a. Determine the fraction of full-load at which the maximum efficiency of the transformer occurs, and compute the efficiency at that load.

b. The load cycle of the transformer operating at a constant power factor is 0.9 full load for 8 hours, 0.5 full load for 12 hours, and no load for 4 hours. Compute the all-day (or energy) efficiency of the transformer under the specified conditions.

Solution

a. Referred to the low-voltage side, the equivalent series impedance of the transformer (based on the solution of Example 4.3.1) is given by

$$|Z_{eq}|_{LV} = \frac{|Z_{eq}|_{HV}}{a^2} = \frac{2.5}{100} = 0.025 \ \Omega$$

$$R_{eq\,LV} = \frac{R_{eq\,HV}}{a^2} = \frac{1.5}{100} = 0.015 \ \Omega$$

$$X_{eq\,LV} = \frac{X_{eq\,HV}}{a^2} = \frac{2.0}{100} = 0.02 \ \Omega$$

For maximum efficiency, the core loss should equal the copper loss. Thus

$$P_{oc} = I_{LV}^2 R_{eq\,LV}$$

where subscript *LV* refers to the low-voltage side. That is,

$$173 = I_{LV}^2(0.015) \quad \text{or} \quad I_{LV} = \sqrt{\frac{173}{0.015}} = 107.4 \text{ A}$$

Full-load current on the low-voltage side, as given by the rating, is

$$I_{FL\,LV} = \frac{50,000}{240} = 208.3 \text{ A}$$

from which the fraction of full-load rating at which the maximum efficiency occurs is given by

$$\frac{107.4}{208.3} = 0.516$$

Alternatively, it may be obtained simply as follows:

$$k^2 P_{sc} = P_{oc}$$

where *k* is the fraction of the full-load rating. Therefore,

$$k = \sqrt{\frac{P_{oc}}{P_{sc}}} = \sqrt{\frac{173}{650}} = 0.516$$

Output power corresponding to this condition is

$$50,000 \times 0.516 \times 0.8 = 20,640 \text{ W}$$

where 0.8 is the power factor given in the problem statement.

$$\text{Core loss} = \text{Copper loss} = 173 \text{ W}$$

The maximum efficiency is then given by

$$\eta_{max} = \frac{\text{output}}{\text{output} + \text{losses}} = \frac{20,640}{20,640 + 173 + 173} = \frac{20,640}{20,986} = 0.9835$$

b. During 24 hours

$$\text{Energy output} = (8 \times 0.9 \times 50 \times 0.8) + (12 \times 0.5 \times 50 \times 0.8) = 528 \text{ kWh}$$

$$\text{Core loss} = 24 \times 0.173 = 4.15 \text{ kWh}$$

$$\text{Copper loss} = (8 \times 0.9^2 \times 0.65) + (12 \times 0.5^2 \times 0.65) = 6.16 \text{ kWh}$$

The all-day (or energy) efficiency of the transformer is given by

$$\eta_{AD} = \frac{528}{528 + 4.15 + 6.16} = 0.9808$$

Regulation

The voltage of a transformer is the change in the magnitude of secondary terminal voltage from no-load to full-load and is usually expressed as a percentage of the full-load value.

$$\text{Percentage voltage regulation} = \frac{V_{2(\text{no load})} - V_{2(\text{full load})}}{V_{2(\text{full load})}} \times 100 \qquad (4.4.5)$$

The voltages in this equation may be referred to either the low-voltage or the high-voltage side.

Referring to the approximate equivalent circuit of Figure 4.2.8c with an equivalent series impedance and to the corresponding phasor diagram for a lagging power-factor case, Figure 4.2.9c,

$$\text{Percentage voltage regulation} = \frac{V_1 - V_2'}{V_2'} \times 100 \qquad (4.4.6)$$

The magnitude of the secondary voltage referred to the primary side (V_2') would rise to a value of V_1, if V_1 were held constant and the load removed. For loads that have a lagging or unity power factor, the regulation is positive, since V_1 is greater (in magnitude) than V_2'. However, for loads with a leading power factor, the regulation may be negative.

From Figure 4.2.9c, for small values of δ, it can be seen that the difference between the magnitudes of $\bar{V}_1$ and $\bar{V}_2'$—that is, $V_1 - V_2'$—is approximately equal to ab given by $(I_1 R_{eq} \cos \theta_L + I_1 X_{eq} \sin \theta_L)$. It is sometimes convenient to express the regulation in per-unit quantities corresponding to rated (i.e., full-load) condition as

$$\varepsilon = \varepsilon_r \cos \theta_L + \varepsilon_x \sin \theta_L \qquad (4.4.7)$$

where ε_r is the per-unit equivalent resistance, or the per-unit resistance drop, or the per-unit full-load I^2R, given by

$$\varepsilon_r = I_1^2 R_{eq}/(V_1 I_1) = I_1 R_{eq}/V_1 = R_{eq}/(V_1/I_1) \qquad (4.4.8)$$

and ε_x is the per-unit equivalent reactance, or the per-unit reactance drop, given by

$$\varepsilon_x = I_1^2 X_{eq}/(V_1 I_1) = I_1 X_{eq}/V_1 = X_{eq}/(V_1/I_1) \qquad (4.4.9)$$

Most loads, such as motors, that are operated near rated voltage and frequency are supplied by transformers that have small values of regulation, of the order of a few percent. On the other hand, loads, such as welding arcs, that operate at nearly constant current are supplied by their own individual transformer, which has a high value of regulation.

Transformers are often designed with taps on one winding so that the turns ratio can be varied over a small range. The *tap-changing* is often done automatically in large power-system transformers so that a sensibly constant secondary-terminal voltage is maintained as the magnitude and power factor of the load vary. Tap-changing is usually done manually for distribution transformers, adjusted to a setting that gives optimum voltage over the projected load cycle. Tap-changing can also be used for compensating the variations in primary terminal voltage due to the feeder impedance.

EXAMPLE 4.4.3

The transformer of Example 4.3.1 is supplying full load at 240 V and 0.8 power factor lagging.

a. Determine the transformer's voltage regulation.

b. Find the approximate change in turns ratio required if the primary terminal voltage is fixed at 2,400 V.

Solution

a. The equivalent circuit of the transformer, referred to the high-voltage (primary) side, neglecting the exciting current of the transformer, and the corresponding phasor diagram are shown in Figure 4.4.1.

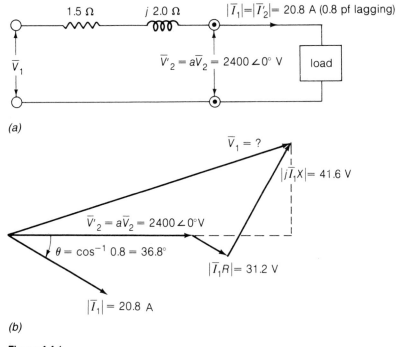

(a)

(b)

Figure 4.4.1

$$\bar{V}_1 = \bar{V}_2' + \bar{I}_1(R + jX)$$
$$= 2,400\angle 0° + 20.8\angle -36.8°(1.5 + j2.0)$$
$$= 2,400 + j0 + (16.64 - j12.48)(1.5 + j2.0)$$
$$= 2,400 + j0 + 49.92 + j14.56 = 2,449.92 + j14.56$$
$$= 2,450\angle 0.34°$$

$$\text{Percentage voltage regulation} = \frac{2,450 - 2,400}{2,400} \times 100 = +2.08\%$$

The per-unit voltage regulation is 0.0208. The regulation can also be approximately calculated by using Eq. (4.4.7).

b. Based on the result of part a, the turns ratio of the transformer must be reduced on load by 2.08% if the secondary terminal voltage is to remain unaltered. That is,

$$\text{Approximate new turns ratio} = \frac{100 - 2.08}{100} \times 10 = 9.792$$

■■■■■■■■■■■■■■■ ||||||||
4.5 Three-Phase Transformers

Most electrical energy is generated and transmitted using a three-phase system that involves several three-phase voltage transformations. For transforming three-phase power, either a bank of three identical single-phase transformers suitably connected or one three-phase transformer can be employed. Under conditions of balanced load and balanced voltages, with identical transformers in a given arrangement, each single-phase transformer will carry one third of the total three-phase load.

A three-phase transformer for a given rating—compared to a bank of three single-phase transformers—weighs less, costs less, requires less floor space, and has somewhat higher efficiency. In addition, the number of external connections is reduced from twelve to six, and so also the number of elaborate bushing arrangements. However, the cost of a spare, stand-by three-phase transformer is much greater than that of a spare single-phase transformer; when one phase breaks down the whole three-phase transformer must be removed for repair.

Cores and windings for all three phases are combined in a single structure in a three-phase transformer. To visualize the development of a three-phase, core-type transformer, let us start with three identical single-phase core-type transformers arranged as in Figure 4.5.1a. Both the primary and secondary windings of each phase have been placed on only one leg of each transformer, and for simplicity, only the primary windings are shown. For balanced three-phase sinusoidal voltages, the fluxes ϕ_a, ϕ_b, and ϕ_c are 120° apart in time phase, and the sum of the three fluxes at any instant is zero. The leg carrying the sum by merging the three legs shown in part a of the figure can be omitted as in part b. In practice, however, the construction is as shown in part c because it is more economical. The magnetic paths of the outer legs are somewhat longer than that of the center leg; but the resultant imbalance in the magnetizing currents is seldom significant. Figure 4.5.2a shows a conventional three-phase core with cruciform section to fit cylindrical coils. In order to reduce the overall height, a five-legged core

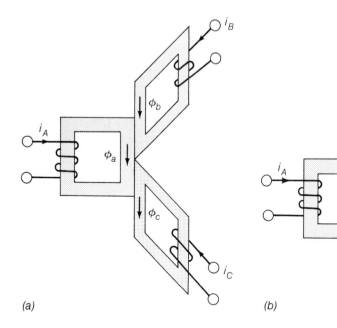

(a)   (b)

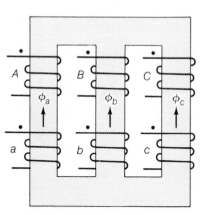

(c)

Figure 4.5.1 Evolution of a three-phase, two-winding, core-type transformer.

is sometimes used, in which the phase coils are wound around the three center legs, as shown in Figure 4.5.2b. See Photo 4.5.1 for a completed core-and coil-assembly with leads of a three-phase core-type transformer.

Next, to visualize the development of a three-phase, shell-type transformer, one can stack three single-phase shell-type units as in Figure 4.5.3a, which shows only the primary windings of the three

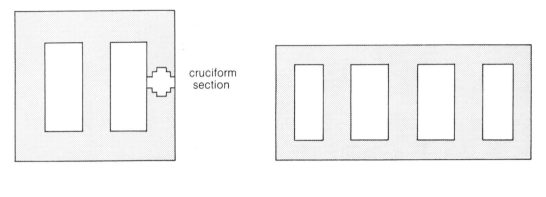

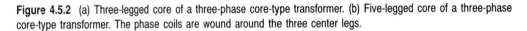

(a)   (b)

Figure 4.5.2 (a) Three-legged core of a three-phase core-type transformer. (b) Five-legged core of a three-phase core-type transformer. The phase coils are wound around the three center legs.

Photo 4.5.1 Completed core-and-coil assembly with leads of a 30-MVA, 69Y: 13.2 Δ-kV, three-phase core-type transformer. (Photo courtesy of Cooper Industries.)

phases for simplicity. The direction of the winding on the center unit is made opposite to that on the outer units. Thus the polarities of both middle windings (primary and secondary) are reversed with respect to those of the outer windings. With this arrangement, for balanced three-phase sinusoidal voltages, the adjacent yoke sections of units a and b carry a combined flux given by

$$\frac{\overline{\Phi}_a}{2} + \frac{\overline{\Phi}_b}{2} = \frac{\overline{\Phi}_a}{2}(1\angle 0° + 1\angle 240°) = \frac{\overline{\Phi}_a}{2}\angle -60° \tag{4.5.1}$$

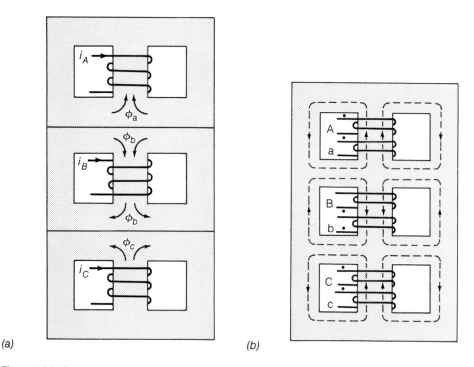

(a) (b)

Figure 4.5.3 Evolution of a three-phase, two-winding shell-type transformer.

where $\overline{\Phi}_b$ is leading $\overline{\Phi}_a$ by 240° for the positive phase sequence *a-b-c*. Because of this it is possible to make the cross-sectional area of the combined adjacent yoke sections the same as that used for the outer legs. Thus Figure 4.5.3b illustrates a conventional three-phase, shell-form core with the coils. The slight imbalance in the magnetic paths among the three phases of the shell-type construction is quite insignificant. See Photo 4.5.2 for an assembly of a typical shell-type transformer.

Three-Phase Transformer Connections

The windings of either core-type or shell-type three-phase transformers may be connected in either wye or delta. Four possible combinations of connections for the three-phase, two-winding transformers are

$$Y-\Delta, \quad \Delta-Y, \quad \Delta-\Delta, \quad \text{and} \quad Y-Y$$

These are shown in Figure 4.5.4 with the transformers assumed to be ideal. The windings on the left are the primaries; those on the right are the secondaries, and a primary winding of the transformer is linked magnetically with the secondary winding drawn parallel to it. With the per-phase primary-to-secondary turns ratio (N_1/N_2), the resultant voltages and currents for balanced applied line-to-line voltages V and line currents I are marked in the figure.

As in the case of three-phase circuits under balanced conditions, only one phase needs to be considered for circuit computations, because the conditions in the other two phases are the same except for the phase displacements associated with a three-phase system. It is usually convenient to carry out the analysis on a per-phase of Y (i.e., line-to-neutral) basis, and in such a case the transformer series impedance can then be added in series with the transmission-line series impedance. In dealing with Y–Δ

Photo 4.5.2 Assembly of a 400-MVA, 345Y:138Y:13.8 Δ-kV shell-type transformer. (Photo courtesy of Cooper Industries.)

and Δ–Y connections, all quantities can be referred to the Y-connected side. For Δ–Δ connections, it is convenient to replace the Δ-connected series impedances of the transformers with equivalent Y-connected impedances by using the relation

$$Z_{\text{per phase of Y}} = \frac{1}{3}Z_{\text{per phase of }\Delta} \qquad (4.5.2)$$

It can be shown that to transfer the ohmic value of impedance from the voltage level on one side of a three-phase transformer to the voltage level on the other side, the multiplying factor is the square of the ratio of line-to-line voltages, regardless of whether the transformer connection is Y–Y, Δ–Y or Δ–Δ.

The Y–Δ connection is commonly used in high-voltage transmission systems to step down the voltage, with the high-voltage side connected in Y and the low-voltage side connected in Δ. A common arrangement in distribution circuits is the 208 (line-to-line)/120 (line-to-neutral)-volt system supplied by the Y-connection on the low-voltage side, with the high-voltage side of the transformer connected in Δ. The neutral point of the Y is usually grounded. The Δ–Y connection is popularly used for stepping up to a high voltage in transmission systems. The Y-connection on the high-voltage side, besides providing a neutral for grounding, reduces insulation requirements.

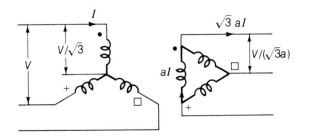

(a)

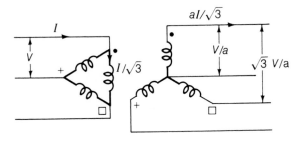

(b)

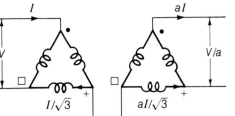

(c)

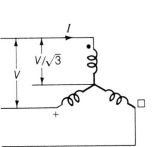

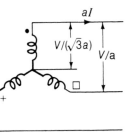

(d)

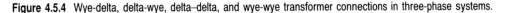

Figure 4.5.4 Wye-delta, delta-wye, delta–delta, and wye-wye transformer connections in three-phase systems.

The Δ–Δ connection is generally used in moderate voltage systems since, the windings operate at full line-to-line voltage. In the case of a bank consisting of three single-phase transformer units, it is an advantage of the Δ–Δ connection that if one transformer breaks down it can be removed from

Photo 4.5.3 Completely assembled 910-MVA, 20.5:500-kV step-up transformer, 39 feet high, 35 feet long, 20 feet wide, weighing 562 tons. (Photo courtesy of Cooper Industries.)

Photo 4.5.4 A 10-MVA, 66:6.6-kV mobile-substation transformer with the high-voltage section, transformer section, and low-voltage section. (Photo courtesy of Westinghouse Electric Corporation.)

the circuit, and reduced three-phase output will be supplied by the other two units connected in *open-delta*. The open-delta connection is also known as *V-connection*. For reasons of economy this connection is sometimes used for instrument transformers and in growing load centers—whose full growth may take several years, at which time a third unit is added to the existing open-delta connection to make it delta–delta connection. The rating of the V–V connection is only about 58 percent of that of the Δ–Δ connection without overheating.

The Y–Y connection may be used in high-voltage applications, because the voltage across the transformer winding is only $1/\sqrt{3}$ of the line-to-line voltage. However, it is seldom used because of the difficulties with the exciting current phenomena. In cases where Y–Y transformation is utilized, it is quite common to incorporate a third winding, known as a *tertiary winding,* connected in delta. See Photos 4.5.3 and 4.5.4 for views of a large step-up power transformer and a mobile-substation transformer.

EXAMPLE 4.5.1

A one-line diagram of part of a three-phase transmission system and necessary details appear in Figure 4.5.5. Draw the equivalent reactance diagram per phase (line-to-neutral) with all reactances marked in per unit, choosing the bases of 30,000 kVA and 120 kV in the transmission line.

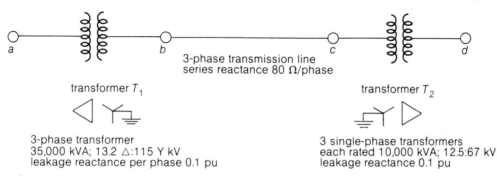

3-phase transmission line
series reactance 80 Ω/phase

transformer T_1

transformer T_2

3-phase transformer
35,000 kVA; 13.2 △:115 Y kV
leakage reactance per phase 0.1 pu

3 single-phase transformers
each rated 10,000 kVA; 12.5:67 kV
leakage reactance 0.1 pu

Figure 4.5.5

Solution
The per-unit leakage reactance of a transformer is based on its own rating. The per-unit reactances of the transformers need to be converted to the chosen bases of 30,000 kVA and 120 kV in the transmission line. Making use of Equation 4.2.34, one obtains

$$\text{for transformer } T_1: \qquad X_l = 0.1\frac{30,000}{35,000}\left(\frac{115}{120}\right)^2 = 0.0787 \text{ pu}$$

$$\text{for transformer } T_2: \qquad X_l = 0.1\frac{30,000}{30,000}\left(\frac{67\sqrt{3}}{120}\right)^2 = 0.0935 \text{ pu}$$

The base impedance per phase of Y of the transmission line is given by Equation 4.2.42:

$$\frac{(120,000)^2}{30,000 \times 10^3} = 480 \ \Omega$$

The per-unit series reactance of the transmission line per phase is

$$\frac{80}{480} = 0.167 \text{ pu}$$

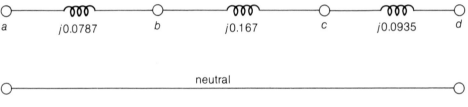

Figure 4.5.6

Figure 4.5.6 gives the desired line-to-neutral reactance diagram with per-unit values. Note that each of the transformers, as well as the transmission line, is represented by a corresponding equivalent series reactance (adjusted properly to the chosen base, of course); only a single-phase circuit needs to be considered when analyzing balanced three-phase systems. In the specifications given for a three-phase transformer, the kVA is the three-phase kVA and the voltages are the line-to-line voltages on either side.

▌▌▌▌▌▌▌▌▌▌▌
EXAMPLE 4.5.2

A one-line diagram of a three-phase distribution system is given in Figure 4.5.7. Determine the line-to-line voltage at the sending end of the high-voltage feeder when the transformer delivers rated load at 240 V (line-to-line) and 0.8 lagging power factor. Neglect the exciting current of the transformer.

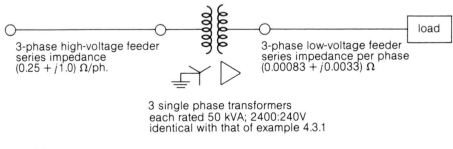

3 single phase transformers
each rated 50 kVA; 2400:240V
identical with that of example 4.3.1

Figure 4.5.7

Solution
The analysis is carried out on a per-phase-of-Y basis by referring all quantities to the high-voltage Y-side of the transformer bank. The impedance of the low-voltage feeder—referred to the high-voltage side by means of the square of the ideal line-to-line voltage ratio of the transformer bank—is

$$(0.00083 + j0.0033)\left(\frac{2{,}400\sqrt{3}}{240}\right)^2 = (0.25 + j1.0) \ \Omega$$

Thus we have, referred to the high-voltage side, the combined series impedance of the high-voltage and low-voltage feeders as

$$Z_{f\,HV} = (0.25 + j1.0) + (0.25 + j1.0) = (0.5 + j2.0) \ \Omega \text{ per phase of Y}$$

The equivalent series impedance of the transformer bank referred to its high-voltage Y-side is given from Example 4.3.1 as

$$Z_{eq\,HV} = (1.5 + j2.0) \; \Omega \text{ per phase of Y}$$

The equivalent circuit per phase of Y referred to the Y-connected primary side is then given by exactly the same circuit shown in Example 4.2.3b. For the conditions given in the problem, the solution on a per-phase basis is exactly the same as the solution of Example 4.2.3b, from which it follows that the voltage at the sending end of the high-voltage feeder is 2,483.5 volts. The actual line-to-line voltage, because of the primary Y-connection, is given by

$$2{,}483.5\sqrt{3} = 4{,}301.4 \text{ V}$$

Per-Unit System and Phase Shift

Balanced three-phase circuits can be solved in per-unit system on a per-phase basis after converting delta-load impedances to equivalent wye-connected impedances. The following relations hold good for a three-phase system, in which the rated complex 3-phase power $|S|_{3\phi}$ and the rated line-to-line voltage V_{LL} are usually selected as base values. (Subscripts 3ϕ, LL, and LN denote three-phase, line-to-line, and line-to-neutral, respectively.)

$$|S|_{\text{base }1\phi} = \frac{|S|_{\text{base }3\phi}}{3} \tag{4.5.3}$$

$$V_{\text{base LN}} = \frac{V_{\text{base LL}}}{\sqrt{3}} \tag{4.5.4}$$

$$|S|_{\text{base }3\phi} = P_{\text{base }3\phi} = Q_{\text{base }3\phi} \tag{4.5.5}$$

$$I_{\text{base}} = \frac{|S|_{\text{base }1\phi}}{V_{\text{base LN}}} = \frac{|S|_{\text{base }3\phi}}{\sqrt{3}V_{\text{base LL}}} \tag{4.5.6}$$

$$|Z|_{\text{base}} = \frac{V_{\text{base LN}}}{I_{\text{base}}} = \frac{V_{\text{base LN}}^2}{|S|_{\text{base }1\phi}} = \frac{V_{\text{base LL}}^2}{|S|_{\text{base }3\phi}} \tag{4.5.7}$$

$$R_{\text{base}} = X_{\text{base}} = |Z|_{\text{base}} = \frac{1}{|Y|_{\text{base}}} \tag{4.5.8}$$

When only one transformer is considered, its nameplate ratings are usually selected as base values. When several transformers and devices are involved, however, the system base values may be different from the nameplate ratings of any particular device. It then becomes necessary to convert the per-unit impedance of a device based on its nameplate ratings to that with the system base values. To convert a per-unit impedance from "old" to "new" base values, the equations are:

$$|Z|_{\text{pu new}} = \frac{|Z|_{\text{actual}}}{|Z|_{\text{base new}}} = \frac{|Z|_{\text{pu old}} |Z|_{\text{base old}}}{|Z|_{\text{base new}}} \tag{4.5.9}$$

or

$$|Z|_{\text{pu new}} = |Z|_{\text{pu old}} \left(\frac{V_{\text{base old}}}{V_{\text{base new}}} \right)^2 \left(\frac{|S|_{\text{base new}}}{|S|_{\text{base old}}} \right) \tag{4.5.10}$$

Three identical single-phase, two-winding transformers can be connected to form a three-phase bank. In the case of a Y–Y or Δ–Δ bank, there is no angular displacement between corresponding

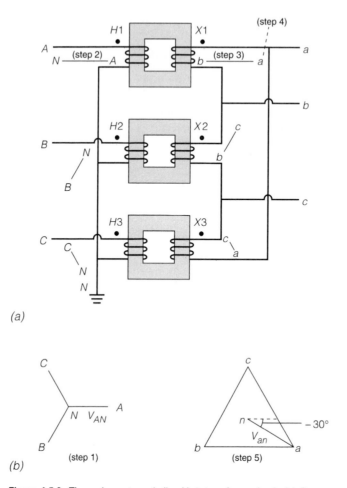

(a)

(b)

Figure 4.5.8 Three-phase, two-winding Y–Δ transformer bank. (a) Core and coil arrangement. (b) Positive-sequence phasor diagram.

primary and secondary line voltages. In the case of a Y–Δ or Δ–Y bank, however, a *phase shift* exists between the corresponding primary and secondary line voltages. Only banks with the same phase shift should be operated in parallel; otherwise, large circulating currents and consequent severe overheating of the windings will occur.

Figure 4.5.8a shows a three-phase, two-winding Y–Δ transformer bank. The American standard notation denotes $H1$, $H2$, and $H3$ on the high-voltage terminals and $X1$, $X2$, and $X3$ on the low-voltage terminals. Also, for identifying phases in this book, uppercase letters A, B, and C and lowercases letters a, b, and c are used on the high-voltage and low-voltage sides of the transformer, respectively. The positive sequence phasor diagram in Figure 4.5.8b can be constructed by following the five steps given here (and indicated in Figure 4.5.8):[3]

---

[3] J. D. Glover and M. S. Sarma, *Power System Analysis and Design,* Boston: PWS-Kent, 1989.

- **Step 1** Assume that positive-sequence voltages are applied to the Y winding. Draw the positive-sequence phasor diagram for these voltages.
- **Step 2** Move phasor A–N next to terminals A–N in Figure 4.5.8a. Identify the ends of this line in the same manner as in the phasor diagram. Similarly, move phasors B–N and C–N next to terminals B–N and C–N in the figure.
- **Step 3** For each single-phase transformer, the voltage across the low-voltage winding must be in phase with the voltage across the high-voltage winding, assuming an ideal transformer. Therefore, draw a line next to each low-voltage winding parallel to the corresponding line already drawn next to the high-voltage winding.
- **Step 4** Label the ends of the lines drawn in Step 3 by inspecting the polarity marks. For example, phase A is connected to dotted terminal H1, and A appears on the *right* side of line A–N. Therefore, phase a, which is connected to dotted terminal X1, must be on the *right* side, and b on the left side of line a–b. Similarly, phase B is connected to dotted terminal H2, and B is *down* on line B–N. Therefore, phase b, connected to dotted terminal X2, must be down on line b–c. Similarly, c is up on line c–a.
- **Step 5** Bring the three lines labeled in Step 4 together to complete the phasor diagram for the low-voltage Δ winding. Note that $\bar{V}_{AN}$ leads $\bar{V}_{an}$ by 30° in accordance with the American standard notation.

Thus, the labeling of the windings and the schematic representation are in accordance with the American standard notation, as per which, in either a Y–Δ or Δ–Y transformer, positive-sequence quantities on the high-voltage side shall lead their corresponding quantities on the low-voltage side by 30°.

It can be shown that the negative-sequence phase shift is the reverse of the positive-sequence phase shift; that is, the high-voltage phasors lag the low-voltage phasors by 30°. (See Problem 4-41.)

In order to develop the per-unit positive sequence model of a three-phase, two-winding transformer, a common $|S|_{base}$ is selected for both the H and X terminals, and the ratio of the voltage bases $(V_{base\ H}/V_{base\ X})$ is selected as the ratio of the rated line-to-line voltages $(V_{rated\ HLL}/V_{rated\ XLL})$. The per-unit positive-phase sequence networks of Y–Y, Y–Δ, and Δ–Δ transformers are shown in Figure 4.5.9. Notice the phase-shifting transformer in the case of Y–Δ. Also, the per-unit impedances do not depend on the winding connections, but the base voltages do.

IIIIIIIIIII▐▬▬▬▬▬▬▬▬▬▬▬▬▬▬▬▬▬▬▬▬▬▬

EXAMPLE 4.5.3

A power system contains four 3-phase Δ–Y transformers T_1, T_2, T_3 and T_4, each rated 1,000 MVA, 500 kV Y/20 kV Δ with a per-unit series reactance of 0.1, and a 3-phase Y–Y transformer T_5, rated 1,500 MVA, 500 kV Y/20 kV Y with a per-unit series reactance of 0.1. The resistance, phase shift, and magnetizing reactance of the transformers are neglected. Using a base of 1,000 MVA and 500 kV on the high-voltage side, determine the per-unit reactances of each transformer in the system.

Solution

$$|S|_{base} = 1000 \text{ MVA}; \qquad V_{base\ H} = 500 \text{ kV}$$

$$X_{T1} = X_{T2} = X_{T3} = X_{T4} = 0.1$$

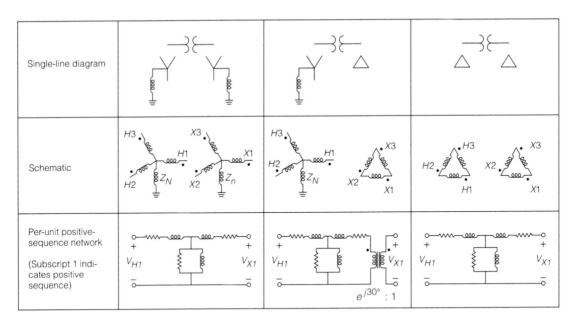

Figure 4.5.9 Per-unit positive-phase sequence networks of Y–Y, Y–Δ, and Δ–Δ transformers.

By using Equation 4.5.10, because of a change of $|S|_{base}$ and no change of V_{base}, one obtains

$$X_{T5} = (0.1)\left(\frac{1,000}{1,500}\right) = 0.0667$$

EXAMPLE 4.5.4

Three single-phase two-winding transformers, each rated 400 MVA, 13.8/199.2 kV, with leakage reactance $X_{eq} = 0.10$ per unit, are connected to form a three-phase bank. Winding resistances and exciting current are neglected. The high-voltage windings are connected in Y. A three-phase load operating under balanced positive-sequence conditions on the high-voltage side absorbs 1000 MVA at 0.90 pf lagging, with $\bar{V}_{AN} = 199.2\angle0°$ kV. Determine the voltage $\bar{V}_{an}$ at the low-voltage bus if the low-voltage windings are connected (a) in Y and (b) in Δ.

Solution

The positive-sequence network is shown in Figure 4.5.10. Using the transformer bank ratings as base quantities, $|S|_{base\ 3\phi} = 1200$ MVA, $V_{base\ HLL} = 345$ kV, and $I_{base\ H} = 1200/(345\sqrt{3}) = 2.008$ kA. The per-unit load voltage and load current are then

$$\bar{V}_{H1} = \bar{V}_{AN} = 1.0\angle0°\ \text{pu}$$

$$\bar{I}_{H1} = \bar{I}_A = \frac{1000/(345\sqrt{3})}{2.008}\angle-\cos^{-1}0.9 = 0.8333\angle-25.84°\ \text{pu}$$

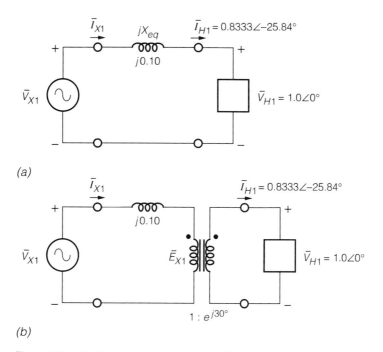

Figure 4.5.10 Positive-sequence network. (a) Y-Connected low-voltage windings. (b) Δ-Connected low-voltage windings.

a. For the Y–Y transformer,

$$\bar{I}_{X1} = \bar{I}_{H1} = 0.8333\angle-25.84° \text{ pu}$$
$$\bar{V}_{X1} = \bar{V}_{H1} + (jX_{eq})\bar{I}_{X1} = 1.0\angle0° + (j0.10)(0.8333\angle-25.84°)$$
$$= 1.0 + 0.08333\angle64.16° = 1.0363 + j0.0750 = 1.039\angle4.139° \text{ pu}$$
$$\bar{V}_{an} = \bar{V}_{X1} = 1.039\angle4.139°$$

Further, since $V_{\text{base } XLN} = 13.8$ kV for the low-voltage Y windings, $V_{an} = 1.039(13.8) = 14.34$ kV, and

$$\bar{V}_{an} = 14.34\angle4.139° \text{ kV}$$

b. For the Δ–Y transformer,

$$\bar{E}_{X1} = e^{-j30°}\bar{V}_{H1} = 1.0\angle-30° \text{ pu}$$
$$\bar{I}_{X1} = e^{-j30°}\bar{I}_{H1} = 0.8333\angle-25.84° - 30° = 0.8333\angle-55.84° \text{ pu}$$
$$\bar{V}_{X1} = \bar{E}_{X1} + (jX_{eq})\bar{I}_{X1} = 1.0\angle-30° + (j0.10)(0.8333\angle-55.84°)$$
$$= 1.039\angle-25.861° \text{ pu}$$
$$\bar{V}_{an} = \bar{V}_{X1} = 1.039\angle-25.861° \text{ pu}$$

Further, since $V_{\text{base } XLN} = 13.8/\sqrt{3} = 7.967$ kV for the low-voltage Δ windings, $V_{an} = (1.039)(7.967) = 8.278$ kV, and

$$\bar{V}_{an} = 8.278\angle{-25.861°} \text{ kV}$$

■■■■■■■■■ ||||||
4.6 Autotransformers

In a two-winding transformer—whether single-phase or three-phase—the primary and secondary windings, though magnetically linked, are isolated electrically. For the case of power-system transformers, in which large transformation ratios are involved and safety plays a dominant role, the electrical isolation of the primary and secondary windings is necessary. However, when the transformation ratio is close to unity and electrical isolation between the two windings is less critical, an *autotransformer*—consisting of a single tapped winding on a transformer core—can be used instead of a two-winding transformer with distinct advantages. The same fundamental considerations discussed for transformers having two separate windings hold true for the autotransformer, which is really just a normal transformer connected in a special way.

Let us consider the two-winding transformer shown in Figure 4.6.1a. We will convert it to an autotransformer arrangement (as shown in Figure 4.6.1b) by connecting the two windings electrically in series so that the polarities are additive. The N_1-turn winding is placed between points a and b, while the N_2-turn winding is placed between points b and c, as illustrated in the figure. Such an arrangement may be used for either stepping up or stepping down the voltage. Several other connections and selections of input and output terminals are possible. However, basic considerations will essentially remain the same for analysis.

Referring to the autotransformer arrangement of Figure 4.6.1b, these relations apply:

$$\bar{V}_H = \bar{V}_1 + \bar{V}_2 = \bar{V}_1 + \bar{V}_X \tag{4.6.1}$$
$$\bar{I}_X = \bar{I}_1 + \bar{I}_2 \tag{4.6.2}$$

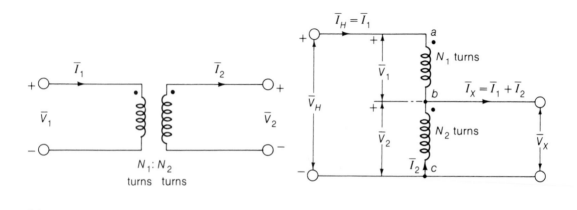

(a)   (b)

Figure 4.6.1 (a) Two-winding transformer, (b) An autotransformer arrangement.

The phase angle between the primary and secondary voltages being very small (usually less than 10°), the phasor sum of Equation 4.6.1 can be regarded as practically equal to the arithmetic sum in magnitude.

$$V_H \simeq V_1 + V_2 \simeq V_1 + V_X \qquad (4.6.3)$$

If the exciting current is neglected, the currents I_H and I_X are in phase, and the magnetomotive forces of the two windings are equal, so that

$$N_1 \bar{I}_1 = N_1 \bar{I}_H = N_2 \bar{I}_2 \qquad (4.6.4)$$

or

$$\bar{I}_H = \bar{I}_1 = \frac{N_2}{N_1} \bar{I}_2 \qquad (4.6.5)$$

The ratio of input and output currents is given by

$$\frac{\bar{I}_H}{\bar{I}_X} = \frac{\bar{I}_1}{\bar{I}_1 + \bar{I}_2} = \frac{N_2}{N_1 + N_2} \qquad (4.6.6)$$

or

$$\frac{\bar{I}_X}{\bar{I}_H} = \frac{\bar{I}_1 + \bar{I}_2}{\bar{I}_1} = \frac{N_1 + N_2}{N_2} = 1 + \frac{N_1}{N_2} = 1 + a \qquad (4.6.7)$$

where a is the turns ratio (N_1/N_2) of the two-winding transformer of Figure 4.6.1a. The ratio of the terminal voltages is given by

$$\frac{\bar{V}_H}{\bar{V}_X} = \frac{\bar{V}_1 + \bar{V}_2}{\bar{V}_2} = \frac{N_1 + N_2}{N_2} = 1 + \frac{N_1}{N_2} = 1 + a \qquad (4.6.8)$$

Thus, the autotransformer with a transformation ratio of $(1 + a)$ can be treated as an equivalent two-winding transformer with a turns ratio of a. The rating of the autotransformer configuration is much higher than the two-winding transformer rating as shown below:

$$\frac{\text{Autotransformer rating}}{\text{Two-winding transformer rating}} = \frac{V_H I_H}{V_1 I_1} = \frac{V_H}{V_1} = 1 + \frac{V_X}{V_1} = 1 + \frac{N_2}{N_1} = 1 + \frac{1}{a} \qquad (4.6.9)$$

This ratio becomes much larger than unity, since for many autotransformer applications a is rather small and the ratio of the terminal voltages $(1 + a)$ is very nearly unity. This is because most of the power in autotransformers is delivered to the secondary through direct *conduction,* and only a small portion is delivered through transformation or induction as it is in a two-winding transformer.

When the ratio of the terminal voltages does not differ too greatly from unity, autotransformers, in comparison with two-winding transformers of the same output, offer lower leakage reactances, lower losses, as well as smaller exciting current at lower cost. Since the autotransformer's primary and secondary share one winding and the current in the common winding is always the difference between the input and output currents, the cross-sectional area of the common winding is much smaller than a two-winding transformer's primary or secondary winding of the same current density. Hence, significant savings in copper result.

For the same output, the autotransformer is smaller in size than a two-winding transformer and has higher efficiency as well as superior voltage regulation due to the reduced resistance drop and lower

Photo 4.6.1 A 367-MVA, 525-kV extra-high-voltage autotransformer. It is 40 feet high and 28 feet wide and weighs 270 tons. (Photo courtesy of Westinghouse Electric Corporation.)

leakage reactance drop. An important disadvantage of the autotransformer is the direct copper connection (i.e., no electrical isolation) between the high- and low-voltage sides. A type of autotransformer commonly found in laboratories is the variable-ratio autotransformer in which the tapped point *b* (shown in Figure 4.6.1b), is movable. It is known as the *variac* ("variable ac").

Although here, for the sake of simplicity, we have considered only the single-phase autotransformer, three-phase autotransformers can be modeled on a per-phase basis and also analyzed just as the single-phase case. In Photo 4.6.1 an extra-high-voltage autotransformer is shown on test at the plant.

▮▮▮▮▮▮▮▮▮▮▮▮▮▬▬▬▬▬▬▬▬▬▬▬▬▬▬▬

EXAMPLE 4.6.1

The single-phase 50-kVA, 2,400:240-volt, 60-Hz, two-winding distribution transformer of Example 4.3.1 is connected as a step-up autotransformer as shown in Figure 4.6.2. Assume that the 240-volt winding is provided with sufficient insulation to withstand a voltage of 2,640 volts to ground.

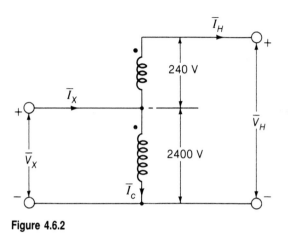

Figure 4.6.2

a. Find V_H, V_X, I_H, I_C, and I_X corresponding to rated (full-load) conditons.

b. Determine the kVA rating as an autotransformer, and find how much of that is the conducted kVA.

c. Based on the data given for the two-winding transformer in Example 4.3.1, compute the efficiency of the autotransformer corresponding to full load, 0.8 power factor lagging, and compare it with the efficiency calculated for the two-winding transformer in part a of Example 4.4.1.

Solution

a. The two windings are connected in series so that the polarities are additive. Neglecting the leakage-impedance voltage drops,

$$V_H = 2,400 + 240 = 2,640 \text{ V}$$

$$V_X = 2,400 \text{ V}$$

The full-load rated current of the 240-V winding, based on the rating of 50 kVA as a two-winding transformer, is

$$\frac{50,000}{240} = 208.83 \text{ A}$$

Since the 240-V winding is in series with the high-voltage circuit, the full-load current of this winding is the rated current on the high-voltage side of the autotransformer.

$$I_H = 208.33 \text{ A}$$

Neglecting the exciting current, the mmf produced by the 2,400-V winding must be equal and opposite to that of the 240-V winding.

$$I_C = 208.33\left(\frac{240}{2,400}\right) = 20.83 \text{ A}$$

in the direction shown in the figure. Then the current on the low-voltage side of the autotransformer is given by

$$I_X = I_H + I_C = 208.33 + 20.33 = 229.16 \text{ A}$$

b. The kVA rating as an autotransformer is

$$\frac{V_H I_H}{1,000} = \frac{(2,640)(208.33)}{1,000} = 550 \text{ kVA}$$

or

$$\frac{V_X I_X}{1,000} = \frac{(2,400)(229.16)}{1,000} = 550 \text{ kVA}$$

which is 11 times that of the two-winding transformer. The transformer must boost the current I_H of 208.33 A through a potential rise of 240 volts; thus the kVA transformed by electromagnetic induction is given by

$$\frac{240 \times 208.33}{1,000} = 50 \text{ kVA}$$

The remaining 500 kVA is the conducted kVA.

c. With the currents and voltages shown for the autotransformer connections, the losses at full load will be 823 W, the same as in Example 4.4.1a. However, the output as an autotransformer at 0.8 power factor is given by

$$550 \times 1,000 \times 0.8 = 440,000 \text{ W}$$

The efficiency of the autotransformer is then calculated as

$$\eta = 1 - \frac{\text{losses}}{\text{input}} = 1 - \frac{823}{440,823} = 0.9981$$

whereas that of the two-winding transformer was calculated as 0.98 in Example 4.4.1a. Because the only losses are those due to transforming 50 kVA, higher efficiency results for the autotransformer configuration complared to that of the two-winding transformer.

4.7 Multiwinding Transformers

For interconnecting three or more circuits that may have different voltages, *multicircuit* or *multiwinding* transformers consisting of three or more windings—rather than of two or more two-winding transformers—are often used for the sake of efficiency and economy. In supplying power to electronic circuits, a transformer with a primary and two secondaries is commonly employed. The distribution transformer with a center tap on the secondary side illustrated in Figure 4.7.1 is also a three-winding

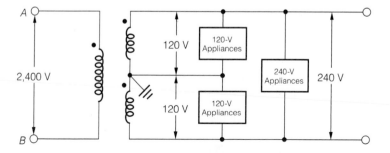

Figure 4.7.1 A three-winding distribution transformer.

transformer. Domestic loads consisting of 120-volt appliances and lighting, divided roughly equally between the two 120-volt secondaries, are connected from line to neutral. Appliances such as electric ranges and water heaters are supplied with 240-volt power from the series-connected secondaries, as shown in Figure 4.7.1.

Large power-system transformers in three-phase systems used for transmission and distribution of power are usually three-winding transformers. While the primary and secondary windings of the three-phase transformer bank are used to interconnect two transmission systems of different voltages, a suitable voltage for auxillary power purposes in the substation or for supplying a local distribution system is provided by the third winding, known as the *tertiary* winding. Whereas the primaries and secondaries are connected in Y–Y, the tertiary windings are connected in Δ to provide a path for the third harmonics and their multiples in the exciting current. The problems associated with multiwinding transformers—such as the effects of leakage impedances on voltage regulation, short-circuit currents, and division of load among different circuits—can be solved by an equivalent-circuit approach similar to the one used in solving the two-winding transformer problems. Since the leakage impedances associated with each pair of windings must be considered, the multiwinding-transformer equivalent circuits tend to be complicated. It is most convenient to express all quantities in per unit, referred to a common base, in these equivalent circuits.

Let us consider the three-winding transformer whose schematic diagram is shown in Figure 4.7.2. For the case of an ideal transformer,

$$\frac{\bar{V}_2}{\bar{V}_1} = \frac{N_2}{N_1} \tag{4.7.1}$$

$$\frac{\bar{V}_3}{\bar{V}_1} = \frac{N_3}{N_1} \tag{4.7.2}$$

$$N_1 \bar{I}_1 = N_2 \bar{I}_2 + N_3 \bar{I}_3 \tag{4.7.3}$$

The equivalent circuit that takes into account the leakage impedance associated with three windings on a common magnetic core and exciting admittance is part b of the figure. All quantities are expressed in per unit and referred to a common base.

The parameters of the equivalent circuit can be determined from open-circuit and short-circuit tests. The data for calculating the exciting admittance are obtained by the open-circuit test, as with a two-winding transformer. Since there are three windings, three short-circuit tests are needed for obtaining the data and determining Z_1, Z_2, and Z_3 of the equivalent circuit. In analyzing the short-circuit test data, the exciting current is neglected. Thus, if Z_{12} is the short-circuit impedance of circuits 1 and 2 with Circuit 3 open (i.e., found by applying voltage to the primary circuit 1 with the secondary circuit 2 short-circuited and the tertiary circuit 3 open), the measured impedance is Z_{12}.

$$Z_{12} = Z_1 + Z_2 \tag{4.7.4}$$

Similarly,

$$Z_{13} = Z_1 + Z_3 \tag{4.7.5}$$

$$Z_{23} = Z_2 + Z_3 \tag{4.7.6}$$

where Z_{13} is the short-circuit impedance of circuits 1 and 3 with secondary circuit 2 open, and Z_{23} is the short-circuit impedance of circuits 2 and 3 with primary circuit 1 open. These impedances should be referred to a common base. From Equations 4.7.4, 4.7.5, and 4.7.6, the leakage impedances Z_1, Z_2, and Z_3 are then given by

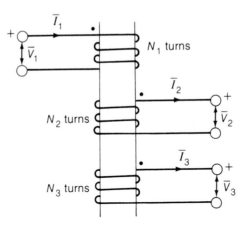

(a)

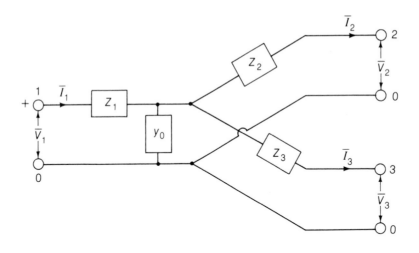

(b)

Figure 4.7.2 Three-winding transformer. (a) Schematic diagram. (b) Equivalent circuit with all quantities expressed in per unit, referred to a common base.

$$Z_1 = \frac{Z_{12} + Z_{13} - Z_{23}}{2} \tag{4.7.7}$$

$$Z_2 = \frac{Z_{23} + Z_{12} - Z_{13}}{2} \tag{4.7.8}$$

$$Z_3 = \frac{Z_{13} + Z_{23} - Z_{12}}{2} \tag{4.7.9}$$

The following example illustrates the computations of the equivalent circuit parameters in per unit referred to a common base.

||||||||||||

EXAMPLE 4.7.1

A three-phase bank consisting of three single-phase, three-winding transformers is used to step down the voltage of a three-phase 66-kV (line-to-line) transmission line. The transformers are connected Y–Y–Δ; the wye-connected secondary and the delta-connected tertiary have line-to-line voltages of 11 kV and 2.2 kV, respectively. Data obtained from short-circuit and open-circuit tests conducted on one of the transformers appear in Table 4.7.1.

Neglecting the resistance of the windings and the core losses, determine the values of reactance and admittance to be used in the equivalent circuit of Figure 4.7.2b, while using a three-phase base of 90,000 kVA (or base of 30,000 kVA per single-phase transformer) for determining the per-unit values.

Table 4.7.1 Data for Example 4.7.1

Short-Circuit Test Data			
Winding		**Measured on the Excited-Winding Side**	
Excited	**Short-Circuited**	**Volts**	**Amperes**
primary(1)	secondary(2)	3,000	800
primary(1)	tertiary(3)	5,800	650
secondary(2)	tertiary(3)	550	1,600

Open-Circuit Test Data		
Winding	**Volts**	**Amperes**
tertiary(3)	2,200	150

Solution

The short-circuit test data yield the following, referred to the primary:

$$Z_{12} = j3,000/800 = j3.75 \ \Omega$$

$$Z_{13} = j5,800/660 = j8.79 \ \Omega$$

$$Z_{23} = (j550/1,600)(N_1/N_2)^2 = (j550/1,600)(66/11)^2 = j12.38 \ \Omega$$

From Equations 4.7.7, 4.7.8, and 4.7.9 we obtain

$$Z_1 = \frac{j3.75 + j8.79 - j12.38}{2} = j0.08 \ \Omega$$

$$Z_2 = \frac{j12.38 + j3.75 - j8.79}{2} = j3.67 \ \Omega$$

$$Z_3 = \frac{j8.79 + j12.38 - j3.75}{2} = j8.71 \ \Omega$$

Referred to the tertiary, the exciting admittance is given by

$$y_0 = -j\frac{150}{2,200} = -j0.068 \text{ S}$$

or, referred to the primary,

$$(-j0.068)(N_3/N_1)^2 = -j0.068 \left(\frac{2.2}{66/\sqrt{3}}\right)^2 = -j0.00023 \text{ S}$$

Base impedance referred to the primary is given by

$$Z_{\text{base pri}} = \frac{(\text{volts})^2}{\text{volt-amperes}} = \frac{(66,000/\sqrt{3})^2}{30,000 \times 1,000} = 48.4 \ \Omega$$

Hence, the per unit values of the equivalent-circuit parameters of Figure 4.7.2b are

$$Z_1 = j0.08/48.4 = j0.0017 \text{ pu}$$
$$Z_2 = j3.67/48.4 = j0.076 \text{ pu}$$
$$Z_3 = j8.71/48.4 = j0.18 \text{ pu}$$
$$y_0 = (-j0.00023)(48.4) = j0.011 \text{ pu}$$

4.8 Some Other Transformer Topics

Several other important transformer topics are discussed in this section.

Instrument Transformers

Instrument transformers are generally of two types, potential transformers and current transformers. These are designed in such a way that the former may be regarded as having an ideal *potential* ratio, and the latter as having an ideal *current* ratio. For instrument transformers it is the accuracy of measurement that is important, rather than the load-carrying capability. Instrument transformers are commonly used in ac power circuits to supply instruments, protective relays, and control circuits.

Potential transformers are employed to step down the voltage to about 110 volts for connecting to voltmeters, wattmeters, relays, and control devices. In three-phase installations, these may be connected line-to-neutral or line-to-line, depending upon the voltage.

Current transformers (connected in series with the line) are used to step down the current to about 5 amperes for metering purposes. Often the primary is not an integral part of the transformer but, rather, is part of the line whose current is being measured. In addition to providing a desirable low current in the metering circuit, the current transformer isolates the meter from the line, which may be at a high potential. The secondary terminals of a current transformer should never be open-circuited under load, because (1) there would then be no secondary mmf to oppose the primary mmf, (2) the net mmf acting on the core would rise to a value many times greater than that for which the core is designed, and (3) as a consequence, very high voltage might be induced in the secondary. Besides endangering the user, this could damage the transformer insulation and cause transformer overheating due to excessive core losses.

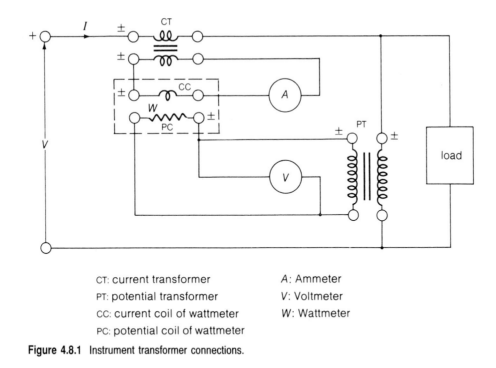

CT: current transformer   A: Ammeter
PT: potential transformer   V: Voltmeter
CC: current coil of wattmeter   W: Wattmeter
PC: potential coil of wattmeter

Figure 4.8.1 Instrument transformer connections.

The connections of instrument transformers that measure current, voltage and power in a single-phase power circuit are shown in Figure 4.8.1. Note the polarity markings indicated by ±, located on primary and secondary terminals of like polarity, as in laboratory transformers. Before connecting to a wattmeter, an instrument transformer's terminal polarity must be known.

Variable-Frequency Transformers

In electronic and communication circuits, audio-frequency transformers are required to operate over the audible frequency range of about 20 to 20,000 Hz. Other examples of variable-frequency operations include variable-frequency sources that contain transformers and supply power to variable-speed motors. The *frequency characteristic* and the *phase characteristic* are the curves usually referred to the variable-frequency operation. The frequency characteristic (also known as the *amplitude-frequency characteristic*) is a curve of the phase angle of the load voltage with respect to the source voltage, plotted as a function of the frequency. A small phase angle is desirable.

Modeling of a circuit or system consisting a transformer depends on the frequency range of operation:

1. In the mid-frequency range, the equivalent circuit reduces to a network of resistances, without any inductance, as shown in Figure 4.8.2a. The variable frequency source (such as an electronic amplifier) is represented by E_S in series with an internal resistance R_S. The load, such as a speaker, is represented by the load resistance R_L. The leakage reactances and the magnetizing current are generally neglected in this range.

2. At low frequencies, leakage reactances are negligible, but the magnetizing current of the transformer becomes significant. The circuit of Figure 4.8.2b applies.

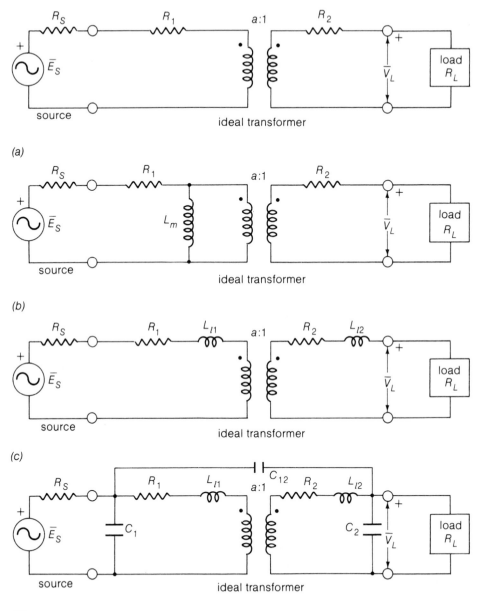

(a)

(b)

(c)

(d)

Figure 4.8.2 Equivalent circuits of a variable-frequency transformer. (a) Mid-frequency-range equivalent circuit. (b) Low-frequency-range equivalent circuit. (c) High-frequency-range equivalent circuit without capacitance effects. (d) High-frequency-range equivalent circuit with capacitances.

3. In the high-frequency range, the transformer's leakage reactances become significant, and the equivalent circuit of Figure 4.8.2c applies. If the effects of distributed capacitances across and between the windings are significant, however, the equivalent circuit of Figure 4.8.2d applies. A lumped circuit containing three capacitances is commonly used to describe the effect of the distributed capacitances. C_1 and C_2 are the primary (winding 1) and secondary (winding 2) capacitances connected across the transformer terminals, and C_{12} is the capacitance between the primary and secondary terminals. The values of these capacitances depend on the winding arrangements. It is preferable to refer all the parameters to one side of the ideal transformer for facilitating analysis of the equivalent circuit.

Accurate analysis of the transient surge potential distribution in high-voltage power-system transformers requires a more elaborate circuit model, one that accounts for the distributed leakage inductance and interwinding capacitances. Analysis of the effects of lightning and switching surges on the power-system transformers is too complex to be discussed here. Electrostatic shielding is usually provided in order to increase the series capacitance of the winding and thereby prevent excessive concentration of transient voltages on the line-end turns.

Pulse Transformers

Pulse transformers are used in radar, television, digital-computer circuits, and linear accelerators. Recently, they have also been used in the gate-firing circuits of power thyristors and several other electronic and control applications. Most pulse transformers are physically small and have relatively few turns so as to minimize the leakage inductance. The cores are made out of ferrites or wound strips of high-permeability alloys such as permalloy. Because the time interval between pulses is longer than the pulse duration, the load-carrying duty of the pulse transformer is rather light; and hence, relatively high pulse-power levels can be handled by a small-size transformer.

Pulse excitation typically consists of a rectangular voltage pulse, though many other pulse shapes also are used. The average transformer power is determined by the pulse magnitude and repetition rate. It is important that the transformer be able to reproduce the input pulse as truly as possible at its secondary terminals. Figure 4.8.3a shows a rectangular pulse exitation, whose pulse width may be a fraction of a microsecond to about 20 microseconds with a relatively long time elapsing between pulses. Part b of the figure is the waveform of the output (either a voltage or current) resulting from a rectangular pulse excitation. Transient, rather than steady-state, analysis is required for determining the output waveform.

Equivalent circuits such as those in Figure 4.8.2 are used for analyzing pulse transformers. Detailed analysis including the transformer nonlinearities requires computer simulation techniques. However, by employing simplified equivalent circuits during different portions of the pulse, it is possible to observe the response. The rise time associated with the leading edge of the output waveform can be found by the capacitive high-frequency equivalent circuit of Figure 4.8.2d. The response to the flat-top portion of the input pulse can be determined by the low-frequency equivalent circuit of Figure 4.8.2b. The fall time and the backswing associated with the trailing edge of the output response can be analyzed by a parallel RLC circuit in which the magnetizing inductance is discharging the energy of the decaying magnetic field through the capacitance and circuit resistance.

Saturable Reactors

Saturable reactors are a means of controlling the power delivered to a load from a constant-potential ac source. They consist of a group of saturable magnetic elements whose core materials may be assumed to have an idealized *B-H* characterisitic with infinite unsaturated permeability and zero saturated

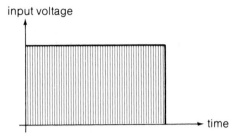

(a)

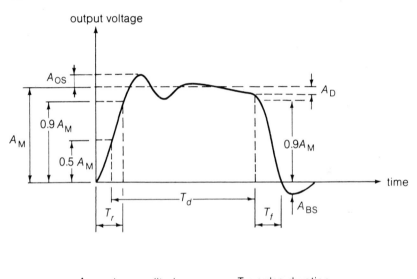

A_M: pulse amplitude

A_{OS}: overshoot

A_D: droop or tilt

A_{BS}: back swing

T_d: pulse duration

T_r: pulse rise time

T_f: pulse fall time

(b)

Figure 4.8.3 Input and output voltage waveforms of a pulse transformer. (a) Rectangular input-voltage pulse excitation. (b) Output-voltage waveform of a pulse transformer.

permeability, as shown in Figure 4.8.4a. Consider an arrangement, shown in Figure 4.8.4(b), of a series saturable reactor, in which the primary windings are connected in series with each other and with the load. The control windings are connected in series opposition and supplied by a direct current I. To ensure that the current is essentially constant and free of ripples, a large inductance is introduced in the control winding, as shown in the figure.

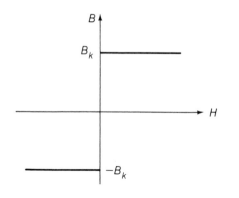

(a)

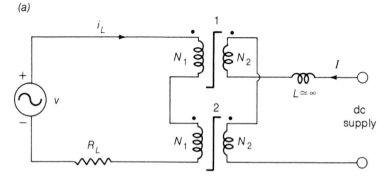

(b)

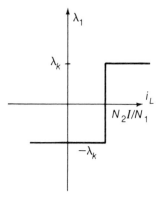

(c)

Continued

Figure 4.8.4 A saturable reactor. (a) Idealized *B-H* characteristic. (b) Series saturable reactor with high-impedance control. (c) Flux linkages λ_1 of core 1 as a function of the current i_L. (d) Flux linkages λ_2 of core 2 as a function of the current i_L. (e) Combined flux linkages $(\lambda_1 + \lambda_2)$ as a function of the current i_L. (f) Waveforms of supply voltage v and the load voltage $(R_L i_L)$ as a function of ωt.

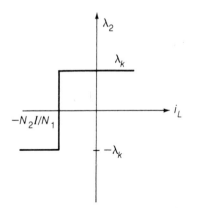

(d)

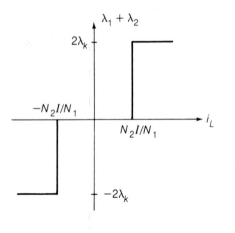

(e)

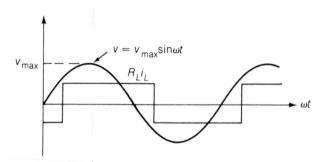

(f)

Figure 4.8.4 *Continued*

With the control current I, core 1 is unsaturated only if

$$i_L = \frac{N_2}{N_1} I \qquad (4.8.1)$$

and Core 2 is unsaturated only if

$$i_L = -\frac{N_2}{N_1} I \qquad (4.8.2)$$

The idealized relations of flux linkages λ_1 of core 1, flux linkages λ_2 of core 2, and combined flux linkages $(\lambda_1 + \lambda_2)$ as a function of the current i_L are shown in parts c, d, and e of Figure 4.8.4, respectively. Since the combined flux linkages $(\lambda_1 + \lambda_2)$ can change only if the current i_L has one of the values given by Equations 4.8.1 and 4.8.2, it follows that the current i_L in the load R_L will be a square wave having the amplitude given by Equations 4.8.1 and 4.8.2 when an ac voltage of $v = v_{max}$ sin ωt is impressed. Thus the saturable reactor acts as a source of constant rectangular-wave current, the magnitude of which is directly proportional to the control current. In practical situations the response is somewhat different from the idealized conditions assumed in this analysis.

A series saturable reactor arranged as in Figure 4.8.4b but with low-impedance control can also be considered. With a sinusoidal applied voltage, the waveforms of λ_1, λ_2, control current i, and load current i_L can be made to assume desired shapes, and the power delivered to the load R_L can be controlled. The control current as well as the load current can be made to consist of only sections of waves. Saturable reactors can also be connected in an alternative arrangement in which the primary windings are in parallel instead of in series.

Inrush Current

When the primary winding of an unloaded transformer is switched on to the normal voltage supply, it acts as a nonlinear inductor. Neglecting the core loss and the primary resistance, the applied voltage must be balanced at every instant by the emf induced by the magnetizing current. The response depends on the magnitude, the polarity, and the rate of change of applied voltage at the instant of switching. It may result in a current peak many times (about eight to ten times) the rated transformer current, or it may be practically unobservable. Such a transient effect is known as the *inrush current* phenomenon. Even though the exciting current eventually returns to its normal value (about 5% or less of the rated transformer current) because of the winding resistance and core losses, the inrush current is a matter of great concern while energizing large power transformers, since it can stress the transformer windings unduly and cause improper operation of protective devices. Thus, it should be looked into.

Consider the transformer core to be initially unmagnetized and the primary winding of the unloaded transformer to be switched on at a voltage peak, as shown in Figure 4.8.5a. Neglecting the core loss and the primary-winding resistance, the voltage and flux need to satisfy

$$v_1 = N_1 \frac{d\phi}{dt} \qquad (4.8.3)$$

Hence, the flux is as shown in the figure and the corresponding magnetizing current can be found from the *B-H* characteristic of the transformer core. Such conditions correspond to those of normal no-load operation, when there is no transient.

Next, consider the switch to be closed at $t = 0$ when the voltage is zero as shown in Figure 4.8.5b, corresponding to which

$$v_1 = \sqrt{2}V \sin \omega t = N\frac{d\phi}{dt} \qquad (4.8.4)$$

The flux is then given by $\qquad \phi = \frac{\sqrt{2}V}{N_1} \int_0^t \sin \omega t \, dt = \frac{\sqrt{2}V}{\omega N_1}(1 - \cos \omega t)$

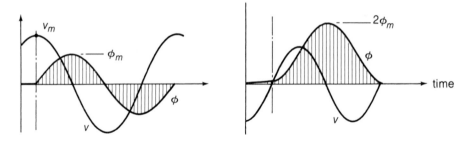

(a)   (b)

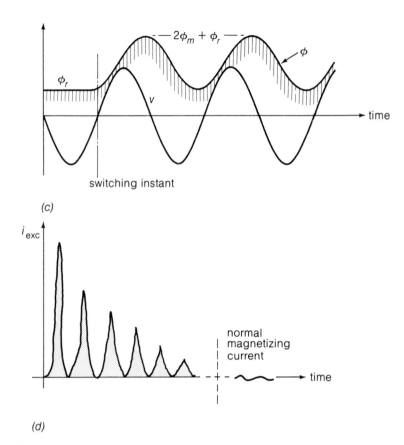

(c)

(d)

Figure 4.8.5 Inrush current phenomenon.

or

$$\phi = -\phi_m \cos \omega t + \phi_m \qquad (4.8.5)$$

The instantaneous flux at $\omega t = \phi$ is then $2\phi_m$, as shown in the figure. The residual flux ϕ_r in the core can aggravate the condition, as shown in Figure 4.8.5c, if the flux increases in the direction of the residual flux. The peak flux in the worst case could then be $(2\phi_m + \phi_r)$. As time progresses, the effect of losses will rapidly reduce the flux waveform to a symmetrical one about the time axis.

Since the maximum steady-state flux occurs around the knee of the saturation curve for most transformers, a flux of two or more times the maximum steady-state flux may demand an exciting current many hundreds of times the normal excitation current because of supersaturation of the core. Depending upon the change in flux path and the tranformer time constant (which may be a few milliseconds for small transformers and a few seconds for large transformers), the inrush current phenomenon may persist for a few seconds, as shown in Figure 4.8.5d. Analytical methods for predicting the inrush current phenomenon are too involved to be addressed here.

Parallel Operation of Transformers

In paralleling single-phase transformers, besides observing the polarities in making physical connections, four conditions are necessary. They are discussed here in the order of their importance:

1. *Equal primary and secondary voltage ratings.* If the voltage ratings are not the same, some windings will be operating at higher and others at lower voltage than that for which they are designed.
2. *Equal ratios of transformation.* If the ratios of transformation are not the same, there will be circulating currents in the transformer windings in addition to the exciting currents, when there is no load on the system. They increase the copper loss, overload the transformer, and reduce the overall permissible load kVA.
3. *Equivalent impedances (in ohms) inversely proportional to their current or kVA ratings.* If the impedances (in ohms) are not inversely proportional to the current outputs, the load does not get divided evenly between the transformers in proportion to their ratings. Such uneven load sharing may lead to overheating or to reduced total rated capacity.
4. *Equal ratios of equivalent resistance to equivalent reactance.* If the ratios of equivalent resistance to equivalent reactance are not equal, the currents delivered by the transformers will not be in phase with each other or with the load current; that is, the transformers will be carrying kilowatt loads that are not proportional to their current load. This will lead to increased copper loss and consequent overheating or to reduced overall kilowatt output.

These conditions apply equally in paralleling three-phase transformers, but with the added condition that the secondary voltages must be at compatible phase angles. Not only must the transformer polarities be observed, but also, the phase sequence must be identical for parallel transformers. Transformers with similar connections on either side can be connected in parallel. Also, three-phase transformer banks connected in Y–Y and Δ–Δ can be paralleled if the line-voltage ratios are identical. The phase shifts resulting from Δ–Y or Y–Δ connections generally preclude the use of either in parallel with a configuration that is different. However, the secondary voltages of a Y–Δ or Δ–Y system are either in phase or are $60°$ out of phase, depending upon the way the connections are made; two such systems can therefore be paralleled with proper connections.

Superconducting Transformer Windings

Superconducting metal films capable of carrying very high current densities when immersed in a magnetic field are being developed. The possibility of operating large power-system transformers under cryogenic conditions has been researched for more than two decades.[4] The superconducting windings would have to conduct in the presence of high leakage fields and carry very high current densities. With these windings, the conventional I^2R loss could be entirely eliminated. In addition, the core loss would be reduced because of the reduction in the core mass. However, refrigeration power would have to be supplied, and capital costs would have to be taken into account. Although small experimental superconducting transformers are being built, the research has not yet led to a commercially feasible, economical proposition.

■■■■■■■■■||||||
4.9 Transformer Design

Let us continue with a transformer-design project in order to show the basic elements and considerations of the design aspect. The specifications for our project are given as:

> 400-kVA, single-phase, 60-Hz, 42,000:2,400-V, self-oil-cooled, core-type transformer that will carry its rated load continuously with a temperature rise that does not exceed 55°C. The full-load efficiency at unity power factor must not be less than 98%, and the ratio of losses (copper to iron) should be about 1.0.

The flux density in the core is chosen as 1.5 T. Using high-resistance 4-percent silicon steel of nominal thickness 0.35 mm, the loss per kg for this flux density can be calculated from the information supplied by the transformer steel manufacturer in the form of iron-loss curves.

$$\text{Iron loss per kg} = 3.1 \times 1.12 = 3.47 \text{ W}$$

where 3.1 is taken from the iron-loss curve corresponding to 1.5 T, and the factor 1.12 is used because the additional loss (due to bending and shearing strains, etc.) in the finished transformer core are taken as 12% of the fundamental-frequency loss.

For an average current density of 2.3 A/mm^2, the copper loss per kg is calculated empirically as

$$\text{Copper loss per kg} = 5.53(2.3 \times 10^6)1.1 \times 10^{-6} = 14 \text{ W}$$

where a stray-loss factor of 1.1 is used. The ratio of core weight to copper weight is then

$$\frac{\text{copper loss per kg}}{\text{iron loss per kg}} \times \frac{\text{iron-core loss}}{\text{total copper loss}} = \frac{14}{3.47} \times 1 = 4.03$$

The net section area of the core can be shown to be related to the kVA, flux density, and current density as follows:

$$\text{Net sectional area of core} = C\sqrt{\frac{\text{kVA} \cdot \frac{\text{core weight}}{\text{copper weight}} \cdot 10^{11}}{\text{flux density} \cdot \text{current density} \cdot \text{frequency}}} \text{ m}^2$$

---

[4] J. M. Feldman, B. Cogbill, and M. S. Sarma, "Superconducting Windings in a Power Transformer (An Old Question with a New Answer) Part II: Some of the Practical Problems," *IEEE Paper* no. A77-020-1, February 1977; and K. J. R. Wilkinson, "Superconductive Windings in Power Transformers," *Proceedings of the IEEE*, vol. 110 (1963), p. 2271.

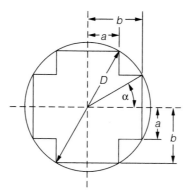

Figure 4.9.1 Cruciform-shaped core section of transformer.

where C is called the *output constant,* which varies over a range depending on the type of transformer—core or shell—and single-or three-phase. Choosing the output constant as (0.46×10^{-4}),

$$\text{Net sectional area of core} = 0.46 \times 10^{-4} \sqrt{\frac{400 \times 4.03 \times 10^{11}}{(1.5)(2.3 \times 10^{6})60}} \ \text{m}^2 = 0.0406 \ \text{m}^2$$

The cruciform-shaped core section as shown in Figure 4.9.1 will be used for the transformer. The diameter of the core, D, is then given by

$$D = \sqrt{\frac{(\text{net area})4}{\pi \ (\text{core space factor})}}$$

where the core space factor is defined as the ratio of the net area of the core to the area of the circumscribed circle.

$$D = \sqrt{\frac{(0.0406)4}{\pi \times 0.7}} = 0.272 \ \text{m}; \qquad \text{choose } 0.275 \ \text{m}$$

The dimensions of the core section are

$$2a = 0.526D = 0.526 \times 0.275 = 0.145 \ \text{m}$$

$$2b = 0.85D = 0.85 \times 0.275 = 0.234 \ \text{m}; \qquad \text{choose } 0.235 \ \text{m}$$

Recalculating D by using $\sqrt{(2a)^2 + (2b)^2}$, one gets $D = \sqrt{(0.145)^2 + (0.235)^2} = 0.276$ m and the net area as

$$\frac{(0.276)^2 \times \pi \times 0.7}{4} = 0.042 \ \text{m}^2$$

Then the total flux for a density of 1.5 T is given by

$$\phi = (0.042)1.5 \simeq 0.06 \text{ Wb}$$

The number of turns for high-voltage and low-voltage windings are then computed as follows by the use of Equation 4.2.10:

$$N_1 = \frac{42,000}{4.44 \times 60 \times 0.06} \simeq 2625 \qquad \text{on the high-voltage side}$$

$$N_2 = \frac{2,400}{42,000} \times 2625 = 150 \qquad \text{on the low-voltage side}$$

The full-load current in the two windings are calculated as

$$I_{HV} = \frac{400 \times 10^3}{42,000} = 9.52 \text{ A}; \qquad I_{LV} = \frac{400 \times 10^3}{2,400} = 166.67 \text{ A}$$

With the assumed average current density of 2.3 A/mm^2, the sectional area of the conductors for the HV and LV windings are then

$$\frac{9.52}{2.3} = 4.14 \text{ mm}^2 \qquad \text{on the HV side}$$

$$\frac{166.67}{2.3} = 72.47 \text{ mm}^2 \qquad \text{on the LV side}$$

The area of the window opening is

$$H_w W_w = \frac{2 \times 4.14 \times 2625}{0.105} \times 10^{-6} = 0.207 \text{ m}^2$$

where 0.105 represents the copper space factor, H_w is the window height, and W_w is the window width. For a ratio of window height to window width equal to 3, the dimensions of the window come out as

$$H_w = \sqrt{3 \times 0.207} = 0.788 \text{ m} \quad \text{or} \quad 78.8 \text{ cm}; \qquad \text{choose 79 cm}$$

$$W_w = \frac{79}{3} = 26.3 \text{ cm}; \qquad \text{choose 27 cm.}$$

Now we are in a position to design low-voltage and high-voltage windings and then to calculate the length of the mean turn for each. The copper and core weight can then be determined. Next the core and copper losses can be calculated. By estimating the resistance and reactance drops with the use of empirical formulae, one can compute the regulation for various load conditions.

The mean length of the flux path is shown by the dotted line in Figure 4.9.2. The ampere-turns necessary to maintain the flux in the iron part of the magnetic circuit can be determined as in Section 3.3 from the standard magnetization curve for 4% silicon steel and the mean length. An approximate adjustment can be made by adding some ampere-turns for the joints in the magnetic circuit. The magnetizing current and the exciting current are then computed.

By estimating the stray-load losses as 10% of the total I^2R losses, the efficiencies for various operating conditions can be calculated as in Section 4.4.

A corrugated sheet-steel tank can then be designed, and the transformer positioned in the tank as shown in Figure 4.9.2. The volume of oil required and the oil weight can be specified. A typical transformer layout is shown in Figure 4.9.2, with the core clamps, coil supports, and so on, omitted.

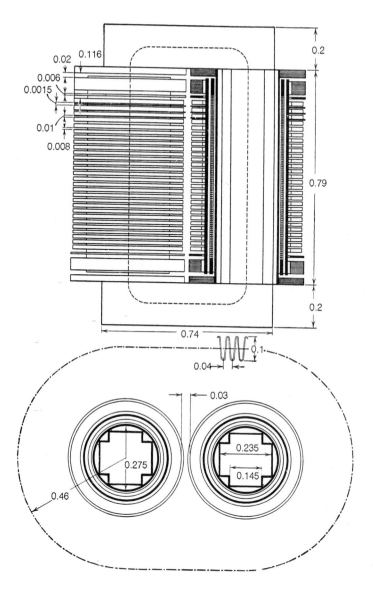

Note: Dimensions in SI System of units

Figure 4.9.2 Typical layout of a single-phase, core-type transformer. (Adapted from John H. Kuhlmann, *Design of Electrical Apparatus,* 3rd ed. New York: John Wiley & Sons, 1950.)

Figure 4.9.3 gives the design details in the form of a typical transformer design sheet. Not all the design details shown on the design sheet have been calculated in this section. However, the student is encouraged to pursue the design and justify the details furnished here.

TRANSFORMER DESIGN SHEET

kVA, 400   Phase, Single   Hertz, 60   Volts $\begin{cases} \text{HV., 42,000} \\ \text{LV., 2,400} \end{cases}$ Line Amperes $\begin{cases} \text{HV., 9.52} \\ \text{LV., 166.7} \end{cases}$

Type, Circular Core   Type of Cooling, Self Oil

CORE

Sheet steel .... 0.35×10^{-3}, 4% Si.
Output constant .... 0.46×10^{-4}

Core leg ...	Center ...	Outside ...
Area	0.042	
Diameter	0.276	
Dimensions	$2a = 0.145, 2b = 0.235$	
Density	1.5	
Weight	507	

Core factor 0.70
Yoke:
  Area.......................... 0.042
  Dimensions................. 0.205×0.235
  Density........................ 1.5
  Weight....................... 473
Copper space factor, f_s 0.1
Window dimensions .... 0.27×0.79
Lamination factor 0.90

Percent:
  Resistance 0.844
  Reactance 4.29
  Impedance 4.37

Power factor	80	100
Regulation	3.38	0.94

Losses:
  Total core 3380
    Stray load....................... 308
  Total copper...................... 3062
Percent:

Load	25	50	75	100
Efficiency	96.5	97.9	98.3	98.3

Square meters per watt 0.002
Ratio of losses 1.0
Ratio of weights...................... 4.41

TANK

Type of tank, corrugated sheet steel, 0.04×0.1
Square meters per watt 0.005
Total wetted surface 32.64
Depth of oil 1.47
Liters of oil 1480
Weight of oil 1245

CORE AND WINDINGS

Mean length of flux path.................. 3
Total ampere–turns 3980
Magnetizing current 1.25
Core loss current.................... 0.0805
Exciting current:
  Amperes 1.25
  Percent 13.1

WINDINGS	HIGH-VOLTAGE	LOW-VOLTAGE
Type of winding	Disc coils	Helical
Connections		
Conductor:		
Dimensions	0.001×0.007	0.005×0.008
Section	4.1×10^{-6}	32.7×10^{-6}
Number in parallel	None	2
Current density	2.33×10^6	2.54×10^6
turns per phase	2660	152
Coils:		
Total number	50	2
Per core leg	25	1
turns:		
Per coil	*	76
Per layer	†	76
Number of layers	2-sections	1
Coil:		
Connections	Series	Series
Dimensions	0.014×0.044	0.0115×0.655
Ducts, number and size	$21 - 0.008; 2 - 0.01; 1 - 0.014$	
Insulations:		
Layer	0.38×10^{-3}	
Core and coils	....	1.6×10^{-3} F. B. $+4.8 \times 10^{-3}$ P. B. $+6.4 \times 10^{-3}$ D.
HV and LV	13×10^{-3} D. $+8 \times 10^{-3}$ P.B. $+13 \times 10^{-3}$ D.	
Voltage per turn	15.8	
Maximum voltage between layers	885	
Length of mean–turn	1.35	0.98
Copper:		
Weight	135.5	86.8
Loss	1910	1460
Resistance, 75° C.	21.1	0.0527
Percent end turns with extra insulation	5.3	

Remarks:
* 21 coils, 56 turns
   1 coil, 50 turns
   1 coil, 44 turns
   1 coil, 36 turns
   1 coil, 24 turns

† 21 coils, 28 turns per sections
   1 coil, 25 turns per section
   1 coil, 22 turns per section
   1 coil, 18 turns per section
   1 coil, 10 and 14 turns per section

Notes: d.c.c. : double cotton covered   P.B. : pressed board   SI system of units is used.
      F.B. : fuller board   D. : duct

Figure 4.9.3 Typical transformer design sheet. (Adapted from John H. Kuhlmann, *Design of Electrical Apparatus*, 3rd ed. New York: John Wiley & Sons, 1950).

■■■■■■■■■■■■**IIII**
BIBLIOGRAPHY

Blume, L. F., et al. *Transformer Engineering,* 2d ed. New York: John Wiley & Sons, 1951.

Fink, D. G., and H.W. Beaty, eds. *Standard Handbook for Electrical Engineers,* 11th ed. New York: McGraw-Hill, 1978.

Glover, J. D., and M. S. Sarma. *Power System Analysis and Design.* Boston: PWS-Kent, 1989.

Grossner, N. R. *Transformers for Electronic Circuits.* New York: McGraw-Hill, 1967.

Institute of Electrical and Electronics Engineers. *Standard No. 389-1978: Test for Electronics Transformers and Inductors.* New York, 1978.

———— *Standard No. 390-1975: Low-Power Pulse Transformers.* New York, 1975.

———— *Standard No. 393-1977: Test Procedures for Magnetic Cores.* New York, 1977.

Kuhlmann, J. H. *Design of Electrical Apparatus,* 3d ed. New York, John Wiley & Sons, Inc., 1950.

Langsdorf, A. S. *Theory of Alternating-Current Machiney,* 2d ed. New York: McGraw-Hill, 1955.

Lawrence, R. R., and H. E. Richards. *Principles of Alternating-Current Machinery,* 4th ed. New York: McGraw-Hill, 1953.

McLyman, W. T. *Magnetic Core Selection for Transformers and Inductors.* New York: Marcel Dekker, Inc., 1987.

————. *Transformer and Inductor Design Handbook,* 2d ed. New York: Marcel Dekker, Inc., 1989.

————. *Transformer and Inductor Design Software for the IBM PC* and *Transformer and Inductor Design Software for the Macintosh.* New York: Marcel Dekker, Inc., 1989.

Say, M. G. *Alternating Current Machines.* New York: John Wiley & Sons, Halsted Press, 1976.

Westinghouse Electric Corporation Central Station Engineers. *Electric Transmission and Distribution Reference Book,* 4th ed. East Pittsburgh, Pa.: Westinghouse Electric Corporation, 1964.

■■■■■■■■■■■■■■■■■■**IIIIIIIIIII**
PROBLEMS

4-1. The sinusoidal flux in the core of a transformer is given by $\phi = 0.002 \sin 377t$ Wb. This flux links a primary winding of 100 turns.

    a. Determine the induced voltage in the coil as a function of time.

    b. Find the rms value of the induced voltage.

4-2. A sinusoidal voltage $v = 100 \sin 377t$ is applied to a 200-turn transformer winding. Determine the rms value of the impressed voltage and the flux produced in the core, neglecting the leakage flux and the winding resistance.

4-3. The primary of a transformer has 200 turns and is excited by a 60-Hz, 220-V source.

    a. Find the maximum value of the core flux.

    b. Let a voltage $v = 155.5 \sin 377t + 15.5 \sin 1131t$ (V) be applied to the transformer primary. Neglecting leakage, determine the instantaneous and rms values of the core flux.

4-4. A transformer is rated 10 kVA, 220:110 V. Consider it as an ideal transformer.

    a. Find the turns ratio and the winding current ratings.

    b. If a 2-Ω load resistance is connected across the 110-V winding, what will the current be in the high-voltage and low-voltage windings?

    c. What is the equivalent load resistance referred to the 220-V side?

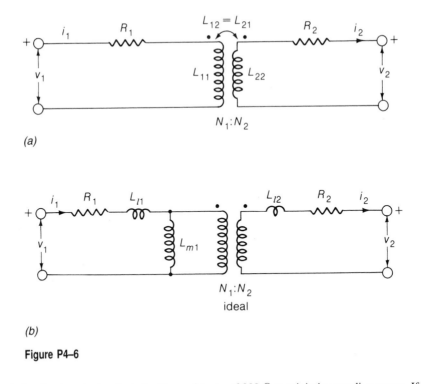

(a)

(b)

Figure P4–6

4–5. A transmission line is to be terminated with a resistance of 200 Ω to minimize standing waves. If the actual load resistance is 500 Ω, what transformer ratio is required to match the actual load to the transmission line?

4–6. The coupled-circuit model shown in Figure P4–6a below is often used to represent a two-winding air-core transformer with negligible core loss. L_{11} and L_{12} are the self-inductances of the primary and secondary windings, respectively; $M = L_{12} = L_{21}$ is the mutual inductance between the two windings.

   a. Write the flux-linkage equations (for λ_1 and λ_2) and the voltage equations (for v_1 and v_2) as a function of the currents (i_1 and i_2) and circuit parameters.

   b. Show that the circuit is equivalent to the linear circuit shown in Figure P4–6b if

$$L_{11} = L_{l1} + L_{m1}; \qquad L_{22} = L_{12} + L_{m2}$$

and

$$L_{m1} = \frac{N_1}{N_2}L_{12}; \qquad L_{m2} = \frac{N_2}{N_1}L_{12}$$

   c. The coupled-circuit model is rarely used with iron-core transformers because the high relative permeability of the core will lead to large values of L_{11}, L_{22}, and L_{12}. What will happen to all inductances in the ideal case of an infinitely permeable core?

4–7. For a single-phase, 60-Hz transformer rated 500 kVA, 2,400:480 V, following the notation of Figure 4.2.7, the equivalent circuit impedances in ohms are

$$R_1 = 0.06 \qquad R_2 = 0.003 \qquad R_c = 2,000$$

$$X_{l1} = 0.3 \qquad X_{l2} = 0.012 \qquad X_m = 500$$

The load connected across the low-voltage terminals draws rated current at 0.8 lagging power factor with rated voltage at the terminals.

a. Calculate the high-voltage winding terminal voltage, current, and power factor.

b. Compute the transformer efficiency for the given condition of load.

c. Determine the transformer series equivalent impedance for the high-voltage and low-voltage sides.

d. Considering the T-equivalent circuit, find the Thévenin equivalent impedance of the transformer under load as seen from the primary high-voltage terminals.

4–8. A 20-kVA, 2,200:220-V, 60-Hz single-phase transformer has these parameters:

Resistance of the 2,200-V winding $R_1 = 2.50\ \Omega$
Resistance of the 220-V winding $R_2 = 0.03\ \Omega$
Leakage reactance of the 2,200-V winding $X_{l1} = 10\ \Omega$
Leakage reactance of the 220-V winding $X_{l2} = 0.1\ \Omega$
Magnetizing reactance on the 2,200-V side $X_m = 25,000\ \Omega$

a. Draw the equivalent circuit of the transformer referred to the high-voltage side and referred to the low-voltage side. Label impedances numerically in ohms.

b. Draw the equivalent circuit in per-unit quantities referred to the high-voltage and low-voltage sides.

c. The transformer is supplying 15 kVA at 220 volts and a lagging power factor of 0.85. Determine the required voltage at the high-voltage terminals of the transformer.

4–9. The parameters of the equivalent circuit of the 150-kVA, 2,400-V/240-V transformer shown in Figure P4–9 (page 162) are $R_1 = 0.2\ \Omega$, $R_2 = 2\ m\Omega$, $X_1 = 0.45\ \Omega$, $X_2 = 4.5\ m\Omega$, $R_c = 10\ k\Omega$ and $X_m = 1.55\ k\Omega$. Using the circuit referred to the primary, determine the

a. voltage regulation.

b. efficiency of the transformer operating at rated load with rated voltage of 240 V and 0.8 lagging power factor.

c. Draw the phasor diagram corresponding to the circuit referred to the primary.

4–10. Corresponding to Figure P4–9a, an approximate equivalent circuit of a transformer is shown in Figure P4–10 (page 162). Using this circuit, repeat the calculations of Problem 4–9 and compare the results. Draw a phasor diagram showing all the voltages and currents of the circuit.

4–11. Using the approximate circuit given in Figure P4–10, determine the secondary current at which the transformer will have maximum efficiency.

4–12. A single-phase, 60-Hz transformer is rated 110/440 V, 2.5 kVA. The leakage reactance of the transformer measured from its low-voltage side is given as $0.08\ \Omega$.

a. Express the leakage reactance in per unit.

b. If the leakage reactance had been measured on the high-tension side, what would be its ohmic value and per-unit value?

4–13. The no-load test on a certain transformer with its secondary terminals open yields this data: 400 V, 2.5 A, 150 W. Neglecting the winding resistance and leakage flux, calculate

a. the core-loss current.

b. the magnetizing current.

c. the no-load power factor.

4–14. The short-circuit test on a certain transformer with its low-voltage winding short-circuited yields this data: 220 V, 1.4 A, 200 W. Neglecting the exciting current of the transformer, determine the series equivalent impedance referred to its high-voltage and low-voltage sides for a voltage rating of 7,200:120 V.

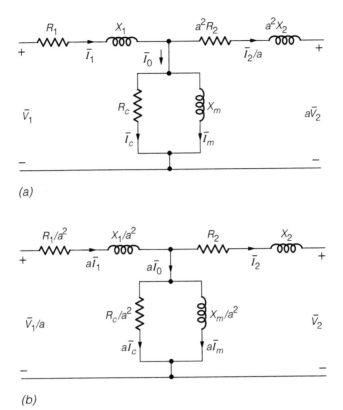

(a)

(b)

Figure P4-9 (a) Referred to primary. (b) Referred to secondary.

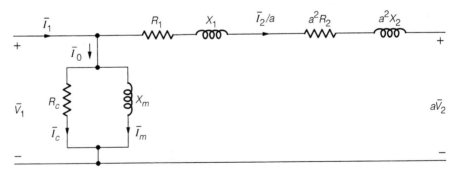

Figure P4-10

4-15.   A 3-kVA, 220:110-V, 60-Hz single-phase transformer yields this test data:

open-circuit test: 200 V, 1.4 A, 50 W
short-circuit test: 4.5 V, 13.64 A, 30 W

Determine the efficiency when the transformer delivers a load of 2 kVA at 0.85 power factor lagging.

4-16. A 75-kVA, 230-V/115-V, 60-Hz transformer was tested with these results:

open-curcuit test: 115 V, 16.3 A, 750 W;   short-circuit test: 9.5 V, 326 A, 1200 W

Determine the

a. Equivalent impendance in high-voltage terms.

b. Equivalent impendace in per unit.

c. Voltage regulation at rated load, 0.8 power factor lagging.

d. Efficiency at rated load, 0.8 power factor lagging and at half-load, unity power factor.

e. Maximum efficiency and the current at which it occurs.

4-17. These data were obtained from tests carried out on a 10-kVA, 2,300:230-V, 60-Hz distribution transformer:

open-circuit test, with the low-voltage winding excited: applied voltage: 230 V
current: 0.45 A
input power: 70 W

short-circuit test, with the high-voltage winding excited: applied voltage: 120 V
current: 4.5 A
input power: 240 W

a. Determine the equivalent circuit of the transformer referred to its high-voltage side and also referred to its low-voltage side, in actual ohmic values for impedances as well as in per unit.

b. Express the exciting current of the transformer in per unit on the basis of the rated full-load current.

c. Compute the efficiency of the transformer when it is delivering full load at 230 V and 0.85 power factor lagging.

d. Find the efficiency of the transformer when it is delivering 7.5 kVA at 230 V and 0.85 power factor lagging.

e. Determine the per-unit rating at which the transformer efficiency is a maximum, and calculate the efficiency corresponding to that load, if the transformer is delivering the load at 230 V and a power factor of 0.85.

f. The transformer is operating at a constant load power factor of 0.85 on this load cycle: 0.85 full load for 8 hours, 0.60 full load for 12 hours, no load for 4 hours. Compute the transformer's all-day (or energy) efficiency.

g. If the transformer is supplying full load at 230 V and 0.8 lagging power factor, determine the voltage regulation of the transformer. Also, find the power factor at the high-voltage terminals.

4-18. A 300-kVA transformer has a core loss of 1.5 kW and a full-load copper loss of 4.5 kW.

a. Calculate its efficiency corresponding to 0.25, 0.5, 0.75, 1.00, and 1.25 per-unit loads at unity power factor.

b. Repeat the efficiency calculations for the loads at power factors of 0.8 and 0.6.

c. Determine the per-unit load for which the efficiency is a maximum, and calculate the corresponding efficiencies for the power factors of unity, 0.8, and 0.6.

4-19. Consider the solution of Example 4.4.3. By means of a phasor diagram, determine the load power factor for which the regulation is maximum (i.e., the poorest), and find the corresponding regulation.

4-20. A 10-kVA, 200:400-V, single-phase transformer gave these test results:

open-circuit test (LV winding supplied):   200 V, 3.2 A, 450 W
short-circuit test (HV winding supplied):   38 V, 25 A, 600 W

a. Determine the per-unit total equivalent series resistance.

b. Compute the efficiency when the transformer delivers half its rated kVA at 0.85 power factor lagging.

4–21. A high-voltage feeder connects a source to the HV terminals of a 50-kVA, 5-kV/220-V, single-phase transformer. The load on the LV terminals of the transformer is 32 kW at 200 V, 0.8 power factor lagging. These impedances are in per unit on the transformer rating as base.

$$\text{feeder:} \quad 0.01 + j0.02$$

$$\text{transformer:} \quad 0.02 + j0.03$$

Working in the per-unit system, compute the source voltage in per unit and in kV.

4–22. A short-circuit test was run at rated current on a 7.5 kVA, 100:300-V, single-phase transformer, with the LV side shorted and the HV side supplied at 20 V. No power reading was taken because the wattmeter was faulty. It is known, however, that the transformer has its maximum efficiency of 94.2% when it delivers a load of 6.0 kW at unity power factor at rated voltage.

The transformer is now loaded so that it delivers 4.5 kW at 100 V and 0.6 power factor lagging. Determine

a. The current in amperes referred to the LV side for the short-circuit test, the maximum efficiency condition, and the load test.

b. R_{eq} and X_{eq} referred to the HV side and to the LV side.

c. The regulation corresponding to the load test, neglecting the effect of the exciting current of the transformer.

4–23. A 10-kVA, 100:200-V, single-phase transformer delivers a load of 5 kW at 0.8 power factor lagging at rated voltage on its HV side. The per-unit total equivalent series resistance and total equivalent leakage reactance, respectively, are 0.04 pu and 0.06 pu. Ignore the effect of the exciting current of the transformer. Working in the per-unit system, compute the input voltage in per unit and in volts.

4–24. A 100-kVA transformer stepping down from 2,000 to 400 V has a primary resistance of 0.17 Ω and a secondary resistance of 0.0068 Ω. The corresponding reactances are 0.25 and 0.01 Ω, respectively. Compute

a. The total equivalent resistance, reactance, and impedance
   (1) in ohms referred to the secondary side.   (2) in per unit.

b. The percent regulation when the transformer delivers rated current at rated voltage and 0.8 power factor lagging.

4–25. The efficiency of a 400-kVA, single-phase, 60-Hz transformer is 98.77% when delivering full-load current at 0.8 power factor, and 99.13% with half-rated current at unity power factor. Calculate

a. The iron loss.   b. The full-load copper loss.   c. The efficiency at 3/4 load, 0.9 pf.

4–26. A transformer has its maximum efficiency of 0.9800 when it delivers 15 kVA at unity power factor. During the day it is loaded as follows:

$$\begin{array}{ll} \text{12 hours:} & \text{2 kW at power factor 0.5} \\ \text{6 hours:} & \text{12 kW at power factor 0.8} \\ \text{6 hours:} & \text{18 kW at power factor 0.9} \end{array}$$

Determine the all-day efficiency.

4–27. A single-phase, 3-kVA, 220:110-V, 60-Hz transformer has a high-voltage winding resistance of 0.3 Ω, a low-voltage winding resistance of 0.06 Ω, a leakage reactance of 0.8 Ω on its high-voltage side, and a leakage reactance of 0.2 Ω on its low-voltage side. The core loss at rated voltage is 45 W, and the copper loss at rated load is 100 W. Neglect the exciting current of the transformer.

a. Find the per-unit voltage regulation when the transformer is supplying full load at 110 V and 0.9 lagging power factor.

b. Calculate the efficiency when the transformer delivers 3 kW at 0.95 power factor lagging.

c. Find the all-day efficiency for this loading: 3.5 kVA, 0.95 pf, for 3 hours; 3.0 kVA, 0.90 pf, for 8 hours; 2.4 kVA, 0.80 pf, for 4 hours; 1.0 kVA, 0.95 pf, for 9 hours.

4–28. A 5-MVA, 66:13.2 kV, three-phase transformer supplies a three-phase resistive load of 4,500 kW at 13.2 kV. What is the load resistance in ohms as measured from the line-to-neutral on the high-voltage side of the transformer, if it is

a. connected in Y–Y.

b. reconnected in Y–Δ, with the same high-tension voltage supplied and the same load resistors connected.

4–29. A three-phase transformer bank consisting of three 10-kVA, 2,300:230-V, 60-Hz, single-phase transformers connected in Y–Δ is used to step down the voltage. The loads are connected to the transformers by means of a common three-phase low-voltage feeder whose series impedance is $(0.005 + j0.01)$ Ω per phase. The transformers themselves are supplied by means of a three-phase high-voltage feeder whose series impedance is $(0.5 + j5.0)$ Ω per phase. The equivalent series impedance of the single-phase transformer referred to the low-voltage side is $(0.12 + j0.24)$ Ω. The star point on the primary side of the transformer bank is grounded. The load consists of a heating load of 2 kW per phase and a three-phase induction-motor load of 20 kVA with a lagging power factor of 0.8, supplied at 230 V line-to-line.

a. Draw a one-line diagram of this three-phase distribution system.

b. Neglecting the exciting current of the transformer bank, draw the per-phase equivalent circuit of the distribution system.

c. Determine the line current and the line-to-line voltage at the sending end of the high-voltage feeder.

4–30. Three single-phase transformers are used to form a three-phase transformer bank rated 18 MVA, 13.8Δ–120Y kV. One side of the transformer bank is connected to a 120-kV transmission line, and the other side is connected to a three-phase load of 10 MVA at 13.2 kV and 0.8 lagging power factor.

a. Determine the MVA and voltage ratings of each of the three single-phase transformers.

b. Using the base in the transmission line of 120 kV and 20 MVA, find the load impedance in per unit to be used in the line-to-neutral impedance diagram.

4–31. A single-line diagram of a three-phase transformer bank connected to a load is given in Figure P4–31. Find the magnitudes of the line-to-line voltages, line currents, phase voltages, and phase currents on either side of the transformer bank. Determine the primary to secondary ratio of the line-to-line voltages and the line currents.

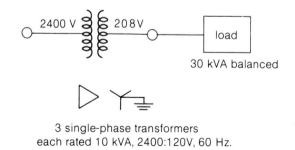

3 single-phase transformers
each rated 10 kVA, 2400:120V, 60 Hz.

Figure P4–31

4–32. A single-line diagram of part of a three-phase transmission system is given in Figure P4–32. Choose a base of 20 MVA, 12.5 kV in the load circuit, and draw the equivalent line-to-neutral impedance diagram in per-unit quantities. Show the base kV in the three parts of the system.

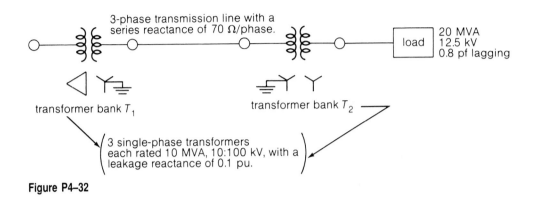

Figure P4–32

4-33. A 400-kVA, 11-kV/415-V, three-phase, Δ–Y-connected transformer on rated load has an I^2R loss of 2.5 kW for its high-voltage windings and 2.0 kW for its low-voltage windings. The total leakage reactance is 0.06 per unit. Neglect the exciting current. Determine the per-phase equivalent circuit of the transformer referred to the high-voltage and the low-voltage sides, both in ohmic values of impedances and in per unit.

4-34. Three single-phase transformers, each rated 15 MVA, 13.2:66 kV, with a leakage reactance of 0.1 pu, are connected to form a three-phase Y–Δ bank, one side of which is connected to a 120-kV transmission line and the other side, to three 12-ohm resistors in delta. The star point of the transformer bank is grounded. The series reactance per phase of the transmission line is 80 Ω. Draw a single-line diagram of the system and obtain the line-to-neutral impedance diagram with per-unit values, for a base voltage of 120 kV in the transmission line and a three-phase voltampere base of 50 MVA.

4-35. A three-phase, Δ–Y, 11-kV/440-V, 60-Hz, 300-kVA transformer has these test data:

no-load test with high-voltage side open: 440 V, 20 A, 1.3 kW

short-circuit test with low-voltage side short-circuited: 630 V, 15.75 A, 3.1 kW

Determine the efficiency and regulation of the transformer when it is delivering full load at 440 V and 0.8 lagging power factor.

4-36. A 2,400-V to 208-V, 300-kVA, three-phase transformer has a core loss at rated voltage of 2.70 kW. If its equivalent resistance is 1.40%, find the efficiency of this transformer for a load power factor of 0.9

a. At full load.   b.   At half load.

4-37. A 200-MVA, 345-kV Δ/34.5-kV Y, substation transformer has an 8% leakage reactance. The transformer acts as a connecting link between 345-kV transmission and 34.5-kV distribution. Transformer winding resistances and exciting current are neglected. The high-voltage bus connected to the transformer is assumed to be an ideal 345-kV positive-sequence source with negligible source impedance. Using the transformer ratings as base values, determine the per-unit magnitudes of transformer voltage drop and voltage at the low-voltage terminals when rated transformer current at 0.8 pf lagging enters the high-voltage terminals.

4-38. a. The leakage reactance of a three-phase, 300-MVA, 230 Y/23 Δ-kV transformer is 0.06 per unit based on its own ratings. Draw the positive sequence network by neglecting the exciting admittance and assuming the American standard phase shift.

b. Choosing system bases to be 240/24 kV and 100 MVA, redraw the sequence networks for part a.

4-39. Three single-phase transformers, each rated 10 MVA, 66.4/12.5 kV, 60 Hz, with an equivalent series reactance of 0.1 per unit divided equally between primary and secondary, are connected in a three-phase bank. The high-voltage windings are Y-connected, and their terminals are directly connected to a 115-kV three-phase bus. The secondary terminals are all shorted together. Find the currents entering the high-voltage terminals and leaving the low-voltage terminals if the low-voltage windings are

a. Y connected.   b. Δ connected.

4-40. A 30-MVA, 13.2-kV, three-phase generator, which has a positive-sequence reactance of 1.5 per unit on the generator base, is connected to a 35-MVA, 13.2Δ/115 Y-kV step-up transformer with a series impedance of $(0.005 + j0.1)$ per unit on its own base.

a. Calculate the per-unit generator reactance on the transformer base.

b. The load at the transformer terminals is 15 MW at unity power factor and at 115 kV. Choosing the transformer high-side voltage as the reference phasor, draw a phasor diagram for this condition.

c. For the condition of part b, find the transformer low-side voltage and the generator internal voltage behind its reactance. Also compute the generator output power and power factor.

4-41. Consider a three-phase, two-winding, Y–Δ transformer bank. Construct the negative-sequence phasor diagram and determine the corresponding phase shift of the transformer.

4-42. A 50-kVA, 60-Hz, 2,400-V Δ/240-V Y transformer is accidentally paralleled with a 50-kVA, 60-Hz, 2,400-V Δ/240-V Y transformer, with no external load connected. The equivalent impedances referred to the respective secondaries are $Z_\Delta = 0.1106\angle60.2°$ Ω and $Z_Y = 0.0369\angle60.2°$ Ω. Determine the circulating current in each transformer secondary, both in magnitude and as a ratio of the rated transformer current of each transformer.

4-43. A single-phase 10-kVA, 2,300:230-V, 60-Hz, two-winding distribution transformer is connected as an autotransformer to step up the voltage from 2,300 V to 2,530 V.

a. Draw a schematic diagram of the arrangement showing all the voltages and currents while delivering full load.

b. Find the permissible kVA rating of the autotransformer, if the winding currents are not to exceed those for full-load operation as a two-winding transformer. How much of that is transformed by electromagnetic induction?

c. Based on the data given for the two-winding transformer in Problem 4–17, compute the efficiency of the autotransformer corresponding to full load and 0.8 lagging power factor. Comment on why the efficiency of the autotransformer is higher than that of the two-winding transformer.

4-44. A two-winding, single-phase transformer rated 3 kVA, 220:110 V, 60 Hz is connected as an autotransformer to transform a line input voltage of 330 volts to a line output voltage of 110 V and to deliver a load of 2 kW at 0.8 lagging power factor. Draw the schematic diagram of the arrangement, label all the currents and voltages, and calculate all the quantities involved.

4-45. Table P4–45 lists the results of three short-circuit tests on a 9,600:2,400:600-V, 60-Hz, single-phase, three-winding transformer. The kVA-ratings of the primary, secondary, and tertiary windings are 1,000, 500, and 500, respectively. Neglect the winding resistances and the exciting current of the transformer.

a. Determine the per-unit values of the equivalent circuit impedances Z_1, Z_2, and Z_3 of Figure 4.7.2b on a 1-MVA rated-voltage base.

b. Three of these transformers are used to form a three-phase, Y–Δ–Δ-connected bank. The wye-connected primaries are connected to a 16.6-kV, three-phase supply. If a three-phase short circuit should occur at the terminals of the tertiary windings, find the per-unit values of the steady-state short-circuit currents and of the voltage at the secondary terminals, on a three-phase, 3-MVA rated-voltage base.

Table P4–45

Winding		Measured on the Excited-Winding Side	
Excited	**Short-Circuited**	**Volts**	**Amperes**
primary (1)	secondary (2)	250	52
primary (1)	tertiary (3)	800	52
secondary (2)	tertiary (3)	200	208

4–46.   The rating of a three-winding, three-phase transformer is given as

primary (1):   Y-connected, 6.6 kV, 15 MVA

secondary (2):   Y-connected, 33 kV, 10 MVA

tertiary (3):   Δ-connected, 2.2 kV, 5 MVA

Neglecting the resistances, the leakage impedances calculated from the short-circuit tests are

$Z_{12} = j0.25 \ \Omega$; $Z_{13} = j0.3 \ \Omega$ (measured from the primary side)

$Z_{23} = j9.0 \ \Omega$ (measured from the secondary side)

Determine the per-unit impedances Z_1, Z_2, and Z_3 of the equivalent circuit of Figure 4.7.2b on a 15-MVA, 6.6-kV base in the primary circuit.

4–47.   The three-phase ratings of a three-winding transformer are given as

primary (1):   Y-connected, 66 kV, 15 MVA

secondary (2):   Y-connected, 13.2 kV, 10 MVA

tertiary (3):   Δ-connected, 2.3 kV, 5 MVA

Neglecting the resistance, the per-unit leakage impedances are:

$Z_{12} = j0.08$ on   15-MVA, 66-kV base

$Z_{13} = j0.10$ on   15-MVA, 66-kV base

$Z_{23} = j0.09$ on   10-MVA, 13.2-kV base

a. Determine the per-unit impedances Z_1, Z_2, and Z_3 of the equivalent circuit of Figure 4.7.2b on a 15-MVA, 66-kV base in the primary circuit.

b. Purely resistive loads of 7.5 MW at 13.2 kV and 5 MW at 2.3 kV are connected to the secondary and tertiary sides of the transformer, respectively. Draw the impedance diagram of the system, showing the per-unit impedances for 66-kV, 15-MVA base in the primary circuit.

4–48.   A three-phase, three-winding, 6,600/400/110-V, Y–Y–Δ-connected transformer with a magnetizing current of 5 A is delivering three-phase loads of 1 MVA at 0.8 lagging power factor connected to the secondary winding and 200 kVA at 0.5 leading power factor connected to the tertiary winding. The per-unit impedances of the equivalent circuit of Figure 4.7.2b on a 6.6-kV, 1-MVA base are

$$Z_1 = 0.005 + j0.03; \qquad Z_2 = 0.006 + j0.025; \qquad Z_3 = 0.008 + j0.033$$

a. Determine the approximate per-unit regulation component associated with the primary alone (ε_1), the secondary alone (ε_2), and the tertiary alone (ε_3).

b. Noting that the resultant of the part-regulations of two circuits together is an addition if power flows from one to the other, and a subtraction otherwise, calculate the per-unit regulations of the primary-secondary (ε_{12}), the primary-tertiary (ε_{13}), the secondary-tertiary (ε_{23}), and the tertiary-secondary (ε_{32}).

4–49. The primary winding of an audio transformer is connected to a source, while its secondary is connected to a speaker that has a resistive impedance of 8 Ω over the frequency range in which it is generally used. The parameters of the transformer equivalent circuit are given in Figure P4–49.

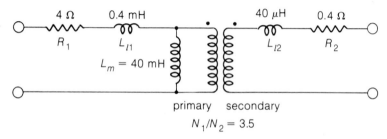

Figure P4–49

a. It is required to supply a power of 60 W to the speaker in the mid-frequency range. Neglecting the internal resistance of the source, compute the applied voltage at the source terminals.

b. Assuming that the same applied voltage is maintained over the whole frequency range, find the frequencies at which the speaker power is reduced to 30 W.

4–50. A single-phase, 500-kVA, 2,400:240-V transformer with negligible resistance and 0.05 per-unit equivalent leakage reactance is operated in parallel with another single-phase 250-kVA, 2,400:240-V transformer with negligible resistance and 0.04 per-unit equivalent leakage reactance. These two transformers in parallel operation are to share a load of 750 kVA at 0.8 lagging power factor.

a. Neglecting the exciting current of the transformers, find the load on each transformer. Which transformer is overloaded and why?

b. In addition to the given data, consider the equivalent resistance of the 500-kVA transformer to be 0.01 per unit and that of the 250-kVA transformer to be 0.015 per unit. How would this change the results of part a?

4–51. A typical layout of a single-phase transformer and its basic dimensions are shown in Figure P4–51. A_i denotes the net cross-sectional area of the ferromagnetic core; A_w denotes window area; k_w is the space-factor allowing for insulation and clearance such that the available area for actual conductor cross section is $k_w A_w$.

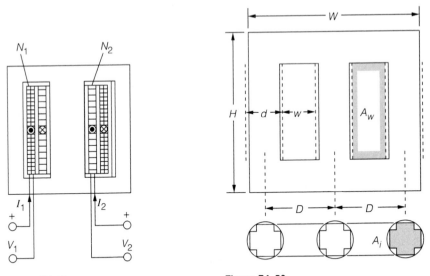

Figure P4–51 Figure P4–52

a. Obtain an expression for the rating of the two-limbed transformer in terms of the peak flux density B_m, frequency f, rms current density J, and the main dimensions given above, assuming that the primary and secondary windings each occupy about half of the available area.

b. With $A_i = 0.03$ m^2, $A_w = 0.07$ m^2, $B_m = 1.375$ T, $J = 2.6$ MA/m$^2 = 2.6$ A/mm^2, $k_w = 0.3$, $f = 60$ Hz, calculate the kVA rating of the two-limbed single-phase transformer.

4–52. In order to design a three-phase, three-limbed, core-type, 400-kVA, 50-Hz, 6,600:450-V, Δ–Y, oil-immersed, natural-cooled transformer, a frame size is considered with a stepped core of net area $A_i = 0.027$ m^2 contained in an overall limb diameter $d = 0.21$ m, with limb centers spaced $D = 0.34$ m. See Figure P4–52. The yoke area is the same as that of a limb. Using $B_m = 1.45$ T and $J = 3$ A/mm^2, a window space factor $k_w = 0.29$, obtain the main dimensions $W(\simeq 2D + 0.9d)$, H, and ℓ, the depth of the windows, as well as the number of primary and secondary turns.

4–53. The project is to design a 1,000-kVA, three-phase, 60-Hz, 7,960-V (primary HV):575-V (secondary LV), core-type power transformer with delta-connected primary and secondary windings. The transformer is to be of the self-oil-cooled type, carrying its rated load continuously with temperature rise not to exceed 50° C. The full-load efficiency at unity power factor is not to be less that 98.2%, and the ratio of the core to copper loss is to be approximately 0.9. Prof. John H. Kuhlmann's[5] transformer design sheet (Figure P4–53a) and the transformer layout (Figure P4–53b, page 172) are given. Pursue the design project and justify the details furnished. Note that the system of units adopted is the American Customary System.[6]

[5] J. H. Kuhlmann, *Design of Electrical Apparatus*, 3rd ed. (New York: John Wiley & Son, 1950).
[6] T. Wildi, *Units and Conversion Charts*, (New York: IEEE Press, 1991).

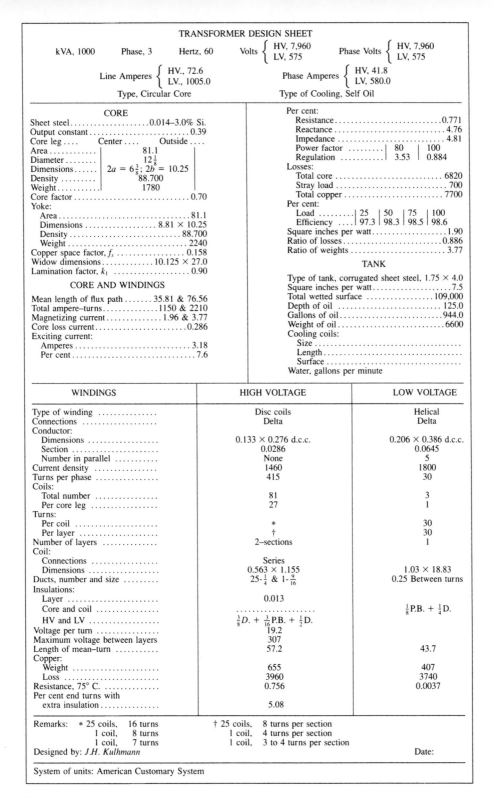

TRANSFORMER DESIGN SHEET

kVA, 1000   Phase, 3   Hertz, 60   Volts { HV, 7,960 / LV, 575   Phase Volts { HV, 7,960 / LV, 575

Line Amperes { HV., 72.6 / LV., 1005.0   Phase Amperes { HV, 41.8 / LV, 580.0

Type, Circular Core   Type of Cooling, Self Oil

CORE

Sheet steel....................0.014–3.0% Si.
Output constant.........................0.39
Core leg....   Center....   Outside....
Area...........   81.1
Diameter........ $12\frac{1}{8}$
Dimensions...... $2a = 6\frac{3}{8} : 2b = 10.25$
Density........   88.700
Weight..........   1780
Core factor.............................0.70
Yoke:
   Area................................81.1
   Dimensions.................. 8.81 × 10.25
   Density............................88.700
   Weight 2240
Copper space factor, f_s 0.158
Widow dimensions............ 10.125 × 27.0
Lamination factor, k_1 .... 0.90

CORE AND WINDINGS

Mean length of flux path .... 35.81 & 76.56
Total ampere–turns.............1150 & 2210
Magnetizing current .... 1.96 & 3.77
Core loss current......................0.286
Exciting current:
   Amperes 3.18
   Per cent..............................7.6

Per cent:
   Resistance...........................0.771
   Reactance 4.76
   Impedance 4.81
   Power factor .... | 80 | 100
   Regulation .... | 3.53 | 0.884
Losses:
   Total core 6820
   Stray load 700
   Total copper 7700
Per cent:
   Load .... | 25 | 50 | 75 | 100
   Efficiency .... | 97.3 | 98.3 | 98.5 | 98.6
Square inches per watt....................1.90
Ratio of losses.......................0.886
Ratio of weights 3.77

TANK

Type of tank, corrugated sheet steel, 1.75 × 4.0
Square inches per watt...................7.5
Total wetted surface 109,000
Depth of oil 125.0
Gallons of oil.........................944.0
Weight of oil 6600
Cooling coils:
   Size ....
   Length ....
   Surface ....
Water, gallons per minute

WINDINGS	HIGH VOLTAGE	LOW VOLTAGE
Type of winding	Disc coils	Helical
Connections	Delta	Delta
Conductor:		
Dimensions	0.133 × 0.276 d.c.c.	0.206 × 0.386 d.c.c.
Section	0.0286	0.0645
Number in parallel	None	5
Current density	1460	1800
Turns per phase	415	30
Coils:		
Total number	81	3
Per core leg	27	1
Turns:		
Per coil	*	30
Per layer	†	30
Number of layers	2–sections	1
Coil:		
Connections	Series	
Dimensions	0.563 × 1.155	1.03 × 18.83
Ducts, number and size	$25\text{-}\frac{1}{4}$ & $1\text{-}\frac{9}{16}$	0.25 Between turns
Insulations:		
Layer	0.013	
Core and coil	....	$\frac{1}{8}$P.B. + $\frac{1}{4}$D.
HV and LV	$\frac{3}{8}$D. + $\frac{3}{16}$P.B. + $\frac{1}{2}$D.	
Voltage per turn	19.2	
Maximum voltage between layers	307	
Length of mean–turn	57.2	43.7
Copper:		
Weight	655	407
Loss	3960	3740
Resistance, 75° C.	0.756	0.0037
Per cent end turns with extra insulation..............	5.08	

Remarks: * 25 coils,   16 turns   † 25 coils,   8 turns per section
          1 coil,   8 turns   1 coil,   4 turns per section
          1 coil,   7 turns   1 coil,   3 to 4 turns per section
Designed by: *J.H. Kulhmann*   Date:

System of units: American Customary System

Figure P4–53a Transformer design sheet.

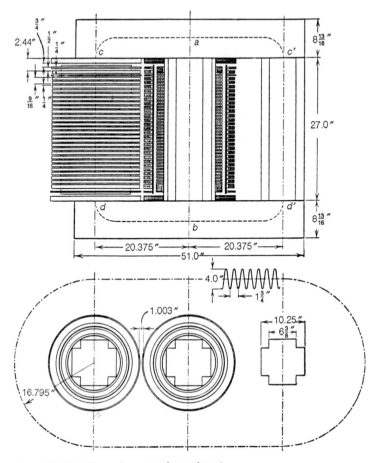

Figure P4–53b Three-phase transformer layout.

5

Principles of Electromechanical Energy Conversion

5.1   Forces and Torques in Magnetic Field Systems
5.2   Singly Excited and Multiply Excited Magnetic Field Systems
5.3   Elementary Concepts of Rotating Machines
5.4   Rotating Magnetic Fields
5.5   Basic Aspects of Electromechanical Energy Converters
5.6   The Design Aspect of AC Machines

The transformer, as we explained in the last chapter, is an electromagnetic device that transmits electrical energy with a change in voltage and current levels from one side to the other. It does not *convert* energy, however. This chapter deals with the principles of electromechanical energy conversion and their application to simple devices. Electromechanical energy conversion involves the interchange of energy between an electrical system and a mechanical system. When the energy is converted from electrical to mechanical form, the device is displaying *motor action*. A *generator action* involves converting mechanical energy into electrical energy. Electromechanical energy converters embody three essential features: (1) an electric system, (2) a mechanical system, and (3) a coupling field.

Both electric and magnetic fields store energy, from which useful mechanical forces can be derived. In air or another gas at normal pressure the dielectric strength of the medium restricts the working electric field intensity to about 3×10^6 V/m and, consequently, the stored electric-energy density to

$$\frac{1}{2} \, \varepsilon_0 \, E^2 = \frac{1}{2} \frac{10^{-9}}{36\pi} (3 \times 10^6)^2 \simeq 40 \text{ J/m}^3$$

where ε_0 is the permitivity of free space, given by $10^{-9}/(36\pi)$ or 8.854×10^{-12} F/m, and E is the electric field intensity. This corresponds to a force density of 40 N/m^2. While there is no comparable restriction on magnetic fields, the saturation of ferromagnetic media required to complete the magnetic circuit limits the working magnetic flux density to about 1.6 T, for which the stored magnetic energy density in air is about

$$\frac{1}{2} \frac{B^2}{\mu_0} = \frac{1}{2} \frac{(1.6)^2}{(4\pi \times 10^{-7})} \simeq 1 \times 10^6 \text{ J/m}^3$$

where μ_0 is the permeability of free space, and B is the magnetic flux density. As this is nearly 25,000 times as much as for the electric field, almost all industrial electric machines are magnetic in principle and are magnetic field devices. Because magnetic poles occur in pairs (north and south) and the movement of a conductor through a natural north–south sequence induces an emf that changes direction in accordance with the magnetic polarity (i.e., an alternating emf), the devices are inherently ac machines.

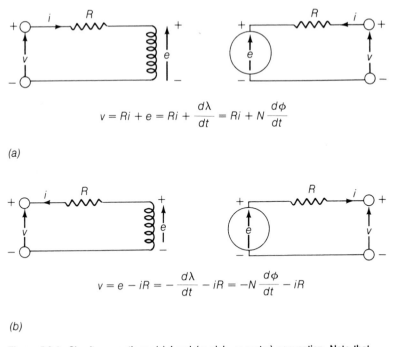

$$v = Ri + e = Ri + \frac{d\lambda}{dt} = Ri + N\frac{d\phi}{dt}$$

(a)

$$v = e - iR = -\frac{d\lambda}{dt} - iR = -N\frac{d\phi}{dt} - iR$$

(b)

Figure 5.0.1 Circuit conventions. (a) Load (or sink, or motor) convention. Note that the power *absorbed* by the circuit is positive when v and i are positive. (b) Source (or generator) convention. Note that the power *delivered* by this circuit to the external circuit is positive when v and i are positive.

Three basic principles associated with an electromagnetic devices are (1) *induction*, (2) *interaction*, and (3) *alignment*. The essentials for producing an electromotive force by magnetic means are electric and magnetic circuits, mutually interlinked. The induced emf is given by Faraday's law of induction:

$$e = -\frac{d\lambda}{dt} = -N\frac{d\phi}{dt}$$

which is the same as Equation 3.1.9, stated earlier. The induced emf acts in the direction of positive current *i* as shown in Figure 5.0.1b with a source (or generator) convention. Sometimes it is more convenient to consider the emf as directed in opposition to positive current, as shown in Figure 5.0.1a with a load (or sink, or motor) convention, in which case

$$e = +\frac{d\lambda}{dt} = +N\frac{d\phi}{dt}$$

which is the same as Equation 4.2.1, stated earlier.

The change of the flux linkage in a coil may occur in one of the following three ways:

a. The flux remaining constant, the coil moves through it.
b. The coil remaining stationary with respect to the flux, the flux varies in magnitude with time.
c. The coil may move through a time-varying flux; that is to say, both changes may occur together.

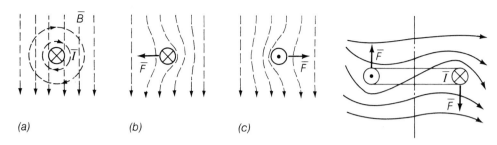

(a)   *(b)*   *(c)*

Figure 5.0.2 Principle of interaction.

Figure 5.0.3 Torque produced by forces caused by interaction of current-carrying conductors and magnetic fields.

ferromagnetic pieces

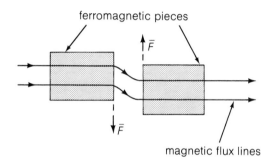

magnetic flux lines

Figure 5.0.4 Principle of alignment.

In *(a)*, the flux-cutting rule given by Equation 3.1.12 or 3.1.13 yields the *motional emf* (or the *speed emf*) that is always associated with the conversion of energy between the mechanical and electrical forms. In *(b)*, since no motion is involved, no energy conversion takes place; Equation 4.2.1 gives the *transformer emf* (or the *pulsational emf*). Both emf components are present in *(c)*.

Current-carrying conductors, when placed in magnetic fields, experience mechanical force, which can be calculated by Equation 3.1.14 or 3.1.15. Consider Figure 5.0.2, where $\bar{B}$ is flux density, $\bar{I}$ is current, and $\bar{F}$ is force. The flux density $\bar{B}$ of an undisturbed uniform field, shown in Figure 5.0.2a, on which the introduction of a current-carrying conductor imposes a corresponding field component, developing the resultant as in Figure 5.0.2b for the case in which the current is directed into and perpendicular to the plane of the paper, as symbolized by the cross in the figure. In the neighborhood of the conductor, as seen in Figure 5.0.2b, the resultant flux density is greater than B on one side and less than B on the other side. Figure 5.0.2c shows the conditions corresponding to the current's being directed out of and perpendicular to the plane of the paper, as symbolized by the dot. The direction of the mechanical force developed is such that it tends to restore the field to its original undisturbed and uniform configuration, as shown in Figures 5.0.2b and c. The force is always in such a direction that the energy stored in the magnetic field is minimized. Figure 5.0.3 shows a one-turn coil in a magnetic field and illustrates how torque is produced by forces caused by interaction between current-carrying conductors and magnetic fields.

Pieces of highly permeable material such as iron, situated in an ambient medium of low permeability such as air in which a magnetic field is established, experience mechanical forces that tend to align them with the field direction in such a way that the reluctance of the system is minimized. Figure 5.0.4

illustrates this principle of alignment and shows the direction of forces. The force is always in a direction that reduces the net magnetic reluctance and shortens the magnetic flux path.

With this general background, we will now evaluate the electromagnetic forces and torques associated with magnetic-field systems.

5.1 Forces and Torques in Magnetic Field Systems

A method of computing the electromagnetic force based on the Lorentz force equation is given by Equations 3.1.14 and 3.1.15. However, this method is applicable only when the flux density is known and the geometry of the system is rather simple. In general, the *energy method* presented here is much more convenient. The energy method of determining electromagnetic forces is based on the principle of conservation of energy:

> Energy can be neither created nor destroyed, even though, within an isolated system, energy may be converted from one form to another form, and transferred from an energy source to an energy sink. The total energy in the system is constant.

Since we will be concerned with low frequencies and low velocities (i.e., velocities much less than that of light), it is fair to assume that no energy is radiated into space, no mass is converted into energy, and no dielectric losses exist.

The energy conversion process involves interchange between electrical and mechanical energy via the stored energy in the magnetic field. This stored energy, which can be determined for any configuration of the system, is a *state function* defined solely by functional relationships between variables and the the final values of these variables. Thus the energy method is a powerful tool for determining the coupling forces of electromechanics.

Figure 5.1.1 is a schematic representation of a practical electromechanical system (with losses) divided into its simpler component parts. Note that all types of dissipation have been excluded to form a *conservative* (or *lossless*) energy-conversion part that can be described by state functions to yield the electromechanical coupling terms in electromechanics. A property of a conservative system is that its energy is a function of its state only and is described by the same independent variables that describe the state. State functions at a given instant of time depend solely on the state of the system at that instant and not on past history; they are independent of how the system is brought to that particular state.

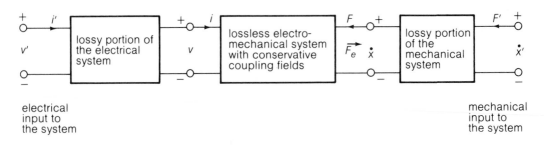

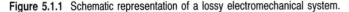

Figure 5.1.1 Schematic representation of a lossy electromechanical system.

For the case of a *sink* of electical energy, such as an electric motor, the principle of conservation of energy allows us to write:

$$
\begin{array}{l}
\text{Electrical input} \\
\text{energy from} \\
\text{source}
\end{array}
=
\begin{array}{l}
\text{mechanical} \\
\text{output energy} \\
\text{to load}
\end{array}
+
\begin{array}{l}
\text{increase in} \\
\text{stored field} \\
\text{energy}
\end{array}
+
\begin{array}{l}
\text{energy loss} \\
\text{converted} \\
\text{to heat}
\end{array}
\qquad (5.1.1)
$$

The energy losses associated with this form of energy conversion are (a) the energy loss due to the resistances of the circuits, (b) the energy loss due to friction and windage associated with motion, and (c) the energy loss associated with the coupling field. Considering the coupling field to be a magnetic field, the field losses are due to hysteresis and eddy-current losses—i.e., the core losses in the magnetic system. Since these losses are usually small, they may be neglected, or their effect may be included in the lossy portion of the electrical system. Then, considering only the conservative (or *lossless*) portion of the system, we have

$$
\begin{array}{l}
\text{Electrical energy from} \\
\text{source } \textit{minus}\text{ electrical} \\
\text{system losses}
\end{array}
=
\begin{array}{l}
\text{mechanical output} \\
\textit{plus}\text{ mechanical} \\
\text{system losses}
\end{array}
+
\begin{array}{l}
\text{increase in stored} \\
\text{magnetic coupling} \\
\text{field energy}
\end{array}
\qquad (5.1.2)
$$

or

$$
\begin{array}{l}
\text{Input electrical energy to the} \\
\text{lossless electromechanical system}
\end{array}
=
\begin{array}{l}
\text{mechanical} \\
\text{work done}
\end{array}
+
\begin{array}{l}
\text{increase in} \\
\text{stored energy}
\end{array}
\qquad (5.1.3)
$$

The increase in energy stored in the magnetic field is considered here, while neglecting the energy stored in the electric field. In incremental form, in time dt, Equation 5.1.3 can be expressed as

$$
dW_e = dW + dW_m \qquad (5.1.4)
$$

or, referring to Figure 5.1.1, it can be written as

$$
vi\,dt = -F\,dx + dW_m \qquad (5.1.5)
$$

where $(-F\,dx)$ corresponds to the mechanical output of the lossless electromechanical system. This may also be expressed as $(F_e\,dx)$, in which F_e is the mechanical force of electrical origin due to magnetic field coupling. Then it follows that

$$
F_e\,dx = vi\,dt - dW_m \qquad (5.1.6)
$$

where dW_m is the increase in energy stored in the magnetic field, and dx is an arbitrary incremental displacement. Since v, which is the same as the induced emf in the lossless electromechanical system, can be expressed in terms of the flux linkage λ by means of Faraday's law of induction as

$$
v = \frac{d\lambda}{dt} \qquad (5.1.7)
$$

Equation 5.1.6 can be rewritten as

$$
F_e\,dx = i\,d\lambda - dW_m \qquad (5.1.8)
$$

In a simple electromechanical system with a singly excited electrical system or a mechanical system consisting of only one-dimensional motion, such as the one in Figure 5.1.2, the independent variable

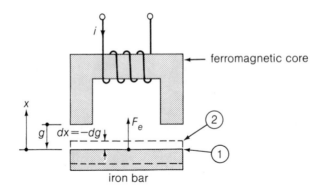

Figure 5.1.2 A simple electromechanical system.

on the electrical side is either the current i or the flux linkage λ. The λ–i characteristic of the nonlinear magnetic system with no core loss is given by a single-valued, nonlinear relationship, typically as shown in Figure 3.4.1. Thus, if the independent variables are the current i and the coordinate x, then the flux linkage λ is a function of both i and x:

$$\lambda = \lambda(i, x) \tag{5.1.9}$$

in which case $d\lambda$ can be expressed as

$$\lambda = \frac{\partial \lambda}{\partial i} di + \frac{\partial \lambda}{\partial x} dx \tag{5.1.10}$$

where di is an arbitrary incremental change in i. Also, since the stored magnetic energy is also a function of i and x, it follows that

$$dW_m = \frac{\partial W_m}{\partial i} di + \frac{\partial W_m}{\partial x} dx \tag{5.1.11}$$

Substituting Equations 5.1.10 and 5.1.11 in Equation 5.1.8, we get

$$F_e \, dx = \left(-\frac{\partial W_m}{\partial x} + i \frac{\partial \lambda}{\partial x} \right) dx + \left(-\frac{\partial W_m}{\partial i} + i \frac{\partial \lambda}{\partial i} \right) di \tag{5.1.12}$$

In order that the force F_e be independent of the change in the current i (and λ) during the arbitrary displacement, the coefficient of di in Equation 5.1.12 must be zero. Consequently, the force F_e is always given by

$$F_e = -\frac{\partial W_m(i, x)}{\partial x} + i \frac{\partial \lambda(i, x)}{\partial x} \tag{5.1.13}$$

in which W_m and λ are functions of independent variables i and x. It is possible to express Equation 5.1.13 in a simpler form in terms of magnetic coenergy W'_m based on Equations 3.4.7 and 3.4.8:

$$F_e = +\frac{\partial W'_m(i, x)}{\partial x} \tag{5.1.14}$$

If, on the other hand, the independent variables are chosen as λ and x, it follows that

$$i = i(\lambda, x) \tag{5.1.15}$$

Table 5.1.1 Mechanical Force of Electrical Origin Caused by the Magnetic Coupling Field

Stored magnetic energy	$W_m = \displaystyle\int_0^\lambda i \, d\lambda$	(3.4.6)
Magnetic coenergy	$W'_m = \displaystyle\int_0^i \lambda \, di$	(3.4.7)
Relation between energy and coenergy	$W_m + W'_m = \lambda i$	(3.4.8)
Conservation of energy principle applied to consevative coupling fields for an arbitrary displacement dx	$F_e \, dx = i \, d\lambda - dW_m$	(5.1.8)

Independent Variables	Electromechanical Coupling Force Evaluated from Stored Magnetic Energy	Electromechanical Coupling Force Evaluated from Magnetic Coenergy
Current i Coordinate x	$F_e = -\dfrac{\partial W_m(i,x)}{\partial x} + \dfrac{\partial \lambda(i,x)}{\partial x}$ (5.1.13)	$F_e = +\dfrac{\partial W'_m(i,x)}{\partial x}$ (5.1.14)
Flux Linkage λ Coordinate x	$F_e = -\dfrac{\partial W_m(\lambda,x)}{\partial x}$ (5.1.19)	$F_e = \dfrac{\partial W'_m(\lambda,x)}{\partial x} - \lambda\dfrac{\partial i(\lambda,x)}{\partial x}$ (5.1.20)

Note: For the case of a rotational electromechanical system, the force F_e and the linear displacement dx are to be replaced by the torque T_e and the angular displacement $d\theta$, respectively.

$$di = \frac{\partial i}{\partial \lambda} \, d\lambda + \frac{\partial i}{\partial x} \, dx \tag{5.1.16}$$

$$dW_m = \frac{\partial W_m}{\partial \lambda} \, d\lambda + \frac{\partial W_m}{\partial x} \, dx \tag{5.1.17}$$

$$F_e \, dx = i \, d\lambda - \frac{\partial W_m}{\partial \lambda} \, d\lambda - \frac{\partial W_m}{\partial x} \, dx \tag{5.1.18}$$

or

$$F_e = -\frac{\partial W_m(\lambda,x)}{\partial x} \tag{5.1.19}$$

Note that W_m in Equation 5.1.19 is a function of independent variables λ and x. Based on Equation 3.4.8, Equation 5.1.9 may also be written as

$$F_e = \frac{\partial W'_m(\lambda,x)}{\partial x} - \lambda\frac{\partial i(\lambda,x)}{\partial x} \tag{5.1.20}$$

Table 5.1.1 summarizes the relations associated with the mechanical force of electrical origin caused by the magnetic coupling field.

Although we have considered a simple system that can be described with only one electrical and one mechanical variable, an electromechanical system in general may need several electrical and mechanical variables for its modeling description, in which case the expressions given in Table 5.1.1 need to be generalized. The results of Table 5.1.1 are completely general and independent of any electrical source variations. A similiar development can be made for determining mechanical forces due to electric field coupling in an electromechanical system.

It should be observed that the force F_e is independent of the variations of λ and i during the arbitrary displacement. However, the changes in λ and i follow the functional relationship given by the

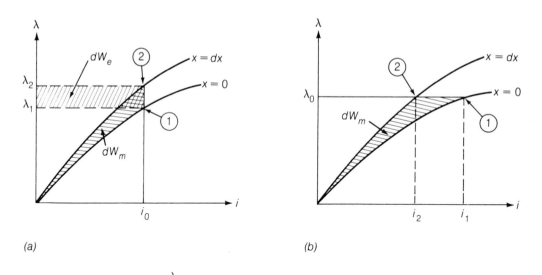

(a)

(b)

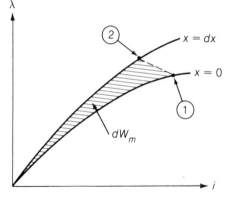

(c)

Figure 5.1.3 Energy balance in a nonlinear electromechanical system. (a) Constant current operation. (b) Constant flux linkage (or voltage) operation. (c) A general case.

λ–i characteristic of the magnetic system, which is also the reason that only one of the two variables (λ or i) can be treated as independent. Since the stored magnetic energy is a state function, it does not matter how the configuration of the system is arrived at.

Figure 5.1.3a illustrates a differential movement of the operating point in a λ–i diagram corresponding to a differential displacement dx of the iron bar of Figure 5.1.2 made at low speed (i.e., at constant current). The increase in coenergy from position 1 to position 2 can be observed, and Equation 5.1.14 can be applied to evaluate F_e readily (with i treated as a constant), when it is more convenient to express λ as a function of i and x.

Figure 5.1.3b illustrates a differential movement of the operating point in the λ–i diagram corresponding to a differential displacement dx of the iron bar of Figure 5.1.2 made at high speed—i.e., at constant flux linkage (or voltage). The decrease in energy from position 1 to position 2 can be observed, and Equation 5.1.19 can be applied to evaluate F_e readily (with λ treated as a constant), when it is more convenient to express the coil current i as a function of λ and x.

It must be emphasized that the holding of i or λ as a constant is a mathematical restriction imposed by the selection of independent coordinates and has nothing to do with the terminal electrical constraints. The generality of the force expressions holds good even when the change from position 1 to position 2 follows a general path (shown in Figure 5.1.3c) in which neither condition of constant i or λ applies.

For a linear magnetic system, however, the λ–i characteristic is a straight line; the magnetic energy and coenergy are always equal in magnitude. Equation 3.5.2 applies:

$$W_m = \frac{1}{2}i\lambda = \frac{Li^2}{2} \tag{3.5.2}$$

from which

$$F_e = \frac{1}{2}i^2\frac{\partial L}{\partial x} = -\frac{1}{2}\lambda\frac{\partial i}{\partial x} \tag{5.1.21}$$

follows from the force expressions given in Table 5.1.1.

It should now be clear that, in order for energy conversion to occur, the electromechanical device must have at least one component that is capable of storing energy and that the amount of energy stored is a function of the space variable. The foregoing analysis is concerned with the force F_e and the linear displacement dx. For the case of a rotational electromechanical system, however, the torque T_e and the angular displacement $d\theta$ must be considered. A few examples will illustrate the application of the force and torque expressions.

EXAMPLE 5.1.1

The λ–i relationship for the electromechanical system shown in Figure 5.1.2 (page 178) is given by

$$\lambda = \frac{0.1i^{1/2}}{g}$$

which holds good within the limits of $0 < i < 4$ A and $0.04 < g < 0.10$ m. Given that the current is maintained at 2 A, find the mechanical force (of electrical origin caused by the magnetic coupling field) on the iron bar for the space variable coordinate $g = 0.06$ m.

Solution
Noting that the λ–i relationship is nonlinear, and that the λ is given as a function of the current i and the space variable coordinate g, it is convenient to apply Equation 5.1.14, for which we need to evaluate the magnetic coenergy W'_m of the system by Equation 3.4.7.

$$W'_m = \int_0^i \lambda\, di = \int_0^i \frac{0.1i^{1/2}}{g}\, di = \frac{0.1}{g}\frac{2}{3}i^{3/2}\ \text{J}$$

$$F_e = \frac{\partial W'_m(i, g)}{\partial g} = \frac{\partial}{\partial g}\left[\frac{0.1}{g}\frac{2}{3}i^{3/2}\right] = -\frac{0.1}{g^2}\frac{2}{3}i^{3/2}$$

For $g = 0.06$ m and $i = 2$ A, we obtain

$$F_e = -\frac{0.1}{0.06^2}\times\frac{2}{3}\times(2)^{3/2} = -52.37\ \text{N}$$

The negative sign indicates that the force F_e acts in a direction that decreases the air-gap length g. Note that a positive displacement dx in Figure 5.1.2 corresponds to a reduction dg in the air-gap length: i.e., $dx = -dg$.

Alternatively, the problem can be solved by expressing i as a function of λ and g, evaluating the magnetic energy W_m, and using Equation 5.1.19:

$$i = \left(\frac{\lambda g}{0.1} \right)^2$$

$$W_m = \int_0^\lambda i \, d\lambda = \int_0^\lambda \frac{g^2}{0.1^2} \lambda^2 \, d\lambda = \frac{g^2}{0.1^2} \frac{\lambda^3}{3}$$

$$F_e = -\frac{\partial W_m(\lambda, g)}{\partial g} = -\frac{\partial}{\partial g} \left(\frac{g^2}{0.1^2} \frac{\lambda^3}{3} \right) = -\frac{\lambda^3}{3} \frac{2g}{0.1^2}$$

When $i = 2$ A and $g = 0.06$ m,

$$\lambda = \frac{0.1 \times 2^{1/2}}{0.06}$$

and

$$F_e = -\frac{0.1^3 \times 2^{3/2}}{0.06^3} \times \frac{1}{3} \times \frac{2 \times 0.06}{0.1^2} = -\frac{0.1}{0.06^2} \times \frac{2}{3} \times 2^{3/2} = -52.37 \text{ N}$$

which is the same as the result obtained before. The selection of the energy or coenergy function as a basis for analysis is a matter of convenience, depending upon the initial description of the system and the desired variables in the result.

EXAMPLE 5.1.2

Let the λ–i relationship for the electromechanical system in Figure 5.1.2 (page 178) be given by

$$i = \lambda^{1/2} + 20\lambda(x - 0.1)^2, \quad x < 0.1 \text{ m}$$

Compute the force on the iron bar at $x = 0.05$ m in terms of λ.

Solution

Since the λ–i relationship is nonlinear, and i is given as a function of λ and x, it is convenient to evaluate $W_m(\lambda, x)$ and apply Equation 5.1.19.

$$W_m = \int_0^\lambda i \, d\lambda = \tfrac{2}{3}\lambda^{3/2} + 10\lambda^2(x - 0.1)^2$$

$$F_e = -\frac{\partial W_m(\lambda, x)}{\partial x} = -20\lambda^2(x - 0.1)$$

At $x = 0.05$ m, $F_e = +\lambda^2$. This shows that the force is proportional to the square of the voltage, assuming that the leakage is neglected, of course. The positive sign indicates that the force F_e acts in such a direction as to increase the coordinate x in its positive direction, or to decrease the air-gap length g. In this problem, it is rather inconvenient to express λ as a function of i and x.

|||||||||||||

EXAMPLE 5.1.3

The magnetic structure shown with dimensions in Figure 5.1.4 is made out of a ferromagnetic material that has negligible reluctance. The rotor is free to rotate about a vertical axis. Neglect leakage and fringing.

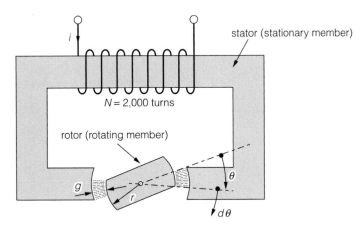

Figure 5.1.4

a. Obtain an expression for the torque acting on the rotor.
b. Calculate the torque for a current of 1.5 amps and the given dimensions.
c. Assuming that the maximum flux density in the air gap is to be limited to 1.5 Wb/m^2 because of saturation in the ferromagnetic structure, compute the maximum torque of the device.

> Axial length perpendicular to the plane of the paper = h = 0.05 m
> Length of a single air gap = g = 0.004 m
> Radius of rotor face = r = 0.04 m

Solution

a. Choosing as independent variables the current i and the space coordinate θ, the expression for the torque is given by

$$T_e = \frac{\partial W'_m(i, \theta)}{\partial \theta}$$

Since the air-gap region is linear, the magnetic energy or coenergy density in the gap is given by Equation 3.4.3:

$$w_m = w'_m = \frac{B_g^2}{2\mu_0} = \frac{\mu_0 H_g^2}{2} \text{ J/m}^3$$

where B_g is the air-gap flux density, H_g is the air-gap field intensity, and μ_0 is the permeability of free space. Noting that the total length of the air gap is $2g$, the volume of the overall air-gap region is calculated as

$$2gh(r + 0.5g)\theta \text{ m}^3$$

where θ is the angle in radians between the stator-pole tip and the adjacent rotor-pole tip, as shown in Figure 5.1.4, and $[(r + 0.5g)\theta]$ is the mean arc length in the air gap. Then

$$W'_m = \mu_0 H_g^2 gh(r + 0.5g)\theta \text{ J}$$

$$T_e = \mu_0 H_g^2 gh(r + 0.5g) = \frac{B_g^2}{\mu_0} gh(r + 0.5g) \text{ N} \cdot \text{m}$$

The torque acts in such a direction as to align the rotor pole faces with the stator pole faces, in the positive direction of θ, as shown. The relationship between the current and the air-gap field intensity H_g is given by

$$Ni = 2gH_g \quad \text{or} \quad H_g = \frac{Ni}{2g}$$

where N is the number of stator-coil turns as shown in Figure 5.1.4. Using the above equation, the mechanical torque T_e of electrical origin caused by the magnetic coupling field can be expressed as

$$T_e = \frac{\mu_0 N^2 i^2}{4g^2} gh(r + 0.5g) \quad \text{or} \quad T_e = \frac{i^2}{2}\left[\frac{\mu_0 N^2}{2g}h(r + 0.5g)\right]$$

For a linear magnetic system, T_e can also be calculated from Equation 5.1.21 by simply replacing F_e and dx by T_e and $d\theta$, respectively:

$$T_e = \frac{i^2}{2} \cdot \frac{dL}{d\theta}$$

For our problem,

$$L = \frac{N^2}{\mathbf{R}} = \frac{\mu_0 A N^2}{l}$$

where the cross-sectional area of the air-gap region, $A = h(r + 0.5g)\theta$, and the total air-gap length, $l = 2g$. The same result for T_e is obtained by substituting and working out the details.

b. The self-inductance L in terms of θ is given by

$$L = \frac{(4\pi \times 10^{-7}) \times 0.05(0.04 + 0.002) \times 2,000^2}{0.008}\theta \text{ H} \quad \text{or} \quad L = 1.32\theta$$

$$T_e = \frac{i^2}{2} \cdot \frac{dL}{d\theta} = \frac{(1.5)^2}{2} \times 1.32 = 1.485 \text{ N} \cdot \text{m}$$

Note that the torque acts to increase the inductance by pulling on the rotor so as to reduce the reluctance of the magnetic path linking the coil.

The air-gap flux density B_g corresponding to a current of 1.5 amps is given by

$$B_g = \mu_0 H_g = \frac{\mu_0 Ni}{2g} = \frac{(4\pi \times 10^{-7}) \times 2,000 \times 1.5}{2 \times 0.004} = 0.47 \text{ T}$$

c. Corresponding to $B_{g\max}$ of 1.5 Wb/m^2, the current is calculated as

$$i_{\max} = \frac{B_g(2g)}{\mu_0 N} = \frac{1.5 \times 0.008}{(4\pi \times 10^{-7})2,000} = 4.77 \text{ A}$$

$$T_{e\max} = \frac{i^2}{2} \cdot \frac{dL}{d\theta} = \frac{4.77^2}{2} \times 1.32 = 15.02 \text{ N} \cdot \text{m}$$

This may also be obtained by directly substituting in the torque expression given in terms of B_g.

5.2 Singly Excited and Multiply Excited Magnetic Field Systems

Before we proceed to analyze magnetic field systems excited by more than one electrical circuit, let us consider an elementary *reluctance machine* that is singly excited—carrying only one winding on its stationary member—called the *stator*. The device is very similar to the one analyzed in Example 5.1.3, but now we will consider sinusoidal excitation to the stator winding, while the rotor is free to rotate about an axis normal to the plane of the paper. For notational convenience, the diagram in Figure 5.1.4 is redrawn as Figure 5.2.1a, showing an elementary rotating reluctance machine. We will assume that the reluctance of the stator and rotor iron are negligible. We will also neglect leakage and fringing.

The stator and rotor poles are so shaped that the reluctance varies sinusoidally about a mean value as shown in Figure 5.2.1b. $\mathbf{R}_d$ is the reluctance of the magnetic system when the rotor is in the direct-axis position ($\theta = 0$), and $\mathbf{R}_q$ is the reluctance when the rotor is in the quadrature-axis position ($\theta = \pi/2$). For each revolution of the rotor, there are two cycles of reluctance. The space variation of inductance is also of double frequency, since the inductance is inversely proportional to the reluctance. The inductance of the stator winding as a function of space coordinate θ measured from the direct axis, as shown in Figure 5.2.1a, is given by

$$L(\theta) = L_0 + L_2 \cos 2\theta \tag{5.2.1}$$

This is sketched in Figure 5.2.1c. Let the stator coil excitation be

$$i_s = I_s \sin \omega_s t \tag{5.2.2}$$

We will investigate the instantaneous and average electromagnetic torques produced because of this sinusoidal excitation, whose angular frequency is ω_s.

The electromagnetic torque can be found by Equation 5.1.14 from the coenergy in the magnetic field of the air-gap region, since the independent variables are the current i and the space coordinate θ. Since the air-gap region is linear, Equations 3.5.2 and 5.1.21 apply:

$$W'_m = W_m = \frac{1}{2}L(\theta)i_s^2 \tag{5.2.3}$$

and

$$T_e = \frac{\partial W'_m(i_s, \theta)}{\partial \theta} = \frac{1}{2}i_s^2 \frac{\partial L(\theta)}{\partial \theta} \tag{5.2.4}$$

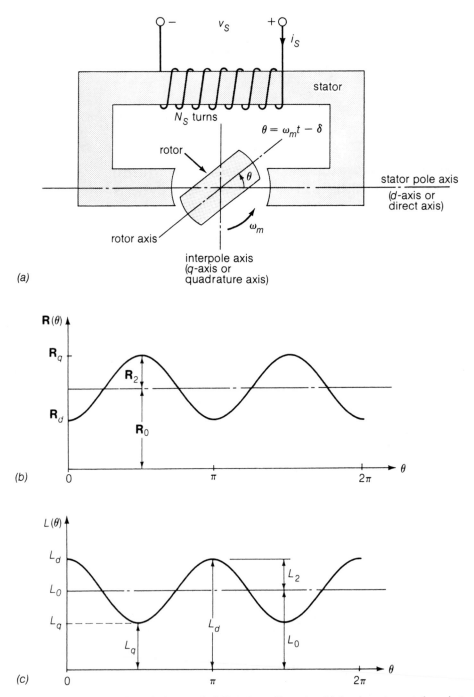

(a)

(b)

(c)

Figure 5.2.1 A singly-excited magnetic field system with a rotor. (a) An elementary rotating reluctance machine. (b) Reluctance variation with rotor position. (c) Inductance variation with rotor position.

Substituting the current and inductance variations, we obtain

$$T_e = -I_s^2 L_2 \sin 2\theta \sin^2 \omega_s t \tag{5.2.5}$$

Let the rotor now be allowed to rotate at an angular velocity ω_m, so that at any instant θ is given by

$$\theta = \omega_m t - \delta \tag{5.2.6}$$

where $\theta = -\delta$ is the angular position of the rotor at $t = 0$, when the current i_s is zero.

Equation 5.2.5 for the instantaneous torque as a function of time can be rearranged in terms of ω_m and ω_s by using the trigonometric identities

$$\sin^2 A = \frac{1 - \cos 2A}{2}$$

and

$$\sin A \cos B = \frac{1}{2} \sin (A + B) + \frac{1}{2} \sin (A - B)$$

The instantaneous electromagnetic torque is then given by

$$T_e = -\frac{1}{2} I_s^2 L_2 \left\{ \sin 2(\omega_m t - \delta) - \frac{1}{2} \sin 2 \left[(\omega_m + \omega_s)t - \delta \right] - \frac{1}{2} \sin 2 \left[(\omega_m - \omega_s)t - \delta \right] \right\} \tag{5.2.7}$$

This torque expression consists of three sinusoidally time-varying terms of frequencies $2\omega_m$, $2(\omega_m + \omega_s)$, and $2(\omega_m - \omega_s)$. The time-average value of these three sine terms is zero unless the coefficient of t becomes zero in one of them. Since $\omega_m \neq 0$, the necessary condition for nonzero time-average torque is then

$$|\omega_m| = |\omega_s| \tag{5.2.8}$$

corresponding to which

$$(T_e)_{av} = -\frac{1}{4} I_s^2 L_2 \sin 2\delta \tag{5.2.9}$$

Expressed in terms of L_d and L_q, the maximum and minimum values of inductance—shown in Figure 5.2.1c—are known as the direct-axis inductance and the quadrature-axis inductance, respectively,

$$(T_e)_{av} = -\frac{1}{8} I_s^2 (L_d - L_q) \sin 2\delta \tag{5.2.10}$$

Several conclusions can be drawn from a close examination of the preceding analysis. The machine develops an average torque only at one particular speed—given by Equation 5.2.8—for either direction of rotation. This speed, known as the *synchronous speed,* exists when the speed of mechanical rotation in radians per second is equal to the angular frequency of the electrical source. Since the torque in this particular electromechanical energy converter is due to the variation of reluctance with rotor position, the device is known as a *synchronous reluctance machine.* As seen from Equation 5.2.10, the torque is zero if $L_d = L_q$; i.e., if there is no inductance or reluctance variation with rotor position.

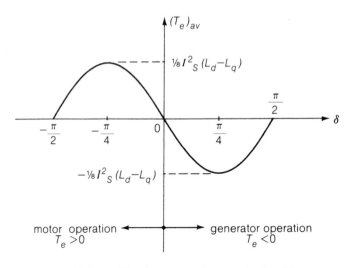

Figure 5.2.2 Variaton of the electromagnetic torque developed by a synchronous reluctance machine.

Figure 5.2.2 shows the variation of the average electromagnetic torque developed by the machine as a function of the angle δ, the *torque angle*. Angle δ gives a measure of the torque. For $\delta < 0$, $(T_e)_{av} > 0$. That is, the developed torque acts in the direction of rotation, and the machine operates as a motor. This torque tends to maintain the speed of the rotor against friction, windage, and any external-load torque applied to the rotor shaft. The maximum torque for motor operation occurs at $\delta = -\pi/4$ and it is known as the *pull-out torque*. Any load requiring a torque greater than the maximum torque results in unstable operation of the machine; the machine pulls out of synchronism and comes to a stand still.

For $\delta > 0$, $(T_e)_{av} < 0$. That is, the developed torque opposes the rotation. An external driving torque drives the machine against the electromagnetically developed torque to maintain the rotor at synchronous speed. The mechanical energy supplied to the system, after meeting the friction and windage losses, is converted to electrical energy. In other words, the machine operates as a generator. However, this can occur only if the stator winding is already connected to an ac source, which becomes a sink as the external driving torque is applied and the machine starts to generate. The maximum torque for the generator operation occurs at $\delta = \pi/4$. If the driving torque externally applied to the shaft exceeds the sum of the developed torque and that due to friction and windage, then the machine tends to be driven above synchronous speed and it may run away unless the speed of the prime mover is controlled. Under such conditions, however, continuous energy conversion ceases. The machine can develop only a certain maximum power, and there is a limit to the rate of energy conversion.

It should be emphasized that the singly excited synchronous reluctance motor cannot start by itself. Also, under excitation from a source of constant amplitude and sinusoidally time-varying voltage, the coil current can be shown to contain a third-harmonic component because of the variation of reluctance with rotor position. Similarly, with sinusoidal current excitation given by Equation 5.2.2, the induced voltage in the stator coil will contain a third-harmonic component. This undesirable feature makes reluctance machines less useful as practical generators and limits their size as motors. However, reluctance machines (modified to develop starting torque) are popularly used to drive electric clocks, record players, and other small mechanisms that need constant speed.

EXAMPLE 5.2.1

In continuation with the analysis of the elementary rotating reluctance machine presented above, and with the assumed current-source excitation of Equation 5.5.2, find the voltage induced in the stator winding at synchronous speed.

Solution

$$e_s = p\lambda_s = \frac{d}{dt}[L(\theta)i_s]$$

where p is the time-derivative operator (d/dt). Noting that L is not a constant but a function of θ, which in turn is a function of time, we have

$$e_s = \frac{d}{dt}[(L_0 + L_2 \cos 2\theta)(i_s)]$$

Applying the product rule for differentiation, we get

$$e_s = (L_0 + L_2 \cos 2\theta)(pi_s) - 2L_2 i_s \sin 2\theta (p\theta)$$

The second term on the right side is the speed–voltage term caused by mechanical motion, proportional to the instantaneous speed. Including the current variation with time, we get

$$e_s = (L_0 + L_2 \cos 2\theta)(\omega_s I_s \cos \omega_s t) - 2L_2(I_s \sin \omega_s t) \sin 2\theta (p\theta)$$

Substituting $\theta = \omega_m t - \delta$, $p\theta = \frac{d\theta}{dt} = \omega_m$, and $\omega_m = \omega_s$,

$$e_s = [L_0 + L_2 \cos 2(\omega_s t - \delta)](\omega_s I_s \cos \omega_s t) - 2L_2 \omega_s I_s \sin \omega_s t \sin 2(\omega_s t - \delta)$$

Rearranging with the aid of these trigonometric identities:

$$\cos A \cos B = \frac{1}{2}\cos(A + B) + \frac{1}{2}\cos(A - B)$$

$$\sin A \sin B = \frac{1}{2}\cos(A - B) - \frac{1}{2}\cos(A + B)$$

we obtain

$$e_s = \omega_s L_0 I_s \cos \omega_s t$$
$$+ \frac{1}{2}\omega_s L_2 I_s[\cos(3\omega_s t - 2\delta) + \cos(\omega_s t - 2\delta)]$$
$$- \omega_s L_2 I_s[\cos(\omega_s t - 2\delta) - \cos(3\omega_s t - 2\delta)]$$

or

$$e_s = \omega_s L_0 I_s \cos \omega_s t - \frac{1}{2}\omega_s L_2 I_s \cos(\omega_s t - 2\delta) + \frac{3}{2}\omega_s L_2 I_s \cos(3\omega_s t - 2\delta)$$

which has a third-harmonic component. The flux will also contain a third-harmonic component.

||||||||||||█████████████████████████████

EXAMPLE 5.2.2

Consider the elementary rotating reluctance machine shown in Figure 5.2.1 (page 186). The reluctance variation is given by

$$\mathbf{R}(\theta) = \mathbf{R}_0 - \mathbf{R}_2 \cos 2\theta$$

Let the stator be excited from a voltage source

$$v_s = V_s \sin \omega_s t$$

The resitance of the winding may be neglected.

a. Develop an expression for the instantaneous electromagnetic torque T_e in terms of the flux ϕ in the magnetic circuit and the reluctance.

b. Let $\theta = \omega_m t + \theta_0$ where ω_m is the angular velocity of the rotor. Obtain the expression for T_e in terms of ω_m, ω_s, and θ_0.

c. Determine the necessary condition for nonzero time-average torque, and find the corresponding expression for the average torque $(T_e)_{av}$.

d. Show that the coil current contains a third-harmonic component when the machine is operating at synchronous speed.

Solution

a. Treating the flux linkage λ and the coordinate θ as independent variables, the electromagnetic torque is given by

$$T_e = -\frac{\partial W_m(\lambda, \theta)}{\partial \theta}$$

Since $W_m = \frac{1}{2}\phi^2 \mathbf{R}(\theta)$ for a linear system, it follows that

$$T_e = -\frac{1}{2}\phi^2 \frac{d\mathbf{R}(\theta)}{d\theta}$$

Compare the above expressions to those of Equations 5.2.3 and 5.2.4, which show that the torque always acts in such a direction that the resulting rotation increases the inductance or decreases the reluctance. For the given reluctance variation, we obtain

$$T_e = -\frac{1}{2}\phi^2 \frac{d}{d\theta}(\mathbf{R}_0 - \mathbf{R}_2 \cos 2\theta) \quad \text{or} \quad T_e = -\phi^2 \mathbf{R}_2 \sin 2\theta$$

b. Neglecting the resistance of the electric circuit, the flux in the magnetic circuit is related to the voltage v_s through the relation

$$v_s = N_s \frac{d\phi}{dt}$$

or, ϕ is given by

$$\phi = -\phi_{max} \cos \omega_s t$$

where $\phi_{max} = V_s/N_s\omega_s$, and the constant of integration may be assumed as zero in the steady-state analysis of linear circuits. Substituting the time variation of ϕ and θ in the expression for T_e, we get

$$T_e = -\phi_{max}^2 \cos^2 \omega_s t \mathbf{R}_2 \sin 2(\omega_m t + \theta_0)$$

This can be rearranged using the trigonometric identities:

$$\cos^2 A = \frac{1 + \cos 2A}{2} \qquad\qquad \cos A \sin B = \frac{1}{2}[\sin(A + B) - \sin(A - B)]$$

Then

$$T_e = -\frac{\mathbf{R}_2 \phi_{max}^2}{2} \left\{ \sin 2(\omega_m t + \theta_0) + \frac{1}{2} \sin 2[(\omega_m + \omega_s)t + \theta_0] + \frac{1}{2} \sin 2[(\omega_m - \omega_s)t + \theta_0] \right\}$$

c. Since

$$(T_e)_{av} = \frac{1}{2\pi} \int_0^{2\pi} T_e \, d\theta$$

the average value of the three sinusoidally time-varying terms is zero unless the coefficient of t is zero in one of them. Because $\omega_m \neq 0$, the necessary condition for nonzero time-average torque is given by

$$|\omega_m| = |\omega_s|$$

corresponding to which

$$(T_e)_{av} = -\frac{\mathbf{R}_2 \phi_{max}^2}{4} \sin 2\theta_0 = -\frac{(\mathbf{R}_d - \mathbf{R}_q)\phi_{max}^2}{8} \sin 2\theta_0$$

d. The current is related to the flux by

$$\phi = N_s i_s / \mathbf{R}$$

For the condition $\omega_m = \omega_s$ (i.e., at synchronous speed),

$$i_s = -\frac{\phi_{max}}{N_s} \cos \omega_s t [\mathbf{R}_0 - \mathbf{R}_2 \cos 2(\omega_s t + \theta_0)]$$

By using the identitiy

$$\cos A \cos B = \frac{1}{2} \cos(A + B) + \frac{1}{2} \cos(A - B)$$

we have

$$i_s = -\frac{\phi_{max} \mathbf{R}_0}{N_s} \cos \omega_s t + \frac{\phi_{max} \mathbf{R}_2}{2N_s} [\cos(\omega_s t + 2\theta_0) + \cos(3\omega_s t + 2\theta_0)]$$

which shows that the coil current contains a third-harmonic component.

Energy conversion implies rotation for machines, because power is the product of a torque and a speed, and energy is converted only when the speed is nonzero. The equation of mechanical motion (or the torque equation) describing a rotating electromechanical energy converter is of the form

$$T = J\ddot{\theta} + \alpha \dot{\theta} + K\theta - T_e \tag{5.2.11}$$

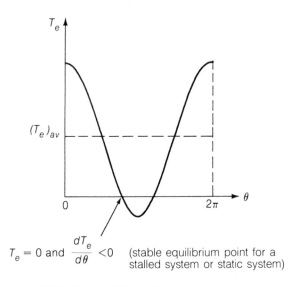

$$T_e = 0 \text{ and } \frac{dT_e}{d\theta} < 0 \quad \text{(stable equilibrium point for a stalled system or static system)}$$

Figure 5.2.3 Torque variation with angle.

where T is the torque applied externally to the shaft, T_e is the electromagnetic torque developed internally and applied to the shaft by the electrical system, J is the rotor moment of inertia, α is the rotor damping factor, K is the torsional spring constant, $\ddot{\theta}$ is $d^2\theta/dt^2$, $\dot{\theta}$ is $d\theta/dt$, and θ is the angular displacement. The power converted from electrical to mechanical form is

$$p_{em} = T_e \dot{\theta} = T_e \omega_m \tag{5.2.12}$$

and the mechanical power flowing into the machine from external mechanical systems attached to the shaft is

$$p_m = T\dot{\theta} = T\omega_m \tag{5.2.13}$$

Considering $\dot{\theta}$ to be positive—that is, for positive rotation—motor action is defined by a conversion of electrical power to mechanical power ($p_{em} > 0$ and $T_e > 0$), and generator action is defined by a conversion from mechanical power to electrical power ($p_{em} < 0$ and $T_e < 0$).

With no externally applied shaft torque T and with zero spring constant K, the condition for *starting* a motor in the positive direction is given by

$$T_e > 0 \quad \text{for} \quad 0 \le \theta \le 2\pi \tag{5.2.14}$$

When Equation 5.2.14 is not satisfied, the machine may come to rest when $T_e = 0$, and stable equilibrium is achieved when $dT_e/d\theta < 0$, as shown in Figure 5.2.3. Since all physical machines have rotor inertia and since kinetic energy is stored in the rotating inertia of the rotor, the condition for successful *running* of a machine is less restrictive than that of starting it (Equation 5.2.14) and is given by

$$(T_e)_{av} = \frac{1}{2\pi} \int_0^{2\pi} T_e \, d\theta > 0 \tag{5.2.15}$$

which implies that the *average* torque be greater than zero over a revolution. Thus a machine with the torque-angle characteristic shown in Figure 5.2.3 will run successfully.

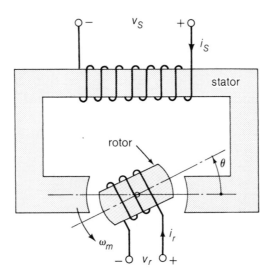

Figure 5.2.4 An elementary doubly excited magnetic-field system.

The synchronous reluctance machine of Figure 5.2.1 would work equally well even if the exciting coil were placed on the rotor instead of on the stator. Singly excited devices are usually employed to produce uncontrolled bulk forces in such devices as relays, solenoids, and force actuators. To obtain forces proportional to electrical signals and signals proportional to forces and velocities, the devices must be *multiply excited* (i.e., having two or more paths for excitation and exchange of energy with sources). Permanent magnets may be used as one of the excitation sources. While one excitation path sets the level of the electrical or magnetic field, the other can be made to handle signals. Loudspeakers, torque motors, and tachometers are examples of signal-handling devices, while most of the known types of motors and generators are examples of continuous energy conversion and power handling.

Multiply Excited Magnetic-Field Systems

Let us consider an elementary multiply excited magnetic system, shown in Figure 5.2.4, with two sets of electrical terminals and one mechanical terminal. This model is the same as in Figure 5.2.1 but with the addition of the rotor coil connected to its electrical source by means of *carbon brushes* bearing on *slip rings* (or *collector rings*). The flux linkages of the stator and rotor windings can be expressed as functions of the coil curents:

$$\lambda_s = L_{ss}i_s + L_{sr}i_r \tag{5.2.16}$$

$$\lambda_r = L_{sr}i_s + L_{rr}i_r \tag{5.2.17}$$

where L_{ss} and L_{rr} are the self-inductances of the stator and rotor coils respectively, and L_{sr} is the stator-rotor mutual inductance. All these inductances are generally functions of the angle θ between the magnetic axes of the stator and rotor windings. The equations of motion for the electrical side (i.e., the volt-ampere equations) are given by

$$v_s = R_s i_s + p\lambda_s = R_s i_s + L_{ss}(pi_s) + i_s(pL_{ss}) + L_{sr}(pi_r) + i_r(pL_{sr}) \tag{5.2.18}$$

$$v_r = R_r i_r + p\lambda_r = R_r i_r + L_{sr}(pi_s) + i_s(pL_{sr}) + L_{rr}(pi_r) + i_r(pL_{rr}) \tag{5.2.19}$$

Given two sets of electrical terminals, two volt-ampere equations result, as in Equations 5.2.18 and 5.2.19. Neglecting the reluctances of the stator- and rotor-iron circuits, the electromagnetic torque can be found from either the energy or the coenergy stored in the magnetic field of the air-gap region:

$$T_e = -\frac{\partial W_m(\lambda_s, \lambda_r, \theta)}{\partial \theta} = +\frac{\partial W'_m(i_s, i_r, \theta)}{\partial \theta} \tag{5.2.20}$$

For a linear system, the energy or coenergy stored in a pair of mutually coupled inductors is given by Equation 3.5.14. (Also see Problem 3–14).

$$W'_m(i_1, i_2, \theta) = \frac{1}{2}L_{ss}i_s^2 + L_{sr}i_s i_r + \frac{1}{2}L_{rr}i_r^2 \tag{5.2.21}$$

The instantaneous electromagnetic torque is then given by

$$T_e = \frac{i_s^2}{2}\frac{dL_{ss}}{d\theta} + i_s i_r \frac{dL_{sr}}{d\theta} + \frac{i_r^2}{2}\frac{dL_{rr}}{d\theta} \tag{5.2.22}$$

The first and third terms on the right-hand side of Equation 5.2.22, involving angular rate of change of self-inductance, are the reluctance-torque terms; the middle term, involving angular rate of change of mutual inductance, is the torque caused by the interaction of fields produced by the stator and rotor currents. It is this mutual-inductance torque that is most commonly exploited in practical rotating machines. Multiply excited systems with more than two sets of electrical terminals can be handled in manner similar to that for two pairs by assigning additional independent variables to the terminals.

If the self-inductances L_{ss} and L_{rr} are independent of angle θ, the reluctance torque is zero and the torque is produced only by the mutual term $L_{sr}(\theta)$, as seen from Equation 5.2.22. Consider such a case in the following example.

███████████
EXAMPLE 5.2.3

Consider an elementary two-pole rotating machine with a uniform (or smooth) air gap as shown in Figure 5.2.5, in which the cylindrical rotor is mounted within the stator made up of a hollow cylinder coaxial with the rotor. The stator and rotor windings are distributed over a number of slots so that their mmf can be approximated by space sinusoids. As a consequence of a construction of this type, we can fairly assume that the self-inductances L_{ss} and L_{rr} are constant, but the mutual L_{sr} is given by

$$L_{sr} = L \cos \theta$$

where θ is the angle between the magnetic axes of the stator and rotor windings. Let the currents in the two windings be given by

$$i_s = I_s \cos \omega_s t \quad \text{and} \quad i_r = I_r \cos(\omega_r t + \alpha)$$

and let the rotor rotate at an angular velocity

$$\omega_m = \dot{\theta} \quad \text{rad/s}$$

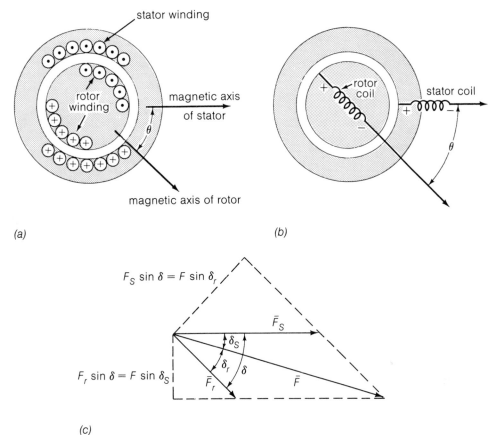

Figure 5.2.5 An elementary two-pole rotating machine with a uniform air gap.

such that the position of the rotor at any instant is given by

$$\theta = \omega_m t + \theta_0$$

Assume that the reluctances of the stator and rotor-iron circuits are negligible, and that the stator and rotor are concentric cylinders neglecting the effect of slot openings.

a. Derive an expression for the instantaneous electromagnetic torque developed by the machine.
b. Find the necessary condition for the developement of an average torque in the machine.
c. Obtain the expression for the average torque corresponding to the following cases:
  (1) $\omega_s = \omega_r = \omega_m = 0$; $\alpha = 0$
  (2) $\omega_s = \omega_r$; $\omega_m = 0$
  (3) $\omega_r = 0$; $\omega_s = \omega_m$; $\alpha = 0$
  (4) $\omega_m = \omega_s - \omega_r$,   where ω_s and ω_r are different angular frequencies

d. With the assumed current-source excitations of part c, determine the voltages induced in the stator and rotor windings at the corresponding angular velocity ω_m at which an average torque results.

Solution

a. Equations 5.2.16 through 5.2.22 apply. With constant L_{ss} and L_{rr} and the variation of L_{sr} as a function of θ substituted, Equation 5.2.22 simplifies to

$$T_e = i_s i_r \frac{dL_{sr}}{d\theta} = -i_s i_r L \sin \theta$$

For the given current variations, the instantaneous electromagnetic torque developed by the machine is given by

$$T_e = -LI_s I_r \cos \omega_s t \cos (\omega_r t + \alpha) \sin (\omega_m t + \theta_0)$$

Using the trigonometric identities, the product of the three trigonometric terms in this equation may be expressed to yield

$$T_e = \frac{-LI_s I_r}{4} \Big\{ \sin \{[\omega_m + (\omega_s + \omega_r)]t + \alpha + \theta_0\} + \sin \{[\omega_m - (\omega_s + \omega_r)]t - \alpha + \theta_0\}$$
$$+ \sin \{[\omega_m + (\omega_s - \omega_r)]t - \alpha + \theta_0\} + \sin \{[\omega_m - (\omega_s - \omega_r)]t + \alpha + \theta_0\} \Big\}$$

b. The average value of each of the sinusoidal terms in the previous equation is zero, unless the coefficient of t is zero in that term. That is, the average torque $(T_e)_{av}$ developed by the machine is zero unless

$$\omega_m = \pm(\omega_s \pm \omega_r)$$

which may be also expressed as

$$|\omega_m| = |\omega_s \pm \omega_r|$$

c. (1) The excitations are direct currents I_s and I_r. For the given conditions of $\omega_s = \omega_r = \omega_m = 0$ and $\alpha = 0$,

$$T_e = -LI_s I_r \sin \theta_0$$

which is a constant. That being so,

$$(T_r)_{av} = -LI_s I_r \sin \theta_0$$

The machine operates as a *dc rotary actuator,* developing a constant torque against any displacement θ_0 produced by an external torque applied to the rotor shaft.

(2) With $\omega_s = \omega_r$, both excitations are alternating currents of the same frequency. For the conditions $\omega_s = \omega_r$ and $\omega_m = 0$,

$$T_e = -\frac{LI_s I_r}{4}[\sin (2\omega_s t + \alpha + \theta_0) + \sin (- 2\omega_s t - \alpha + \theta_0)$$
$$+ \sin (- \alpha + \theta_0) + \sin (\alpha + \theta_0)]$$

The machine operates as an *ac rotary actuator,* and the developed torque is fluctuating. The average value of the torque is

$$(T_e)_{av} = -\frac{LI_s I_r}{2} \sin \theta_0 \cos \alpha$$

Note that α becomes zero if the two windings are connected in series, in which case $\cos \alpha$ becomes unity.

(3) With $\omega_r = 0$, the rotor excitation is a direct current I_r. For the conditions $\omega_r = 0, \omega_s = \omega_m$, and $\alpha = 0$,

$$T_e = -\frac{LI_sI_r}{4}[\sin(2\omega_st + \theta_o) + \sin(\theta_0) + \sin(2\omega_st + \theta_0) + \sin(\theta_0)]$$

or

$$T_e = -\frac{LI_sI_r}{2}[\sin(2\omega_st + \theta_0) + \sin\theta_0]$$

The device operates as an idealized *single-phase synchronous machine,* and the instantaneous torque is pulsating. The average value of the torque is

$$(T_e)_{av} = -\frac{LI_sI_r}{2}\sin\theta_0$$

since the average value of the double-frequency sine term is zero. If the machine is brought up to *synchronous* speed ($\omega_m = \omega_s$), an average unidirectional torque is established. Continuous energy conversion takes place at synchronous speed. Note that the machine is not self-starting , since an average unidirectional torque is not developed at $\omega_m = 0$ with the specified electrical excitations.

(4) With $\omega_m = \omega_s - \omega_r$, the instantaneous torque is given by

$$T_e = -\frac{LI_sI_r}{4}[\sin(2\omega_st + \alpha + \theta_0) + \sin(-2\omega_rt - \alpha + \theta_0) \\ + \sin(2\omega_st - 2\omega_rt - \alpha + \theta_0) + \sin(\alpha + \theta_0)]$$

The machine operates as a *single-phase induction machine,* and the instantaneous torque is pulsating. The average value of the torque is

$$(T_e)_{av} = -\frac{LI_sI_r}{4}\sin(\alpha + \theta_0)$$

If the machine is brought up to a speed of $\omega_m = \omega_s - \omega_r$, an average unidirectional torque is established, and continuous energy conversion takes place at the *asynchronous* speed of ω_m. Again, note that the machine is not self-starting, since an average unidirectional torque is not developed at $\omega_m = 0$ with the specified electrical excitations.

The pulsating torque, which may be acceptable in small machines, is in general an undesirable feature in a rotating machine working as either a generator or a motor, since it may result in speed fluctuation, vibration, noise, and waste of energy. In magnetic-field systems excited by single-phase alternating sources, the torque pulsates while the speed is relatively constant; consequently, pulsating power becomes a feature. This calls for improvement; in fact, by employing polyphase windings and polyphase sources, constant power is developed in a balanced system.

d. The flux-linkage relations are given by

$$\lambda_s = L_{ss}i_s + L_{sr}(\theta)i_r = L_{ss}i_s + Li_r\cos\theta; \qquad \lambda_r = L_{sr}(\theta)i_s + L_{rr}i_r = Li_s\cos\theta + L_{rr}i_r$$

in which L_{ss}, L_{rr}, and L are constants. The volt-ampere equations are then given by

$$v_s = R_si_s + p\lambda_s; \qquad v_r = R_ri_r + p\lambda_r$$

where R_s and R_r are the winding resistances, and p is the time-derivative operator d/dt.

Substituting for λ_s and λ_r, and recognizing that θ is a variable and a function of time, we obtain

$$v_s = R_s i_s + L_{ss} p i_s + L \cos \theta (p i_r) - L i_r \sin \theta (p\theta)$$
$$v_r = R_r i_r + L_{rr} p i_r + L \cos \theta (p i_s) - L i_s \sin \theta (p\theta)$$

where $p\theta$ is the instantaneous speed ω_m. The fourth term on the right-hand side of each equation is caused by mechanical motion and is proportional to the instantaneous speed. These are the speed-voltage terms, which are the coupling terms relating the interchange of power between electrical and mechanical systems.

The voltages induced in the stator and rotor windings will now be found for each of the parts in (c).

(1) $i_s = I_s$; $i_r = I_r$; $\omega_m = 0$ so that

$$e_s = 0 \quad \text{and} \quad e_r = 0.$$

(2) $i_s = I_s \cos \omega_s t$; $i_r = I_r \cos (\omega_s t + \alpha)$; $\omega_m = 0$ so that

$$e_s = -\omega_s L_{ss} I_s \sin \omega_s t - \omega_s L I_r \cos \theta_0 \sin (\omega_s t + \alpha)$$
$$e_r = -\omega_s L_{rr} I_r \sin (\omega_s t + \alpha) - \omega_s L I_s \sin \omega_s t \cos \theta_0$$

(3) $i_s = I_s \cos \omega_s t$; $i_r = I_r$; $\omega_m = \omega_s$ so that

$$e_s = -\omega_s L_{ss} I_s \sin \omega_s t - \omega_s L I_r \sin (\omega_s t + \theta_0)$$
$$e_r = -\omega_s L I_s \sin \omega_s t \cos (\omega_s t + \theta_0) - \omega_s L I_s \cos \omega_s t \sin (\omega_s t + \theta_0)$$

or

$$e_r = -\omega L I_s \sin (2\omega_s t + \theta_0)$$

Note that the stator current induces a double-frequency voltage in the rotor circuit.

(4) $i_s = I_s \cos \omega_s t$; $i_r = I_r \cos (\omega_s t + \alpha)$; $\omega_m = \omega_s - \omega_r$ so that

$$e_s = -\omega_s L_{ss} I_s \sin \omega_s t - \omega_r L I_r \sin (\omega_r t + \alpha) \cdot \cos [(\omega_s - \omega_r)t + \theta_0]$$
$$-(\omega_s - \omega_r) L I_r \cos (\omega_r t + \alpha) \cdot \sin [(\omega_s - \omega_r)t + \theta_0]$$

or

$$e_s = -\omega_s L_{ss} I_s \sin \omega_s t$$
$$-\frac{\omega_s L I_r}{2} \{\sin (\omega_s t + \alpha + \theta_0) - [\sin (-\omega_s + 2\omega_r)t + \alpha - \theta_0]\}$$
$$\cdots \omega_r L I_r \sin [(-\omega_s + 2\omega_r)t + \alpha - \theta_0]$$
$$e_r = -\omega_r L_{rr} I_r \sin (\omega_r t + \alpha)$$
$$-\omega_s L I_s \sin \omega_s t \cdot \cos [(\omega_s - \omega_r)t + \theta_0]$$
$$-(\omega_s - \omega_r) L I_s \cos \omega_s t \sin [(\omega_s - \omega_r)t + \theta_0]$$

or

$$e_r = - \omega_r L_{rr} I_r \sin(\omega_r t + \alpha)$$

$$+ \frac{\omega_r LI_s}{2} \{\sin[(2\omega_s - \omega_r)t + \theta_0] - \sin(\omega_r t - \theta_0)\}$$

$$- \omega_s LI_s \sin[(2\omega_s - \omega_r)t + \theta_0]$$

Example 5.2.3 considered a two-pole rotating machine. When a machine has more than two poles, only a single pair of poles needs to be considered because the electric, magnetic, and mechanical conditions associated with every other pole-pair are repetitions of those for the pole-pair under consideration. The angle subtended by one pair of poles in a P-pole machine (or one cycle of flux distribution) is defined to be 360 *electrical degrees* or 2π *electrical radians*. So the relationship between the mechanical angle θ_m and the angle θ in electrical units is given by

$$\theta = \frac{P}{2}\theta_m \tag{5.2.23}$$

because one complete revolution has $P/2$ complete wavelengths (or cycles). In view of this relationship, Equation 5.2.20 for the electromagnetic torque must be modified as

$$T_e = \frac{\partial W'_m(i_s, i_r, \theta_m)}{\partial \theta_m} = \frac{\partial W'_m(i_s, i_r, \theta)}{\partial \theta} \frac{d\theta}{d\theta_m} \tag{5.2.24}$$

where T_e is the electromagnetic torque acting in the positive direction of θ_m. The derivative is to be taken with respect to the actual mechanical angle θ_m because we are dealing with mechanical variables. Thus, if Example 5.2.3 were a P-pole machine, the expression for torque would be

$$T_e = -\frac{P}{2}Li_s i_r \sin\left(\frac{P}{2}\theta_m\right) \tag{5.2.25}$$

The negative sign means that the electromagnetic torque acts in a direction that brings the magnetic fields of the stator and rotor into alignment.

The voltage and torque equations for the idealized elementary machine of Example 5.2.3 with a uniform air gap have been derived on the basis of the *coupled-circuit viewpoint*. These equations can also be obtained from the *magnetic-field viewpoint*, which we explore next. Since the mmf waves of the stator and rotor are considered spatial sine waves, they can be represented by the space vectors $\bar{F}_s$ and $\bar{F}_r$, drawn along the magnetic axes of the stator and rotor mmf waves, as in Figure 5.2.5c, with the phase angle δ (in electrical units) between their magnetic axes. The resultant mmf $\bar{F}$ acting across the air gap is also a sine wave, given by the vector sum of $\bar{F}_s$ and $\bar{F}_r$, so that

$$F^2 = F_s^2 + F_r^2 + 2F_s F_r \cos\delta \tag{5.2.26}$$

where F's are the peak values of the mmf waves. Assuming the air-gap field to be entirely radial, the resultant $\bar{H}$-field is a sinusoidal space wave whose peak is given by

$$H_{\text{peak}} = F/g \tag{5.2.27}$$

where g is the radial length of the air gap (or the clearance between the rotor and stator). This radial length is considered small compared to the radius of either the stator or the rotor. The currents in machine windings produce magnetic flux in the air gap, and the flux paths are completed through the stator and rotor iron.

Equation 3.4.3. gives the energy or coenergy density in the air-gap region at a point where the magnetic field intensity is H

$$w_m = w'_m = \frac{1}{2}\mu_0 H^2 \tag{5.2.28}$$

The average coenergy density obtained by averaging over the volume of the air-gap region is

$$(w'_m)_{av} = \frac{\mu_0}{2}(\text{average value of } H^2) = \frac{\mu_0}{2}\frac{H^2_{peak}}{2} = \frac{\mu_0}{4}\frac{F^2}{g^2} \tag{5.2.29}$$

since the average value of the square of a sine wave is one-half of the square of its peak value. The total coenergy for the air-gap region is then given by

$$(W'_m) = (w'_m)_{av} (\text{ volume of air-gap region}) = \frac{\mu_0}{4}\frac{F^2}{g^2}\pi Dlg \tag{5.2.30}$$

where D is the average diameter at the air gap, and l is the axial length of the machine. Equation 5.2.30 may be rewritten as follows by using Equation 5.2.26:

$$W'_m = \frac{\pi Dl}{4g}\mu_0(F^2_s + F^2_r + 2F_s F_r \cos \delta) \tag{5.2.31}$$

The torque in terms of the interacting magnetic fields is obtained by taking the partial derivative of the field coenergy with respect to the angle δ. For a two-pole machine, such torque is given by

$$T_e = \frac{\partial W'_m}{\partial \delta} = -\frac{\pi Dl}{2g}\mu_0 F_s F_r \sin \delta = -KF_s F_r \sin \delta \tag{5.2.32}$$

in which K is a constant determined by the dimensions of the machine. The torque for a P-pole cylindrical machine with a uniform air gap is then

$$T_e = -\frac{P}{2}KF_s F_r \sin \delta \tag{5.2.33}$$

Equations 5.2.32 and 5.2.33 have shown that the torque is proportional to the peak values of the interacting stator and rotor mmfs and also to the sine of the space-phase angle δ between them (expressed in electrical units). The interpretation of the negative sign is the same as before, in that the fields tend to align themselves by decreasing the displacement angle δ between the fields. If the rotor-mmf axis is fixed relative to the rotor winding, the angle δ between the mmf axes is the same as the angle θ that describes the rotor position in Figure 5.2.3.

Equation 5.2.33 shows that it is possible to obtain a constant torque, varying neither with time nor with rotor position, provided that the two mmf waves are of constant amplitude and have constant angular displacement from each other. While it is easy to conceive of the two mmf waves having constant amplitudes, the question would then be how to maintain a constant angle between the stator and rotor-mmf axes if one winding is stationary and the other is rotating. Three possible answers arise: (1) If the stator-mmf axis is fixed in space, the rotor-mmf axis must also be fixed in space, even when the rotor winding is physically rotating. (2) If the rotor-mmf axis is fixed relative to the rotor, the stator-mmf axis must rotate at the rotor speed relative to the stationary stator windings. (3) The two mmf axes must rotate at such speeds relative to their windings that they remain stationary with respect to each other.

■IIIIIIIIIIIII
5.3 Elementary Concepts of Rotating Machines

The most widely used electromechanical device is the magnetic field rotating machine. The main purpose of most rotating machines is to correct electromechanical energy, i.e., to convert energy between electrical and mechanical systems, either for electric power generation (as in generators or sources) or for the production of mechanical power to perform useful tasks (as in motors or sinks). Rotating machines range in size and capacity from small motors that consume only a fraction of a watt to large generators that produce several hundred megawatts. In spite of the wide variety of types, sizes, and methods of construction, all such machines operate on the same principle, namely, the tendency of two magnets to align themselves. An analysis of an idealized structure of one electric machine will thus provide the essential concepts necessary to understand the operation of most practical machines.

The study of electric machines from the coupled-circuit viewpoint is based on the fundamental consideration that machines can be viewed in terms of sets of linear lumped circuits in relative motion. Relative motion exists between the two magnetic members of the electric machine: the stator, which is the stationary member, and the rotor, which is the rotating member. In general, electrical machines consist of two sets of windings (or coils) in which one set can rotate with respect to the other. The mechanical motion between these two sets of coils is generally restricted to one degree of freedom. This mechanical motion is, in most instances, rotary motion. The annular (ring-shaped) space between the inner surface of the stator and the outer surface of the rotor is known as the air gap. The radial length of the air gap is always kept very small compared to the radial dimension of the outer surface of the rotor in order to produce a large magnetic field for a given current. The two windings in relative motion are known as the *field winding,* which produces the flux density, and the *armature winding,* in which the working emf is induced. The field structure on which the field circuit is located can, in general, be physically situated on either the stationary member (stator) or the rotating member (rotor) of the machine, depending on constructional convenience.

While permanent magnets may be used in small machines as the primary sources of flux, in the majority of machines the field is electromagnetic, and field coils carrying the field current are wound on a magnetic structure. The iron core forming this structure is laminated in order to reduce the field iron losses if the field current is alternating or contains an alternating component.

The field circuit uses two types of construction. In the *salient-pole* arrangement, the field coils are *concentrated* and wound around the protruding poles. This form of construction is used only for machines with direct-current field supply. The air gap around the periphery of the machine is characteristically nonuniform in this arrangement. The second type is the *nonsalient-pole* construction for smooth (or uniform) air-gap machines, in which the field coils are *distributed* in slots cut into a cylindrical magnetic structure. This arrangement is commonly used on certain forms of large ac generators known as turbo-alternators.

Table 5.3.1 summarizes the classes of rotating machines we consider in this book.

The winding in which the voltage is to be induced is known as the armature winding, and the structure containing it is called an *armature.* In general, the electrical circuit representing the armature of the machine consists of coils distributed in slots cut into a cylindrical magnetic structure. To provide the most effective flux path, iron cores are employed for magnetic circuits of both the armature and the field. The armature currents are always alternating, so the armature iron is subjected to a varying flux,

Table 5.3.1 Classes of Rotating Machines

Type of Machine	Stator		Rotor		Typical Examples
	Number of Circuits	Normal type of excitation	Number of Circuits	Normal type of excitation	
Cylindrical Stator and Rotor Structures					
synchronous machines (single and polyphase)	one or more than one (symmetrical winding)	single or polyphase balanced	one	dc	alternators synchronous motors
polyphase induction machines	more than one (symmetrical winding)	polyphase balanced or unbalanced	more than one (symmetrical winding)	short-circuited	3-phase induction motors, 2-phase servo-motors
commutator machines	one or more than one	dc, single or polyphase	commutated winding	short-circuited	amplidynes, metadynes, Schrage motors, repulsion motors, dc machines
single-phase induction machines	one or two (unsymmetrical winding	single-phase	more than one (symmetrical winding)	short-circuited	split-phase, capacitor, shaded-pole motors, ac tachometers, synchros
Saliency on Either Stator or Rotor, but Not Both					
synchronous machines (single and polyphase)	one or more than one (symmetrical winding)	single or polyphase balanced	one (salient rotor)	dc	salient-pole alternators, reluctance motors
commutator machines	one or more than one (salient stator)	ac or dc	commutated winding	ac or dc	conventional dc machines, universal motors, metadynes, rotary amplifiers

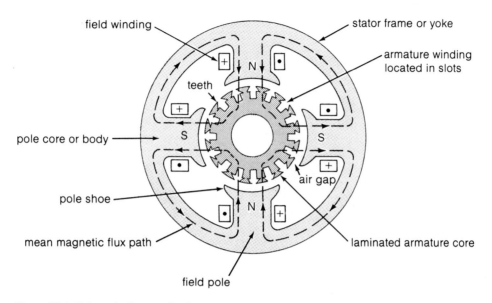

Figure 5.3.1 Schematic diagram of a dc machine.

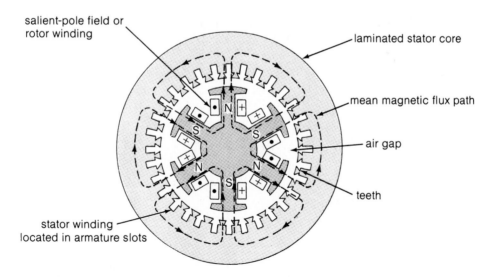

Figure 5.3.2 Schematic diagram of a salient-pole ac machine (synchronous machine).

inducing eddy currents in it. To minimize the eddy-current loss, the armature iron is built up of thin laminations. Figures 5.3.1, 5.3.2, and 5.3.3 are schematic diagrams showing the flux paths in a four-pole dc machine, in a salient-pole ac machine with six poles, and in a two-pole synchronous machine with a cylindrical rotor, respectively. Photos 5.3.1 and 5.3.2 show typical stators of a dc machine and a synchronous machine, respectively. Photos 5.3.3 and 5.3.4 illustrate the construction of salient-pole and nonsalient-pole rotors of synchronous machines.

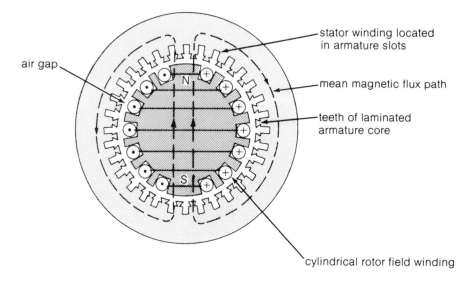

Figure 5.3.3 Schematic diagram of a nonsalient-pole synchronous machine.

Photo 5.3.1 Typical stator of a large dc machine. (Photo courtesy of Westinghouse Electric Corporation.)

In general, either the field or the armature can rotate. When the armature of the machine is on the rotor, the machine is known as a rotating armature machine. The form of different rotating machines (ac or dc) is governed by external constraints, such as the form of electrical supply connected to the field circuit or the mechanical method of connection to the rotating member. For a dc machine, as shown in Figure 5.3.1, the armature is the rotating member, or rotor; the field structure is the stationary member,

Photo 5.3.2 Typical stator of a synchronous machine. (Photo courtesy of Westinghouse Electric Corporation.)

Photo 5.3.3 Eight-salient-pole rotor of a synchronous machine. (Photo courtesy of Westinghouse Electric Corporation.)

Photo 5.3.4 Cutaway view of a two-pole nonsalient rotor of a synchronous generator. (Photo courtesy of General Electric Company.)

or stator. For most ac machines, on the other hand, the armature windings are located on the stator, and the field winding is on the rotor, which may be of salient-pole or nonsalient-pole construction, as in Figures 5.3.2 and 5.3.3.

Currents in the machine windings produce magnetic flux in the air gap between the stator and rotor, as shown in Figures 5.3.1, 5.3.2, and 5.3.3, while the flux path is completed through the stator and rotor iron. This situation is equivalent to the appearance of magnetic poles on both stator and rotor, the number of poles depending on the specific winding design. The concept of interaction between magnetic fields shows that electromagnetic torque cannot be obtained by using unequal numbers of rotor and stator poles, and that all rotating machines must have the same number of poles on the stator as on the rotor. In motor and generator action, then, the magnetic fields tend to line up, pole to pole. In the case of a generator, the electromagnetic torque opposes the mechanical torque applied from the prime mover, which is the source of mechanical energy. In a motor, the direction of rotation is determined by the electromagnetic torque, and the speed voltages that are produced act in opposition to the applied voltages. Thus, the electromagnetic torque and rotational voltages, essential for electromechanical energy conversion, are produced in both generators and motors. In their construction, generators and motors of the same type differ only in details necessary for the best adaptation of a machine for its intended use. In general, any machine can be used as either a motor or a generator for energy conversion.

Machines have another possible mode of operation, known as *braking*. These three modes of operation are schematically shown in Figure 5.3.4.

a. *Motoring mode* has electrical power input and mechanical power output. The electromagnetic torque T_e drives the machine against the load torque T. The input voltage v drives the current into the winding against the generated emf e.

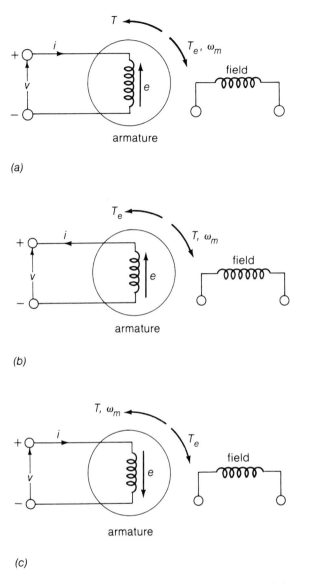

Figure 5.3.4 Three modes of operation of a rotating electrical machine. (a) Motoring mode. (b) Generating mode. (c) Braking mode.

b. *Generating mode* has mechanical power input and electrical power output. The torque T applied externally to the shaft drives the machine against the electrically developed torque T_e. The generated emf e drives current out of the winding against the terminal voltage v.

c. *Braking mode* has both mechanical and electrical energy input. The total input is dissipated as heat. The machine is driven by the externally applied torque T, while the electromagnetic torque T_e is opposing T, thereby braking the machine. The electric braking of motor drives is achieved by causing the motor to act as a generator, receiving mechanical energy from the moving parts and converting it to electrical energy, which is dissipated in a resistor or pumped back into the power line. Note that the applied voltage v and the generated emf e do not oppose each other.

Many forms of electrical machines operate from a three-phase ac supply. The armature winding of such a machine consists of a three-phase distributed winding for which the three separate phase windings are identically wound and displaced by 120 electrical degrees from each other. The phase windings may be connected in wye or delta. We will see in Section 5.4 that, in general, a rotating field of constant amplitude and sinusoidal space distribution of mmf around the periphery of the stator is produced by a q-phase winding ($q > 1$) located on the stator and excited by balanced q-phase currents when the respective phase windings are wound $2\pi/q$ electrical radians apart in space. However, for a balanced two-phase case, the two phase windings are displaced $\pi/2$ electrical radians in space, and the phase currents in the two windings are phase-displaced by $\pi/2$ electrical radians in time. The constant amplitude is $q/2$ times the maximum contribution of any one phase and the speed is $\omega = 2\pi f$ electrical radians per second, where f is the frequency of the electrical supply in hertz. For a P-pole machine, the rotational speed of the mmf is the synchronous speed, given by

$$\omega_m = \frac{2}{P}\omega \text{ rad/s} \tag{5.3.1}$$

or

$$n = \frac{120f}{P} \text{ rpm} \tag{5.3.2}$$

Thus, a polyphase winding excited by balanced polyphase currents produces an effect equivalent to that of spinning a permanent magnet about an axis perpendicular to the magnet or to that of the rotation of the dc-excited field poles. The armature, in general, may be located either on the stator or on the rotor of the machine; the speed of rotation of the rotating mmf in space is then the algebraic sum of the speed of the mmf relative to the structure plus the mechanical speed of the structure itself.

Special mechanical arrangements must be provided when making electrical connections to the rotating member. Such connections are usually made through carbon brushes bearing on either a *slip ring* or a *commutator,* mounted on—but insulated from—the rotor shaft and rotating with the rotor. A slip ring is a continuous ring, usually made of brass, to which only one electrical connection is made. For example, two slip rings are used to supply direct current to the field winding on the rotor of a synchronous machine. A commutator, on the other hand, is an elegant mechanical switch consisting of a cylinder formed of hard-drawn copper segments separated and insulated from each other by mica. See Photo 5.3.5.

In the conventional dc machine, for example, full-wave rectification of the alternating voltage induced in individual armature coils is achieved by means of a commutator, which makes a unidirectional voltage available to the external circuit through the stationary carbon brushes held against the commutator surface. The armature windings of dc machines are located on the rotor because of this necessity for commutation. The winding connected to the commutator, called the commutator winding, can be viewed as a pseudostationary winding because it produces a stationary flux when carrying a direct current, as a stationary winding would. The direction of the flux axis is determined by the position of the brushes. Later, in the detailed analysis of commutator action in dc machines, we will see that the flux axis corresponds to the brush axis (the line joining the two brushes). The brushes are located so that commutation (i.e., reversal of current in the commutated coil) occurs when the coil sides are in the neutral zone, midway between the field poles. The axis of the armature mmf is then 90 electrical degrees from the field (or direct) axis of field poles, i.e., in the quadrature axis. Figure 5.3.5 shows the schematic

Photo 5.3.5 Commutator of a dc machine on the rotor shaft. (Photo courtesy of Westinghouse Electric Corporation.)

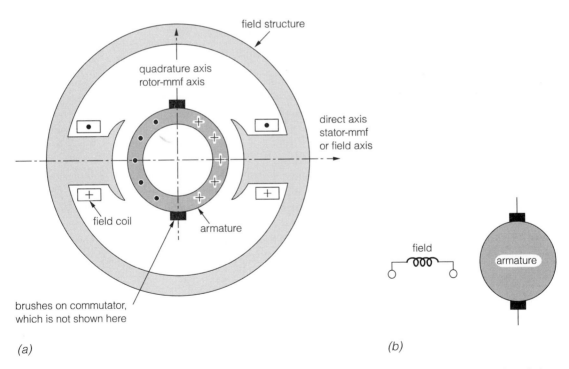

(a)   *(b)*

Note: The geometrical position of the brushes in an actual machine is approximately 90 electrical degrees from their position in the schematic diagram because of the shape of the end connections to the commutator.

Figure 5.3.5 Schematic representations of a dc machine. (a) Schematic arrangement. (b) Circuit representation.

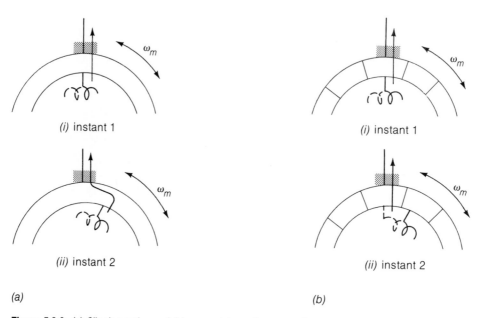

Figure 5.3.6 (a) Slip-ring action and (b) commutator action connections.

representations of a dc machine. The commutator is thus a device for changing the connections between a rotating closed winding and an external circuit at the instants when the individual coil-generated voltages reverse. In a dc machine, then, this arrangement enables a constant and unidirectional output voltage. The armature mmf axis is fixed in space because of the switching action of the commutator (even though the closed armature winding on the rotor is rotating), so the commutator winding becomes pseudostationary.

The action of slip rings and that of a commutator differs in only one way: The conducting coil connected to the slip ring is always connected to the brush, regardless of the mechanical speed ω_m of the rotor and the rotor position, but, with the commutator, the conducting coil conducts current only when it is physically under the commutator brush, i.e., when it is stationary with respect to the commutator brush. This difference is illustrated in Figure 5.3.6. It follows then that the voltage frequency at the slip-ring brush must be equal to that of the rotor mmf relative to the rotor, i.e., $\omega_r = \pm(\omega_s - \omega_m)$, and the voltage frequency at the commutator brush must be that of the rotor mmf in space, i.e., $\omega_s = \omega_m \pm \omega_r$, where $\omega_s \, (= 2\pi f_s)$ is the angular speed of the mmf set up when polyphase ac of frequency f hertz is applied to the stator.

As seen from Equation 5.2.33 and the discussion that followed, to maintain torque production when the machine rotates, the relative current-sheet patterns giving rise to the mmfs and the angle δ between their magnetic axes must be maintained. The manner in which this configuration is achieved depends on the form (dc or ac) of the stator and rotor currents and on the windings in which they circulate. Machine windings are excited so as to develop particular current-sheet and mmf patterns. A winding for dc excitation of a working flux is usually concentrated, while for an ac flux it is commonly distributed to reduce the leakage. As mentioned earlier and as we will discuss in detail later, polyphase windings can be arranged to yield, a close degree of approximation to sinusoidally distributed current sheets and rotating mmfs. Transformers and rotating machines can then be classified depending on (1) the kind of

Table 5.3.2 Classification of Electromagnetic Devices ($\omega_s = \omega_m \pm \omega_r$)

| Air-Gap flux | Electric Supply and Windings | | Name of the Device |
	Member 1 (primary or stator)	Member 2 (secondary or rotor)	
Fixed-Axis			
alternating	ac(ω_s) concentrated	ac(ω_s) concentrated	transformer (not an energy-conversion device)
alternating	ac(ω_s) phase	ac($\omega_r = \omega_s - \omega_m$) phase	1-phase induction machine
constant	ac($\omega_s = \omega_m$) phase	dc($\omega_r = 0$) concentated	1-phase synchronous machine
constant	dc($\omega_s = 0$) concentrated	dc($\omega_r = 0$) commutator	dc commutator machine
alternating	ac(ω_s) distributed	ac(ω_r) commutator	1-phase commutator machine
Rotating-Wave (or traveling-wave, in case of a linear form that the following can take)			
constant	ac($\omega_s = \omega_m$) polyphase	dc($\omega_r = 0$) concentrated or distributed	polyphase synchronous machine, salient-pole or nonsalient-pole type (3-phase machines are most common)
constant	ac(ω_s) polyphase	ac($\omega_r = \omega_s - \omega_m$) polyphase	polyphase induction machine (3-phase machines are most common)

electrical supply (dc or ac) to which each of its windings is connected, and (2) the type of connections made between the winding and its supply. Such connections might be by phase or tapped windings or by switched connections, as in a commutator winding. Table 5.3.2 lists a number of possible combinations, giving the names by which the machines are generally known.

Elementary Synchronous Machines

Figure 5.3.7 shows an elementary single-phase, two-pole synchronous machine. In almost all cases, the armature winding of a synchronous machine is on the stator and the field winding is on the rotor, because it is constructionally advantageous to have the low-power field winding on the rotating member. The field winding is excited by direct current that is supplied by a dc source connected to carbon brushes bearing on slip rings (or collector rings). The armature windings, though distributed in the slots around

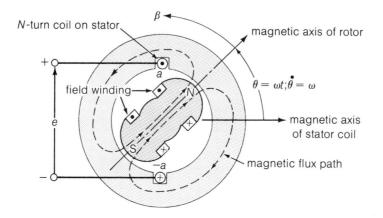

Figure 5.3.7 Elementary single-phase two-pole synchronous machine.

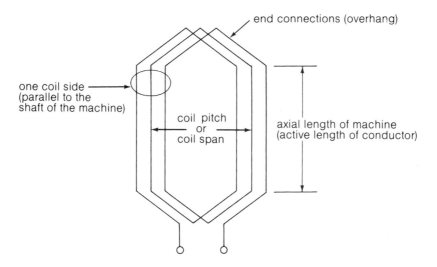

Figure 5.3.8 A three-turn single coil of armature winding.

the inner periphery of the stator in an actual machine, are shown for simplicity as consisting of a single coil of N turns (see Figure 5.3.8), indicated in cross section by the two sides a and $-a$ placed in diametrically opposite narrow slots. The conductors forming these coil sides are placed in slots parallel to the machine shaft and connected in series by means of the end connections, as shown in Figure 5.3.8. The coil in Figure 5.3.7 spans 180 degrees (or a complete *pole pitch*, which is the peripheral distance from the center line of a north pole to the center line of an adjacent south pole) and is hence denoted as a *full-pitch* coil. For simplicity and convenience, Figure 5.3.7 shows only a two-pole synchronous machine with salient-pole construction; the flux paths are shown by dotted lines.

The space distribution of the radial air-gap flux density around the air-gap periphery can be made to approximate a sinusoidal distribution by properly shaping the rotor-pole faces facing the air gap:

$$B = B_m \cos \beta \tag{5.3.3}$$

where B_m is the peak value at the rotor-pole center, and β is measured in electrical radians from the rotor-pole axis (or the magnetic axis of rotor), as shown in Figure 5.3.7. The air-gap flux per pole is the integral of the flux density over the pole area. For a two-pole machine,

$$\phi = \int_{-\pi/2}^{+\pi/2} B_m \cos \beta \, lr \, d\beta = 2B_m lr \tag{5.3.4}$$

and for a P-pole machine,

$$\phi = \frac{2}{P} 2B_m lr \tag{5.3.5}$$

where l is the axial length of the stator, r is the average radius at the air gap. For a P-pole machine, the pole area is $2/P$ times that of a two-pole machine of the same length and diameter.

The flux linkage with stator coil is $N\phi$ when the rotor poles are in line with the magnetic axis of the stator coil. If the rotor is turned at a constant speed ω by a source of mechanical power connected to its shaft, the flux linkage with the stator coil varies as the cosine of the angle θ between the magnetic axes of the stator coil and rotor:

$$\lambda = N\phi \cos \omega t \tag{5.3.6}$$

where time t is arbitrarily taken as zero when the peak of the flux-density wave coincides with the magnetic axis of the stator coil. By Faraday's law, the voltage induced in the stator coil is given by

$$e = -\frac{d\lambda}{dt} = \omega N\phi \sin \omega t - N\frac{d\phi}{dt} \cos \omega t \tag{5.3.7}$$

The minus sign associated with Faraday's law in Equation 5.3.7 implies generator reference directions, as explained earlier and shown in Figure 5.3.7. Considering the right-hand side of Equation 5.3.7, the first term is a speed voltage caused by the relative motion of the field and the stator coil. The second term is a transformer voltage, which is negligible in most rotating machines under normal steady-state operation because the amplitude of the air-gap flux wave is fairly constant. The induced voltage is then given by the speed voltage itself:

$$e = \omega N\phi \sin \omega t \tag{5.3.8}$$

The above equation may alternatively be obtained by the application of the cutting-of-flux concept given by Equation 3.1.13, from which the motional emf is given by the product of B_{coil} times the total active length of conductors l_{eff} in the two coil sides times the linear velocity of the conductor relative to the field, provided that these three are mutually perpendicular. For the case under consideration, then,

$$e = B_{coil} l_{eff} v = (B_m \sin \omega t)(2lN)(r\omega_m)$$

or

$$e = (B_m \sin \omega t)(2lN)(r2\omega/P) = \omega N\frac{2}{P} 2B_m lr \sin \omega t \tag{5.3.9}$$

which is the same as Equation 5.3.8 when the expression for ϕ given by Equation (5.3.5) is substituted.

The resulting coil voltage is thus a time function having the same sinusoidal waveform as the spatial distribution B. The coil voltage passes through a complete cycle for each revolution of the two-pole machine of Figure 5.3.7. So its frequency in hertz is the same as the speed of the rotor in revolutions per second; that is, the electrical frequency is synchronized with the mechanical speed of rotation. Thus a two-pole synchronous machine, under normal steady-state conditions of operation, revolves at 60 rps or 3,600 rpm in order to produce 60-Hz voltage. For a P-pole machine in general, however, the coil voltage

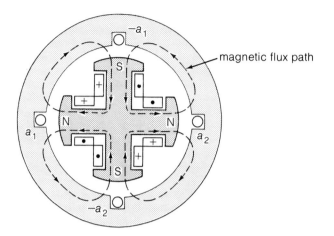

Figure 5.3.9 Elementary single-phase four-pole synchronous machine.

passes through a complete cycle every time a pair of poles sweeps, i.e., $P/2$ times in each revolution. The frequency of the voltage wave is then given by

$$f = \frac{P}{2} \cdot \frac{n}{60} \text{ Hz} \tag{5.3.10}$$

where n is the mechanical speed of rotation in rpm. The synchronous speed in terms of the frequency and the number of poles is given by

$$n = \frac{120f}{P} \text{ rpm} \tag{5.3.11}$$

The radian frequency ω of the voltage wave in terms of ω_m the mechanical speed in radians per second, is given by

$$\omega = \frac{P}{2}\omega_m \tag{5.3.12}$$

which is in accordance with Equation 5.2.23. Figure 5.3.9 shows an elementary single-phase synchronous machine with four salient poles; the flux paths are shown by dashed lines. Two complete wavelengths (or cycles) exist in the flux distribution around the periphery, since the field coils are connected so as to form poles of alternate north and south polarities. The armature winding now consists of two coils $(a_1, -a_1)$ and $(a_2, -a_2)$, connected in series by their end connections. The span of each coil is one-half wavelength of flux, or 180 electrical degrees for the full-pitch coil. Since the generated voltage now goes through two complete cycles per revolution of the rotor, the frequency is then twice the speed in revolutions per second, consistent with Equation 5.3.10.

The field winding may be concentrated around the salient poles, as shown in Figures 5.3.7 and 5.3.9, or distributed in slots around the cylindrical rotor, as in Figure 5.3.3 (shown for a two-pole machine). By properly shaping the pole faces in the former case and by appropriately distributing the field winding in the latter case, an approximately sinusoidal field is produced in the air gap.

A salient-pole rotor construction is best suited mechanically for hydroelectric generators because hydroelectric turbines operate at relatively low speeds, and a relatively large number of poles is required in order to produce the desired frequency (60 Hz in the United States), in accordance with Equation 5.3.11. Salient-pole construction is also employed for most synchronous motors.

Photo 5.3.6 A salient-pole three-phase synchronous generator. (Photo courtesy of Marathon Electric Manufacturing Company.)

The nonsalient-pole (smooth or cylindrical) rotor construction is preferred for turbine-driven alternators (known also as turbo-alternators or turbine generators), which are usually of two or four poles driven by steam turbines or gas turbines, operating best at relatively high speeds. The rotors for such machines may be made either from a single steel forging or from several forgings shrunk together on the shaft.

Going back to Equation 5.3.8, the maximum value of the induced voltage is

$$E_{max} = \omega N \phi = 2 \pi f N \phi \qquad (5.3.13)$$

and its rms value is

$$E_{rms} = \frac{2\pi}{\sqrt{2}} f N \phi = 4.44 f N \phi \qquad (5.3.14)$$

which are identical in form to the corresponding emf equations for a transformer. Consistent with the discussion that followed Equations 3.1.12 and 3.1.13, the effect of a time-varying flux in association with stationary transformer windings is the same as that of relative motion of a coil and a constant-amplitude spatial flux-density wave in a rotating machine. The space distribution of flux density is transformed into a time-variation of voltage because of the time element introduced by mechanical rotation. The induced voltage is a single-phase voltage for single-phase synchronous machines of the nature discussed so far. As pointed out at the end of the solution of Example 5.2.3, part c, to avoid the pulsating torque, the designer could employ polyphase windings and polyphase sources to develop constant power under balanced conditions of operation.

In fact, with very few exceptions, three-phase synchronous machines are most commonly used for power generation. In general, three-phase ac power systems, including power generation, transmission, and usage, have grown most popular because of their relative economic advantages. Photos 5.3.6 and 5.3.7 show typical salient-pole and nonsalient-pole three-phase synchronous generators. An elementary

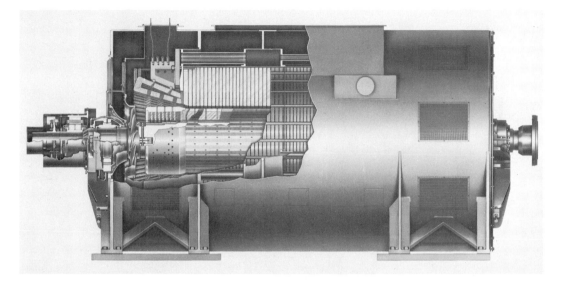

Photo 5.3.7 A nonsalient-pole three-phase synchronous generator. (Photo courtesy of Electric Machinery Manufacturing Company.)

three-phase, two-pole synchronous machine with one coil per phase (chosen for simplicity) is shown in Figure 5.3.10a. The coils are displaced by 120 electrical degrees from each other in space so the three-phase voltages of positive phase sequence *a-b-c*, displaced by 120 electrical degrees from each other in time, could be produced. Figure 5.3.10b shows an elementary three-phase, four-pole synchronous machine with one slot per pole per phase. It has 12 coil sides or 6 coils in all; 2 coils belong to each phase, which may be connected in series either in wye or delta, as shown in Figures 5.3.10c and d. Equation 5.3.14 can be applied to give the rms voltage per phase when *N* is treated as the total series turns per phase. The coils may also be connected in parallel to increase the current rating of the machine. In actual ac machine windings, instead of concentrated full-pitch windings, distributed fractional-pitch armature windings are commonly used to make better use of iron and copper and to make waveforms of the generated voltage (in time) and the armature mmf (in space) as nearly sinusoidal as possible. That is to say, the armature coils of each phase are distributed in a number of slots, and the coil span may be shorter than a full pitch. In such cases, Equation 5.3.14 is modified to be

$$E_{rms} = 4.44 k_w f N_{ph} \phi \text{ volts per phase} \tag{5.3.15}$$

where k_w is a winding factor (less than unity, usually about 0.85 to 0.95), and N_{ph} is the number of series turns per phase.

The polyphase synchronous machine operates with direct current supplied to the field winding (assumed to be on the rotor, which is usually the case) through two slip rings and with polyphase ac supplied to the armature (assumed to be on the stator). The rotor mmf, which is obtained from a dc source, is stationary with respect to the rotor structure. When carrying balanced polyphase currents, as stated earlier and shown later in Section 5.4, the armature winding produces a magnetic field in the air gap rotating at synchronous speed (Equation 5.3.11 or 5.3.12), as determined by the system frequency and the number of poles in the machine. But the field produced by the dc rotor winding revolves with the rotor. To produce a steady unidirectional torque, the rotating fields of stator and rotor must be traveling at the same speed; therefore, the rotor must turn precisely at the synchronous speed.

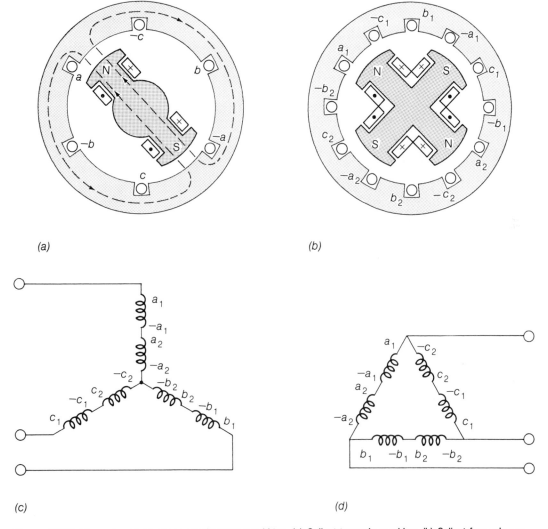

Figure 5.3.10 Elementary three-phase synchronous machines. (a) Salient two-pole machine. (b) Salient four-pole machine. (c) Phase windings connected in wye. (d) Phase windings connected in delta.

Such conditions satisfy the second possibility for the development of a constant torque suggested in the discussion following Equation 5.2.33.

The rotor-mmf axis of the sinusoidally distributed rotor-mmf wave of constant amplitude in the air gap remains at a constant angle to that of the stator. The resulting physical system is shown in Figure 5.3.11, in which the poles correspond to peak values of gap mmf components produced by the rotor and stator windings. The *polyphase synchronous machine* is one in which the rotor rotates in synchronism with the rotating mmf wave produced by the stator. The actual flux-density distribution in the air gap is produced by the resultant mmf, the maximum of which occurs at some angle between the mmf axes. Since the rotor is rotating at the same speed as the resultant flux wave produced in the air gap, no emf is induced in the rotor winding because there is no variation in the associated flux linkage. As explained

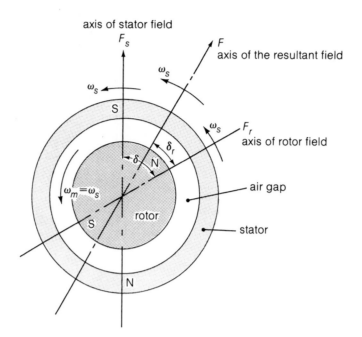

Figure 5.3.11 Simplified two-pole synchronous machine with nonsalient poles.

earlier, the essential condition for energy conversion in the cylindrical machine with single stator and rotor windings is expressed by

$$\omega_m = \pm(\omega_s \pm \omega_r) \tag{5.3.16}$$

where the $\pm$ outside the parentheses implies that the direction of rotation is immaterial in such single-winding machines. In a machine with polyphase stator winding, however, direction is no longer immaterial, since the rotor-mmf axis must rotate in the same direction as that of the stator in order to satisfy the condition for constant torque deduced from Equation 5.2.33. Thus, for machines with polyphase stator winding, Equation 5.3.16 should be modified as

$$\omega_m = \omega_s \pm \omega_r \tag{5.3.17}$$

Equation 5.2.33 applies for the electromagnetic torque produced by the nonsalient-pole (or cyclindrical-rotor) machine. The torque can also be expressed in terms of the resultant mmf wave $\bar{F}$ (see Problem 5–20):

$$T_e = -\frac{P}{2}KF_s F_r \sin \delta \tag{5.2.33}$$

or

$$T_e = -\frac{P}{2}KFF_s \sin \delta_s \tag{5.3.18}$$

or

$$T_e = -\frac{P}{2}KFF_r \sin \delta_r \tag{5.3.19}$$

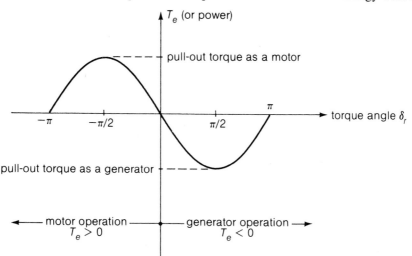

Figure 5.3.12 Torque-angle characteristic curve of a cylindrical-rotor synchronous machine (for a given field current and a fixed terminal voltage, i.e., for a constant resultant air-gap flux).

where δ_s and δ_r are the angles, as shown in Figure 5.2.5c, between $\bar{F}$ and $\bar{F}_s$, and $\bar{F}$ and $\bar{F}_r$, respectively. Alternatively, the torque can be expressed in terms of the resultant flux ϕ per pole produced by the combined effect of the stator and rotor mmfs (see Problem 5–20d):

$$T_e = -\frac{\pi}{2}\left(\frac{P}{2}\right)^2 \phi F_r \sin \delta_r \tag{5.3.20}$$

where δ_r is the angle between the rotor-mmf axis and the resultant-flux or mmf axis. When the armature terminals are connected to a balanced polyphase *infinite bus* (which is a high-capacity, constant-voltage, constant-frequency system), the resultant air-gap flux ϕ is approximately constant, independent of shaft load. Under normal operating conditions, the armature-resistance voltage drop is negligible, and the armature-leakage flux is relatively small compared to the resultant air-gap flux ϕ, which is given by Equation 5.3.15 as

$$\phi = \frac{\text{terminal phase voltage}}{4.44 k_w f N_{ph}} \tag{5.3.21}$$

The rotor mmf F_r determined by the dc field current is also a constant under normal operating conditions. So, as seen from Equation 5.3.20, any variation in the torque requirements of the load has to be accounted for entirely by variation of the angle δ_r, which is why δ_r is known as the *torque angle* (or the *load angle*) of a synchronous machine. The effect of salient poles on the torque-angle characteristic is discussed later in detail in Chapter 8 on synchronous machines.

The torque-angle characteristic curve of a cylindrical-rotor synchronous machine is shown in Figure 5.3.12 as a function of the angle δ_r. For $\delta_r < 0, T_e > 0$, the developed torque is positive and acts in the direction of rotation: the machine operates as a motor. If, on the other hand, the machine is driven by a prime mover so that δ_r becomes positive, the torque is then negative, and the machine operates as a generator. Note that, at standstill, i.e., when $\omega_m = 0$, no average undirectional torque is developed by the synchronous machine; such a synchronous motor is not capable of self-starting because it has no starting torque. The designer must provide a method for bringing the machine up to synchronous speed.

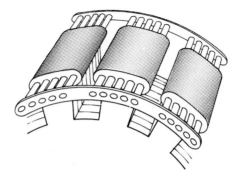

Figure 5.3.13 A sketch of damper bars located on the salient-pole shoes of a synchronous machine.

Because a synchronous machine operates only at synchronous speed under steady-state conditons, the machine cannot operate at synchronous speed during the load transition when the load on the synchronous machine is to be changed; the readjustment process is in fact dynamic. In the case of a generator connected to an infinite bus, an increase in the electrical output power of the generator is brought about by increasing the mechanical input power supplied by the prime mover. The speed of the rotor increases momentarily during the process, and the axes of the rotor field mmf advances relative to the axes of both the armature mmf and the resultant air-gap mmf. The increase in torque angle results in an increase in electrical power output. The machine then locks itself into synchronism and continues to rotate at synchronous speed until the load changes further. The same general argument can be made for the operation of a synchronous motor, except that an increase in mechanical load decreases the speed of the rotor during the transition period, so that the axis of the rotor-field mmf falls behind that of the stator mmf or the resultant air-gap mmf by the required value of the load angle.

When the load on a synchronous machine changes, the load angle changes from one steady value to another. During this transition, oscillations in the load angle and consequent associated mechanical oscillations (known as *hunting*) occur. To damp out these oscillations, it is common practice to provide an additonal short-circuited winding on the field structure, made out of copper or brass bars located in pole-face slots on the pole shoes of a salient-pole machine and connected together at the ends of the machine. This additional winding (known as a *damper* or an *amortisseur winding),* is shown in Figure 5.3.13. This winding is also useful in getting a synchronous motor started, as explained later. When the machine is operating under steady-state conditions at synchronous speed, however, this winding has no effect. Because there is no rate of change of flux linkage, no voltage is induced in it.

As seen in Figure 5.3.12, when δ_r is $\pm\pi/2$ or $\pm90°$, the maximum torque or power (called the *pull-out torque* or *pull-out power)* is reached for a fixed terminal voltage and a given field current. If the load requirements exceed this value, the motor slows down because of the excess shaft torque; synchronous-motor action is lost because the rotor and stator fields are no longer stationary with respect to each other. Any load requiring a torque greater than the maximum torque results in unstable operation of the machine; the machine pulls out of synchronism, known as *pulling out of step* or losing synchronism. The motor is usually disconnected from the electric supply by automatic circuit breakers, and the machine comes to a standstill. Note that the pull-out torque can be increased by increasing either the field current or the terminal voltage. In the case of a generator connected to an infinite bus, synchronism will be lost if the torque applied by the prime mover exceeds the maximum generator pull-out torque; the speed will then increase rapidly unless the quick-response governor action comes into play on the prime mover to control the speed.

Elementary DC Machines

A dc machine operates with direct current applied to its field winding (generally located on the salient-pole stator of the machine) and to a commutator (via the brushes) connected to the armature winding situated inside slots on the cylindrical rotor, as shown schematically in Figures 5.3.1 and 5.3.5. In terms of the discussion that followed Equation 5.2.33 for development of a constant torque, the first possibility is satisfied: In a direct current machine, the stator-mmf axis is fixed in space, and the rotor-mmf axis is also fixed in space, even when the rotor winding is physically rotating, because of the commutator action, which was briefly discussed earlier. Thus the dc machine will operate under steady-state conditions, whatever the rotor speed ω_m. The dc motor is capable of producing starting torque. The armature current in the armature winding is alternating; the action of the commutator is to change the armature current from a frequency governed by the mechanical speed of rotation to zero frequency at the commutator brushes connected to the external circuit.

For the case of a dc machine with a flux per pole ϕ, the total flux cut by one conductor in one revolution is given by ϕP, where P is the number of poles of the machine. If the speed of rotation is n rpm, the emf generated in a single conductor is given by

$$e = \frac{\phi P n}{60} \tag{5.3.22}$$

For an armature with Z conductors and α parallel paths, the total generated armature emf E_a is given by

$$E_a = \frac{P \phi n Z}{60 \alpha} \tag{5.3.23}$$

Since the angular velocity ω_m is given by $2\pi n/60$, Equation 5.3.23 becomes

$$E_a = \frac{PZ}{2\pi \alpha} \phi \omega_m = K_a \phi \omega_m \tag{5.3.24}$$

where K_a is the design constant given by $PZ/2\pi\alpha$. The value obtained is the speed voltage appearing across the brush terminals in the quadrature axis (see Figure 5.3.5) due to the field excitation producing ϕ in the direct axis. For this reason, in the schematic circuit representation of a dc machine, the field axis and the brush axis are shown in quadrature, i.e., perpendicular to each other. The generated voltage as observed from the brushes is the sum of the rectified voltages of all the coils in series between brushes. If the number of coils is sufficiently large, the ripple in the waveform of the armature voltage (as a function of time) becomes very small, thereby making the voltage direct or constant in magnitude.

The instantaneous electrical power associated with the speed voltage should be equal to the instantaenous mechanical power associated with the electromagnetic torque, the direction of power flow being determined by whether the machine is operating as a motor or generator.

$$T_e \omega_m = E_a I_a \tag{5.3.25}$$

With the aid of Equation 5.3.23, it follows that

$$T_e = K_a \phi I_a \tag{5.3.26}$$

which is created by the interaction of the magnetic fields of stator and rotor. As mentioned earlier, if the machine is acting as a generator, this torque opposes rotation; if the machine is acting as a motor, the electromagnetic torque acts in the direction of rotation.

Given a large number of conductors on the rotor, if a commutated dc source is adopted to excite the rotor, the variation of armature mmf around the periphery is not sinusoidal; in fact, it is very nearly triangular, as discussed in Section 6.7. This is not important, however, as the dc machine is not to be connected to an ac system. For the same reason, the distribution of stator mmf around the air gap need not be sinusoidal. In fact, the air-gap flux distribution for a dc machine usually approximates a flat-topped wave (nearly rectangular) rather than the sine wave found in ac machines. The location of brushes on the commutator arrangement connected to the armature winding ensures that the rotor and stator-mmf axes are at all times at right angles to one another, as shown in Figure 5.3.5. The expression for torque given by Equation 5.2.33 is applicable to the dc machine, provided the constant K is adjusted for the nonsinusoidal mmf distributions and $\sin \delta$ is made equal to unity. Not only are the conditions for constant torque fulfilled but also the condition for maximum torque and hence for maximum energy conversion is satisfied, since $\delta = \pm\pi/2$, depending upon whether generator or motor action occurs for a given direction of rotation.

Since both armature and field circuits carry direct current in the case of a dc machine, they can be connected either in series or in parallel. When the armature and field circuits are connected in parallel, the machine is known as a *shunt machine*. In the shunt machine, the field coils are wound with a large number of turns carrying a relatively small current. When the circuits are connected in series, the machine is known as a *series machine*. The field winding in the series machine carries the full armature current and is wound with a smaller number of turns. A dc machine provided with both a series and a shunt field winding is known as a *compound machine*. In the compound machine, the series field may be connected either *cumulatively,* so that its mmf adds to that of the shunt field, or *differentially,* so that the mmf opposes. The differential connection is very rarely used. The voltage of both shunt and compound generators (or the speed, for motors) is controlled over reasonable limits by means of a field rheostat in the shunt field. The machines are said to be *self-excited* when the machine supplies its own excitation of the field windings as in the cases we have just defined. The field windings may be *separately excited* from an external dc source, however. A small amount of power in the field circuit can control a large amount of power in the armature circuit. The dc generator may then be viewed as a power amplifier. Some possible field-circuit connections of dc machines are shown in Figure 5.3.14.

For self-excited generators, residual magnetism must be present in the ferromagnetic circuit of the machine in order to start the self-excitation process. For a dc *generator*, the relationship between the steady-state generated emf E_a and the terminal voltage V_t is given by

$$V_t = E_a - I_a R_a \tag{5.3.27}$$

where I_a is the armature current *output,* and R_a is the armature circuit resistance. For a dc *motor,* the relationship is given by

$$V_t = E_a + I_a R_a \tag{5.3.28}$$

where I_a is now the armature current *input*. Under steady-state conditions, *volt-ampere* characteristic curves are of interest for dc generators, and *speed-torque* characteristics are of interest for dc motors. Depending on the method of excitation of the field windings, a wide variety of operating characteristics can be obtained. These possibilities make the dc machine both versatile and adaptable for control. Photo 5.3.8 shows a cross-sectional view of a typical dc machine.

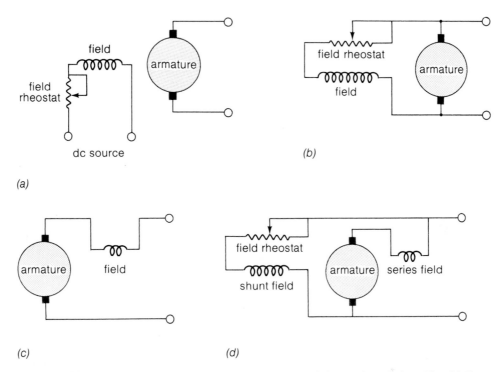

Figure 5.3.14 Field-circuit connections of direct-current machines. (a) Separately excited machine. (b) Shunt machine. (c) Series machine. (d) Compound machine.

Elementary Induction Machines

In our discussion that followed Equation 5.2.33, the third possible method of producing constant torque was to cause the mmf axes of stator and rotor to rotate at such speeds relative to their windings that they remain stationary with respet to each other. If the stator and rotor windings are polyphase and carry polyphase ac, then both the stator-mmf and rotor-mmf axes may be caused to rotate relative to their windings. Such a machine will have polyphase stator ac excitation at ω_s, polyphase rotor ac excitation at ω_r, and the rotor speed ω_m satisfying Equation 5.3.17. Let us consider the rotor speed $\omega_m = \omega_s - \omega_r$ and the same phase sequence of sources. A rotating magnetic field of constant amplitude, rotating at ω_s rad/s relative to the stator, is produced because of polyphase stator excitation. A rotating magnetic field of constant amplitude, rotating at ω_r rad/s relative to the rotor, is also produced because of polyphase rotor excitation. The speed of rotation of the rotor magnetic field relative to the stator is $(\omega_m + \omega_r)$ or ω_s if the rotor is rotating with a positive speed of rotation ω_m in the direction of rotating fields. If so, the condition for energy conversion at constant torque is satisfied. Such a situation is diagramatically illustrated in Figure 5.3.15. The machine under these conditions is operating as a double-fed polyphase machine. Normally, in an induction machine with polyphase stator and rotor windings, only a source to excite the stator is employed, and the rotor excitation at the appropriate frequency is induced from the stator winding. The device is thus known as an induction machine.

In an induction machine, the stator winding (Photo 5.3.9) is essentially the same as that of a synchronous machine. Equation 5.3.15 and the considerations leading to it hold good here as in the case

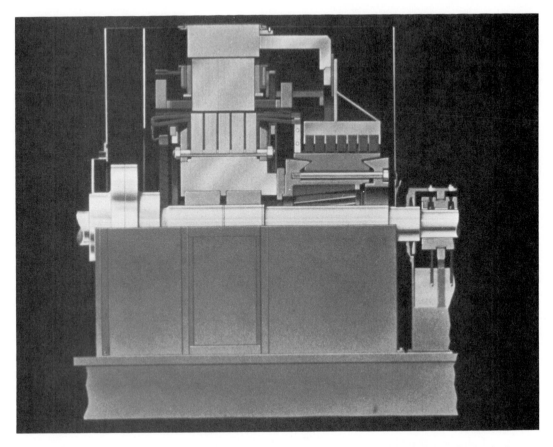

Photo 5.3.8 Cross-sectional view of a typical dc machine. (Photo courtesy of Westinghouse Electric Corporation.)

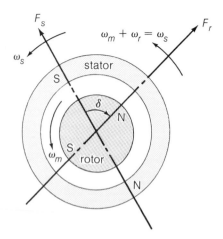

Figure 5.3.15 Mmf-axes of an induction machine.

Photo 5.3.9 Stator of an induction motor. (Photo courtesy of General Electric Company.)

of a synchronous machine. When excited from a balanced polyphase source, the polyphase windings produce a magnetic field in the air gap rotating at a synchronous speed determined by the number of poles and the applied stator frequency given by Equation 5.3.11. On the rotor, the winding is electrically closed on itself and often has no external terminals.

The induction machine rotor may be one of two types: the *wound rotor* or the *squirrel-cage rotor.* A wound rotor has a polyphase winding similiar to and wound for the same number of poles as the stator winding. The terminals of the rotor winding (wye- or delta-connected, in the case of three-phase machines) are brought to insulated slip rings mounted on the shaft. Carbon brushes bearing on these slip rings make the rotor terminals available to the circuitry external to the motor. The rotor winding is usually short-circuited through external resistences that can be varied. A squirrel-cage rotor (Photo 5.3.10) has a winding consisting of conducting bars of copper or aluminum embedded in slots cut in the rotor iron and short-circuited at each end by conducting end rings. The squirrel-cage induction machine is the electromagnetic machine most widely used as a motor because of its extreme simplicity and ruggedness. Although the induction machine in motor mode is the most common of all motors, the induction machine is very rarely used as a generator because its performance characteristics as a generator are not satisfactory for most applications. The induction machine with a wound rotor is also used as a *frequency changer.*

The polyphase induction motor (Photo 5.3.11) operates with polyphase ac applied to the primary winding, usually located on the stator of the polyphase machines. Three-phase motors are most popularly used in practice, while two-phase motors are used in control systems. The induction machine has emf (and consequently current) induced in the short-circuited secondary (or rotor) winding by virtue of the primary rotating mmf. Such a machine is then singly excited. The induction machine may be regarded as a generalized transformer in which energy conversion takes place, and electric power is

Photo 5.3.10 A squirrel-cage rotor of a three-phase induction motor. (Photo courtesy of Westinghouse Electric Corporation.)

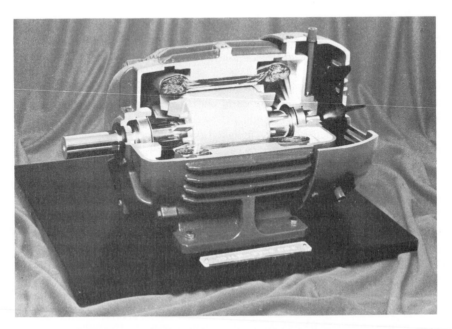

Photo 5.3.11 Totally enclosed fan-cooled 10-hp, three-phase, 60-Hz, 230/460-V squirrel-cage induction motor. (Photo courtesy of Siemens Energy and Automation, Inc.)

transformed between the stator and rotor along with a change of frequency and a flow of mechanical power.

Let us assume that the rotor is turning at a steady speed of n rpm in the same direction as the rotating stator field. Let the synchronous speed of the stator field be n_1 rpm, as given by Equation 5.3.11, corresponding to the applied stator frequency f_s Hz or ω_s rad/s. It is convenient to introduce the concept of a *per-unit slip* S given by

$$S = \frac{\text{synchronous speed - actual rotor speed}}{\text{synchronous speed}} = \frac{n_1 - n}{n_1} = \frac{\omega_s - \omega_m}{\omega_s} \qquad (5.3.29)$$

The rotor is then traveling at a speed $(n_1 - n)$ or $n_1 S$ rpm in the backward direction with respect to the stator field. The relative motion of the flux and rotor conductors induces voltages of frequency (Sf_s), known as the *slip frequency,* in the rotor winding. Thus the induction machine is similiar to a transformer in its electrical behavior but with an additional feature of frequency change. The frequency f_r Hz of the secondary (or rotor) currents is then given by

$$f_r = \frac{\omega_r}{2\pi} = Sf_s \qquad (5.3.30)$$

At standstill, $\omega_m = 0$ so that the slip $S = 1$ and $f_r = f_s$; that is, the machine then acts as a simple transformer with an air gap and a short-circuited secondary winding. A steady starting torque is produced because the condition for energy conversion at constant torque is satisfied; hence the polyphase induction motor is self-starting. At synchronous speed, however, $\omega_m = \omega_s$ so that the slip $S = 0$ and $f_r = 0$; no induction takes place because there is no relative motion between the flux and rotor conductors. Thus, at synchronous speed, the value of the secondary mmf is zero, and no torque is produced; that is, the induction motor cannot run at synchronous speed. The no-load speed of the induction motor is usually on the order of 99.5 percent of synchronous speed so that the no-load per-unit slip is about 0.005, and the full-load per-unit slip is on the order of 0.05. Thus the polyphase induction motor is effectively a constant-speed machine.

An induction machine connected to a polyphase exciting source on its stator side can be made to generate (i.e., with the power flow reversed compared to that of a motor) if its rotor is driven mechanically by an external means at above synchronous speed, so that $\omega_m > \omega_s$ and the slip becomes negative. If the machine is driven mechanically in the direction opposite to its primary rotating mmf, then the slip is greater than unity and the machine acts as a brake. For example, let the machine be operating normally as a loaded motor: if two of the three-phase supply lines to the stator are reversed, the direction of the stator-rotating mmf will reverse (to be explained in detail in Section 5.4). The rotor will then be rotating in the direction opposite that of the rotating mmf, so the machine will act as a brake and the speed will rapidly come to zero, at which time the electric supply can be removed from the machine. Such a reversal of two supply lines of the three-phase system, a useful method of stopping the motor rapidly, is generally referred to as *plugging* or *plug-braking*. If the electric supply is not removed at zero speed, however, the machine will reverse its direction of rotation because of the change of phase sequence of the supply resulting from the interchange of the two stator leads. The general form of the torque-speed curve (or torque-slip characteristic) for a polyphase induction machine between rotor-speed limits of $-\omega_s \leq \omega_m \leq 2\omega_s$, corresponding to a range of slips of $2 \leq S = -1$, is shown in Figure 5.3.16.

In general, concerning the polyphase operation of induction machines, a rotor with any number of phases will develop torque in a stator of the same or any other number of phases (excepting only

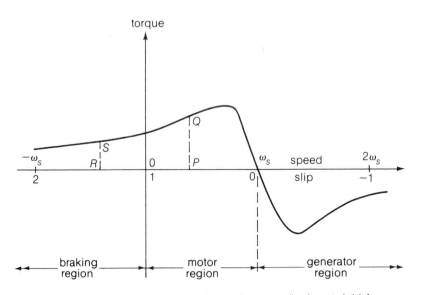

Figure 5.3.16 General form of torque-speed curve (or torque-slip characteristic) for a polyphase induction machine.

single-winding stators). The student should be able to reason this out from the necessary condition for producing the torque. However, the number of poles in the stator and rotor must be the same for torque production, as mentioned earlier. For a wound-rotor induction machine, the rotor must be wound for the same number of poles as the stator; for a squirrel-cage machine, however, an equal number of poles is induced in the rotor. The emfs induced in squirrel-cage conductors produce currents that circulate through the bars and the end rings, thereby effectively forming a rotor winding with a large number of phases. Machines of small size sometimes employ solid iron rotors as well as rotors with a thin cylinder of conductor (such as copper) enclosing the rotor magnetic core. Also, in composite-rotor induction machines, alternate layers of copper and iron in the form of thin cylinders enclosing the rotor-iron structure are used for high-speed and high-frequency operation.

The torque that exists at any mechanical speed other than synchronous speed is known as an *asynchronous torque.* The induction machine is also known as an *asynchronous machine,* since no torque is produced at synchronous speed and the machine runs at a speed other than synchronous speed. In fact, as a motor, the machine runs only at a speed that is less than synchronous speed with positive slip. The factors influencing the general shape of the torque-speed characteristic (shown in Figure 5.3.16) can be appreciated in terms of the torque equation (5.3.20). Noting that the resultant air-gap flux ϕ is nearly constant when the stator-applied voltage and frequency are constant as seen by Equation 5.3.21, and that the motor mmf F_r is proportional to the rotor current I_r, the torque may be expressed as

$$T_e = K_1 i_r \sin \delta_r \tag{5.3.31}$$

where K_1 is a constant and δ_r is the angle between the rotor-mmf axis and the resultant-flux or mmf axis. The rotor current I_r is determined by the rotor-induced voltage (proportional to slip) and rotor impedance. Since the slip is small under normal running conditions, as already mentioned, the rotor frequency $f_r = \mathbf{S}f_s$ is very low (on the order of 3Hz in 60-Hz motors of a per-unit slip of 0.05). Hence, in this range the rotor impedance is largely resistive, and the rotor current is very nearly proportional

Photo 5.3.12 Single-phase squirrel-cage induction motor. (Photo courtesy of Marathon Electric Manufacturing Company.)

to and in phase with the rotor voltage; that is, the rotor current is very nearly proportional to slip. An approximately linear torque-speed relationship can be observed in the range of low values of slip in Figure 5.3.16. Further, with the rotor-leakage reactance being very small compared with rotor resistance, the rotor-mmf wave lags approximately 90 electrical degrees behind the resultance flux wave, and therefore $\sin \delta_r$ is approximately equal to unity.

As slip increases, the rotor impedance increases because of the increasing effect of rotor-leakage inductance; the rotor current is then somewhat less than proportional to slip. The rotor current lags further behind the induced voltage, and the rotor-mmf wave lags further behind the resultant-flux wave, so that $\sin \delta_r$ decreases. The torque increases with increasing values of slip up to a point and then decreases as shown in Figure 5.3.16 for the motor region. The maximum torque that the machine can produce is sometimes referred to as the *breakdown torque,* because it limits the short-time overload capabilitiy of the motor. Higher starting torque can be obtained by inserting external resistances in the rotor circuit, as usually done in the case of the wound-rotor induction motor. These resistances can be cut out for the normal running conditions in order to operate the machine with a higher efficiency. Recall that a synchronous motor has no starting torque. It is usually provided with a damper or amortisseur winding located in the rotor-pole faces. Such a winding acts like a squirrel-cage winding to make the synchronous motor start as an induction motor and come up almost to synchronous speed, with the dc field winding unexcited. If the load and inertia are not too large, the motor will pull into synchronism and act as a synchronous motor when the field winding is energized from a dc source.

So far the discussion of induction machines applies only to machines operating from a polyphase supply. Of particular interest is the single-phase induction machine (Photo 5.3.12), which is widely used as a fractional-horsepower ac motor supplying the motive power for all kinds of equipment in

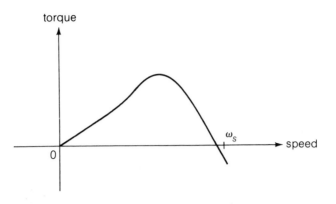

Figure 5.3.17 Approximate shape of the torque-speed curve for a
single-phase induction motor.

the home, office, and factory. For the sake of simplicity, let us consider a single-phase induction motor
with a squirrel-cage rotor and a stator carrying a single-phase winding, connected to a single-phase ac
supply. The primary mmf cannot be rotating, but, in fact, it is pulsating in phase with the variations in
the single-phase primary current. It can be shown, however, that any pulsating mmf can be resolved in
terms of two rotating mmfs of equal magnitude, rotating in synchronism with the supply frequency but
in opposite directions (see Section 5.4). The wave rotating in the same direction as the rotor is known
as the *forward-rotating wave,* while the one rotating in the opposite direction is the *backward-rotating
wave.* Then the slip, S_f, of the machine with respect to the forward-rotating wave is given by

$$S_f = \frac{\omega_s - \omega_m}{\omega_s} \tag{5.3.32}$$

which is the same as Equation 5.3.29. The slip S_b of the machine with respect to the backward-rotating
wave, however, is given by

$$S_b = \frac{-\omega_s - \omega_m}{\omega_s} = 2 - S_f \tag{5.3.33}$$

S_f and S_b are known as the forward (or positive-sequence) slip and the backward (or negative-sequence)
slip, respectively. Assuming that the two components mmfs exist separately, the frequency and magnitude
of the component emfs induced in the rotor by their presence will, in general, be different because S_f is
not equal to S_b. Then the machine can be thought of as producing a steady total torque as the algebraic
sum of the component torques. At standstill, however, $\omega_m = 0$ and the component torques are equal
and opposite; no starting torque is produced. Thus it is clear that a single-phase induction motor is not
capable of self-starting, but it will continue to rotate once started in any direction. In practice, additional
means are provided to get the machine started (usually as an asymmetrical two-phase motor) from a
single-phase source, and the machine is then run as a single-phase motor. An approximate shape of
the torque-speed curve for the single-phase motor can readily be obtained from that of a three-phase
machine shown in Figure 5.3.16. Corresponding to a positive slip of $S_f = OP$ in Figure 5.3.16, the
positive-sequence torque is PQ; then, corresponding to $S_b = 2 - S_f = OR$, the negative-sequence torque
is RS. The resultant torque is given by $(PQ - RS)$. This procedure can be repeated for a range of slips
$(1 \leq S \leq 0)$ to give the general form of the torque-speed characterisitc of a single-phase induction
motor, as shown in Figure 5.3.17.

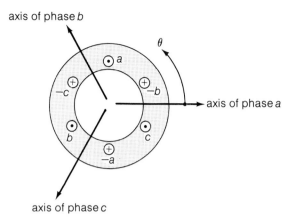

axis of phase b

θ

axis of phase a

axis of phase c

Figure 5.4.1 Simple two-pole, three-phase winding arrangement on a stator.

▮▮▮▮▮▮▮▮ |||||||
5.4 Rotating Magnetic Fields

In this section we set out to show, as stated earlier, that a rotating field of constant amplitude and sinusoidal space distribution of mmf around a periphery of the stator is produced by a three-phase winding located on the stator and excited by balanced three-phase currents when the respective phase windings are wound $2\pi/3$ electrical radiance (or 120 electrical degrees) apart in space. Let us consider the two-pole, three-phase winding arrangement on a stator shown in Figure 5.4.1. The windings of the individual phase are displaced by 120 electrical degrees from each other in space around the air-gap periphery. The reference directions are given for positive phase currents. The concentrated full-pitch coils, shown here for simplicity and convenience, do in fact represent the actual distributed windings producing sinusoidal mmf waves centered on the magnetic axes of the respective phases. Thus these three sinusoidal mmf waves are displaced by 120 electrical degrees from each other in space. Let a balanced three-phase excitation be applied with phase sequence a-b-c:

$$i_a = I \cos \omega_s t; \qquad i_b = I \cos(\omega_s t - 120°); \qquad i_c = I \cos(\omega_s t - 240°) \qquad (5.4.1)$$

where I is the maximum value of the current, and the time $t = 0$ is arbitrarily chosen when the a-phase current is a positive maximum. Each phase current is an ac wave varying in magnitude sinusoidally with time. Hence the corresponding component mmf waves vary sinusoidally with time. The sum of these components yields the resultant mmf.

Analytically, the resultant mmf at any point at an angle θ from the axis of phase a is given by

$$F(\theta) = F_a \cos \theta + F_b \cos(\theta - 120) + F_c \cos(\theta - 240°) \qquad (5.4.2)$$

But the mmf amplitudes vary with time according to the current variations:

$$F_a = F_m \cos \omega_s t; \qquad F_b = F_m \cos(\omega_s t - 120°); \qquad F_c = F_m \cos(\omega_s t - 240°) \qquad (5.4.3)$$

Then, on substitution, it follows that

$$F(\theta, t) = F_m \cos \theta \cos \omega_s t + F_m \cos(\theta - 120°) \cos(\omega_s t - 120°)$$
$$+ F_m \cos(\theta - 240°) \cos(\omega_s t - 240°) \qquad (5.4.4)$$

By the use of the trigonometric identity

$$\cos \alpha \cos \beta = \frac{1}{2} \cos (\alpha - \beta) + \frac{1}{2} \cos (\alpha + \beta)$$

and noting that the sum of three equal sinusoids displaced in phase by 120° is equal to zero, Equation 5.4.4 can be simplified as

$$F(\theta, t) = \frac{3}{2} F_m \cos (\theta - \omega_s t) \tag{5.4.5}$$

which is the expression for the resultant mmf wave. It has a constant amplitude $(3/2)F_m$, is a sinusoidal function of the angle θ, and rotates in synchronism with the supply frequency; hence it is called a rotating field. The constant amplitude is 3/2 times the maximum contribution F_m of any one phase. The angular velocity of the wave is $\omega_s = 2\pi f_s$ electrical radians per second, where f_s is the frequency of the electrical supply in hertz. For a P-pole machine, the rotational speed is given by

$$\omega_m = \frac{2}{P} \omega_s \text{ rad/s} \quad \text{or} \quad n = \frac{120 f_s}{P} \text{ rpm} \tag{5.4.6}$$

which is the synchronous speed.

The same result may be obtained graphically, as shown in Figure 5.4.2, which shows the spatial distribution of the mmf of each phase and that of the resultant mmf (given by the algebraic sum of the three components at any given instant of time). Part a applies for that instant when the a-phase current is a positive maximum; part b refers to that instant when the b-phase current is a positive maximum; the intervening time corresponds to 120 electrical degrees. It can be seen from Figure 5.4.2 that, during this time interval, the resultant sinusoidal mmf waveform has traveled (or rotated through) 120 electrical degrees of the periphery of the stator structure carrying the three-phase winding. That is to say, the resultant mmf is rotating in synchronism with time variations in current, with its peak amplitude remaining constant at 3/2 times that of the maximum phase value. Note that the peak value of the resultant stator mmf wave coincides with the axis of a particular phase winding when that phase winding carries its peak current. The graphical process can be continued for different instants of time to show that the resultant mmf is in fact rotating in synchronism with the supply frequency.

Although the analysis here is carried out only for a three-phase case, it holds good for any q-phase ($q > 1$; i.e., polyphase) winding excited by balanced q-phase currents when the respective phases are wound $2\pi/q$ electrical radians apart in space. However, in a balanced two-phase case, note that the two phase windings are displaced 90 electrical degrees in space, and the phase currents in the two windings are phase-displaced by 90 electrical degrees in time. The constant amplitude of the resultant rotating mmf can be shown to be $q/2$ times the maximum contribution of any one phase. Neglecting the reluctance of the magnetic circuit, the corresponding flux density in the air gap of the machine is then given by

$$B_g = \frac{\mu_0 F}{g} \tag{5.4.7}$$

where g is the length of the air gap.

Production of Rotating Fields from Single-Phase Windings

In this subsection, we show that a single-phase winding carrying alternating current produces a stationary pulsating flux that can be represented by two counter-rotating fluxes of constant and equal magnitude.

Let us consider a single-phase winding, shown in Figure 5.4.3a, carrying alternating current $i = I \cos \omega t$. This winding will produce a flux-density distribution whose axis is fixed along the axis of the

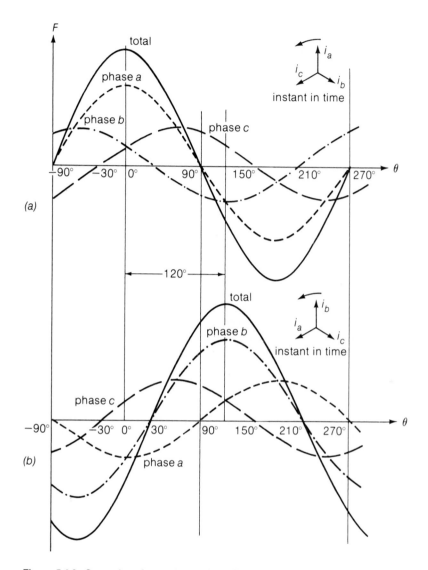

Figure 5.4.2 Generation of a rotating mmf. (a) Spatial mmf distribution at the instant in time when the *a*-phase current is a maximum. (b) Spatial mmf distribution at the instant in time when the *b*-phase current is a maximum.

winding and that pulsates sinusoidally in magnitude. The flux density along the coil axis is proportional to the current and is given by $B_m \cos \omega t$, where B_m is the peak flux density along the coil axis.

Let the winding be on the stator of a rotating machine with uniform air gap, and let the flux density be sinusoidally distributed around the air gap. Then the instantaneous flux density at any position θ from the coil axis can be expressed as

$$B(\theta) = (B_m \cos \omega t) \cos \theta \tag{5.4.8}$$

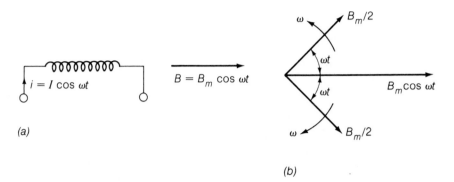

(a)

(b)

Figure 5.4.3 Single-phase winding carrying ac, producing a stationary pulsating flux or equivalent rotating flux components.

which may be rewritten by the use of the trigonometric identity as

$$B(\theta) = \frac{B_m}{2} \cos(\theta - \omega t) + \frac{B_m}{2} \cos(\theta + \omega t) \tag{5.4.9}$$

The sinusoidal flux-density distribution given by Equation 5.4.8 can be represented by a vector $B_m \cos \omega t$ of pulsating magnitude on the axis of the coil, as shown in Figure 5.4.3b. Alternatively, as suggested by Equation 5.4.9, this stationary pulsating flux-density vector can be represented by two counter-rotating vectors of constant magnitude $B_m/2$, as shown in Figure 5.4.3b. While Equation 5.4.8 represents a standing space wave varying sinusoidally with time, Equation 5.4.9 represents the two rotating components of constant and equal magnitude, rotating in opposite directions at the same angular velocity given by $d\theta/dt = \omega$. The vertical components of the two rotating vectors in Figure 5.4.3b always cancel, and the horizontal components always yield a sum equal to $B_m \cos \omega t$, the instantaneous value of the pulsating vector.

As indicated earlier, this principle is often used in the analysis of single-phase machines. The two rotating fluxes are considered separately, as if each represents the rotating flux of a three-phase machine, and the effects are then superimposed. If the system is linear, the principle of superposition holds and yields correct results. In a nonlinear system with saturation, however, one must be careful in reaching conclusions since the results are not as obvious as in a linear system.

Phase Splitting

Phase splitting is a technique to obtain two currents in a two-phase winding equal in magnitude and 90° displaced in phase, by using only a single-phase supply. The effect is the same as if the two identical coils were supplied by a two-phase supply.

Let us consider two windings, as shown in Figure 5.4.4a, displaced by 90° in space and connected to the same single-phase supply of voltage V. Let the inductive impedance of each coil be $Z_L = R_L + jX_L$, in which R_L is usually much less than X_L. Let a resistance R_S and a capacitance of reactance X_C be connected in series with coil B, as shown. The total impedance of the circuit B is then given by

$$Z_B = R_B + jX_B = (R_L + R_S) - j(X_C - X_L) \tag{5.4.10}$$

If the values are chosen so that

$$R_B = X_L; \qquad X_B = R_L \tag{5.4.11}$$

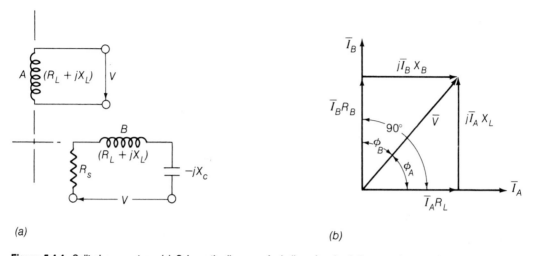

(a)

(b)

Figure 5.4.4 Split-phase system. (a) Schematic diagram of windings for simulating two-phase performance from a single-phase supply. (b) Phasor diagram for part (a).

then $|Z_B| = |Z_L|$, which gives the two currents the same magnitude but a phase difference of 90°, as seen from the phasor diagram for the split-phase system in Figure 5.4.4b. A two-phase system is simulated, and a rotating flux of constant magnitude is produced, as discussed earlier.

The system described above (with some modifications) is often used in a single-phase induction motor so it can start itself. If the values used for R_S and X_C are such that the two currents $\bar{I}_A$ and $\bar{I}_B$ are not exactly equal in magnitude, or not exactly 90° out of phase, then the system becomes an unbalanced two-phase system.

5.5 Basic Aspects of Electromechanical Energy Converters

As one goes deeper into the study of rotating machines, detailed differences and particularly challenging problems emerge among various machine types. In spite of a number of special problems peculiar to a given class of machines, one group of interrelated problems are common to all machine types. This section briefly touches on these interrelated problems, including losses and efficiency, ventilation and cooling, machine ratings, magnetic saturation, leakage and harmonic fluxes, and problems generally associated with machine applications. Matters relating to machine ratings, insulation, and allowable temperature rise, as well as determination of losses, are the subject of standardization by professional organizations like the Institute of Electrical and Electronics Engineers (IEEE), the National Electrical Manufacturers Association (NEMA), the American National Standards Institute (ANSI), and the International Electrotechnical Commission (IEC).

The limitations on the performance of a machine stem from the properties of the materials of which it is made. Much of the great progress made over the years in electric machinery is due to the improvements in the quality and characteristics of steel and insulating materials and to innovative cooling methods. In fact, advancement in cooling methods is the single most important factor making it possible to increase power ratings of turboalternators for a given physical size. The steady increase in the horsepower rating from 7.5 to 120 hp, obtained for a given size of an induction-motor frame over a period of the last eight decades, illustrates the effect of continued research in science and technology.

Losses and Efficiency

Losses are an inevitable part of the workings of any energy-transforming device, even though loss plays essentially no role in the energy-conversion process. For transformers, the efficiency of a machine is given by

$$\text{Efficiency} = \frac{\text{output}}{\text{input}} = \frac{\text{input} - \text{losses}}{\text{input}} = 1 - \frac{\text{losses}}{\text{input}} = \frac{\text{output}}{\text{output} + \text{losses}} \tag{5.5.1}$$

The efficiency of an electric machine is commonly determined by measuring losses instead of directly measuring the input and output under load conditions. Many definitions and methods of measuring losses and specifying efficiencies, as well as other performance characteristics, are standardized by international professional agencies; refer to these standards as necessary.

Typical values of full-load efficiencies for rotating machines are about 50% for fractional-horsepower motors, 75% for 1-hp motors, 90% for 100-hp motors, and 98% for 100-MVA turboalternators. Rotating machines generally operate quite efficiently except at light loads. For transformers, maximum efficiency occurs at a load for which variable losses are equal to constant losses. Because most practical machines operate at about 80–90% of full load (thus allowing for growth in demand) or reserve capacity in emergencies, it is common practice to design machines with maximum efficiency at about 80% of their full load.

Besides affecting the efficiency and hence the operating cost of the machine, losses determine machine heating and consequently the rating or power output that a machine can produce without overheating and deteriorating its insulation over a reasonable period. Also, the current components for supplying the losses and the associated voltage drops affect regulation and other performance characteristics of a machine. The losses in a machine are classified in terms of physical phenomena and methods of measurement and testing. These fall into four basic classes:

1. *No-load* or *open-circuit core losses* consist of the hysteresis and eddy-current losses discussed in Section 3.2 under "Iron Losses." No-loaded core losses occur because of the periodic magnetic reversals in the ferromagnetic parts of the machine. In dc and synchronous machines, the principal core losses due to the main flux (ignoring the effect of spatial harmonics in flux density due to slots and nonsinusoidal current density) are confined to the armature side of the air gap. In induction machines, these losses are largely confined to the stator.

   Additional core losses result from harmonic components of flux (such as slot harmonics), harmonic components of mmf, and end-leakage flux caused in the end portions of the windings. Losses due to the dominating alternating effect in the teeth are known as *tooth losses,* which may be quite pronounced in induction machines. In the case of dc and synchronous machines, the rotational effect is predominant at the surface of the pole, causing *surface losses.* Laminating the iron in the pole shoes can minimize these losses. The losses due to harmonic components of mmf and end-leakage flux generally depend upon the load and increase with the load on the machine.

   The principal core losses (hysteresis and eddy-current losses) are, for the most part, constant at a given voltage. No-load core losses are considered the common loss of the machine, because they are nearly independent of the load. Other core losses are usually included in *stray-load losses,* discussed in item 4.

2. *Mechanical losses* are *friction* and *windage losses* caused by friction of bearings, brushes, and windage. Mechanical losses are generally functions of the machine speed. They are usually assumed to be practically constant for small speed variations. The sum of the mechanical losses plus the no-load core losses is called *no-load rotational losses,* which are effectively constant.

3. *Copper losses* are losses that consist principally of the I^2R losses in the windings, where R is taken to be the dc resistance of the winding, usually corrected to 75°C to allow for the effect of the usual

temperature rise under normal operation. The effect of the brush-contact resistance in dc machines is conventionally taken into account by asssuming a full-load drop of 2 volts in series with the armature circuit. The principal copper losses, which depend on the square of the load current, vary much more widely than the no-load rotational losses.

Additional copper losses result from the increase in winding resistance caused by the skin effect and the consequent nonuniform current-density distribution in the conductors. Also, the mmf and slot harmonics produce I^2R lossses in induction machines and in the damper windings of synchronous machines. These additional copper losses are generally included with the stray-load losses discussed in item 4.

4. *Stray-load losses* arise from the nonuniform current distribution in the conductors and the additional core losses produced in the iron by the distortion of the magnetic flux distribution caused by the load currents. It is usually difficult to determine the stray-load losses accurately; estimates are based on tests, experience, and judgment. Although these losses may vary from 0.5 percent of the output in large machines to 5 percent of the output in medium-size machines, it is customarily assumed to be 1 percent of the output in dc machines.

Ventilation and Cooling

In electrical machines, as in any other energy-conversion device, some of the energy is dissipated as heat during operation. The problem of transferring heat without increasing the temperature beyond reasonable limits in parts of the electrical machine can be very complex. The difficulty generally increases with increasing size, because the surface area from which the heat is to be carried away increases approx- imately as the square of the dimensions, whereas the heat developed by the losses increases roughly proportionally to the cube of the dimensions.

Machines have two types of ventilating systems: the self-ventilating system, in which ventilating devices such as fans are mounted on the machine shaft, and the separate (or independent) ventilating system, in which ventilating devices, independent of the actual machine, are externally driven. Both systems can be open circuit or closed circuit. The stator and rotor cores are usually provided with radial ducts for the passage of air or any other cooling medium, such as hydrogen. The flow of the ventilating medium can be axial in small machines, radial in large machines, or both in larger machines. The draft of the cooling medium can be induced or forced. Closely associated with ventilating systems are the types of enclosures commonly employed for the machines: open, protected, or totally enclosed. The type of enclosure is determined by considerations of safety and the environment in which the machine is to be used. For example, machines in coal mines and chemical plants are totally enclosed to protect the machine parts from particles of corrosion.

The cooling problem is particularly serious in large turboalternators where economy, mechanical requirements, transportation limitations, and erection and assembly problems demand compact config- uration, especially for the rotor forging. Rather elaborate systems of cooling ducts must be provided to ensure that the cooling medium effectively removes most of the heat arising from the losses. Closed ventilating systems are commonly used for moderately sized machines. For these, the cooling medium (such as air or hydrogen), after being heated by its passage through the machine, is passed through a heat exchanger (with air or water as the cooling agent) and then recirculated through the machine. Modern large turboalternators quite commonly have water-cooled stators (with hollow passages in their conductors through which cooling water circulates in direct contact with the copper conductors) and hydrogen-cooled rotors. Water-cooling is employed in both the stator and the rotor in a few cases. Typical cooling arrangements for large turbogenerators are shown in Photos 5.5.1 and 5.5.2. Direct- conductor or inner cooling, either by liquid or gas under 1 to 5 atmospheres of pressure, is the most

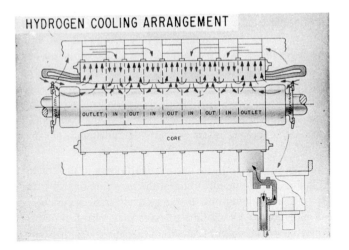

Photo 5.5.1 Hydrogen cooling arrangement for a large two-pole turbogenerator. (Photo courtesy of General Electric Company.)

Photo 5.5.2 Ventilating scheme for a large turboalternator. (Photo courtesy of Electric Machinery Manufacturing Company.)

important single improvement in modern cooling. It is now possible to double the output of a turbine generator for a physical size and increase the efficiency of the unit to more than 99%. A rating of 1,500 MVA or more for a single unit is quite common these days for machines used for power generation in electric utility systems. The constructional details of machines greatly depend on the type of cooling methods employed. Complex problems may arise in appropriately sealing the bearings when the coolant is hydrogen under pressure. For hydroelectric generators the design of thrust bearings may pose serious problems because water-turbine construction requires vertical mounting instead of the usual horizontal mounting.

Machine Ratings

The temperature rise resulting from the losses in a machine plays an important role in rating a machine, because machine life should not be unduly shortened by overheating and consequent deterioration of the insulating materials surrounding the current-carrying conductors from the core. Life expectancy of a large industrial electric machine ranges from 10 to 50 years or more. It can be a matter of minutes in military or missile applications, and it can be on the order of a few thousand hours in aircraft and electronic equipment. Insulating materials employed in electrical machinery are classified according to the maximum allowable temperature rise that can safely be withstood. Table 5.5.1 gives the IEEE classification of electrical insulating materials by tolerance to heat.

For each class of insulation, the permissible rise of temperature depends on the type of enclosure and cooling method used. A *general-purpose motor* must have a lower permissible temperature than a *special purpose motor,* to allow a greater factor of safety in view of the unknown service conditions. The general-purpose motor has a standard rating up to 200 hp (with 450 rpm or more) with standard operating characteristics and mechanical construction for use under normal service conditions without restriction to a particular application or type of application. A special-purpose motor is designed either to achieve some particularly desirable operating characteristics or to satisfy a particular mechanical requirement, or both.

The ASA (American Standards Association) recommends that a general-purpose motor should be able to operate successfully at 115% of its rating continuously as long as (1) the voltage does not vary more than $\pm 10\%$ from the rated value, (2) the frequency does not vary more than $\pm 5\%$ from the rated value, and (3) the sum of these two percentages is not more than 10 at any one time. *Service factor* is another rating term. It is a multiplier which, when applied to the rated output, gives a permissible loading that can be carried continuously under the specified conditions for that service factor. Thus, for general-purpose motors, a service factor of 1.15 is allowed. Consult the NEMA and IEEE standards for the allowable temperature rise for machine types.

Table 5.5.1 Thermal Classification of Electrical Insulating Materials

Class	Maximum Permissible Temperature Rise in °C Beyond the Ambient Temperature of 40°C	Materials
O	50	paper, cotton, silk
A	65	cellulose, phenolic resins
B	90	mica, glass, asbestos with organic binder
F	115	same as above, with suitable binder
H	140	mica, glass, asbestos with slicone binder, silicone resin, teflon
C	>180	mica, porcelain, glass

The rating of a machine gives its working capabilities under specified electrical and environmental conditions. The most common machine rating is called the *continuous rating,* which gives the output (in kW for dc generators, in kVA at a specified power factor for ac generators, and in hp for motors) that can be carried indefinitely without exceeding the specified temperature rise. In contrast, the *short-time rating* is given to some machines, defining the load that can be carried without overheating for a specified time due to intermittent, periodic, or varying duty. Standard periods for short-time ratings are 5, 15, 30, and 60 minutes. The duty cycle is usually represented by the horsepower–time curve, and a motor rating is normally chosen on the basis of the rms value given by

$$\text{rms hp} = \sqrt{\frac{\sum \text{hp}^2(\text{time})}{\text{running time} + (\text{standstill time}/k)}} \tag{5.5.2}$$

in which k is a constant accounting for reduced ventilation at standstill; $k = 3$ for an open motor. In practice, the result is rounded off to the next higher commercially available motor size. Special features, like frequent starting or reversal and high torque peaks in the duty cycle, must be given adequate consideration in the selection of a motor. Special-purpose motors with short-time ratings are, in general, designed to have better torque-producing ability than continuously rated motors that produce the same power output; however, the special-purpose motors have a lower thermal capacity.

In general, every machine has a nameplate attached to the frame inscribed with relevant information regarding the voltage, current, power, power factor, speed, frequency, phases, and allowable temperature rise. The nameplate rating is the continuous rating, unless otherwise specified. Ac generators and transformers are rated in terms of kVA rather than kW because their losses and heating are approximately determined by the voltage and current, regardless of power factor. The physical size and cost of ac power-system apparatus are roughly proportional to the kVA rating.

Magnetic Saturation

From the study of ferromagnetic materials and their properties in Section 3.2, we have seen that the magnetization B-H characteristic is generally linear up to a point (on the knee of the curve), beyond that it is generally nonlinear. When a magnetic material is saturated, the mmf required to produce a given flux is greater than that required under unsaturated conditions. In other words, a given mmf will produce less flux under saturation than under unsaturated conditions. For specified mmfs in the windings of a rotating machine, the fluxes depend not only on the reluctance of the air gaps but also on that of the iron portions of the magnetic circuits that may be saturated. In order to fully utilize the magnetic properties of the iron and optimize the machine design, the machine iron is worked at fairly saturated levels of flux density, such that the normal operating point on the open circuit is near the knee of the *open-circuit characteristic* (or the *no-load saturation curve,* which is similiar to the magnetization B-H characteristic) given by the plot of the open-circuit geometry (involving air and iron) of the machine. Saturation may therefore influence the machine performance characteristic to a considerable degree.

As a first crude approximation, neglect the reluctance of the iron portions of the magnetic circuit, and only consider air gaps for mmf calculations. As a next step towards refinement, assume that iron portions have constant permeability and can be included in the calculations. As a further step, depending on the actual working saturated conditions of the machine, permeabilities of the iron can be accounted for. This is analytically rather difficult, since saturation affects different parts of the machine to varying degrees; in addition, the load saturation curve is not the same as the open-circuit (no-load) characteristic. However, serious research efforts are focusing on reproducing the magnetic conditions at the air gap correctly by increasing the air-gap length to a sufficient degree to account for the effects of saturation and air-gap nonuniformities due to slots and ventilating ducts. Final experimental tests results will show the validity of the approach adopted. From an engineering point of view, such an approach may be adequate for a number of situations. In other cases, however, where there is too much discrepancy between the test and computed results, more refined methods of calculation are needed.

Several empirical and approximate methods to account for the effects of magnetic saturation in machines do work well within certain limits under specified conditions for certain classes of machines. Linear mathematical models and circuit techniques, for instance, will always retain their important place for analyzing and understanding rotating machines. Substantially better understanding of the fundamental phenomena (including saturation) awaits exploration of the determination of flux distribution within a machine at various conditions of operation. But, field-theory methods of analysis are quite complex.

Leakage and Harmonic Fluxes

The mutual flux linking the stator and rotor windings has been assumed to be sinusoidally distributed in space in our analysis so far. In real machines, we find fluxes that do not link both the windings; they are called *leakage fluxes*. Examples of these include slot-leakage flux due to slots, end-leakage flux due to end connections of the winding, and pole-to-pole leakage in salient-pole structures. The effect of the leakage flux, for transformers, for instance, is accounted for by defining leakage inductances that will cause the inductive voltage drop of fundamental frequency induced in ac windings. The paths that the leakage fluxes take in a practical machine are very complicated; the load and saturation also affect these paths considerably. Exact methods of calculating leakage inductances are usually too involved to be used, but empirical formulas have been developed for particular situations to yield reasonable solutions.

In addition to leakage fluxes, the air gap has *harmonic fluxes*. That is, the flux-density distribution produced by a winding contains not only a fundamental component but also several other harmonics in space. As we will see later, in Chapter 6 on machine windings, it is possible, however, to reduce harmonic fluxes and even to completely eliminate one or more undesirable harmonic components. Harmonic fluxes, in addition to causing harmonics in induced emf, produce secondary effects in the energy-conversion process itself. They develop undesirable *parasitic torques* that may become responsible for vibration and noise. In induction motors, for example, the problem may become rather serious: For certain stator and stator slot numbers, slot-harmonic fluxes may cause the machine to *cog*. The seventh harmonic flux will develop a torque that may keep the machine, when started from standstill, from rising above one-seventh the normal speed and make the machine *crawl*. A detailed discussion of harmonic fluxes and their effects on machine performance is outside the scope of this book.

General Nature of Machine-Application Problems

At the outset, we can state that several important, interesting, challenging, and complex engineering problems spring from the design, development, and manufacture of rotating machines—most of them beyond the scope of this book. We would like to touch upon the general nature of machine-application considerations, however.

For motors the major consideration is the torque-speed characteristics; for generators it is the volt-ampere or voltage-load characteristics. Also vital are the limits between which these characteristics can be varied and how to obtain such variations. Further, we need to investigate relevant economic features, like the efficiency, power factor, relative costs, and the effect of losses on the heating and machine rating.

The motor, which is generally supplied with electric power from a constant-voltage source, drives a mechanical load whose torque requirements vary with the speed at which the motor is driven. The point at which the electromagnetic torque supplied with the motor equals the mechanical torque that the load can absorb is the steady-state operating speed. The requirements of motor loads generally vary from one application to another. In ordinary hydraulic pumps, the speed remains approximately constant while the load varies. Compact disc players require absolutely constant speed. Cranes and traction-type drives need a varying speed characteristic with heavy torques at low speeds and light torques at high speeds. Machine tool drives can demand constant speeds adjustable over a wide range, while still other machines may need adjustable varying speed. For any motor application, the starting torque, maximum torque, and running characteristics, along with the current requirements, should be addressed.

Similiar considerations apply for machines operating as generators. A generator's terminal voltage and power output are fixed by the characteristics of both the generator and its load. When the electrical load is connected to the generator, an operating point is attained such that the generator gives exactly what the load can take. Normally the terminal voltage must remain substantially constant over a wide range of load. On the other hand, the terminal voltage might be required to vary with load to provide greater flexibility and better control in certain applications.

A generator, even of the size of 1,500 MVA or more, is only one component in a complex modern power system. Similarly, a motor may be only one component in an involved electromechanical system for an industrial application demanding dynamic controls of great accuracy and rapid response. In such system-related applications, the electromechanical transient behavior of the system as a whole becomes a major consideration. Thus it becomes necessary for us to study not only the steady-state but also the transient (or dynamic) analysis and behavior of rotating machinery. It is one thing to study the detailed dynamic behavior of a rotating machine as an individual unit; it is quite different when the same machine is a component in a major system. The required degree of representational detail, the mathematical models, and the pertinent simplifying assumptions all depend on the particular problem at hand. This interdependence underscores how essential a thorough understanding of both the fundamentals and the orders of magnitude of the dynamic effects can be for choosing the right model with the available data.

5.6 The Design Aspect of AC Machines

The design process is fundamentally that of compromise. Nowhere is this more amply demonstrated than in the design of rotating electrical machines. The electrical, mechanical, and thermal objectives—which must be simultaneously satisfied—require that the constraints imposed by one aspect of the design specification be considered while other aspects are optimized.

Innovative design and development remain the cornerstones of progress. With the advent of computational robots and efficient computer software over the last several decades, the procedures of machine design have significantly changed. Design optimization has become a realistic objective in a good number of cases. Today the design engineer is more often concerned with arbitrating design calculations than with actually performing them. This shift naturally demands that the designer have a thorough understanding of the design process, since a design that appears to be the most technically proficient may not yield optimal reliability or economy. The final decision obviously requires high skill and experience. The primary considerations in designing an ac machine are given below.

Output Coefficient

The output coefficient is defined as

$$C = W/D^2 L \omega_s \tag{5.6.1}$$

where W = air-gap power (W)
   D = stator bore diameter (m)
   L = stator core length (m)
   ω_s = synchronous angular speed (rad/s)

Noting that $D^2 L$ is closely related to the volume of the machine and that W/ω_s is related to the torque at rated output, C then gives a measure of rated torque per unit volume. The higher C is, the better utilized the material of the machine is. Although the output coefficient is an important design criterion, due consideration must also be given to the costs of materials and manufacture, temperature rises, efficiency, mechanical integrity, and other application requirements.

Electric Loading

Electric loading is a measure of the armature strength and is expressed as the amperes per unit of air-gap circumference in the armature winding. The electric loading (A/m) is the amperes per conductor times the conductors around the circumference, all divided by the circumferential length. It is also equal to the total current in an armature slot divided by the slot pitch. The upper limit for the electric loading is usually determined by the ability of the machine to dissipate the heat generated.

Magnetic Loading

Magnetic loading is normally expressed as the flux density in the air gap, as either average or maximum density. The average density is the total flux per pole divided by the pole area, whereas the maximum density is the density at the center of the pole. The magnetic loading is to be specified in relation to the excitation, saturation conditions, core loss, and power factor. It determines the field strength or magnetization requirements of the machine and can only vary over a restricted range.

   The magnetic and electric loadings compete for space in a given frame size. It can be seen that an increase in one requires a reduction in the other, which sometimes give rise to comparisons between a "copper design" and an "iron design." In view of the relatively high cost of copper, an "iron design" is generally preferred.

   The output coefficient C can be shown to be proportional to the product of electric and magnetic loadings and is thus a direct measure of how effectively the space has been used to achieve the product. Having decided the proportion of electric and magnetic loadings that will yield the flux per pole, the designer can calculate the number of conductors per phase required to give that flux by the use of the voltage equation (Equation 5.3.15), determine the number of slots to give a balanced winding, and calculate the number of slots per pole per phase.

   From the viewpoint of mechanical considerations and various reactances, the designer must decide the air-gap length. The magnetic circuit can then be designed and analyzed for satisfactory flux-density levels in various parts.

EXAMPLE 5.6.1

The mean air-gap flux density in a rotating machine is given as 0.50 T. The slots are spaced with a slot pitch of 40 mm. Each slot contains a conductor of section 35 mm × 12 mm, with a current density of 3.3 A/mm^2. Calculate the tangential force per unit length of air-gap periphery and the per-unit axial length of the machine.

Solution
The specific electric loading is

$$(35 \times 12)3 \cdot 3 \times 1000/40 = 35 \text{ kA/m}$$

The specific magnetic loading is 0.5 T. Hence the specific force is the product of electric and magnetic loadings, given by

$$0.5 \times 35,000 = 17,500 \text{ N/m}^2$$

which is typical of a medium-rated machine. The torque depends on the diameter and axial length of the machine.

BIBLIOGRAPHY

Ames, Robert L. *A.C. Generators: Design and Application.* New York: John Wiley & Sons, Inc., 1990.

Andreas, J. C. *Energy-Efficient Electric Motors: Selection and Application.* New York: Marcel Dekker, Inc., 1982.

Chalmers, Brian, and Alan Williamson. *A.C. Machines: Electromagnetics and Design.* New York: John Wiley and Sons, Inc., 1991.

Fitzgerald, A. E.; C. Kingsley, Jr.; and A. Kusco. *Electric Machinery,* 3d ed. New York: McGraw-Hill, 1971.

Kuhlmann, John H. *Design of Electrical Apparatus,* 3d ed. New York: John Wiley and Sons, Inc., 1950.

Majmudar, H. *Electromechanical Energy Converters.* Boston: Allyn and Bacon, 1965.

Matsch, L. W. *Electromagnetic and Electromechanical Machines.* New York: Intext Educational Publishers, 1972.

Morgan, A. T. *General Theory of Electrical Machines.* London: Heyden & Son, 1979.

Nasar, S. A., and L. E. Unnewehr. *Electromechanics and Electric Machines.* New York: John Wiley & Sons, 1979.

Say, M. G. *Alternating Current Machines.* New York: John Wiley and Sons, Inc., 1976.

Slemon, G. R., and A. Straughen. *Electric Machines.* Reading, Mass.: Addison Wesley, 1980.

White, D. C., and H. H. Woodson. *Electromechanical Energy Conversion.* New York: New York: John Wiley & Sons, 1959.

Woodson, H. H., and J. R. Melcher. *Electromechanical Dynamics—Part I: Discrete Systems.* New York: John Wiley & Sons, 1968.

Yamamura, S. *A.C. Motors for High-Performance Applications: Analysis and Control.* New York: Marcel Dekker, Inc., 1987.

PROBLEMS*

5-1. Table 5.1.1 summarizes the relations associated with the mechanical force of electrical origin caused by the magnetic field coupling. Develop similiar relations for determining mechanical forces due to electric field coupling in an electromechanical system. Choose as the independent variables the space coordinate x on the mechanical side, and either the voltage v or the charge q on the electrical side.

5-2. Consider the electromagnetic plunger shown in Figure P5–2. The λ–i relationship for the normal working range is experimentally found to be

$$\lambda = \frac{Ki^{2/3}}{x + t}$$

where K is a constant. Determine the electromagnetic force on the plunger by applying Equations 5.1.13, 5.1.14, 5.1.19, and 5.1.20. Interpret the significance of the sign that you obtain in the force expression.

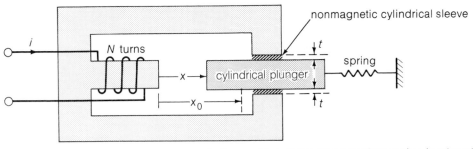

x_0 is the position of the plunger when spring is relaxed.

Figure P5–2

5-3. In Problem 5–2, neglect the saturation of the core, leakage, and fringing. Neglect also the reluctance of the ferromagnetic circuit. Assuming that the cross-sectional area of the center leg is twice the area of the outer legs, obtain expressions for the inductance of the coil and the electromagnetic force on the plunger.

5-4. Consider the system shown in Figure 5.1.2 and assume the λ–i characteristic of the system to be linear.

a. Draw a diagram similar to Figure 5.1.3a, and show that the electrical energy input is divided equally between increasing stored energy and doing mechanical work.

b. Draw a diagram similar to Figure 5.1.3b, and show that the mechanical work done equals the reduction in stored energy.

5-5. In the magnetic field structure shown in Figure P5–5 the rotor core is displaced 1 cm in an axial direction from its correct position. Assuming the air-gap length under the pole shoes to be constant, compute the axial force tending to bring the rotor to its correct position.

    Length of single air-gap = 0.25 cm;   Air-gap flux density = 0.8 T

    Rotor diameter = 25 cm;   Angle subtended by each pole shoe = 120°

* Asssume SI units unless otherwise stated.

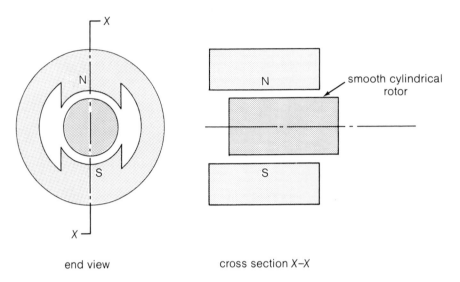

end view

cross section X–X

Figure P5–5

5–6. For the electromagnet shown in Figure P5–6, the λ–i relationship for the normal working range is given by

$$i = a\lambda^2 + b\lambda(x - d)^2$$

where a and b are constants. Determine the force applied to the plunger by the electrical system.

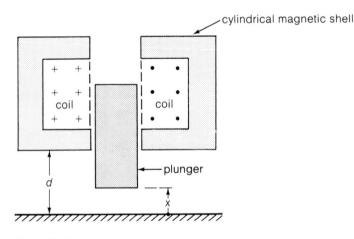

Figure P5–6

5–7. A solenoid of cylindrical geometry is shown in Figure P5–7.

   a. If the exciting coil carries a dc steady current I, derive an expression for the force on the plunger.

   b. For the numerical values $I = 10$ A, $N = 500$ turns, $g = 5$ mm, $a = 20$ mm, $b = 2$ mm, and $l = 40$ mm, find the magnitude of the force. Assume infinite permeability of the core, and neglect leakage.

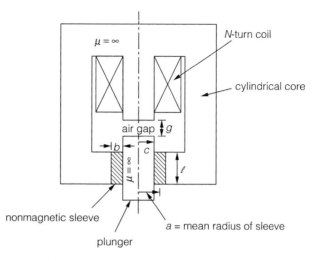

Figure P5–7

5–8.   Let the solenoid of Problem 5–7 carry an alternating current of 10 A (rms) at 60 Hz instead of the dc current.

a.   Find an expression for the instantaneous force.

b.   For the numerical values of N, g, a, b and l given in problem 5–7b compute the average force. Compare it to that of Problem 5–7b.

5–9.   Consider the solenoid with a core of square cross section shown in Figure P5–9.

a.   For a coil current of I (dc), derive an expresion for the force on the plunger.

b.   Given $I = 10$ A, $N = 500$ turns, $g = 5$ mm, $a = 20$ mm, and $b = 2$ mm, calculate the magnitude of the force.

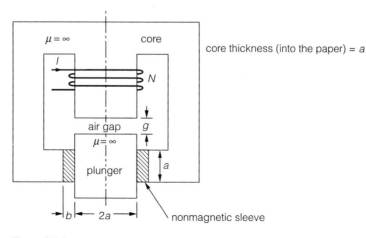

Figure P5–9

5-10   Let the coil of the solenoid of Problem 5–9 have a resistance R and be excited by a voltage $v = V_m \sin \omega t$. Consider a plunger displacement of $g = g_0$.

   a. Obtain the expression for the steady-state coil current.

   b. Obtain the expression for the steady-state electrical force.

5-11   A rotating machine of the form shown in Figure 5.2.1 has a coil inductance that can be approximated by

$$L(\theta) = 0.02 - 0.04 \cos 2\theta - 0.03 \cos 4\theta \text{ H}$$

A current of 5 A (rms) at 60 Hz is passed through the coil, and the rotor is driven at a speed, which can be controlled, of ω_m rad/s.

   a. Find the values of ω_m at which the machine can develop average torque.

   b. At each of the speeds obtained in part a, determine the maximum value of the average torque and the maximum mechanical power output.

5-12.   Consider an elementary rotating reluctance machine of the type shown in Figure 5.2.1, but with four rotor poles instead of two. The poles are so shaped that the reluctance of the magnetic system is given by

$$\mathbf{R}(\theta) = (4 \times 10^5 - 3 \times 10^5 \cos 4\theta) \text{ A/Wb}$$

The stator coil has 100 turns and negligible resistance. An ac voltage of 110 V (rms) at 60 Hz is applied to the coil terminals.

   a. Sketch the function of $\mathbf{R}$ versus θ.

   b. Determine the synchronous speed of the rotor and the maximum average torque that the machine can develop.

5-13.   A reluctance motor of the type illustrated in Figure 5.2.1 has a direct-axis reluctance $\mathbf{R}_d$ of 20×10^6 A/Wb and a quadrature-axis reluctance $\mathbf{R}_q$ of 60×10^6 A/Wb. Assume that the reluctance varies sinusoidally with rotor position as in Figure 5.2.1b. The stator winding has 5,000 turns and is excited by a 230-V, 50-Hz supply. Neglect the winding resistance.

   a. Determine the speed at which the machine can develop an average torque.

   b. Find the maximum average motoring torque and the power developed.

   c. Compute the rms value of the coil current.

5-14.   An electromagnetic relay may be modeled by its lumped-parameter system, as shown in Figure P5–14. Assume no externally applied mechanical force. Neglect saturation of the ferromagnetic circuit, which is considered to be infinitely permeable. Ignore leakage and fringing fluxes. Assume the friction force to be linearly proportional to the velocity, and the spring force to be linearly proportional to the elongation. Let the resistance of the electrical circuit be R and the inductance

$$L(x) = \frac{A}{B + x}$$

where A and B are constants. Obtain the equations of motion for the electrical and mechanical sides of the electromechanical system.

5-15.   A two-winding system has its inductances given by

$$L_{11} = \frac{k_1}{x} = L_{22}; \qquad L_{12} = L_{21} = \frac{k_2}{x}$$

where k_1 and k_2 are constants. Neglecting the winding resistances, derive an expression for the electrical force when both windings are connected to the same voltage source, $v = V_m \sin \omega t$. Comment on its dependence of x.

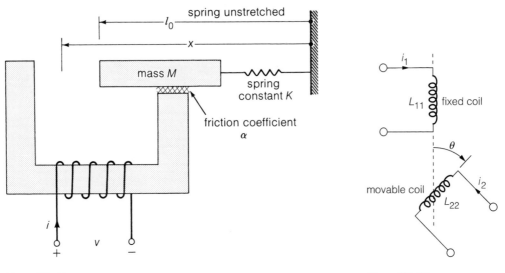

Figure P5-14   **Figure P5-16**

5-16   Two mutually coupled coils are shown in Figure P5–16. The inductances of the coils are $L_{11} = A$, $L_{22} = B$, and $L_{12} = L_{21} = C \cos \theta$. Find the electrical torque for

a. $i_1 = I_0$, $i_2 = 0$

b. $i_1 = i_2 = I_0$

c. $i_1 = I_m \sin \omega t$, $i_2 = I_0$

d. $i_1 = i_2 = I_m \sin \omega t$

e. Coil 1 short-circuited and $i_2 = I_0$

5-17.   Consider an elementary cylindrical-rotor two-phase synchronous machine with uniform air gap as illustrated in the schematic diagram in Figure P5–17. It is similar to that of Figure 5.2.5 except that Figure P5–17 has two identical stator windings in quadrature instead of one. The self-inductance of the rotor or field winding is a constant given by L_{ff} H; the self-inductance of each stator winding is a constant given by $L_{aa} = L_{bb}$; the mutual inductance between the stator windings is zero since they are in space quadrature; the mutual inductance between a stator winding and the rotor winding depends on the angular position of the rotor:

$$L_{af} = L \cos \theta; \qquad L_{bf} = L \sin \theta$$

where L is the maximum value of the mutual inductance, and θ is the angle between the magnetic axes of the stator a-phase winding and the rotor field winding.

a. Let the instantaneous currents be i_a i_b, and i_f in the respective windings. Obtain a general expression for the electromagnetic torque T_e in terms of these currents, angle θ, and L.

b. Let the stator windings carry balanced two-phase currents given by

$$i_a = I_a \cos \omega t; \qquad i_b = I_a \sin \omega t$$

and the rotor winding be excited by a constant direct current I_f. Let the rotor revolve at synchronous speed so that its instantaneous angular position θ is given by $\theta = \omega t + \delta$. Derive the torque expression under these conditions and describe its nature.

c. For conditions of part b, neglect the resistance of the stator windings. Obtain the volt-ampere equations at the terminals of stator phases a and b, and identify the speed-voltage terms.

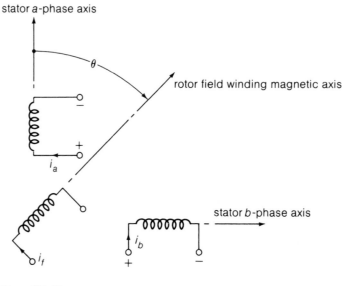

stator a-phase axis

rotor field winding magnetic axis

i_a

stator b-phase axis

i_b

i_f

Figure P5–17

5–18. Consider the machine configuration of Problem 5–17. Let the rotor be stationary and constant direct currents I_a, I_b, and I_f be supplied to the windings. Further, let I_a and I_b be equal. If the rotor is now allowed to move, will it rotate continuously or will it tend to come to rest? If the latter, find the value of θ for stable equilibrium.

5–19. Let us consider an elementary salient-pole, two-phase, synchronous machine with nonuniform air gap. The schematic representation is the same as for Problem 5–17. Structurally the stator is similar to that of Figure 5.2.5 except that this machine has two identical stator windings in quadrature instead of one. The salient-pole rotor with two poles is similar to that of Figure 5.2.4, carrying the field winding connected to slip rings.

The inductances are given as:

$$L_{aa} = L_0 + L_2 \cos 2\theta; \qquad L_{af} = L \cos \theta$$

$$L_{bb} = L_0 - L_2 \cos 2\theta; \qquad L_{bf} = L \sin \theta$$

$$L_{ab} = L_2 \sin 2\theta; \qquad L_{ff} \text{ is a constant, independent of } \theta.$$

L_0, L_2, and L are positive constants, and θ is the angle between the magnetic axes of the stator a-phase winding and the rotor field winding.

a. Let the stator windings carry balanced two-phase currents given by

$$i_a = I_a \cos \omega t; \qquad i_b = I_a \sin \omega t$$

and the rotor winding be excited by a constant direct current I_f. Let the rotor revolve at synchronous speed so that its instantaneous angular position θ is given by

$$\theta = \omega t + \delta$$

Derive an expression for the torque under these conditions and describe its nature.

b. Compare the torque with that of Problem 5–17b.

c. Can the machine be operated as a motor? As a generator? Explain.

d. Suppose that the field current I_f is brought to zero. Will the machine continue to run?

5–20. Consider the analysis leading up to Equation 5.2.33 for the torque of an elementary cylindrical machine with uniform air gap.

   a. Express the torque in terms of F, F_s, and δ_s, where δ_s is the angle between $\bar{F}$ and $\bar{F}_s$.

   b. Express the torque in terms of F, F_r, and δ_r, where δ_r is the angle between $\bar{F}$ and $\bar{F}_r$.

   c. Neglecting magnetic saturation, obtain the torque in terms of B, F_r, and δ_r, where B is the peak value of the resultant flux-density wave.

   d. Let ϕ be the resultant flux per pole given by the product of the average value of the flux density over a pole and the pole area. Express the torque in terms of ϕ, F_r, and δ_r, where ϕ is the resultant flux produced by the combined effect of the stator and rotor mmfs.

5–21. An electromagnetic structure is characterized by the following inductance:

$$L_{11} = L_{22} = 4 + 2\cos 2\theta; \qquad L_{12} = L_{21} = 2 + \cos\theta$$

Neglecting the resistances of the windings, find the torque as a function of θ when both windings are connected to the same ac voltage source such that

$$v_1 = v_2 = 220\sqrt{2}\sin 314t$$

5–22. By using the concept of interaction between magnetic fields, show that the electromagnetic torque cannot be obtained by using a four-pole rotor in a two-pole stator.

5–23. For each of the following devices, is a reluctance torque produced when their coils carry direct current?

   a. Salient-pole stator carrying a coil, and salient-pole rotor.

   b. Salient-pole rotor carrying a coil, and salient-pole stator.

   c. Salient-pole stator carrying a coil, and cylindrical rotor.

   d. Salient-pole rotor carrying a coil, and cylindrical stator.

   e. Cylindrical stator carrying a coil, and cylindrical rotor.

   f. Cylindrical rotor carrying a coil, and cylindrical stator.

   g. Cylindrical stator carrying a coil, and salient-pole rotor.

   h. Cylindrical stator carrying a coil, and salient-pole stator.

5–24. Consider an elementary doubly excited magnetic-field system, such as the one shown in Figure 5.2.4. The stator and rotor inductances are given by

$$L_{ss} = 0.75 + 0.25\cos 2\theta; \qquad L_{rr} = 0.45 + 0.15\cos 2\theta; \qquad L_{sr} = 0.8\cos\theta$$

Let both windings carry a constant current of 1 A.

   a. Calculate the torque when $\theta = 45°$.

   b. If the rotor moves slowly from $\theta = 90°$ to $\theta = 0°$, find
   (1) the work done
   (2) the change in stored energy
   (3) the electrical input

   c. If the rotor rotates at a speed of 200 rad/s with constant winding currents of 1 A, determine the generated emfs e_s and e_r at the instant when $\theta = 45°$.

5–25. An elementary two-pole rotating machine with uniform air gap, as shown in Figure 5.2.5, has a stator-winding self-inductance L_{ss} of 50 mH, a rotor-winding self-inductance L_{rr} of 50 mH, and a maximum mutual inductance L of 45 mH. If the stator were excited from a 60-Hz source, and the rotor were excited from a 25-Hz source, at what speed or speeds would the machine be capable of converting energy?

5–26. A rotating electrical machine with uniform air gap has a cylindrical-rotor winding with inductance $L_2 = 1$ H and a stator winding with inductance $L_1 = 3$ H. The mutual inductance varies sinusoidally with the angle θ between the winding axes, with a maximum of 2 H. Resistances of the windings are negligible. Compute the mean torque if the stator current is 10 A (rms), the rotor is short-circuited, and the angle between the winding axes is 45°.

5-27. The flux-density distribution produced in a 2-pole synchronous generator by an ac-excited field winding is

$$B(\theta, t) = B_m \sin \omega_1 t \cos \theta$$

Find the nature of the armature voltage induced in an N-turn coil if the rotor (or field) rotates at ω_2 rad/s. Comment on the special case $\omega_1 = \omega_2 = \omega$.

5-28   The flux-density distribution in the air gap of a 60 Hz, 2-pole, salient-pole machine is sinusoidal, having an amplitude of 0.6 T. Calculate the instantaneous and rms values of the voltage induced in a 150-turn coil on the armature, if the axial length of the armature and its inner diameter are both 100 mm.

5-29. Consider an elementary three-phase, four-pole alternator with a wye-connected armature winding, consisting of full-pitch concentrated coils, as shown in Figures 5.3.10b and c. Each phase coil has three turns, and all the turns in any one phase are connected in series. The flux per pole, sinusoidally distributed in space, is 0.1 Wb. The rotor is driven at 1,800 rpm.

a. Calculate the generated rms voltage in each phase.

b. If a voltmeter were connected across the two line terminals, what would it read?

c. For the a-b-c phase sequence, take $t = 0$ at the instant when the flux linkages with a-phase are a maximum.
1. Express the three phase voltages as functions of time.
2. Draw a corresponding phasor diagram of these voltages with the a-phase voltage as reference.
3. Represent the line-to-line voltage on the phasor diagram in part c2.
4. Obtain the time functions of the line-to-line voltages.

5-30. A wye-connected, three-phase, 50-Hz, six-pole synchronous alternator develops a voltage of 1,000 V (rms) between the lines when the rotor dc field current is 3 A. If this alternator is to generate 60-Hz voltages, find the new synchronous speed, and calculate the new terminal voltage for the same field current.

5-31. Determine the synchronous speed in rpm and the useful torque in newton-meters of a 200-hp, 60-Hz, six-pole synchronous motor operating at its rated full load.

5-32. From a three-phase, 60-Hz system, through a motor-generator set consisting of two directly coupled synchronous machines, electrical power is supplied to a three-phase, 50-Hz system.

a. Determine the minimum number of poles for the motor.

b. Determine the minimum number of poles for the generator.

c. With the number of poles decided, find the speed in rpm at which the motor-generator set will operate.

5-33. Two coupled synchronous machines are used as motor-generator set to link a 25-Hz system to a 60-Hz system. Find the three highest speeds at which this linkage would be possible.

5-34. A 10-turn square coil of side 200 mm is mounted on a cylinder 200 mm in diameter. If the cylinder rotates at 1,800 rpm in a uniform 1.1-T field, determine the maximum value of the voltage induced in the coil.

5-35. A four-pole dc machine with 728 active conductors and 30 mWb flux per pole runs at 1,800 rpm.

a. If the armature winding is lap-wound, find the voltage induced in the armature, given that the number of parallel paths is equal to the number of poles for lap windings.

b. If the armature is wave-wound, calculate the voltage induced in the armature, given that the number of parallel paths is equal to 2 for wave windings.

c. If the lap-wound armature is designed to carry a maximum line current of 100 A, compute the maximum electromagnetic power and torque developed by the machine.

d. If the armature were to be reconnected as wave-wound, while limiting per-path current to the same maximum as in part c, will either the maximum developed power or torque be changed from those obtained in part c?

5-36. A four-pole, lap-wound armature has 144 slots with two coil sides per slot, each coil having two turns. If the flux per pole is 20 mWb and the armature rotates at 720 rpm, calculate the induced voltage, given that the number of parallel paths is equal to the number of poles for lap windings.

5-37. A four-pole dc generator is lap wound with 326 armature conductors. It runs at 650 rpm on full load, with an induced voltage of 252 V. If the bore of the machine is 42 cm in diameter, its axial length is 28 cm, and each pole subtends an angle of 60°, determine the air-gap flux density. (Note that the number of parallel paths is equal to the number of poles for lap windings.)

5-38. A dc shunt machine has an armature winding resistance of 0.12 Ω and a shunt field winding resistance of 50 Ω. The machine may be run on 250-V mains as either a generator or a motor. Find the ratio of the speed of the generator to the speed of the motor when the total line current is 80 A in both cases.

5-39. A 100-kW, dc shunt generator connected to 220-V mains is belt-driven at 300 rpm, when the belt suddenly breaks and the machine continues to run as a motor, taking 10 kW from the mains. The armature winding resistance is 0.025 Ω; the shunt field winding resistance is 60 Ω. Determine the speed at which the machine runs as a motor.

5-40. A dc shunt motor runs off a constant 200-V supply. The armature winding resistance is 0.4 Ω and the field winding resistance is 100 Ω. When the motor develops rated torque, it draws a total line current of 17.0 A.

   a. Determine the electromagnetic power developed by the armature under these conditions, in watts and in horse power.

   b. Find the total line current drawn by the motor when the developed torque is one half of the rated value.

5-41. A dc machine operating as a generator develops 400 V at its armature terminals, corresponding to a field current of 4 A, when the rotor is driven at 1,200 rpm and the armature current is zero.

   a. If the machine produces 20 kW of electromagnetic power, find the corresponding armature current and the electromagnetic torque produced.

   b. If the same machine is operated as a motor supplied from a 400-V dc supply, and if the motor is delivering 30 hp to the mechanical load, determine (1) the current taken from the supply, (2) the speed of the motor, and (3) the electromagnetic torque, if the field current is maintained at 4 A. Assume that the armature circuit resistance and the mechanical losses are negligible.

   c. If this machine is operated from a 440-V dc supply, determine the speed at which this motor will run and the current taken from the supply, assuming no mechanical load, no friction and windage losses, and that the field current is maintained at 4 A.

5-42. Consider the operation of a dc shunt motor that is affected by the following changes in its operating conditions. Explain the corresponding approximate changes in the armature current and speed of the machine for each change in operating conditions.

   a. The field current is doubled, with the armature terminal voltage and the load torque remaining the same.

   b. The armature terminal voltage is halved, with the field current and load torque remaining the same.

   c. The field current and the armature terminal voltage are halved, with the horsepower output remaining the same.

   d. The armature terminal voltage is halved, with the field current and horsepower output remaining the same.

   e. The armature terminal voltage is halved and the load torque varies as the square of the speed, with the field current remaining the same.

5-43.   A three-phase, 50-Hz induction motor has a full-load speed of 700 rpm and a no-load speed of 740 rpm.

    a.   How many poles does the machine have?

    b.   Find the slip and the rotor frequency at full load.

    c.   What is the speed of the rotor field at full load
        (1)   With respect to the rotor?
        (2)   With respect to the stator?

5-44.   A three-phase, 60-Hz induction motor runs at almost 1,800 rpm at no-load and at 1,710 rpm at full load.

    a.   How many poles does the motor have?

    b.   What is the per-unit slip at full load?

    c.   What is the frequency of rotor voltages at full load ?

    d.   At full load, find the speed of
        (1)   The rotor field with respect to the rotor
        (2)   The rotor field with respect to the stator
        (3)   The rotor field with respect to the stator field

5-45.   A four-pole, three-phase induction motor is energized from a 60-Hz supply. It is running at a load condition for which the slip is 0.03. Determine

    a.   Rotor speed, in rpm

    b.   Rotor current frequency, in Hz

    c.   Speed of the rotor rotating magnetic field with respect to the stator frame, in rpm

    d.   Speed of the rotor rotating magnetic field with respect to the stator rotating magnetic field, in rpm

5-46.   Consider a three-phase induction motor with a normal torque-speed characteristic. Neglecting the effects of stator resistance and leakage reactance, discuss the approximate effect on the characteristic

    a.   If the applied voltage and frequency are halved.

    b.   If only the applied voltage is halved, but the frequency is at its normal value.

5-47.   Induction motors are often braked rapidly by a technique known as *plugging,* which is the reversal of the phase sequence of the voltage supplying the motor. Assume that a motor with four poles is operating at 1,750 rpm from an infinite bus (a load-independent voltage supply) at 60 Hz. Two of the stator supply leads are suddenly interchanged.

    a.   Find the new slip.

    b.   Calculate the new rotor current frequency.

5-48.   A four-pole, three-phase, wound-rotor induction machine is to be used as a variable frequency supply. The frequency of the supply connected to the stator is 60 Hz.

    a.   Let the rotor be driven at 3,600 rpm in either direction by an auxiliary synchronous motor. If the slip-ring voltage is 20 V when the rotor is at standstill, what frequences and voltages can be available at the slip rings?

    b.   If the slip-ring voltage is 400 V when the rotor frequency is 120 Hz, at what speed must the rotor be driven in order to give 150 Hz at the slip-ring terminals? What will the slip-ring voltage be in this case?

5-49.   A three-phase wound-rotor induction machine, with its shaft rigidly coupled to the shaft of a three-phase synchronous motor, is used to change balanced 60-Hz voltages to other frequencies at the wound-rotor terminals brought out through slip rings. Both machines are electrically connected to the same balanced three-phase, 60-Hz source. Let the synchronous motor, which has four poles, drive the interconnecting shaft in the clockwise direction, and let the eight-pole balanced three-phase stator winding of the induction machine produce a counterclockwise rotating field, i.e., opposite that of the synchronous motor. Determine the frequency of the rotor voltages of the induction machine.

5-50. For a balanced two-phase stator supplied by balanced two-phase currents, carry out the steps leading up to an equation such as Equation 5.4.5 for the rotating mmf wave.

5-51. A split-phase induction motor has two stator windings of impedance $(5 + j50)$ Ω, displaced by 90 electrical degrees. Compute the necessary resistance and capacitance in series with one winding in order to obtain a simulated balanced two-phase current system when both windings are suppplied from a 60-Hz, 110-V single-phase supply.

5-52. Consider the rectangular coil $PQRS$ shown in Figure P5–52, moving in the x-direction in the xy-plane with a velocity U m/s in a traveling-wave magnetic field distribution given by the flux density function

$$B = B_m \cos(\beta x - \omega t) \text{ Wb/m}^2$$

directed in the z-direction perpendicular to the plane of the coil. Determine the total induced voltage, identifying the transformer-emf and speed-emf components

a. By the method of relative velocity.   b. By applying Faraday's law of induction.

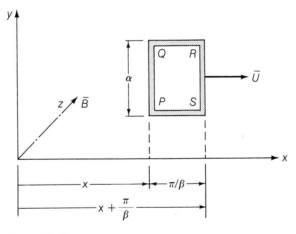

Figure P5–52

5-53. A certain 10-hp, 230-V motor has a rotational loss of 600 W, a stator copper loss of 350 W, a rotor copper loss of 350 W, and a stray load loss of 50 W. It is not known whether the motor is an induction, synchronous, dc machine.

a. Calculate the full-load efficiency.

b. Compute the efficiency at half-full load, assuming that the stray-load and rotational losses do not change with load.

5-54. A synchronous motor operates continuously on the following duty cycle: 50 hp for 8 minutes, 100 hp for 8 minutes, 150 hp for 10 minutes, 120 hp for 20 minutes, and no-load for 14 minutes. Specify the required continuous-rated hp of the motor.

5-55. A 2,500-kVA, unity-power factor, 225-rpm, three-phase, 60-Hz, 2,400-V, salient-pole synchronous generator is to be designed. For an output coefficient of 3.8×10^4 J/m^3, obtain values for diameter D and length L of the machine for these values of L/τ: 1.00, 1.50, 1.75, 2.00, and 2.25. (Note that τ is the pole pitch.)

5-56 A synchronous motor is to be designed for the following rating: 200 hp, 440 V, three-phase, 60 Hz, 900 rpm. For an output coefficient of 2.2×10^4 J/m^3, determine the diameter D and length L of the machine for these values of L/τ: 0.5, 0.6, 0.7, 0.8, 0.9, and 1.00. (Note that τ is the pole pitch.)

6

Machine Windings

6.1 Basic Winding Arrangements in Rotating Machines
6.2 DC Field Windings
6.3 DC Armature Windings
6.4 AC Armature Windings
6.5 Winding Factors
6.6 EMF Produced by an Armature Winding
6.7 MMF Produced by Windings

In an electromagnetic rotating machine, energy conversion takes place through the medium of the magnetic field, an electromotive force (emf) being induced in any coil that experiences a change of flux linkage. In practice, a machine usually has several coils, connected in series or in series-parallel circuits to form a winding. Machine windings are the means by which theory is translated into practice; their importance thus warrants a short chapter of their own. Windings can also be viewed as a unifying link among different kinds of machines.

The machine designer must determine the best way of placing the coils to collect their emfs in the most effective manner. The stationary windings of a transformer present no problem, as they can readily be wound in close proximity on the iron core to reduce the leakage, as discussed earlier. The situation is rather different for rotating machines, however. In fact, the study of armature windings is rather specialized; the many design considerations lead to several possibilities for each situation. Since all possibilities cannot be dealt with fully in this book, we will present a simplified treatment of winding problems to give practical support to theoretical analysis.

6.1 Basic Winding Arrangements in Rotating Machines

Machine windings can be classified as either *field windings* or *armature windings*. Armature windings may further be subdivided into *dc* armature windings (also known as *commutator windings*) and *ac* armature windings. Such windings are generally formed by wire-wound coils of one or more *turns* (see Figure 5.3.8) placed in slots arranged to form either *single-layer* or *double-layer* windings.

In a single-layer winding, one side of a coil, known as a *coil side,* occupies the whole of one slot, whereas, in a double-layer winding, any one slot has two separate coil sides. Most polyphase ac and dc armature windings are double-layer windings with multiturn coils. In general, a double-layer winding has a lower leakage reactance and produces a better waveform than the corresponding single-layer winding.

When one coil side is under the influence of a north pole and the other under that of a south pole, the emfs induced are in opposite directions. To add these two emfs around the coil circuit, a back-to-back connection is made. In such a case, the coil is known as a *full-pitch coil,* with the *coil span* equal to

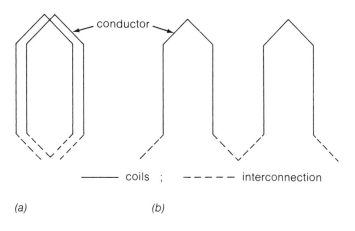

Figure 6.1.1 Methods of interconnection of armature coils. (a) Lap. (b) Wave.

one pole pitch. The coils are sometimes *chorded (short-pitched);* i.e., the coil sides are not exactly one pole pitch apart, with the coil span somewhat shorter than a pole pitch. The coils are laid in succession, i.e., *distributed,* around the periphery of the machine.

For mechanical reasons the coils are insulated and embedded in *slots* that may be *open, semiclosed,* or *closed.* The reactances are affected by the shape and relative dimensions of these slots. When the slot opening equals the slot width, the slots are said to be open; when the slot opening is less than the slot width, the slots are said to be semiclosed; and when there is no slot opening at all, the slots are said to be closed. Semiclosed and closed slots are normally used in induction motors, while open slots are most frequently used in large synchronous machines.

The separate coils of winding can be interconnected in a variety of ways. The real differences among windings lie in their interconnections. The common features are that, in progressing from one coil to another in series, the emfs are additive, and any time-phase displacement between the coil emfs is usually kept small by suitable choice of connection. The two most usual methods of interconnection are the *lap* connection and the *wave* connection, illustrated in Figure 6.1.1. Machine armature windings for direct current are always of the closed type of double-layer lap or waveform connected to a commutator. The commutator itself, as mentioned earlier, consists of many copper segments, insulated from and running parallel to one another, clamped around the periphery of an insulated cylinder mounted on the machine shaft. Each segment is extended to a coil junction, so that there are as many segments as there are coils.

By referring to Figure 6.1.2, let us familiarize ourselves with the usual armature winding terms and symbols:

- *Conductor:* the active length of wire or strip in the slot along the axial length of the machine core. The conductor can be laminated and transposed (i.e., occupying different physical locations) along its length to reduce eddy currents.

- *Coil:* two conductors separated by a pole pitch or nearly so, connected in series to form a single-turn coil. More conductors placed in virtually the same position magnetically can be formed into multiturn coils. Figure 5.3.8 shows a three-turn lap coil with six conductors.

- *Coil side:* one-half of the coil, lying in one slot, the coil being either single-turn or multiturn. Only one coil side per slot for a single-layer winding; at least two coil sides in any one slot for

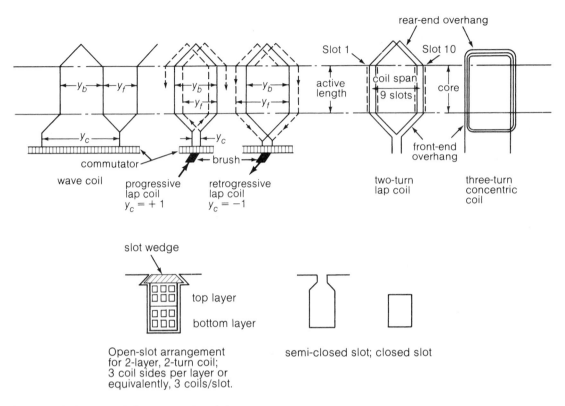

Figure 6.1.2 Armature winding terms and symbols.

a double-layer winding. A commutator winding may have several coil sides in one slot. Figure 6.1.2 shows one slot arrangement with two-turn coils and three coil-sides per layer, giving twelve conductors per slot insulated from one another. All the conductors in one layer in the slot are usually taped together to share a common insulation to earth or iron core.

- *Pitch:* The various pitches of interest can be seen in Figure 6.1.2. The coil-pitch or coil-span (also known as the slot span of the coil) and the pole pitch have already been referred to. For the dc armature winding, the *back pitch* y_b (also known as coil pitch), the *front pitch* y_f, and the *commutator pitch* y_c (also known as coil-end pitch) can be seen from Figure 6.1.2 These pitches or distances are usually measured or specified in terms of the commutator segments between the end connections of a coil. Thus, for a single commuator lap winding, y_c is ± 1, the positive sign indicating a progression to the right when tracing through the winding and the negative sign indicating a retrogression to the left. In all cases, the commutator pitch is equal to the algebraic sum of the front pitch and the back pitch of the coil. The commutator pitch for a *multiplex* lap winding of degree m will be m commutator segments.

With this background, let us move on to briefly discuss some classes of machine windings. See Photos 6.1.1 and 6.1.2 for a laminated (stranded) armature-coil construction and the details of open-slot arrangements.

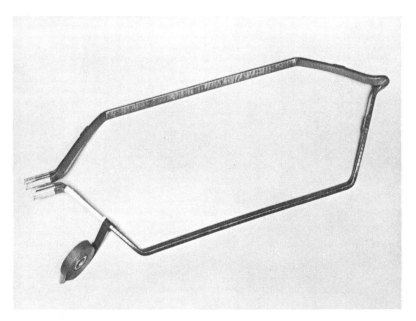

Photo 6.1.1 Typical laminated armature coil consisting of several strands of copper wire in parallel. (Photo courtesy of Westinghouse Electric Corporation.)

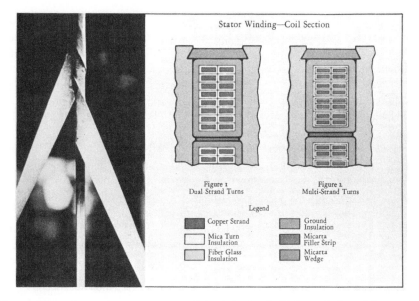

Photo 6.1.2 Cross sections of open-slot arrangements with insulation for double-layer windings. (Photo courtesy of Westinghouse Electric Corporation.)

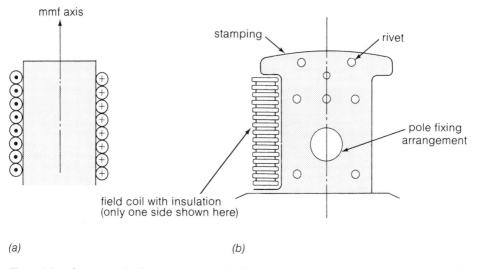

Figure 6.2.1 Concentrated coil on a salient pole. (a) Schematic arrangement. (b) Realistic arrangement.

6.2 DC Field Windings

Synchronous machines and dc machines require dc-excited field windings. Two forms of construction are used for the field circuit: (1) a salient-pole arrangement with a concentrated winding, used for either dc or ac machines, and (2) a distributed arrangement on the nonsalient (cylinderical) rotors of ac machines. Figure 6.2.1 shows a concentrated coil on a salient pole. In order to obtain the indicated current pattern, the turns of the concentrated coil must be connected in series. It makes no difference in which order the coils are connected, except perhaps from the viewpoint of constructional convenience. Photo 6.2.1 shows the dc machine field windings wound on a pole, and Photo 6.2.2 illustrates the salient-pole field arrangement for a synchronous machine.

For high-speed rotors of turbogenerators, which are usually run at 3,600 or 1,800 rpm (for 60-Hz systems), the centrifugal forces are too great to be sustained by a salient-pole construction, so their field winding is distributed into slots, as shown in Figure 5.3.3 for a two-pole machine. A distributed two-pole winding with six coils (12 conductors) is shown schematically in Figure 6.2.2a to illustrate the principle: 1 to 1', 2 to 2', etc., are connected to form coils 1, 2, etc., respectively. The required current distribution is produced by simply connecting the six coils in series. The winding is called a concentric winding because all six coils are arranged in a concentric fashion. Concentric windings are most commonly employed for the field windings of synchronous machines where the field winding is on the rotor.

The winding arrangement may easily be modified to other schemes, like those shown in Figure 6.2.2 b, c, and d, while still yielding the mmf axis in the same position.

Liquid-helium-cooled superconducting rotor field windings are feasible; they are used in large cryogenerators or superconductive turboalternators.

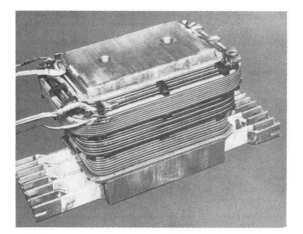

Photo 6.2.1 DC machine field windings wound-on-pole along with pole-face compensating winding bars. (Photo courtesy of Westinghouse Electric Corporation.)

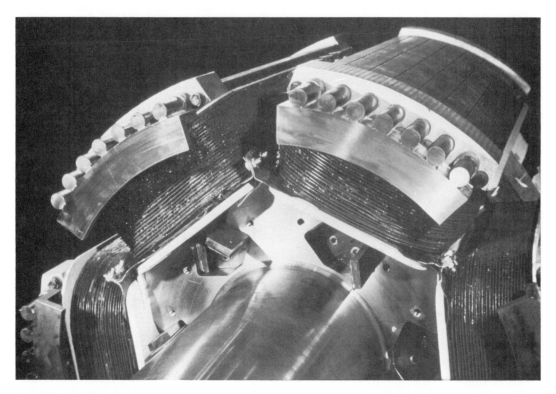

Photo 6.2.2 Salient-pole field arrangement being manufactured for a synchronous machine. (Photo courtesy of Westinghouse Electric Corporation.)

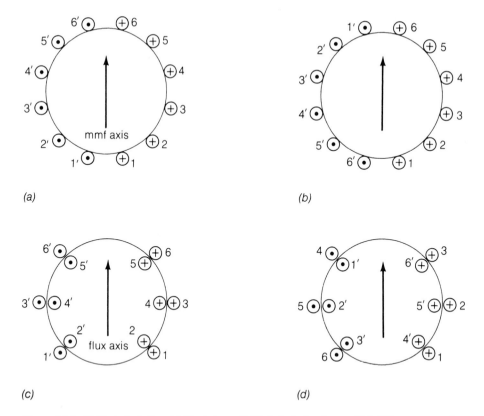

Figure 6.2.2 Distributed winding. (a) Single-layer distributed winding (concentric). This arrangement is most commonly used for nonsalient-pole synchronous machine rotor field windings. (b) Single-layer distributed winding with 180° coil span. (c) Double-layer distributed winding (concentric). (d) Distributed double-layer winding with 180° coil span.

▐▌ ▌▌▌▌▌
6.3 DC Armature Windings

As mentioned already, these windings (also known as commutator windings) are always of the closed continuous type of double-layer lap or waveform. The simplex lap winding has as many parallel paths as there are poles, while the simplex wave winding, always has two parallel paths. Photo 6.3.1 shows a dc armature being wound.

Before proceeding further, let us try to justify a statement made earlier in the text that a rotating commutator winding can be called a pseudostationary winding because it produces a stationary flux when carrying a direct current. Consider a six-coil closed continuous two-pole winding connected to a six-segment commutator, as shown in Figure 6.3.1a, producing the illustrated current distribution when a current is passed into the winding through the brushes located on the commutator. Note that the flux axis is along the brush axis.

Next, consider an instant when the commutator and the winding have both rotated by 60°, but with the same position of brushes, as shown in Figure 6.3.1b. It can be seen that coils 3 and 6 have moved

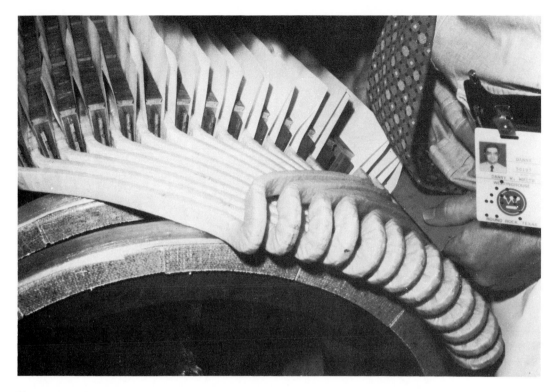

Photo 6.3.1 DC armature being wound. (Photo courtesy of Westinghouse Electric Corporation.)

from one parallel circuit to the other; but the current directions in them are changed and thus their physical position, such that the resultant spatial current distribution and the position of the flux axis remain the same as they were in Figure 6.3.1a. By considering successive 60° movements, we can see that the resultant total mmf (or flux) is a constant in the direction fixed by the brush axis, irrespective of the position of the individual coils. Hence, the commutator winding is viewed as a pseudostationary winding.

In a practical machine with a large number of coils, intervals of much less than 60° can be considered and the same conclusions reached. In practice, however, the axis of flux oscillates forward and backward by an angle equal to that between adjacent commutator segments. By keeping the winding as in Figure 6.3.1a but moving the position of brushes by 60°, as shown in Figure 6.3.1c, we see that the flux axis has also been moved by 60°. Thus we have seen that the flux axis corresponds to the brush axis.

Lap Windings

A *simplex lap winding* is one in which the coil sides are placed in slots approximately one pole pitch apart, the coil leads are connected to adjacent commutator segments, and the number of parallel paths in the armature is equal to the number of poles. This class of winding is best illustrated by considering an example.

Let us consider a four-pole, two-layer, simplex lap winding with 20 slots. Let the number of conductors per coil side be one (i.e., single-turn coils), for the sake of simplicity. The number of coils

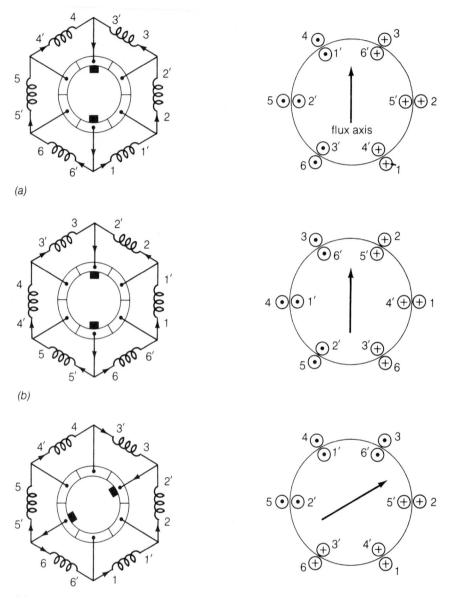

Figure 6.3.1 Commutator winding as a pseudo-stationary winding. (a) Six-coil closed continuous two-pole winding connected to a commutator; corresponding current distribution and the mmf axis. (b) Winding in (a) rotated by 60°, but with the same position of brushes; corresponding current distribution and the mmf axis. (c) Winding in (a) with brushes rotated by 60°; corresponding current distributin and the mmf axis.

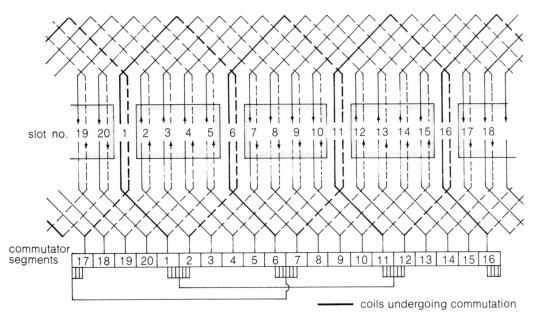

Figure 6.3.2 Developed diagram of a four-pole, two-layer simplex lap dc armature winding with 20 slots and 20 single-turn coils.

as well as the number of commutator segments is 20, which is the same as the number of slots. Instead of showing a polar diagram of the winding located in the slots of an armature of circular cross section, a developed diagram is shown in Figure 6.3.2 for better clarity.

The slots per pole being 20/4 = 5 in our case, the following pitches are selected in terms of the slots or slot pitches:

$$y_b = 5; \qquad y_f = -4; \qquad y_c = y_b + y_f = +1 \qquad (6.3.1)$$

The above choice of $y_c = +1$ makes the winding progressive instead of retrogressive. The development of the winding is rather straightforward, and four parallel paths through the armature can be observed in Figure 6.3.2. The brushes on the commutator are located as nearly possible on the center of the poles so that the coil undergoing commutation (i.e., reversal of current directions) is situated in the interpolar region. Corresponding to the four parallel armature paths, the lap winding must have four brush sets, two of which have the same polarity. The polarity of the brushes can be assigned here arbitrarily, because it is determined definitely only when the pole polarities and direction of rotation are known.

The generated voltages in each of the parallel paths of a lap winding should be equal if the machine is geometrically and magnetically symmetrical. In practice, however, since these voltages are not exactly the same and hence cause circulating currents through the armature and brushes, most machines with lap windings are provided with *equalizers* (copper straps of large cross-sectional area) joining points on the winding that are 360 electrical degrees apart. The number of equalizer bars is given by the number of slots per pole pair, 10 in our example; each bar joins two coils together. Equalization prevents excessive brush currents and helps commutation.

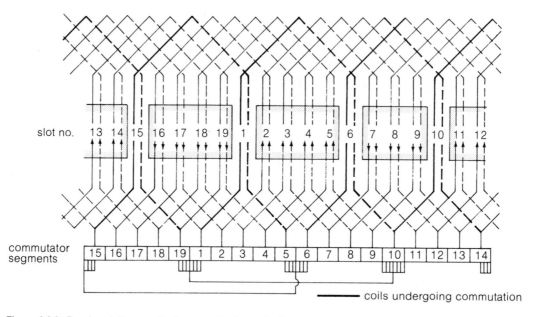

slot no.

commutator segments

———— coils undergoing commutation

Figure 6.3.3 Developed diagram of a four-pole, two-layer simplex wave dc armature winding with 19 slots and 19 single-turn coils. (Note: only two brush sets are really necessary.)

Wave Windings

A simplex wave winding is one in which the coil sides are placed in slots approximately one pole pitch apart, the coil leads are connected to commutator bars approximately two pole pitches apart, and the number of parallel paths in the armature is always two. Again, we can best illustrate by considering an example. The commutator pitch y_c for a simplex wave winding in terms of commutator segments is given by

$$y_c = y_b + y_f = \frac{K \pm 1}{\text{pole pairs}} \tag{6.3.2}$$

where K is the number of commutator segments. A simplex wave winding can be obtained only if y_c is an integer. In some cases in which y_c is not an integer, a so-called dummy coil is introduced into the winding to maintain the mechanical symmetry of the armature, and electrically the coil is left on open circuit.

Let us consider a four-pole, two-layer, simplex wave winding with 19 slots, and let the number of conductors per coil side be one (i.e., single-turn coils) for simplicity. The number of coils as well as the number of commutator segments is 19, the same as the number of slots. A developed winding diagram is shown in Figure 6.3.3. The commutator pitch y_c is given by $(19 \pm 1)/2 = 9$ or 10; choosing $y_c = 9, y_b = 5$, and $y_f = 4$, the winding arrangement is completed as shown.

Only two parallel paths through the armature can be observed in Figure 6.3.3. Only two brush sets are really necessary, although, large machines usually employ the same number of brush sets as poles.

Let us consider the location of the brushes. As seen in Figures 6.3.2 and 6.3.3, the brushes lie physically on the center lines of the poles. Considering the active parts of the conductors to which the brushes are connected, however, the brushes lie on the interpolar (quadrature) axis, which is 90 electrical degrees away from the direct-pole axis. In view of the greater number of parallel paths available, lap windings are normally used for dc machines with large current-carrying capacity. In practice, any one slot usually has more than two coil sides, so that the number of commutator segments is greater than the number of armature slots. For a given output voltage, such a pratice leads to reduced voltage between adjacent commutator segments, which in turn reduces the risk of flashover between segments. Designers rarely use multiplex windings that have two or more (depending on the degree of multiplicity) times as many parallel paths for the same number of poles as the simplex windings.

6.4 AC Armature Windings

When associated with a heteropolar magnetic field, a q-phase ac armature winding develops emfs in all of its phases. When the phases are wound $2\pi/q$ electrical radians apart in space, these emfs are normally equal in magnitude and displaced in time-phase relationship by $2\pi/q$ electrical radians, except in the case of a two-phase winding when a displacement of $\pi/2$ electrical radians for both space and time. The winding is composed of conductors in slots distributed around the periphery of the air gap, connected at the ends, and grouped to form separate phase windings. Polyphase windings are usually double-layer, with the two layers arranged one above the other in the slot. The single-layer polyphase winding is used in the wound rotors—but seldom in the stators—of small induction motors. Polyphase windings can be of lap or wave type. Lap windings are most commonly used; wave windings are employed mainly in wound rotors of medium-sized and large induction motors. Photo 6.4.1 shows an induction-motor stator being wound.

Photo 6.4.1 Induction-motor stator being wound. (Photo courtesy of General Electric Company.)

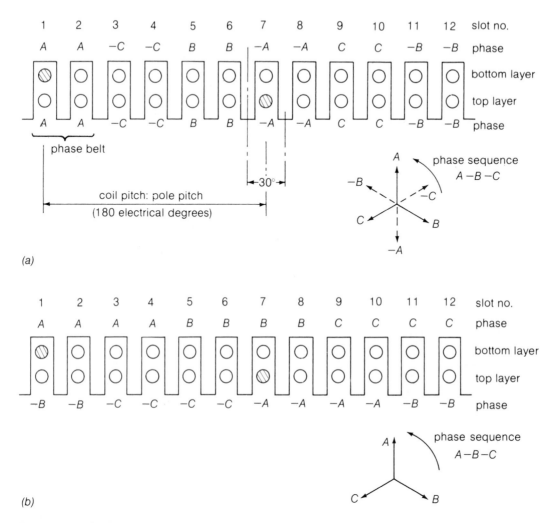

Figure 6.4.1 A flattened layout of a three-phase, two-phase, two-pole, two-layer, full-pitch lap ac armature winding with two slots per pole per phase. (a) 60-electrical-degree phase spreading. (b) 120-electrical-degree phase spreading. (c) Coil interconnections shown in detail for part (a). (Note: The signs indicate the instantaneous directions of current flow, in or out of the plane of paper.)

The ac armature windings can be considered in terms of several separate sections, known as *phase belts*. A three-phase, two-layer winding can be completed with either three phase belts per pole pair (i.e., with 120 electrical-degree spread) or six phase belts per pole pair (i.e., with 60 electrical-degree spread). The more common arrangement is to use a 60° spread for three-phase windings. Note that a three-phase winding is effectively a six-phase winding. Two-phase machines normally have double-layer windings with four phase belts per pole pair, i.e., a 90° spread.

A developed sectional diagram for a three-phase, two-pole, two-layer, full-pitch lap ac armature winding with two *slots per pole per phase* (abbreviated spp), (for a total of $3 \times 2 \times 2 = 12$ slots) is shown in Figure 6.4.1a, along with the instantaneous directions of current flow that make the phase sequence

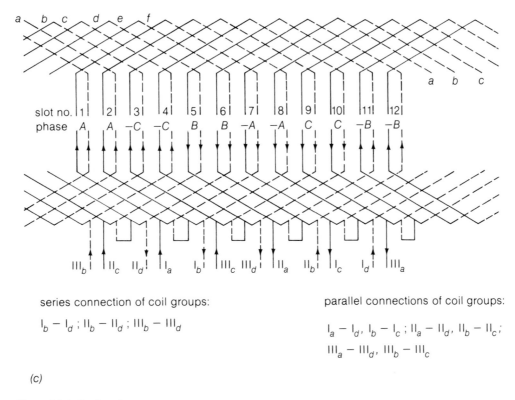

series connection of coil groups:

$$I_b - I_d \; ; \; II_b - II_d \; ; \; III_b - III_d$$

parallel connections of coil groups:

$$I_a - I_d, \; I_b - I_c \; ; \; II_a - II_d, \; II_b - II_c \; ;$$
$$III_a - III_d, \; III_b - III_c$$

(c)

Figure 6.4.1 *Continued*

come out as *A-B-C*. The coil groups assigned to each phase, shown in Figure 6.4.1c with detailed coil intersections, are connected either in series or in parallel according to the voltage to be produced (in the case of a generator) or the voltage impressed (in the case of a motor). The phase end leads are brought out for wye or delta connection.

All the coils shown in Figure 6.4.1 are full-pitch coils. Short-pitch (chorded) windings are used quite often in polyphase armature windings. Using chorded coils reduces the length of the coil-end connections (thereby saving a little copper) and also, as we will see later, significantly reduces the magnitude of certain harmonics in the waveform of the generated emf and mmf. In most ac machines, the coil pitch is generally between 2/3 and 1 full pole pitch, and the coils are usually short-pitched as a standard practice. Figure 6.4.2 shows the flattened layout of a three-phase, two-pole, two-layer, fractional-pitch (5/6 pitch coils) lap ac armature winding with spp (number of slots per pole phase)= 2 and 60° spread.

It can be seen from Figure 6.4.1a that the coil sides occupying the top and bottom layers of each slot belong to the same phase for full-pitch windings. Typical of fractional-pitch windings, shown in Figure 6.4.2, the coil sides occupying the top and bottom layers of some slots are of different phases. Individual phase groups are still displaced by 120 electrical degrees from the groups in other phases, so three-phase voltages are produced.

The ac armature windings may be classified as *integral slot* and *fractional slot windings,* depending on whether spp is an integer or an improper fraction. The fractional slot windings offer great flexibility of design and manufacture and allow a wider latitude in the choice of the number of slots that can be used for the armature punchings. We will not consider either of these windings here.

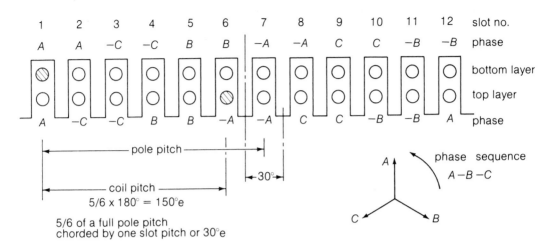

Figure 6.4.2 A flattened layout of a three-phase, two-pole, two-layer, fractional-pitch (5/6 pitch coils) lap ac armature winding with spp = 2 and 60° spread.

Squirrel-cage windings used for rotors of most induction motors have solid, uninsulated bars in the slots and are connected on each end of the rotor by a short-circuiting ring, as mentioned already. In small rotors of induction motors of up to about 50 hp, the bars and rings are usually diecast of aluminum; in large motors the bars are usually made of copper.

▮▮▮▮▮ ▮▮▮▮▮
6.5 Winding Factors

Two winding factors are discussed in this section: (1) the distribution factor, also known as the breadth factor, and (2) the pitch factor, also known as the coil-span or coil-pitch factor.

Distribution Factor

The coils of an ac armature winding of a practical machine are distributed in space so that the induced voltages in the coils are not in phase but are displaced from each other by the slot angle α. The effect of distributing the winding is taken into account by introducing the *distribution factor* k_d, which is defined as the ratio of the actual resultant emf E_r, given by the phasor sum, to the arithmetic sum of the individual coil emfs, each given by E. Thus, for an ac armature winding with m coils per pole per phase, the distribution factor is given by

$$k_d = E_r/mE \qquad (6.5.1)$$

For the particular arrangement with $m = 3$, illustrated in Figure 6.5.1, it follows that

$$E_r = 2 \times \sin(m\alpha/2); \qquad E = 2 \times \sin(\alpha/2) \qquad (6.5.2)$$

and

$$k_d = \frac{\sin(m\alpha/2)}{m\sin(\alpha/2)} \qquad (6.5.3)$$

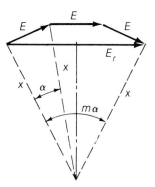

Figure 6.5.1 Coil voltage
phasors and phasor sums
to calculate the fundamental
distribution factor.

where α is measured in electrical degrees. For common three-phase, two-layer windings with 60° spread, it follows that $m\alpha/2$ is 30 electrical degrees, so that

$$k_d = \frac{\sin(\pi/6)}{m\sin(\pi/6m)} = \frac{0.5}{m\sin(\pi/6m)} \tag{6.5.4}$$

Equations 6.5.3 and 6.5.4 give the fundamental distribution factor; they apply for integral slot windings where m is the number of slots per phase belt in which the winding is distributed.

The above discussion assumes a sinusoidal-generated emf. If the emf contains harmonics, then the distribution factor differs for each harmonic. The phase difference between the nth harmonic of adjacent coil voltages in $n\alpha$; the distribution factor for the nth harmonic is given by

$$k_{dn} = \frac{\sin(nm\alpha/2)}{m\sin(n\alpha/2)} \tag{6.5.5}$$

With $n = 1$ (i.e., for the fundamental), Equation 6.5.5 reduces to Equation 6.5.3. If the number of series coils in a phase winding is sufficiently large, then α becomes rather small, and an approximation can be made as follows:

$$k_{dn} \simeq \frac{\sin(nm\alpha/2)}{nm\alpha/2} \tag{6.5.6}$$

in which α is expressed in electrical radians.

Distributing the winding results in a marked reduction in the harmonic content of the flux-density distribution and consequently the generated voltage, as shown later. If we assume that the generated voltages are sinusoidal, the output voltage from a distributed winding of N turns will be equal in magnitude to that of a concentrated winding of $(k_d N)$ turns. Thus, the effective number of turns of the distributed winding is $k_d N$.

The fundamental distribution factor for double-layer, integral-slot windings of common three-phase machines with 60° spread can be found by the application of Equation 6.5.4. For values of m (the number of coils per phase belt) ranging from 1 to 6, k_{d1} varies from 1.0 to 0.956. Thus, distributed windings lower the fundamental generated voltage to some extent, since the voltage expression given by Equation 5.3.14 is to be multiplied by k_{d1}.

Equation 6.5.6 shows that k_{dn} decreases continually as n (the order of the harmonic) increases; it follows that, if α could be made very small, the higher harmonics could be virtually eliminated. Practically, we can achieve this effect by *skewing* the slots by a full slot pitch, in which case, some part of the conductor occupies every angular position over a slot pitch. Skewing is thus equivalent to an infinite number of infinitesimally short lengths displaced by infinitesimally small angles. The winding can then be said to be uniformly distributed, so Equation 6.5.6 can be applied. Skewing introduces constructional difficulties, but it is often used in induction motors to smooth out the variations in air-gap permeance and obtain a more uniform torque and a quieter motor. If the slots are not skewed by a full slot pitch, it is necessary to introduce another factor, called the *skew factor,* which must be applied to Equation 6.5.5 to account for the effect of skewing.

Let us now consider an example that shows the effects of distributing the winding and skewing the slots.

||||||||||||

EXAMPLE 6.5.1

A machine has nine slots per pole.

a. Calculate the fundamental distribution factor for these cases:

   1.   One winding distributed in all the slots.
   2.   One winding using only the first two-thirds of the slots per pole, i.e., 120° groups.
   3.   Three equal windings placed sequentially in 60° groups.

b. Considering case 3, determine the distribution factors for the 5th, 17th, and 19th harmonics for straight slots as well as for slots skewed by one slot pitch.

Solution

a. In each case, the slot angel $\alpha = 180/9 = 20$ electrical degrees. The values of m (the number of slots in a group) are 9, 6, and 3, respectively, for cases 1, 2, and 3. Equation 6.5.5—or, approximately, Equation 6.5.6—can be applied with $n = 1$ for the fundamental, giving

1. $\quad k_{d1} = \dfrac{\sin(9 \times 20°/2)}{9 \sin(20°/2)} = 0.64,$ from Equation 6.5.5

   or approximately, from Equation 6.5.6, $\quad k_{d1} = \dfrac{\sin(\pi/2)}{\pi/2} = 0.637$

2. $\quad k_{d1} = \dfrac{\sin(6 \times 20°/2)}{6 \sin(20°/2)} = 0.831 \quad$ or approximately $\quad k_{d1} = \dfrac{\sin(\pi/3)}{\pi/3} = 0.827$

3. $\quad k_{d1} = \dfrac{\sin(3 \times 20°/20)}{3 \sin(20°/2)} = 0.960 \quad$ or approximately $\quad k_{d1} = \dfrac{\sin(\pi/6)}{\pi/6} = 0.955$

These results illustrate one reason for using three or more separate windings or phases. Dividing the winding into six groups per pole pair instead of three increases the voltage, and hence the power output for the same number of conductors, by 15%, from a factor of 0.831 to one of 0.96. As stated earlier, these six groups can form a six-phase winding, or three groups of two sections each can form a three-phase winding, as shown in Figure 6.4.1a. This example further illustrates why only two-thirds of the slots are usually employed for a single-phase winding. The additional slots would only increase the output by $(9 \times 0.64)/(6 \times 0.831) = 1.15$ times, while the increase in the amount of copper would be 50%.

For a simple two-phase winding, this slotting arrangement is not suitable because an exact displacement of 90° is not possible with an odd number of slots. Considering eight slots per pole as an example, the distribution factor for two equal windings is

$$\frac{\sin\left[4 \times 180°/(8 \times 2)\right]}{4 \sin\left[180°/(8 \times 2)\right]} = 0.907$$

which is not quite as good as for the three-phase winding with 60° groups.

b. $k_{d5} = \dfrac{\sin\left(3 \times 5 \times 20°/2\right)}{3 \sin\left(5 \times 20°/2\right)} = 0.216$, from Equation 6.5.5, and for skewed slots,

$$k_{d5} = \frac{\sin 150°}{150 \times \pi/180} = 0.191, \text{ from Equation 6.5.6.}$$

$k_{d17} = \dfrac{\sin\left(3 \times 17 \times 20°/2\right)}{3 \sin\left(17 \times 20°/2\right)} = 0.959 \quad$ and for skewed slots,

$$k_{d17} = \frac{\sin 510°}{510 \times \pi/180} = \frac{\sin 150°}{510 \times \pi/180} = 0.056.$$

$k_{d19} = \dfrac{\sin\left(3 \times 19 \times 20°/2\right)}{3 \sin\left(19 \times 20°/2\right)} = 0.959 \quad$ and for skewed slots,

$$k_{d19} = \frac{\sin 570°}{570 \times \pi/180} = \frac{\sin 210°}{570 \times \pi/180} = -0.05.$$

With straight slots, some harmonics are reduced, but, unfortunately, the pronounced tooth harmonics caused by armature slottings are not reduced. For example, the 17th and 19th harmonics arise wih nine slots per pole. With skewed slots, however, there is a progressive reduction in the higher harmonics as seen from the previous example. To take full advantage, the degree of skewing should be equivalent to a full slot pitch.

Pitch Factor

If the individual coils of a winding are short-pitched, that is, if the coil span is less than a pole pitch, the coil emf is reduced, and the *pitch factor* k_p is defined as

$$k_p = \frac{\text{emf of short-pitched coil}}{\text{emf of full-pitched coil}} \tag{6.5.7}$$

This equation is equivalent to the ratio of the phasor sum of coil-side emfs to the arithmatic sum of coil-side emfs. When the coil pitch is less than a pole pitch (180 electrical degrees), the flux linking the coil and hence the induced voltage is less than that for a full-pitched coil. Considering the situation

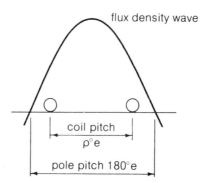

Figure 6.5.2 Fractional-pitch coil in a sinusoidal field.

shown in Figure 6.5.2, the coil pitch ρ in electrical degrees of the fractional-pitch coil is less than 180 electrical degrees; and the coil is located in a sinusoidal flux-density field wave. The ratio of the flux linking with the short-pitched coil to the flux that would have linked with a full-pitch coil is given by

$$\frac{\int_{-\rho/2}^{+\rho/2} \cos\theta\, d\theta}{\int_{-\pi/2}^{+\pi/2} \cos\theta\, d\theta} = \frac{2\int_0^{\rho/2} \cos\theta\, d\theta}{2\int_0^{\pi/2} \cos\theta\, d\theta} = \frac{[\sin\theta]_0^{\rho/2}}{[\sin\theta]_0^{\pi/2}} = \sin\frac{\rho}{2} \tag{6.5.8}$$

where θ is in electrical degrees. The expression obtained in Equation 6.5.8 also gives the pitch factor since the emfs are simply proportional to the flux linkages. Thus, we have

$$k_p = \sin\frac{\rho}{2} = \sin\left\{\frac{\text{coil span in electrical degrees}}{2}\right\} \tag{6.5.9}$$

If the field is nonsinusoidal, containing odd harmonics (and note that even harmonics need not be considered in ac machines because of the symmetry), the magnitude of the pitch factor for the nth harmonic is given by

$$k_{pn} = \sin\frac{n\rho}{2} \tag{6.5.10}$$

For odd harmonics, the sign and magnitude of the pitch factor are given in terms of the short pitch or pitch deficiency $(180 - \rho)$ by

$$k_{pn} = \cos\frac{n(180 - \rho)}{2} \tag{6.5.11}$$

The effect of pitch on the phase of any harmonic in a fractional-pitch winding, as compared with its phase in a full-pitch winding, depends on the order of the harmonic. All odd harmonics are either not changed in phase or reversed. The magnitudes of the harmonics in fractional-pitch windings, as compared with their magnitudes in a full-pitch winding having the same number of turns, are given in Table 6.5.1.

The voltage waveform can be improved by using short-pitched coils. Any harmonic can be eliminated from the voltage generated in an ac armature coil by choosing a pitch that makes the pitch factor zero for that harmonic. To eliminate the nth harmonic,

$$k_{pn} = \sin\left(\frac{n\rho}{2}\right) = 0, \quad \rho = \frac{2k\pi}{n} \tag{6.5.12}$$

Table 6.5.1 Magnitudes of Harmonics in Fractional-Pitch Windings

Pitch	Harmonic				
	1	3	5	7	11
2/3	0.866	0.000	0.866	0.866	0.866
4/5	0.951	0.588	0.000	0.588	0.951
5/6	0.966	0.707	0.259	0.259	0.966
6/7	0.975	0.782	0.434	0.000	0.782

where k is any integer. For example, to eliminate the 5th harmonic, coil pitches of 2/5, 4/5, 6/5, and so forth, can be used. However, since any departure from full pitch reduces the fundamental by an amount that increases progressively with the departure of the pitch form unity, practical use requires the fractional pitch that is nearest to unity, in this case 4/5. Pitches greater than unity require more copper for the coil-end connections than do shortened pitches and possess no compensating advantage. The 7th harmonic is totally eliminated (chorded out) with the choice of 6/7 pitch. A 5/6 pitch is particularly satisfactory for reducing harmonics, as it cuts significantly both the 5th and 7th harmonics.

It can be shown that 3rd harmonic voltages (and multiples of the 3rd harmonic, like the 9th and 15th) caused by corresponding harmonic components in the flux-density wave do not appear in the line-to-line voltages of normally designed either wye- or delta-connected three-phase machines. Such triplex harmonics in the line-to-neutral voltages or wye-connected machines may occasionally cause some difficulty, in which case 2/3 pitch may be used. The most significant harmonics to be minimized by the use of fractional-pitch windings are usually the 5th and 7th. Higher harmonics than the 9th are so small that little attention is required except in rare cases.

For single-layer windings, only the distribution factor is effective. Even though individual coils may be short-pitched or over-pitched, the coils are full-pitched on the average because the coil sides of a phase under neighboring poles occupy the same group of slots relative to the poles.

‖‖‖‖‖‖‖‖‖‖‖▬▬▬▬▬▬▬▬▬▬▬▬▬▬▬
EXAMPLE 6.5.2

An armature of an ac machine has 18 slots per pole. It is wound with a two-layer, three-phase winding consisting of single-turn coils. Each coil spans 15 slots. Calculate the pitch factors for the fundamental k_{p1}, for the 5th harmonic k_{p5}, and for the 7th harmonic k_{p7}.

Solution
The slot angle $\alpha = 180/18 = 10$ electrical degrees. The coil span is $\rho = 15 \times 10 = 150$ electrical degrees. The coil is short-pitched by $180 - \rho = 180 - 150 = 30$ electrical degrees. From Equation 6.5.11, it follows that

$$k_{p1} = \cos \frac{30°}{2} = 0.966; \qquad k_{p5} = \cos \frac{5 \times 30°}{2} = 0.259; \qquad k_{p7} = \cos \frac{7 \times 30°}{2} = -0.259$$

The signs are associated with the relative phase angles of the harmonics in the resultant emf wave. The signs are usually not of practical significance, so Equation 6.5.10 can also be applied.

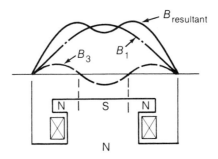

Figure 6.6.1 Space harmonics in the flux-density wave. (*Note:* For simplicity, only third-harmonic is shown.)

6.6 EMF Produced by an Armature Winding

The time variation of emf for a single conductor corresponds to the spatial variation of air-gap flux density. This variation is not purely sinusoidal, and it often contains significant lower-order odd harmonics. By suitable winding design through distributing and chording, the harmonics can be appreciably reduced, while the fundamental is reduced by only a little. Consequently, the waveform of the generated emf more nearly approaches a pure sine shape. The winding factors discussed in Section 6.5 express mathematically the per-unit reduction of the fundamental (as well as of each harmonic) that happens as a result of distribution and chording. In practice, the harmonics are often reduced to negligible proportions, while the fundamental is reduced by only about 10%. The same winding factors also apply to the space distribution of mmf, and so reduce the mmf space harmonics.

For an ac machine, considering only the fundamental, Equation 5.3.14 has beeen developed for the induced voltage. The rms-generated voltage per phase of a concentrated winding having N_{ph} turns in series per phase is given by

$$E_1 = 4.44 \, fN_{ph}\phi_1 \text{ rms volts per phase} \tag{6.6.1}$$

where subscript 1 denotes the fundamental, f the frequency, and ϕ_1 the fundamental field flux per pole. When both the fundamental distribution factor and the pitch factor for the distributed, fractional-pitch winding are applied, the rms voltage per phase is modified to

$$E_1 = 4.44 k_{w1} \, fN_{ph}\phi_1 \text{ rms volts per phase} \tag{6.6.2}$$

which is the same as Equation 5.3.15, and in which k_{w1} is the winding factor for the fundamental given by

$$k_{w1} = k_{d1} \cdot k_{p1} \tag{6.6.3}$$

This result is slightly less than unity for a polyphase machine.

Time harmonics in the emf are produced by space harmonics in the flux-density wave. Figure 6.6.1 shows the flux-density distribution consisting of a fundamental and an inphase third harmonic. Imagine it as being produced by three harmonic poles superimposed on each fundamental pole as shown. Harmonic voltages can be calculated if the field form is analyzed into its fundamental and harmonic flux densities, given by

$$B = B_1 \sin \theta + B_3 \sin 3\theta + B_5 \sin 5\theta + B_7 \sin 7\theta + \cdots \qquad (6.6.4)$$

For the harmonics, the frequency of flux changes is increased three, five, seven, and so forth, times, depending on the order of the harmonics, which tends to bring about an increase of voltage. But corresponding reductions in harmonic pole-pitch area and flux components, as seen in Figure 6.6.1, tend to offset this increase of voltage. Thus, harmonic voltages can be obtained by reducing the fundamental voltage in the ratio of the corresponding flux density and winding factors. That is to say, for the nth harmonic,

$$E_n \text{ per phase} = E_1(B_n/B_1)(k_{dn}/k_{d1})(k_{pn}/k_{p1}) \qquad (6.6.5)$$

or equivalently

$$E_n \text{ per phase} = 4.44 \, k_{dn} k_{pn} k_{fn} f N_{ph} \phi_1 \qquad (6.6.6)$$

where ϕ_1 is the fundamental flux per pole and $k_{fn} = (B_n/B_1)$ is a coefficient, depending on the nature of the surfaces facing the air gap, corresponding to the nth harmonic.

The ratio of the voltage magnitude of any harmonic to the fundamental is given by

$$\frac{E_n}{E_1} = \frac{k_{dn} k_{pn} k_{fn}}{k_{d1} k_{p1}} \qquad (6.6.7)$$

For normal designs and machine proportions, the ratios (k_{dn}/k_{d1}), (k_{pn}/k_{p1}), and the coefficient k_{fn} are all much less than unity, so their product results in a very small quantity indeed. Hence the harmonics in the terminal voltage may usually be neglected under normal conditions.

The rms phase value of the complete voltage wave is given by

$$E_{ph} = \sqrt{E_1^2 + E_3^2 + E_5^2 + \cdots} = E_1 \sqrt{1 + (E_3^2/E_1^2) + (E_5^2/E_1^2) + \cdots} \qquad (6.6.8)$$

Since the value of the radical $\sqrt{1 + (E_3^2/E_1^2) + (E_5^2/E_1^2) + \cdots}$ is very nearly unity, the rms phase value of the complete voltage is very nearly the same as the rms value of the fundamental alone, when the harmonics are relatively small compared to the fundamental. Thus, for practical purposes, the rms value of the generated voltage per phase is given by

$$E_{ph} = 4.44 \, k_{d1} k_{p1} f N_{ph} \phi_1 \text{ rms volts} \qquad (6.6.9)$$

which is the same as Equation 5.3.15. The rms line-to-line voltage for a balanced wye connection is given by $\sqrt{3}$ times the phase value, but excluding the 3rd harmonic, if the harmonic components are considered. This relationship is true because, for the wye connection, the line voltage is obtained from the difference of the phase voltages, and in-phase components cancel, so the line voltage is free from 3rd and multiples of 3rd (triplex) harmonics.

Let us next consider the case of a conventional dc machine with a closed continuous commutator winding on its armature. Equations 5.3.22 through 5.3.24 have already been obtained for the generated emf. The same equations can also be derived from a different point of view, bringing out a better understanding of the interrelationship between the ac and dc machine windings.

The induced voltages in a commutator winding are inherently alternating, and between any two points the resultant can be obtained by vectorial summation. Winding factors that we have already

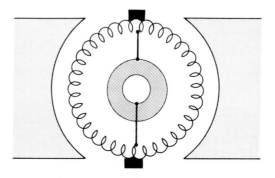

Figure 6.6.2 A closed continuous two-pole commutator winding. (*Note:* The commutator brushes are shown, for convenience, making direct contact with the armature winding.)

discussed may also be applied; but the slot angle is usually small, such that Equation 6.5.3 with the denominator replaced by $m\alpha/2$ in electrical radians is used for the distribution factor, and the chording factor may usually be neglected (i.e., a pitch factor of unity is taken) because the chording angle is small. Let us consider the two tappings opposite each other on a two-pole commutator winding shown in Figure 6.6.2. Assuming a sinusoidal distribution of flux, the vectorial summation is proportional to the diameter, while the arithmetic sum is proportional to one-half of the periphery; that is to say,

$$k_d = \frac{\text{chord}}{\text{arc}} = \frac{2}{\pi} \tag{6.6.10}$$

This equation may also be obtained by the application of Equation 6.5.3, making an approximation for the denominator:

$$k_d = \frac{\sin(\pi/2)}{\pi/2} = \frac{2}{\pi}$$

If the diametrical tappings are connected to two slip rings as shown in Figure 6.6.2, the instantaneous ac voltage for the position indicated will be at its maximum value, and the rms value of the slip-ring voltage will be 0.707 times that maximum value. If the winding is connected to a commutator on which two brushes are located on the quadrature axis, as shown schematically in Figure 6.6.2, the voltage across the brushes will be that maximum voltage at a substantially constant value as a function of time. In practice, however, a small superimposed high-frequency ripple occurs because of the finite number of commutator bars and slots. Thus we see that, when a commutator winding rotates in a magnetic field, a unidirectional emf is obtained whose magnitude depends on the brush position; further, the magnitude of the emf is the same as the instantaneous emf that would be obtained from a slip-ring winding when the rotor position with the tappings corresponds with the commutator brush axis.

As stated already, for a dc machine, a sinusoidal flux waveform is not needed. The flux density is arranged to be as high as possible over the pole pitch in order to get the maximum output voltage. Of course, other considerations, such as commutation, need to be addressed. This investigation leads to a field form that approaches a rectangular shape. In view of the large number of space harmonics in such a field form, the vectorial addition, although feasible, becomes rather complicated. A simpler method is usually employed for calculating the generated emf. The average voltage per conductor is calculated on the basis of the average flux density, because the dc voltage (which is an average value) is of interest.

The average voltage per conductor, when multiplied by the number of conductors in series, gives the total dc voltage across the brushes.

In a conventional dc machine, the brush axis is fixed relative to the stator. If, however, there is continuous relative motion between the poles and brushes, the voltage at the brushes will in fact be alternating. This principle is made use of in ac commutator machines.

IIIIIIIIIIIII▬▬▬▬▬▬▬▬▬▬▬▬▬▬▬▬
EXAMPLE 6.6.1

The field form of a wye-connected, three-phase, 60–Hz, 10-pole alternator has a spatial flux-density distribution given by

$$B = \sin \theta + 0.3 \sin 3\theta + 0.2 \sin 5\theta \text{ T}$$

The machine has 180 slots, lap-wound with double-layer, three-turn coils in 60° groups. Each coil spans 15 slots. The armature diameter is 1.5 m and the core length is 0.50 m. Determine the following:

a. Instantaneous emf per conductor

b. Instantaneous emf per coil

c. RMS phase and line-to-line voltages

Solution

a. Surface area per pole pitch = (pole pitch)(axial length of the core)

$$= \frac{\pi \times 1.50}{10} \times 0.50 = 0.2357 \text{ m}^2$$

Fundamental flux per hole $= \phi_1 = B_{1av} \times \text{area} = 1 \times (2/\pi) \times 0.2357 = 0.15$ Wb where $2/\pi$ is the factor for the average of a sinusoidal wave with an amplitude of unit magnitude.

The maximum or peak value of fundamental voltage per conductor is given by $\pi f \phi$, as seen from Equation 5.3.13, in which N is the number of turns of a coil, or one-half of the number of conductors; i.e., $2N$ is the number of conductors per coil.

Thus, peak value of fundamental conductor voltage equals $\pi \times 60 \times 0.15 = 28.3$ V. The peak harmonic conductor voltages are proportional to their corresponding peak flux densities: for the 3rd harmonic, $0.3 \times 28.3 = 8.49$ V; for the 5th harmonic, $0.2 \times 28.3 = 5.66$ V. The instantaneous conductor voltage is then given by

$$(28.3 \sin \omega t + 8.49 \sin 3\omega t + 5.66 \sin 5\omega t) \text{ V}$$

where $\omega = 2\pi \times 60 = 377$ rad/s, and $t = 0$ when the voltage is zero. In other words, t is measured from the point of zero voltage.

b. The number of slots and the coil span are the same as in Example 6.5.2. The pitch factors for the fundamental and 5th harmonic are calculated in that example:

$$k_{p1} = \cos \left(\frac{30°}{2} \right) = 0.966$$

$$k_{p3} = \cos \left(\frac{3 \times 30°}{2} \right) = 0.707$$

$$k_{p5} = \cos \left(\frac{5 \times 30°}{2} \right) = 0.259$$

Because a three-turn coil has six conductors, it follows that the peak fundamental coil voltage = $6 \times 28.3 \times 0.966 = 164$ V; the peak 3rd harmonic coil voltage = $6 \times 8.49 \times 0.707 = 36$ V; and the peak 5th harmonic coil voltage = $6 \times 5.66 \times 0.259 = 8.8$ V. The instantaneous coil voltage is then given by

$$(164 \sin \omega t + 36 \sin 3\omega t + 8.8 \sin 5\omega t) \text{ V}$$

c. The distribution factors can be calculated from Equation 6.5.5 for each harmonic. The slot angle α being $180/18 = 10$ electrical degrees, and the number of slots per group m being $18/3 = 6$ for $60°$ grouping, then

$$k_{d1} = \frac{\sin(60°/2)}{6 \sin(60°/2)} = \frac{0.5}{6 \times 0.087} = 0.958$$

$$k_{d3} = \frac{\sin(180°/2)}{6 \sin(30°/2)} = \frac{1}{6 \times 0.259} = 0.644$$

$$k_{d5} = \frac{\sin(300°/2)}{6 \sin(50°/2)} = \frac{0.5}{6 \times 0.423} = 0.197$$

Total number of coils per phase = $180/3 = 60$, since two coil sides occupy each slot in a double-layer winding. The rms phase emf may now be calculated:

$$\text{rms fundamental phase emf} = (164/\sqrt{2}) \times 60 \times 0.958 = 6{,}666.7 \text{ V}$$

$$\text{rms 3rd harmonic phase emf} = (36/\sqrt{2}) \times 60 \times 0.644 = 983.8 \text{ V}$$

$$\text{rms 5th harmonic phase emf} = (8.8/\sqrt{2}) \times 60 \times 0.197 = 73.6 \text{ V}$$

These emfs can also be calculated by the direct application of Equation 6.6.6, in which N_{ph} is the number of series turns per phase, given by 180 in our example because of the 60 three-turn coils per phase. The rms phase voltage is then given by Equation 6.6.8:

$$E_{ph} = \sqrt{6{,}666.7^2 + 983.8^2 + 73.6^2} = 6{,}739.3 \text{ V}$$

The rms line-to-line voltage = $\sqrt{3}\sqrt{6{,}666.7^2 + 73.6^2} = \sqrt{3} \times 6{,}667.1 = 11{,}547.4$ V

Note that the 3rd-harmonic component is excluded in the calculation of the line-to-line voltage.

‖‖‖‖‖‖‖‖‖ ▬▬▬▬▬▬▬▬▬▬▬▬

EXAMPLE 6.6.2

The armature of a four-pole dc machine has a simplex-lap wound commutator winding with 120 two-turn coils. If the flux per pole is 0.02 Wb, calculate the dc voltage appearing across the brushes located on the quadrature axis when the machine is running at 1,800 rpm.

Solution

This example can be solved by the direct application of Equation 5.3.23:

$$E_a = \frac{P\phi nZ}{a \times 60}$$

For our example,

$$P = 4; \qquad \phi = 0.02; \qquad n = 1,800$$

$$Z = \text{number of conductors} = 120 \times 2 \times 2 = 480$$

$$a = \text{number of parallel paths} = 4, \text{ for the simplex-lap}$$
$$\text{winding (same as number of poles)}$$

Therefore,

$$E_a = \frac{4 \times 0.02 \times 1,800 \times 480}{4 \times 60} = 288 \text{ V dc}$$

━━━━━━━━━━━━━━━━━━■ ❚❚❚❚❚❚❚❚
6.7 MMF Produced by Windings

Having discussed the formation of windings, we examine in this section the mmf and flux-density distributions that the windings produce. An understanding of the flux-density distribution wave-forms is important, because they determine the waveforms of generated emfs in the machine, the winding inductances, and the inductance variations as functions of rotor position.

Most of the analysis in this book is based on inductances and the coupled-circuit viewpoint. We have already derived the general expressions for emf and torque based on self-inductances and mutual inductances of the machine windings. Emf and torque calculations can, of course, be performed directly in terms of the magnetic fields, as we briefly saw earlier. The concepts of self- and mutual inductances make it possible, however, to express the magnetic flux linkages in terms of the inductances and the winding currents. Such an approach is quite convenient because the subsequent work can be performed by the familiar circuit analysis, which in fact becomes the basis for the generalized theory of electrical machines. While harmonics can be taken into account with the inductance method, saturation effects, if present, cannot easily be addressed. Based on the mmf and flux-density distributions produced by windings, it is possible to develop expressions for the winding inductances in cylindrical and salient-pole machines by evaluating the appropriate flux linkages per ampere of the appropriate current.

Cylindrical Machine with a Single Coil

Let us start by considering a simple case of a two-pole machine with a cylindrical stator and rotor having a single coil of N turns on the rotor, as shown in Figure 6.7.1, carrying a current of I amperes. The flux produced by the coil, with the indicated direction of current, divides itself into two parallel paths, as illustrated by the dotted flux loops. The mmf applied to each of these parallel parts is NI ampere-turns per pole. Ampere's circuital law can be applied around the closed path to obtain the magnetic field intensity and hence the magnetic flux density, recognizing that each flux loop contains two air gaps in series. In other words, the mmf at each pole is $NI/2$ ampere-turns per pole. We can neglect the reluctance of the iron compared with that of air gaps and consequently assume that all the applied mmf acts across the air gaps alone.

With the convention that mmf at the air gap is positive if acting radially outward, and negative if acting inward, we obtain the mmf distribution, F_θ, as a function of the position around the air gap, as shown in Figure 6.7.2 in a developed form laid out flat. The flux density at any position θ is given by $B_\theta = \mu_0 H_\theta = \mu_0 F_\theta / g$, where g is the single radial air-gap length. The air gap is constant at all values

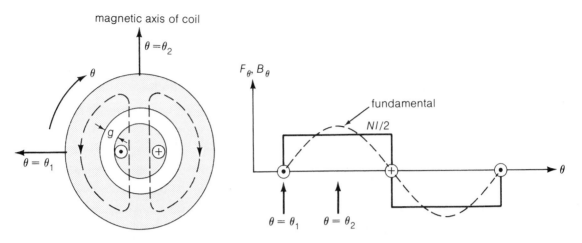

Figure 6.7.1 A two-pole cylindrical machine with a stator and rotor, having a single coil of N turns on the rotor.

Figure 6.7.2 The mmf and flux-density distribution corresponding to Figure 6.7.1 in a flattened (developed) layout.

of θ, and the flux density B_θ for any position θ is proportional to the mmf. Hence it follows that Figure 6.7.2 also represents the flux-density distribution as a function of θ to a different scale.

The wave of mmf distribution is shown in Figure 6.7.2 by the steplike distribution of amplitude $\pm NI/2$. On the basis of narrow slot openings, the mmf wave jumps abruptly by NI in crossing from one side to the other of the coil. The rectangular waveshape of the mmf shown in Figure 6.7.2 can be analyzed by Fourier analysis into harmonic components. Measuring θ from zero position θ_1 gives

$$F_\theta = \frac{4F}{\pi} \sum_{n \text{ odd}} \frac{\sin(n\theta)}{n} \tag{6.7.1}$$

where $F = NI/2$.

If θ is measured from zero position θ_2, i.e., the magnetic axis of the coil, one obtains

$$F_\theta = \frac{4F}{\pi} \sum_{n \text{ odd}} \pm \frac{\cos(n\theta)}{n} \tag{6.7.2}$$

where $F = NI/2$, and the sign alternates between + and −, starting with the positive sign for the fundamental. Explicitly,

$$F_\theta = \frac{4F}{\pi}[\cos\theta - \frac{1}{3}\cos 3\theta + \frac{1}{5}\cos 5\theta - \cdots]$$

The fundamental is sketched in Figure 6.7.2, the amplitude of which is given by $[(4/\pi)NI/2]$. The peak of the fundamental sinusoidal space wave is aligned with the magnetic axis of the coil.

Distributed Winding

Let us next consider the case of a two-pole cylindrical machine with a uniformly distributed winding, consisting of 10 conductors as shown in Figure 6.7.3a. Each conductor carries a current of I amperes in the directions indicated. Assume that pairs of conductors are connected to form coils as shown, 1-1′,

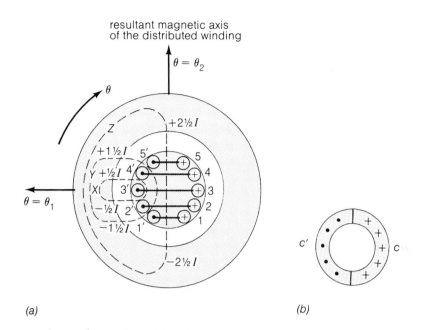

Figure 6.7.3 A two-pole cylindrical machine with uniformly distributed winding of ten conductors in (a), and its equivalent continuous conductor arrangement in (b).

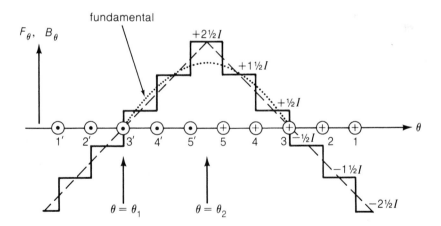

Figure 6.7.4 The mmf and flux-density distribution corresponding to Figure 6.7.3 in a flattened (developed) layout.

2-2', and so forth. Considering the three flux loops sketched on either side of the magnetic axis of the distributed winding, the mmf appplied to the flux loop X is I ampere-turns, the mmf applied to the flux loop Y is $3I$ ampere-turns, and the mmf applied to the flux loop Z is $5I$ ampere-turns. Notice that, at any position of the air gap between two adjacent conductors, the mmf remains at the same constant value. Therefore, the mmf at various parts of the air gap as a function of θ can be determined by inspection. The mmf distribution showing the variation of mmf around the air gap is given in Figure 6.7.4 by a series of steps, each of height I ampere-conductors in the slot, in the flattened (developed) layout. If the radial air gap is a constant, the value of the flux density at any position θ is simply proportional to the mmf. That being so, the waveshape of Figure 6.7.4 also represents the flux-density distribution to a different scale.

As we can see, the distributed winding produces a closer approximation to a sinusoidal mmf wave than the concentrated coil of Figure 6.7.1. Thus, distributing the winding results in a marked reduction in the harmonic content of the flux-density distribution and hence causes the induced voltage to be more nearly sinusoidal. The resultant space-fundamental component of the mmf wave of a distributed winding is less than the sum of the fundamental components of the individual coils, because the magnetic axes of the individual coils are not aligned with the resultant.

If the conductors 1 through 5 and $1'$ through $5'$ are replaced by continuous conductors C and C', as shown in Figure 6.7.3b, carrying a total current of $5I$ amperes, then the mmf will increase and decrease linearly around the air gap, as shown by the dotted triangular waveform in Figure 6.7.4. This assumption is equivalent to assuming that the winding comprises an infinite number of conductors Z with an infinitesimally small circumferential width, each carrying $(1/Z)$ of the total current. In practice, since the number of conductors is normally a much larger number than five, the triangular waveform with linear variation can be taken as a reasonable approximation.

The triangular waveshape of the mmf shown in Figure 6.7.4 can be analyzed by Fourier analysis into harmonic components. If θ is measured from zero position θ_1, then

$$F_\theta = \frac{8F}{\pi^2} \sum_{n \text{ odd}} \left[\pm \frac{\sin(n\theta)}{n^2} \right] \tag{6.7.3}$$

where $F = NI/2$, N being 5 in Figure 6.7.4. Equivalently,

$$F_\theta = \frac{4F}{\pi} \sum_{n \text{ odd}} [\pm k_n \sin(n\theta)] \tag{6.7.4}$$

where $k_n = 2/n^2\pi$. The sign alternates between $+$ and $-$, starting with the positive sign for the fundamental. If θ is measured from zero position θ_2, i.e., the resultant magnetic axis of the distributed winding, then

$$F_\theta = \frac{8F}{\pi^2} \sum_{n \text{ odd}} \frac{\cos(n\theta)}{n^2} = \frac{4F}{\pi} \sum_{n \text{ odd}} k_n \cos(n\theta) \tag{6.7.5}$$

where F is again $NI/2$ and k_n again equals $2/n^2\pi$.

The fundamental is sketched in Figure 6.7.4; its amplitude is given by $8F/\pi^2$. The peak of the fundamental sinusoidal space wave is aligned with the resultant magnetic axis of the distributed winding.

Partially Distributed Winding

Instead of a distributed winding's fully covering the rotor surface, as in Figure 6.7.3, consider a rotor winding in which the conductors, carrying currents as shown in Figure 6.7.5, are distributed over an angle β. The case of $\beta = 180$ electrical degrees corresponds to that of Figure 6.7.3; $\beta = 0$ represents the single coil of Figure 6.7.1. For three-phase windings, each phase occupies one-third of the air-gap periphery; for the usual $60°$ groupings, β is equal to 60 electrical degrees.

Following the procedure indicated for the previous cases, we can obviously see that the mmf and flux-density distributions for the partially distributed winding are illustrated in Figure 6.7.6. Note that

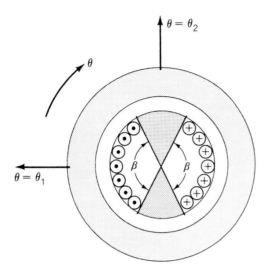

Figure 6.7.5 A two-pole cylindrical machine with partially distributed winding over a portion of the rotor.

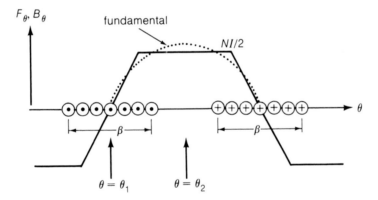

Figure 6.7.6 The mmf and flux density distribution corresponding to Figure 6.7.5 in a flattened (developed) layout.

the flat tops with constant mmf correspond to the parts of the periphery where there are no conductors, and a straight-line approximation is used over the rest. The trapezoidal waveshape of the mmf shown in Figure 6.7.6 can be analyzed by Fourier analysis into harmonic components. If θ is measured from zero position θ_1, then

$$F_\theta = \frac{4F}{\pi} \sum_{n \text{ odd}} k_n \sin(n\theta) \tag{6.7.6}$$

where $F = NI/2$ and

$$k_n = \frac{\sin(n\beta/2)}{(n^2\beta/2)}$$

If θ is measured from the zero position θ_2, i.e., the resultant magnetic axis of the distributed winding, then

$$F_\theta = \frac{4F}{\pi} \sum_{n \text{ odd}} [\pm k_n \cos(n\theta)] \qquad (6.7.7)$$

where F and k_n are the same as given for Equation 6.7.6. The sign alternates between $+$ and $-$, starting with the positive sign for the fundamental.

The magnitude of any harmonic component of mmf produced by the distributed winding is nk_n times that produced by the concentrated winding, as seen from Equations 6.7.1 and 6.7.6 or Equations 6.7.2 and 6.7.7. Thus,

$$nk_n = \frac{\sin(n\beta/2)}{n\beta/2} = k_{dn} \qquad (6.7.8)$$

is called the distribution factor for the nth harmonic, which is the same as given by Equation 6.5.6. The distributed winding of N turns is equivalent to a concentrated winding of $(k_{dn}N)$ turns, as far as the nth harmonic is concerned.

The fundamental with the amplitude of $4Fk_1/\pi$ is aligned with its peak occurring on the resultant magnetic axis of the distributed winding. In general, for a P-pole machine with a distributed winding of N_{ph} series turns per phase, the amplitude of the space-fundamental mmf wave is given by

$$F_{a1} = \frac{4}{\pi} k_1 \frac{N_{ph}}{P} i_a \qquad (6.7.9)$$

where i_a is the current in phase a; the factor $4/\pi$ arises from the Fourier-series analysis of the rectangular mmf wave of a concentrated full-pitch coil; and k_1 accounts for the distribution of the winding. The effects of space harmonics in ac machines, as we learned earlier, can be made small by using distributed windings with fractional-pitch coils.

The above space-fundamental component of the mmf wave produced by current in the distributed phase-a winding is equivalent to the mmf wave produced by a finely divided sinusoidally distributed current sheet placed on the outer periphery of the rotor (or inner periphery of the stator, if the winding is located in the slots of the stator). The mmf is a standing wave for which the fundamental spatial distribution around the periphery is sinusiodal, described by $\cos\theta$ or $\sin\theta$ depending on the zero-position chosen for θ. As stated before, its peak is along the magnetic axis of phase-a winding, and its peak amplitude is proportional to the instantaneous current i_a as given by Equation 6.7.9. If the phase current i_a is a function of time given by $i_a = I \cos \omega t$, the time maximum of the peak is given by

$$F_m = \frac{4}{\pi} k_1 \frac{N_{ph}}{P} I \qquad (6.7.10)$$

The effect of balanced three-phase currents in all three phases was covered in Section 5.4.

Thus far we have seen that the waveshapes of the mmf and flux-density distributions produced by windings around the air-gap periphery can be analyzed by Fourier analysis into harmonic components. Depending on the reference chosen for $\theta = 0$:

$$F_\theta = \sum F_n \sin(n\theta) \text{ or } \sum F_n \cos(n\theta) \qquad (6.7.11)$$
$$B_\theta = \sum B_n \sin(n\theta) \text{ or } \sum B_n \cos(n\theta) \qquad (6.7.12)$$

where $B_n = \mu_0 F_n/g$. Note that there are no even harmonics because of the particular symmetry of the

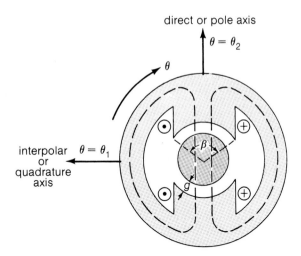

Figure 6.7.7 A two-pole machine with cylindrical rotor and salient-pole stator fitted with concentrated stator winding.

waveforms, i.e, $F(\theta + \pi) = -F(\theta)$. Discussions are often restricted to only the fundamental sinusoidal component of mmf or flux-density distribution because

- This quite often reasonable mathematical approximation also happens to be convenient for analysis.
- When we suitably distribute and chord the winding, we can significantly reduce the effect of harmonics.
- If, in fact, the approximation is not close enough, we can include the effects of each harmonic by following the same general principles of analysis developed for the fundamental.

Thus, in introductory machine analysis, we are often sufficiently accurate if we neglect all but the fundamental components.

Concentrated Coils on Salient Poles

Consider a two-pole machine with a cylindrical rotor and a salient-pole stator, as shown in Figure 6.7.7, with concentrated stator coils of N total number of turns on salient poles, carrying I amperes. The flattened layout of the stator in a developed form is given in Figure 6.7.8 along with the corresponding mmf wave-shape. Maximum mmf of constant value ($NI/2$) occurs over the polar region, and no mmf exists in the interpolar region. If the air gap in the polar region is a constant, the flux density under the pole is proportional to the mmf, and hence the waveform of Figure 6.7.8 also gives the waveshape of the flux-density distribution to a different scale. The maximum constant value of B is given by $\mu_0 F/g$, where F is $NI/2$ and g is the radial air-gap length under the pole. Fourier analysis of the waveforms with zero position of θ as θ_1, leads to the following result:

$$F_\theta = \frac{4F}{\pi} \sum_{n \text{ odd}} [\pm k_n \sin (n\theta)] \tag{6.7.13}$$

where $k_n = [\sin (n\beta/2)]/n$ and β is the angle subtended by the pole-face at the center, as shown in Figure

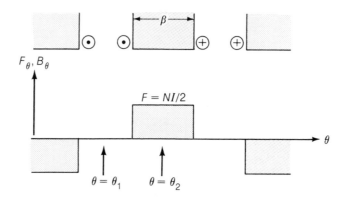

Figure 6.7.8 The mmf and flux-density distribution corresponding to Figure 6.7.7 in a flattened (developed) layout.

6.7.7, and the signs alternate starting with the positive sign for the fundamental. With zero position of θ as θ_2,

$$F_\theta = \frac{4F}{\pi} \sum_{n \text{ odd}} \cos(n\theta) \tag{6.7.14}$$

For salient-pole ac machines, where the sinusoidal nature of the flux-density distribution around the air-gap periphery is desirable as in synchronous machines, the body of the pole structure is usually fitted with a wider poleshoe, which is so shaped that the air gap is shortest under the pole center and increases in radial length toward the pole tips. This process is referred to as *pole-chamfering*. Because the reluctance of the air gap under the poleshoe is not uniform, the flux-density distribution is not the same as the mmf distribution.

Distributed Winding on the Rotor with Saliency on Stator

Let us next consider a two-pole machine with saliency on its stator and a cylindrical rotor with a distributed winding of 10 conductors, as shown in Figure 6.7.9, carrying currents as indicated. Two positions of the rotor-mmf axis with respect to the stator are considered: Figure 6.7.9a in line with the stator-pole axis, and Figure 6.7.9b at right angles to the stator-pole axis. Figure 6.7.10 shows the corresponding developed diagrams for the mmf and flux-density waveforms. The mmf waveshape is triangular, as discussed previously, assuming linear variation instead of steps as we would for a much larger number of conductors than five. But here, the flux density around the air-gap periphery is not proportional to the mmf throughout because the air-gap length is not a constant.

For a given position of the rotor, a rough sketch of the flux-density variation can be drawn. First locate a triangular curve corresponding to the uniform air gap equal to that under the poles, and then reduce the value of the flux density in those portions where the air gap is larger than the uniform value. Dotted curves in Figure 6.7.10 illustrate the approximate waveshapes of the flux-density distribution for the two positions of the rotor considered in Figure 6.7.9. Notice that the flux-density waveform is now different for different positions of the rotor with respect to the stator. Even though the situation looks rather complicated because of the varibale air gap, reasonable approximations can be made to calculate the flux-density distribution in terms of two mutually perpendicular components in the polar (direct)

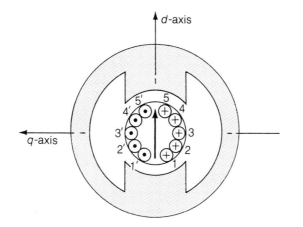

(a)

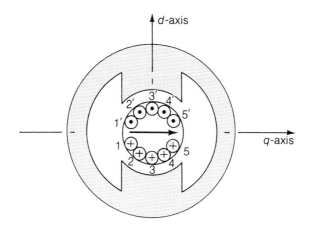

(b)

Figure 6.7.9 A two-pole machine with salient-pole stator and cylindrical rotor carrying a distributed winding. (a) Rotor-mmf axis in line with the stator-pole axis, (b) rotor-mmf axis in quadrature to the stator-pole axis.

and interpolar (quadrature) axes. The physical structure and symmetry of the salient-pole arrangement suggest that this example may be more amenable to mathematical analysis if quantities are resolved into two orthogonal components along the direct and quadrature axes, which correspond to the physical polar and interpolar axes on the machine, that are always 90 electrical degrees away from each other. In fact, such a technique works well, as we see later in the analysis of a salient-pole synchronous machine.

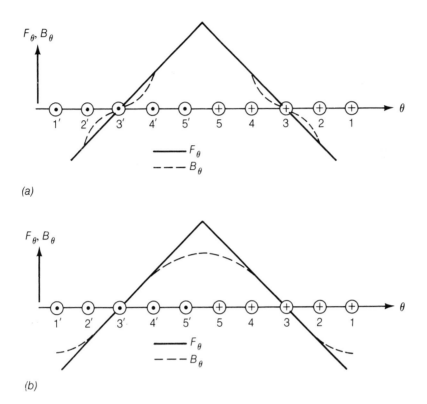

Figure 6.7.10 The mmf and flux-density distributions corresponding to Figure 6.7.9 (a) for the case when the rotor-mmf axis is in line with the stator-pole axis, (b) for the case when the rotor-mmf axis is in quadrature to the stator-pole axis.

BIBLIOGRAPHY

Daniels, A. R. *Introduction to Electrical Machines.* London: Macmillan, 1976.

Garik, M. L., and C. C. Whipple. *Alternating-Current Machines,* 2d ed. New York: D. Van Nostrand, 1961.

Hindmarsh, J. *Electrical Machines and Their Applications,* 2d ed. Oxford: Pergamon Press, 1970.

Kloefffler, R. G.; R. M. Kerchner; and J. L. Brenneman. *Direct-Current Machinery.* New York: Macmillan, 1948.

Lawrence, R. R., and H. E. Richard. *Principles of Alternating-Current Machinery,* 4th ed. New York: McGraw-Hill, 1953.

Morgan, A. T. *General Theory of Electrical Machines.* London: Heyden & Son, 1979.

Sarma, M. S. *Synchronous Machines.* New York: Gordon and Breach Science Publishers, 1979.

Say, M. G. *Alternating Current Machines.* New York: John Wiley & Sons, Halsted Press, 1976.

Siskind, C. S. *Direct-Current Machinery.* New York: McGraw-Hill, 1952.

Veinott, C. A. *Spatial Harmonics (One-Phase and Polyphase).* Sarasota, Fla.: C. A. Veinott, 1991.

Whipple, C. C. *Electric Machinery, Volume 1: Fundamentals and D. C. Machines.* New York: D. Van Nostrand, 1946.

▬▬▬▬▬▬▬▬▬ ||||||||||||

PROBLEMS

6–1. Draw a developed winding diagram for a two-pole, double-layer, dc armature winding in 16 slots for a simplex-lap progressive connection. Indicate the position of the brushes and the poles. You may assume single-turn coils for simplicity.

6–2. Draw a developed view of the dc commutator simplex-lap winding that has the following particulars:

> Number of slots: 27
> Number of poles: 6
> Total number of coils: 54
> Number of coils per slot: 2
> $y_c = +1$; $y_b = 8$;   slot span $= 4$

Determine the number of equalizer bars needed for 100% equalization, and give their corresponding tappings.

6–3. Draw a developed winding diagram of a four-pole, two-layer, simplex-wave dc armature winding with 17 slots and 17 single-turn coils. Locate the position of the brushes and the poles.

6–4. Draw a developed view of the dc commutator simplex-wave winding that has the following particulars:

> Number of slots: 32
> Number of poles: 6
> Total number of coils: 64
> Number of coils per slot: 2
> $y_c = 21$ (retrogressive); $y_b = 10$;   slot span $= 5$

6–5. Calculate the armature current per path of a four-pole, 50-kW, 250-V dc generator if the commutator winding is (a) simplex wave, and (b) simplex lap. You may neglect the shunt field current.

6–6. A six-pole dc machine is required to generate an emf of 260 V when rotating at 450 rpm. The flux per pole is not to exceed 0.02 Wb. If the armature has 74 slots, determine the particulars of a suitable wave winding.

6–7. Draw a flattened layout of a two-phase, two-pole, two-layer, full-pitch lap ac armature winding in 12 slots with 90° spread.

6–8. a. Show that, if entry points displaced by 120 mechanical degrees are to be made to the phases of a three-phase machine with a double-layer winding of 60 electrical degree spread with $(6K + 2)$ poles, where K is an integer (i.e., 2, 8, 14, 20, ... poles), they must be made in the order of the phase sequence $A - B - C$.

  b. Show that, for a machine with $(6K + 4)$ poles (i.e., 4, 10, 16, 22, ... poles), it is never possible to make entry points displaced by 120 mechanical degrees.

6–9. Develop a flattened layout of a three-phase, four-pole, two-layer lap ac armature winding with 60 electrical degree spread in 24 slots with

  a. Full-pitch coils

  b. 5/6 fractional-pitch coils

Show the detailed coil interconnections. Calculate the winding distribution factor and pitch factor for a. and b. for the fundamental.

6–10. Draw a flattened layout of a three-phase, four-pole, two-layer, lap ac armature winding with 60 electrical degree spread in 36 stator slots with

  a. full-pitch coils

b. fractional-pitch coils short-pitched by (1) one slot; (2) two slots; (3) three slots
Determine the winding distribution factor and pitch factor for each of the above cases for the fundamental and the 5th and 7th harmonics.

6–11. Calculate the distribution, pitch, and winding factors for the distributed fractional-pitch winding of Figure P6–11.

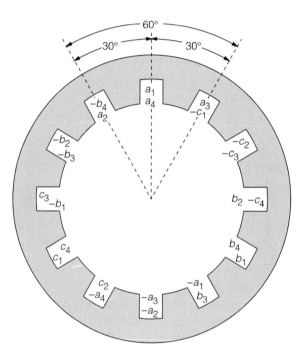

Figure P6–11

6–12. A three-phase, 60-Hz, wye-connected armature winding of a generator has 6 slots per pole per phase. The coil pitch is 16 slots. The winding is a 60° spread and double layered, with 30 turns per phase. If the air-gap flux is sinusoidally distributed, what must be its maximum value to give 600 V across the lines?

6–13. A three-phase, eight-pole, 60-Hz, wye-connected, salient-pole synchronous generator has 96 slots and a 60° spread double-layer winding, with four conductors per slot connected in series in each phase. The coil pitch is 10 slots. If the maximum value of the air-gap flux is 60 mWb and the flux-density distribution in the air gap is sinusoidal, determine (a) the rms phase voltage and (b) the rms line-to-line voltage. (c) If each phase is capable of carrying 650 amperes, find the kVA rating of the machine.

6–14. A 72-slot stator is to be wound for three-phase operation. If the coil pitch is six slots, derive a suitable form for the winding arrangement as an eight-pole winding, and draw the developed view showing the position of the phase belts in each layer of the winding for (a) 60 electrical degree spread, and (b) 120 electrical degree spread.

Calculate the winding distribution and pitch factors for the fundamental and the 5th and 7th harmonics for each of the above cases.

6–15. A three-phase, wye-connected, 60-Hz, six-pole alternator has a fundamental flux per pole of 0.02 Wb. The stator has 90 slots with four conductors per slot, and each stator coil spans 12 slots. If the flux in the air gap contains a 30% 3rd harmonic and a 20% 5th harmonic with respect to the fundamental, determine the rms values of the phases and line-to-line induced emfs.

6–16. A three-phase, four-pole, star-connected synchronous machine, driven at 1,800 rpm, has a 48-slot stator carrying a 60° spread double-layer winding with 5/6 fractional-pitch coils. Each slot has 10 conductors. If the maximum value of the sinusoidally distributed field flux is 0.1 Wb, calculate the fundamental distribution and pitch factors, and the phase and line rms values of the generated emf.

6–17. The field form of a salient-pole synchronous machine has been analyzed for the relative magnitudes of the harmonics. The ratios of the harmonics to the fundamental are

$$k_{f1} = 1.0; \qquad k_{f3} = 0.061; \qquad k_{f5} = 0.004; \qquad k_{f7} = 0.035$$

The double-layer, wye-connected armature winding of the machine has three spp, with a fractional pitch of 8/9.

Determine the effect of the harmonics on the rms values of (a) the line-to-neutral voltage, and (b) the line-to-line voltage.

6–18. A six-pole, double-layer dc armature winding in 28 slots has five turns per coil. If the field flux is 0.025 Wb per pole and the speed of the rotor is 1,200 rpm, find the value of the induced emf when the winding is (a) lap-connected, and (b) wave-connected.

6–19. A four-pole, dc series motor has a lap-connected, two-layer armature winding with a total of 400 conductors. Calculate the gross torque developed for a flux per pole of 0.02 Wb and an armature current of 50 A.

6–20. A three-phase, 60-Hz, six-pole, wye-connected ac generator has 972 conductors distributed in 54 slots. The coils are short-pitched by one slot. Determine the fundamental-frequency line voltage on no-load (i.e., open circuit) when the fundamental flux per pole is 0.01 Wb.

If the rms current per conductor is 100 A, calculate the peak value of the fundamental and the 5th and 7th harmonic components of the armature mmf wave in ampere-turns per pole.

PART
2

Steady-State Theory and Performance

7

Induction Machines

In this chapter we want to analyze the polyphase induction machines introduced in Chapter 5 in sufficient detail that we can develop equivalent-circuit models for predicting the machines' steady-state performance. Corresponding to the number of phases in practical power systems, the polyphase induction motors used in industrial applications are almost without exception three-phase. The stator winding is connected to the ac source, and the rotor winding is either short-circuited, as in *squirrel-cage* machines, or closed through external resistances, as in *wound-rotor* machines. The cage machines are also known as *brushless* machines; the wound-rotor machines are also called *slip-ring* machines.

The polyphase induction motor, invented by Nikola Tesla in 1886, is the most widely used, least expensive industrial motor. An estimated 50 million or more of them are now in use in American industry, totaling some 150 million horsepower. Each normal year's production adds about a million motors to that number. In addition, another 20 million single-phase fractional-horsepower induction motors are used in domestic appliances like electric fans, refrigerators, washing machines, and farm appliances.

The polyphase induction motor requires nothing for its excitation other than the ac source. Because the induction motor has no inherent means for producing its excitation, it requires reactive power and draws a lagging current. While the power factor at rated load may generally be above 0.8, its is quite low at light loads. In order to limit the reactive power, the magnetizing reactance must be high, and therefore the air gap must be kept relatively short compared to that of a synchronous motor of the same size and rating, except in a small motor. The minimum length of the air gap used in an induction motor is determined by mechanical considerations and also by such factors as noise and magnetic losses in the tooth faces.

We recall from Chapter 5 that ac in a polyphase induction motor is supplied to the stator winding directly and to the closed rotor winding by induction from the stator. The balanced polyphase stator and rotor currents produce stator-component and rotor-component mmf waves of constant amplitude that rotate in the air gap at synchronous speed and are therefore stationary with respect to each other regardless of the mechanical speed of the rotor. As a consequence of these mmfs a resultant air-gap flux-density wave is produced; the interaction of the flux wave and the rotor-mmf wave gives rise to torque. At all speeds other than synchronous speed, the conditions necessary for the production of a steady torque are fulfilled. Because no torque is produced at synchronous speed and the induction motor runs below synchronous speed, induction machines are also known as *asynchronous machines*.

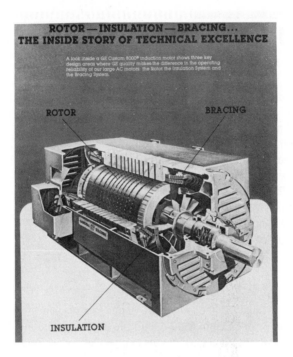

ROTOR — INSULATION — BRACING...
THE INSIDE STORY OF TECHNICAL EXCELLENCE

A look inside a GE Custom 8000® induction motor shows three key
design areas where GE quality makes the difference in the operating
reliability of our large AC motors: the Rotor, the Insulation System and
the Bracing System.

ROTOR   BRACING

INSULATION

Photo 7.1.1 A cross-sectional view of a typical medium-sized three-phase squirrel-cage induction motor. (Photo courtesy of General Electric Company.)

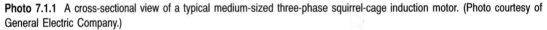

7.1 Constructional Features of Polyphase Induction Machines

A polyphase induction motor has five essential features:

- A laminated stator core carrying a polyphase winding
- A laminated rotor core carrying either a squirrel cage or a polyphase winding with shaft-mounted slip rings
- A stiff shaft to maintain a necessarily short air gap
- A frame for the stator housing to carry the bearings
- A cooling system

See Photos 7.1.1 and 7.1.2 for typical cross-sectional views of a medium-sized and a small-sized three-phase squirrel-cage induction motor.

The stator core is made up of laminations that are usually 0.3 to 0.6 mm thick; slightly thicker laminations can be used for small motors or in cases when low core loss is not so important. The laminations are insulated from one another either by an oxide coating produced by heat treatment or by varnish coating. The built-up laminations are held by flanges in the yoke. Ventilating ducts are provided along the length of the core, spaced every 5 or 7 cm by spacers placed between laminations. The rotor core is built up of laminations of the same material, say, sheet metal, as used for the stator core, but thicker laminations can be used because of the lower frequencies of the rotor flux. The lamination consists of a single piece in small motors; in large motors, the laminations are usually segmented and dovetailed to a center spider. When ventilating ducts are provided for the stator core, they are also

Photo 7.1.2 Cross-sectional view of a drip-proof, fan-cooled, 10-hp, three-phase, 60-Hz, 230/460-V squirrel-cage induction motor. (Photo courtesy of Marathon Electric Manufacturing Company.)

provided in the rotor core in equal number. To force cooling air through the machine, fan blades are usually added to the rotor core ends.

Double-layer windings are most frequently used for the stator windings of the polyphase motors because they are easier to manufacture, assemble, and repair, and also because the coils are all alike. The windings are distributed and are generally short-pitched. The short pitch and the distribution are effective in limiting the magnitudes of the space harmonics in the air-gap flux. Wherever possible, designers use integral-slot windings (for which the number of slots per pole per phase is an integer). To be able to use standard punchings in as wide a variety of machines as possible, however, the manufacturer can use fractional-slot windings for convenience. See Photo 7.1.3 for a wound stator of a medium-sized three-phase induction motor.

Since the air-gap reluctance is a variable at different points on the stator as the rotor turns, and since this pulsating reluctance results in pulsating exciting current, irregular torque, increased tooth loss, and noise, the general tendency is to use a large number of slots to reduce the effect of variable air-gap reluctance. On the other hand, a large number of slots results in narrow teeth and increased manufacturing costs. Making the number of slots on the rotor different from that on the stator and skewing the rotor slots create more uniform reluctance. Partly closed slots on the stator or the rotor, or both, also help make the air-gap reluctance more uniform. However, such slots complicate the problem of winding, since the coils must be fed through the narrow slot, turn by turn, or, if copper bars are used, as in squirrel-cage rotors, the coils must be inserted from the ends of the rotor. While open and partly closed slots are both widely used for the stator, partly closed slots are almost always used for the rotor. Once the number of slots on the stator is fixed, a suitable (different) number of rotor slots is chosen to avoid magnetic locking and obtain a smooth torque-speed characteristic.

For squirrel-cage rotor windings, copper, brass, or aluminum bars, short-circuited on the ends by rings, are used as the rotor conductors. The bars are welded, brazed, or bolted to the end rings, or, after the rotor core is built, aluminum alloy bars can be cast into the slots with the end rings as integral parts. No insulation is necessary between the bars and the laminated rotor core. For wound-rotor machines, the same types of windings used for the stators can be used for the rotors. The number

Photo 7.1.3 Wound stator of a medium-size three-phase induction motor. (Photo courtesy of Westinghouse Electric Corporation.)

Photo 7.1.4 Squirrel-cage rotor of a medium-sized three-phase induction motor. (Photo courtesy of General Electric Company.)

of poles on the stator and rotor must be the same for torque production, while operation is only slightly affected by a change in the number of phases provided on the rotor windings, as long as it is greater than one. Three-phase rotors are commonly used for three-phase and two-phase motors, and also sometimes for single-phase motors. Bar, strap, or wire is used for rotor windings; wire is best for cases in which many turns are needed. Using a large number of rotor turns increases the secondary voltage and decreases the current flowing through the slip rings. The secondary voltage and the current influence the value of the resistance to be used across the slip rings for starting or speed control. The secondary voltage also determines the necessary insulation. See Photos 7.1.4

Photo 7.1.5 Wound rotor of a large three-phase induction motor. (Photo courtesy of Westinghouse Electric Corporation.)

and 7.1.5 for typical views of a squirrel-cage rotor and a wound rotor of three-phase induction machines.

The windings, air gap, and slot details must be selected so that the exciting current and machine reactances conform to the desired performance. The air-gap length should be made as small as possible to reduce the magnetizing current necessary to set up the air-gap flux; yet too small an air gap can increase motor noise and tooth-face loss and can even prevent the motor from accelerating to its rated speed.

7.2 Equivalent Circuit of a Polyphase Induction Machine

A review of the material presented under "Elementary Induction Machines" in Section 5.3 can be helpful at this stage to recall the operating principle of polyphase induction machines.

As stated in that section, the induction machine may be regarded as a generalized transformer in which energy is converted and electric power is transferred between the stator and the rotor, along with a change of frequency and a flow of mechanical power. At standstill, however, the machine acts as a simple transformer with an air gap and a short-circuited secondary winding; the frequency of the rotor-induced emf is the same as the stator frequency as standstill. At any value of the slip under balanced steady-state operation, the rotor current reacts on the stator winding at the stator frequency because the rotating magnetic fields caused by the stator and rotor are stationary with respect to each other.

The induction machine may thus be viewed as a transformer with an air gap and a variable resistance in the secondary: the stator of the induction machine corresponds to the transformer primary, and the rotor corresponds to the secondary. For analysis of the balanced steady-state, it is sufficient to proceed on a per-phase basis with some phasor concepts, so we will now develop an equivalent circuit on a per-phase

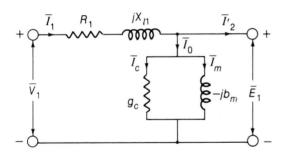

Figure 7.2.1 Equivalent circuit for a stator phase of a polyphase induction motor.

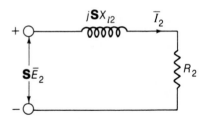

Figure 7.2.2 Slip-frequency equivalent circuit for a rotor phase of a polyphase induction motor.

basis. Only machines with symmetrical polyphase windings excited by balanced polyphase voltages are considered. As in other discussions of polyphase devices, let us think of three-phase machines as wye-connected, so that currents are always line values and voltages are always line-to-neutral values.

The resultant air-gap flux is produced by the combined mmfs of the stator and rotor currents. For the sake of conceptual and analytical convenience, the total flux is divided into a mutual flux (linking both the stator and the rotor) and leakage fluxes (including slot-leakage flux, tooth-top leakage flux, and coil-end-connection leakage flux, among others). An appropriate leakage reactance component can be assigned to each leakage flux component.

Considering the conditions in the stator, the synchronously rotating air-gap wave generates balanced polyphase counter emfs in the phases of the stator. The volt-ampere equation for the phase under consideration *in phasor notation* is given by

$$\bar{V}_1 = \bar{E}_1 + \bar{I}_1 \left(R_1 + jX_{l1} \right) \tag{7.2.1}$$

where $\bar{V}_1$ is the stator terminal voltage, $\bar{E}_1$ is the counter emf generated by the resultant air-gap flux, $\bar{I}_1$ is the stator current, R_1 is the stator effective resistance, and X_{l1} is the stator-leakage reactance.

As in a transformer, the stator (primary) current can be resolved into two components: a load component $\bar{I}_2'$ (which produces an mmf that exactly counteracts the mmf of the rotor current) and an exciting component $\bar{I}_0$ (required to create the resultant air-gap flux). This exciting component itself can be resolved into a core-loss component $\bar{I}_c$ in phase with $\bar{E}_1$ and a magnetizing component $\bar{I}_m$ lagging $\bar{E}_1$ by 90°. A shunt branch formed by the core-loss conductance g_c and magnetizing susceptance b_m in parallel, connected across $\bar{E}_1$, will account for the exciting current in the equivalent circuit, as shown in Figure 7.2.1 along with the positive directions in a motor.

Thus far, the equivalent circuit representing the stator phenomena is exactly like that of the transformer primary. Because of the air gap, however, the value of the magnetizing reactance tends to be relatively low compared to that of a transformer, and the leakage reactance is larger in proportion to the magnetizing reactance than it is in transformers. To complete the equivalent circuit, the effects of the rotor must be incorporated, which we do by referring the rotor quantities to the stator.

The Rotor Equivalent Circuit

Because the frequency of the rotor voltages and currents is the slip frequency, the magnitude of the voltage induced in the rotor circuit is proportional to the slip, as can be verified readily by the voltage equation (4.2.10). Also, in terms of the standstill per-phase rotor leakage reactance X_{l2}, the leakage

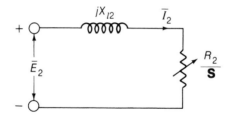

Figure 7.2.3 Alternate form of the per-phase rotor equivalent circuit.

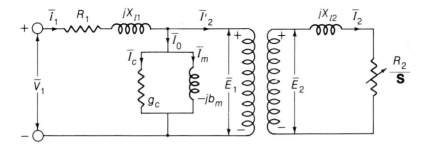

Figure 7.2.4 Per-phase equivalent coupled circuit of a polyphase induction motor.

reactance at a slip $\mathbf{S}$ is given by $\mathbf{S}X_{l2}$. With R_2 as the per-phase resistance of the rotor, the slip-frequency equivalent circuit for a rotor phase is shown in Figure 7.2.2, in which E_2 is the per-phase voltage induced in the rotor at standstill. The rotor current I_2 is given by

$$I_2 = \frac{SE_2}{\sqrt{R_2^2 + \left(SX_{l2}\right)^2}} \tag{7.2.2}$$

which may be rewritten as

$$I_2 = \frac{E_2}{\sqrt{\left(\frac{R_2}{S}\right)^2 + X_{l2}^2}} \tag{7.2.3}$$

resulting in the alternate form of the per-phase rotor equivalent circuit shown in Figure 7.2.3.

The Complete Per-Phase Equivalent Circuit

All rotor electrical phenomena, when viewed from the stator, become stator-frequency phenomena because the stator winding sees the mmf and flux waves traveling at synchronous speed. Returning to the analogy of a transformer, and considering that the rotor is coupled to the stator in the same way the secondary of a transformer is coupled to its primary, we may draw the equivalent circuit as shown in Figure 7.2.4.

Now we refer the rotor quantities to the stator. When we include the effect of the stator and rotor winding distributions, the ratio of the stator to the rotor voltage is given by

$$\frac{E_1}{E_2} = \frac{k_{w1}N_1}{k_{w2}N_2} \tag{7.2.4}$$

where k_{w1} and k_{w2} are the winding factors of the stator and rotor windings, respectively; N_1 and N_2 are the numbers of series-connected turns per phase of the stator and rotor windings, respectively. For a cage rotor, however, note that k_{w2} equals unity, so the number of turns per phase N_2 is given by $(1/2)(P/2)$, where P is the number of poles.

If m_1 and m_2 are the numbers of phases on the stator and rotor, respectively, the rotor volt-amperes per phase referred to the stator must be the same as the original rotor volt-amperes:

$$m_1 E_2' I_2' = m_2 E_2 I_2 \tag{7.2.5}$$

where the primed quantities are the rotor quantities referred onto the stator side. For the wound rotor, m_2 is usually equal to m_1; and for the cage rotor, the number of phases m_2 is equal to the number of rotor bars per pole-pair. The rotor current I_2', referred to the stator, is given by

$$I_2' = \frac{m_2 k_{w2} N_2}{m_1 k_{w1} N_1} I_2 \tag{7.2.6}$$

Using Equation 7.2.6, E_2' can be solved from Equation 7.2.5 and be seen to equal E_1:

$$E_2' = \frac{k_{w1} N_1}{k_{w2} N_2} E_2 = E_1 \tag{7.2.7}$$

Noting that the rotor $I^2 R$ losses must be invariant, that is,

$$m_1 \left(I_2' \right)^2 R_2' = m_2 I_2^2 R_2 \tag{7.2.8}$$

it follows with the aid of Equation 7.2.6 that the rotor resistance per phase referred to the stator is given by

$$R_2' = \frac{m_1}{m_2} \left(\frac{k_{w1} N_1}{k_{w2} N_2} \right)^2 R_2 \tag{7.2.9}$$

For the wound rotor, R_2 is the per-phase resistance of the rotor winding. For the cage rotor, R_2 is approximately the resistance of one bar.

The rotor leakage reactance X_{l2}', referred to the stator, is similarly given by

$$X_{l2}' = \frac{m_1}{m_2} \left(\frac{k_{w1} N_1}{k_{w2} N_2} \right)^2 X_{l2} \tag{7.2.10}$$

which leads to the result that the magnetic energy stored in the standstill rotor leakage reactance remains unchanged:

$$\frac{1}{2} m_1 \left(\frac{X_{l2}'}{2 \pi f_s} \right) \left(I_2' \right)^2 = \frac{1}{2} m_2 \left(\frac{X_{l2}}{2 \pi f_s} \right) I_2^2 \tag{7.2.11}$$

where f_s is the stator frequency.

Having thus referred the rotor quantities to the stator, we can now draw the per-phase equivalent circuit of the polyphase induction motor, as shown in Figure 7.2.5a. The combined effect of the shaft

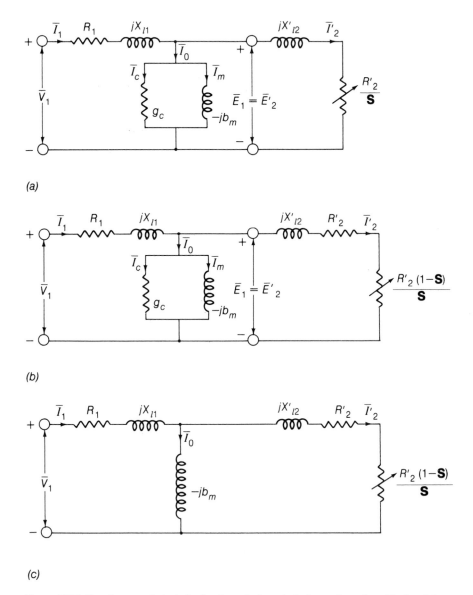

Figure 7.2.5 Per-phase equivalent circuits of a polyphase induction motor, referred to the stator.

load and the rotor resistance appears as a reflected resistance R'_2/S, which is a function of slip and therefore of the mechanical load. The quantity R'_2/S may conveniently be split into two parts:

$$\frac{R_2}{S} = R'_2 + R'_2\frac{(1-S)}{S} \tag{7.2.12}$$

and the equivalent circuit may be redrawn as in Figure 7.2.5b. R'_2 is the per-phase standstill rotor resistance referred to the stator, and $R'_2[(1-S)/S]$ is a dynamic resistance that depends on the rotor speed and corresponds to the load on the motor. (See the discussion after Equation 7.4.2.)

When power aspects need to be emphasized, the equivalent circuit is frequently redrawn as in Figure 7.2.5c, in which the shunt conductance g_c is omitted; the core losses can be included in efficiency calculations along with the friction, windage, and stray load losses.

Recall that, in static-transformer theory, analysis of the equivalent circuit is often simplified either by neglecting the exciting shunt branch entirely or by adopting the approximation of moving it out directly to the primary terminals. For the induction machine, however, such approximations might not be permissible under normal running conditions because the air gap leads to a much higher exciting current (30 to 50% of full-load current) and relatively higher leakage reactances.

7.3 Equivalent Circuit from Test Data

The parameters of the equivalent circuit of an induction machine can be obtained from the *no-load* and *blocked-rotor* tests. These tests correspond to the no-load and short-circuit tests on the transformer.

No-Load Test

Rated balanced voltage at rated frequency is applied to the stator, and the motor is allowed to run on no-load. Input power, voltage, and current are measured and then reduced to *per-phase* values, denoted by P_0, V_0, and I_0, respectively. When the machine runs on no-load, the slip is close to zero, and the circuit to the right of the shunt branch in Figure 7.2.5a is taken to be an open circuit. Thus the equivalent circuit corresponding to the no-load test conditions is given in Figure 7.3.1a or equivalently in Figure 7.3.1b, in which the shunt branch is replaced by an equivalent series impedance to facilitate evaluation of the motor parameters.

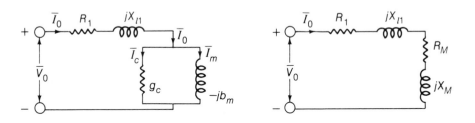

(a)   (b)

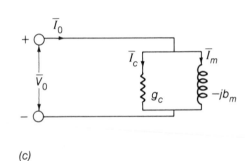

(c)

Figure 7.3.1 Per-phase equivalent circuits of a polyphase induction motor corresponding to the no-load test conditions ($S \simeq 0$).

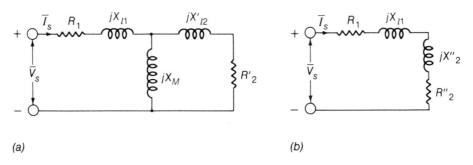

Figure 7.3.2 Per-phase equivalent circuits of a polyphase induction motor corresponding to the blocked-rotor test conditions ($S = 1$).

The conductance g_c in Figure 7.3.1a takes into account not only losses from the stator core but also from windage and friction. Because of the relatively low value of rotor frequency, the rotor core loss is practically negligible at no-load. From Figure 7.3.1b, it follows that

$$R_0 = R_1 + R_M = P_0/I_0{}^2 \tag{7.3.1}$$

$$Z_0 = V_0/I_0 \tag{7.3.2}$$

$$X_0 = X_{l1} + X_M = \sqrt{Z_0^2 - R_0^2} \tag{7.3.3}$$

$$\text{No-load power factor} = \cos \phi_0 = P_0/(V_0 I_0) \tag{7.3.4}$$

in which R_1 is the stator resistance per phase; the series resistance $R_M \ll X_M$; and $X_M \simeq 1/b_M$. The rotational losses given by the sum of the friction, windage, and core losses are found on a per-phase basis by subtracting the stator copper loss from the no-load power input:

$$P_{r0} = P_0 - I_0^2 R_1 \tag{7.3.5}$$

An approximate per-phase equivalent circuit, shown in Figure 7.3.1c is sometimes used for the no-load test conditions, in which case, the calculation of shunt-branch parameters becomes much simpler—more like a transformer.

Blocked-Rotor Test

In this test, the rotor of the induction motor is blocked so that the slip is equal to unity, and a reduced voltage is applied to the machine stator terminals so that the rated current flows through the stator windings. The input power, voltage, and current values are recorded and reduced to *per-phase values,* denoted respectively by P_s, V_s, and I_s. The iron losses are assumed to be negligible in this test. The equivalent circuit corresponding to the blocked-rotor test conditions is then given by Figure 7.3.2a or equivalently by Figure 7.3.2b. If we consider the shunt branch of the circuit shown in Figure 7.3.2a to be absent, as in a transformer short-circuit test, the calculations become much simpler because X_2'' and R_2'' are then equal to X_{l2}' and R_2', respectively. From Figure 7.3.2b, it then follows that

$$R_{eq} = R_1 + R_2'' = P_s/I_s^2 \tag{7.3.6}$$

$$Z_{eq} = V_s/I_s \tag{7.3.7}$$

$$X_{eq} = X_{l1} + X_2'' = \sqrt{Z_{eq}^2 - R_{eq}^2} \tag{7.3.8}$$

$$\cos \phi_s = P_s/(V_s I_s) \tag{7.3.9}$$

For a more complete discussion of tests on induction motors, refer to published test codes and procedures (listed in the bibliography at the end of the chapter), in which the empirical proportions for stator and rotor leakage reactances are given for three-phase induction motors by class. When the classification of the motor is not known, assume that

$$X_{l1} = X'_{l2} = 0.5X_{eq} \tag{7.3.10}$$

The magnetizing reactance X_M can now be evaluated from Equation 7.3.3.

The value of R'_2 requires a closer approximation than that of X'_{l2} because, in the running range, $(R'_2 S) \gg (X_{l1} + X'_{l2})$ and R'_2 has a correspondingly greater effect on the performance of the motor within that range. From the equivalent circuits of Figure 7.3.2a and b, it follows that

$$R''_2 + jX''_2 = \frac{(R'_2 + jX'_{l2})jX_M}{R'_2 + j(X'_{l2} + X_M)} \tag{7.3.11}$$

Equating the real parts of both sides, it can be shown that

$$R''_2 = \frac{R'_2 X_M^2}{(R'_2)^2 + (X'_{l2} + X_M)^2} \tag{7.3.12}$$

Since $R'_2 \ll (X'_{l2} + X_M)$, Equation 7.3.12 can be approximated as

$$R''_2 \simeq \frac{R'_2 X_M^2}{\left(X'_{l2} + X_M\right)^2} \tag{7.3.13}$$

Substituting from Equation 7.3.6 that $R''_2 = R_{eq} - R_1$, we get

$$R'_2 = (R_{eq} - R_1) \frac{\left(X'_{l2} + X_M\right)^2}{X_M^2} \tag{7.3.14}$$

If X_M is much larger than X'_{l2}, R'_2 is nearly equal to $(R_{eq} - R_1)$ or R''_2; otherwise, R'_2 is somewhat larger than R''_2.

▋▋▋▋▋▋▋▋▋▋▋ EXAMPLE 7.3.1

The results of the no-load and blocked-rotor tests on a three-phase, wye-connected, 10-hp, 440-V, 14-A, 60-Hz, eight-pole induction motor with a single squirrel-cage rotor are given below:

> No-load test: Line-to-line voltage = 440 V
>   Total input power = 350 W
>   Line current = 6 A
> Blocked-rotor test: Line-to-line voltage = 95 V
>   Total input power = 900 W
>   Total input power = 14 A

The dc resistance of the stator measured immediately after the blocked-rotor test yields an average value of 0.75 ohm per phase. Calculate the parameters of the equivalent circuit shown in Figure 7.2.5c, and compute the no-load rotational losses.

Solution

From the no-load test data and Equations 7.3.1 to 7.3.3, we get

$$R_0 = \frac{350}{3} \times \frac{1}{6^2} = 3.24 \ \Omega$$

$$Z_0 = \frac{440}{\sqrt{3}} \times \frac{1}{6} = 42.34 \ \Omega$$

$$X_0 = \sqrt{(42.34)^2 - (3.24)^2} = 42.22 \ \Omega$$

From the blocked-rotor test data and Equations 7.3.6 to 7.3.8. we get

$$R_{eq} = \frac{900}{3} \times \frac{1}{14^2} = 1.53 \ \Omega$$

$$Z_{eq} = \frac{95}{\sqrt{3}} \times \frac{1}{14} = 3.92 \ \Omega$$

$$X_{eq} = \sqrt{3.92^2 - 1.53^2} = 3.61 \ \Omega$$

Taking $X_{l1} = X'_{l2} = 0.5 \times 3.61 = 1.805$ ohms, we get from Equation 7.3.3

$$X_M = X_0 - X_{l1} = 42.22 - 1.805 = 40.415 \ \Omega$$

From Equation 7.3.14, we get

$$R'_2 = (1.53 - 0.75) \left(\frac{1.805 + 40.415}{40.415} \right)^2 = 0.78 \times \left(\frac{42.22}{40.415} \right)^2 = 0.851 \ \Omega$$

The per-phase equivalent circuit is given in Figure 7.3.3. The no-load rotational losses are given by Equation 7.3.5, so we get

$$3 \left[\tfrac{350}{3} - \left(6^2 \times 0.75 \right) \right] = 350 - \left(3 \times 6^2 \times 0.75 \right) = 269 \ W$$

Knowing the equivalent-circuit constants, we can now compute the machine performance.

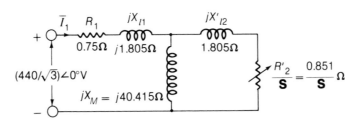

Figure 7.3.3

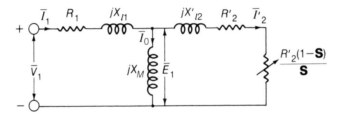

Figure 7.4.1 Per-phase equivalent circuit of a polyphase induction motor used for performance calculations.

▌▌▌▌▌▌▌▌▌▌▌▌

7.4 Polyphase Induction Machine Performance

Some of the important steady-state performance characteristics of a polyphase induction motor include the variation of current, speed, and losses as the load-torque requirements change, and the starting and maximum torque. Performance calculations can be made from the equivalent circuit. All calculations can be made on a *per-phase basis,* assuming balanced operation of the machine. Total quantities can be obtained by using an appropriate multiplying factor.

The equivalent circuit of Figure 7.2.5c, redrawn for convenience in Figure 7.4.1, is usually employed for the analysis. The core losses, most of which occur in the stator, as well as friction, windage, and stray-load losses, are included in efficiency calculations. The power-flow diagram for an induction motor is given in Figure 7.4.2, in which m_1 is the number of stator phases, ϕ_1 is the power-factor angle between $\bar{V}_1$ and $\bar{I}_1$, ϕ_2 is the power-factor angle between $\bar{E}_1$ and $\bar{I}_2$, T is the internal electromagnetic torque developed, ω_s is the synchronous angular velocity in mechanical radians per second, and ω_m is the actual mechanical rotor speed given by $\omega_s(1 - S)$.

The total power P_g transferred across the air gap from the stator is the difference between the electrical power input P_i and the stator copper loss. P_g is thus the total rotor input power, which is dissipated in the resistance R_2'/S of each phase so that

$$P_g = m_1 \left(I_2'\right)^2 R_2'/S = T\omega_s \tag{7.4.1}$$

where T is the internal electromagnetic torque developed by the machine, and ω_s is the synchronous angular velocity in mechanical radians per second. Subtracting the total rotor copper loss, which is $m_1(I_2')^2 R_2'$ or SP_g, from Equation 7.4.1 for P_g, we get the internal mechanical power developed:

$$P_m = P_g (1 - S) = T\omega_m = m_1 \left(I_2'\right)^2 R_2' \frac{(1 - S)}{S} \tag{7.4.2}$$

This much power is absorbed by a resistance of $R_2' \frac{(1-S)}{S}$, which corresponds to the load. For this reason, the resistance term R_2'/S has been split into two terms as in Equation 7.2.12 and shown in the equivalent circuit of Figure 7.4.1. From Equation 7.4.2, we can see that, of the total power delivered to the rotor, the fraction $(1 - S)$ is converted to mechanical power and the fraction S is dissipated as rotor copper loss. We can conclude then that an induction motor operating at high slip values will be inefficient.

The total rotational losses including the core losses can be subtracted from P_m to obtain the mechanical power output P_0 that is available in mechanical form at the shaft for useful work:

$$P_0 = P_m - P_{rot} = T_0\omega_m \tag{7.4.3}$$

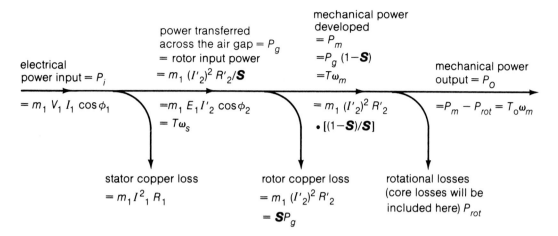

Figure 7.4.2 Power flow in an induction motor.

The per-unit efficiency of the induction motor is then given by

$$\eta = P_0/P_i \tag{7.4.4}$$

Let us now illustrate the above procedure and the analysis of the equivalent circuit in the following example.

EXAMPLE 7.4.1

The parameters of the equivalent circuit given in Figure 7.4.1 for a three-phase, wye-connected, 220-V, 10-hp, 60-Hz, six-pole induction motor are given below in ohms per phase referred to the stator:

$$R_1 = 0.3; \quad R'_2 = 0.15; \quad X_{l1} = 0.5; \quad X_{l2'} = 0.2; \quad X_M = 15$$

The total friction, windage, and core losses can be assumed to be constant at 400 W, independent of load. For a per-unit slip of 0.02, when the motor is operated at rated voltage and frequency, calculate the stator input current, the power factor at the stator terminals, rotor speed, output power, output torque, and efficiency.

Solution
From the equivalent circuit of Figure 7.4.1, the total impedance per phase as viewed from the stator input terminals is given by

$$Z_t = R_1 + jX_{l1} + \frac{jX_M \left(\frac{R'_2}{S} + jX'_{l2} \right)}{\frac{R'_2}{S} + j\left(X_M + X'_{l2} \right)} = 0.3 + j0.5 + \frac{j15\,(7.5 + j0.2)}{7.5 + j(15 + 0.2)}$$

$$= (0.3 + j0.5) + (5.87 + j3.10) = 6.17 + j3.60 = 7.14\angle 30.26°\ \Omega$$

Phase voltage $= 220/\sqrt{3} = 127$ V

Stator input current $= 127/7.14 = 17.79$ A

Power factor $= \cos 30.26° = 0.864$

Synchronous speed $= \frac{120 \times 60}{6} = 1{,}200$ rpm

Rotor speed $= (1 - 0.02)\, 1{,}200 = 1{,}176$ rpm

Total input power $= \sqrt{3} \times 220 \times 17.79 \times 0.864 = 5{,}856.8$ W

Stator copper loss $= 3 \times 17.79^2 \times 0.3 = 284.8$ W

Power transferred across the air gap $= P_g = 5{,}856.8 - 284.8 = 5{,}572$ W

P_g can also be obtained as follows:

$$P_g = m_1 \left(I_2'\right)^2 R_2'/S = m_1 I_1^2 R_f$$

where R_f is the real part of the parallel combination of jX_M and $(R_2'/S + jX_{l2}')$. Thus,

$$P_g = 3 \times 17.79^2 \times 5.87 = 5{,}573 \text{ W}$$

Internal mechanical power developed $= P_g (1 - S) = 0.98 \times 5{,}572 = 5{,}460$ W

Total mechanical power output $= 5{,}460 - 400 = 5{,}060$ W or $\dfrac{5{,}060}{745.7} = 6.8$ hp

Total output torque = output power $\omega_m = \dfrac{\text{output power}}{(1 - S)\, \omega_s}$

Since $\omega_s = \dfrac{4\pi f}{\text{poles}} = \dfrac{4\pi \times 60}{6} = 40\pi = 125.7$ mechanical radians per second, it follows that

Total output torque $= \dfrac{5{,}060}{(0.98)(125.7)} = 41.08 \text{ N} \cdot \text{m}$

Efficiency $= \dfrac{5{,}060}{5{,}856.8} = 0.864$ or 86.4%

The efficiency may alternatively be calculated from the losses:

Total stator copper loss $= 284.8$ W

Rotor copper loss $= m_1 \left(I_2'\right)^2 R_2' = SP_g = (0.02)(5{,}572) = 111.4$ W

Friction, windage, and core losses $= 400$ W

Total losses $= 284.8 + 111.4 + 400 = 796.2$ W

Output $= 5{,}060$ W

Input $= 5{,}060 + 796.2 = 5{,}856.2$ W

Efficiency $= 1 - \dfrac{\text{Losses}}{\text{Input}} = 1 - \dfrac{796.2}{5{,}856.2} = 1 - 0.136 = 0.864$ or 86.4%

By repeating these calculations for other values of slip ranging from 0 to 1, using a procedure similar to that given in Example 7.4.1, the complete performance characteristics of the induction motor can be determined from its equivalent circuit. Typical induction-motor characteristics are shown in Figure 7.4.3.

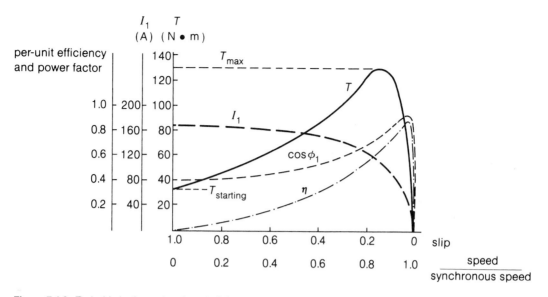

Figure 7.4.3 Typical induction-motor characteristics.

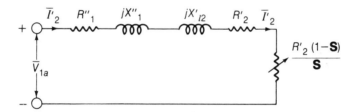

Figure 7.4.4 Another form of per-phase equivalent circuit for the polyphase induction motor shown in Figure 7.4.1.

Because the torque-slip characteristic is one of the most important aspects of the induction motor, we will now develop an expression for torque as a function of slip and other equivalent circuit parameters. Recalling Equation 7.4.1 and the equivalent circuit of Figure 7.4.1, we will obtain an expression for I'_2. To that end, let us redraw the equivalent circuit as in Figure 7.4.4. By applying Thévenin's theorem, we have the following from Figures 7.4.1 and 7.4.4:

$$\bar{V}_{1a} = \bar{V}_1 - \bar{I}_0 \left(R_1 + jX_{l1} \right) = \bar{V}_1 \frac{jX_M}{R_1 + j \left(X_{l1} + X_M \right)} \tag{7.4.5}$$

$$R''_1 + jX''_1 = \frac{\left(R_1 + jX_{l1} \right) jX_M}{R_1 + j \left(X_{l1} + X_M \right)} \tag{7.4.6}$$

$$I'_2 = \frac{V_{1a}}{\sqrt{\left[R''_1 + \left(R'_2/S \right) \right]^2 + \left(X''_1 + X'_{l2} \right)^2}} \tag{7.4.7}$$

$$T = \frac{1}{\omega_s} \frac{m_1 V_{1a}^2 \left(R'_2/S \right)}{\left[R''_1 + \left(R'_2/S \right) \right]^2 + \left(X''_1 + X'_{l2} \right)^2} \tag{7.4.8}$$

Neglecting the stator resistance in Equation 7.4.5 results in negligible error for most induction motors. If X_M of the equivalent circuit shown in Figure 7.4.1 is sufficiently large that the shunt branch need not be considered, calculations become much simpler; R_1'' and X_1'' is then equal to R_1 and X_{l1}, respectively; also, V_{1a} is then equal to V_1.

The general shape of the torque-speed or torque-slip characteristic is shown in Figure 5.3.16, in which the motor region ($0 < \mathbf{S} \leq 1$), the generator region ($\mathbf{S} < 0$), and the breaking region ($\mathbf{S} > 1$) are included for completeness. The performance of an induction motor can be characterized by such factors as efficiency, power factor, starting torque, starting current, pull-out (maximum) torque, and maximum internal power developed. Starting conditions are those corresponding to $\mathbf{S} = 1$.

The maximum internal (or breakdown) torque T_{max} occurs when the power delivered to $R_2'/\mathbf{S}$ in Figure 7.4.4 is a maximum. Applying the familiar impedance-matching principle of the circuit theory, this power will be a maximum when the impedance $R_2'/\mathbf{S}$ equals the magnitude of the impedance between that and the constant voltage V_{1a}. That is to say, the maximum occurs at a value of slip $\mathbf{S}_{\max T}$ for which the following condition is satisfied:

$$\frac{R_2'}{\mathbf{S}_{\max T}} = \sqrt{\left(R_1''\right)^2 + \left(X_1'' + X_{l2}'\right)^2} \tag{7.4.9}$$

The same result can also be obtained by differentiating Equation 7.4.8 with respect to $\mathbf{S}$, or, more conveniently, with respect to $R_2'/\mathbf{S}$, and setting the result equal to zero. This calculation has been left to the enterprising student. The slip corresponding to maximum torque, $\mathbf{S}_{\max T}$, is thus given by

$$\mathbf{S}_{\max T} = \frac{R_2'}{\sqrt{\left(R_1''\right)^2 + \left(X_1'' + X_{l2}'\right)^2}} \tag{7.4.10}$$

and the corresponding maximum torque from Equation 7.4.8 results in

$$T_{max} = \frac{1}{\omega_s} \frac{0.5 m_1 V_{1a}^2}{R_1'' + \sqrt{\left(R_1''\right)^2 + \left(X_1'' + X_{l2}'\right)^2}} \tag{7.4.11}$$

which can be verified by the reader. Equation 7.4.11 shows that the maximum torque is independent of the rotor resistance. The slip corresponding to the maximum torque is directly proportional to the rotor resistance R_2', however, as seen from Equation 7.4.10. Thus, when the rotor resistance is increased by inserting external resistance in the rotor of a wound-rotor induction motor, the maximum internal torque is unaffected, but the speed or slip at which it occurs is increased, as shown in Figure 7.4.5 Also note that maximum torque and maximum power do not occur at the same speed. The student is encouraged to work out the reason.

A conventional induction motor with a squirrel-cage rotor has about 5% drop in speed from no-load to full load and is thus substantially a constant-speed motor. Employing a wound-rotor motor and inserting external resistance in the rotor circuit achieves speed variation but resutls in poorer efficiency. Variation of starting torque (at $\mathbf{S} = 1$) with rotor-circuit resistance can also be seen from Figure 7.4.5. As stated in Section 5.3, we can obtain higher starting torque by inserting external resistances in the

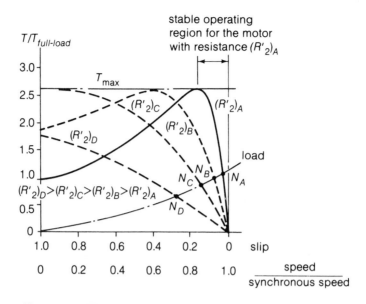

Figure 7.4.5 Effect of changing rotor-circuit resistance on the torque-slip characteristic of a polyphase induction motor.

rotor circuit and then cutting them out eventually for the normal running conditions in order to operate the machine at a higher efficiency. Creating a sufficiently large rotor-circuit resistance might make it possible to achieve an almost linear torque-slip relationship for the slip range of 0 to 1. For instance, two-phase servo motors (used as output actuators in feedback control systems) are usually designed with very high rotor resistance to ensure a negative slope for the torque-speed characteristic over the entire operating range.

The resistances R_1 and R_2' are kept very small to reduce the copper losses and thereby increase the efficiency; the motor is most compatible with a load running at the highest possible speed with a low value of slip. From Equations 7.4.8 and 7.4.11, we see that the leakage reactance $(X_1'' + X_{l2}')$ must be low to assure good starting torque as well as adequate maximum torque. Power factor can be improved by decreasing the leakage reactances and increasing the magnetizing reactance. Since the starting current of the motor is essentially limited by the leakage reactances, however, they should not be reduced below a certain value. Attempting to decrease the core losses beyond a limit by reducing the working flux density results in an increase in the copper losses for a given load because the torque (which is determined by the load) is dependent on the product of the flux density and I_2'.

Since the stator resistance is quite low and has only a negligible influence, let us set $R_1 = R_1'' = 0$, in which case, from Equations 7.4.8 and 7.4.11, we can see that

$$\frac{T}{T_{max}} = \frac{2}{(S/S_{max\,T}) + (S_{max\,T}/S)} \qquad (7.4.12)$$

where $\mathbf{S}$ and $\mathbf{S}_{max\,T}$ are the slips corresponding to T and T_{max}, respectively.

||||||||||||

EXAMPLE 7.4.2

For the motor specified in Example 7.4.1, compute the following:

a. The load component I_2' of the stator current, the internal torque T, and the internal power P_m for a slip of 0.02

b. The maximum internal torque, and the corresponding slip and speed

c. The internal starting torque and the corresponding stator-load current I_2'

Solution

Let us first reduce the equivalent circuit of Figure 7.4.1 to its Thévenin-equivalent form shown in Figure 7.4.4. With the aid of Equations 7.4.5 and 7.4.6, we obtain

$$\bar{V}_{1a} = \bar{V}_1 \frac{jX_M}{R_1 + j(X_{l1} + X_M)} \simeq \bar{V}_1 \frac{X_M}{X_{l1} + X_M} = \frac{220}{\sqrt{3}} \frac{15}{0.5 + 15} \angle 0° = 122.9\angle 0°\text{ V}$$

$$R_1'' + jX_1'' = \frac{(0.3 + j0.5)\,j15}{0.3 + j(0.5 + 15)} = 0.281 + j0.489$$

a. Corresponding to a slip of 0.02, $\frac{R_2'}{S} = \frac{0.15}{0.02} = 7.5$. From Equation 7.4.7, we get

$$I_2' = \frac{122.9}{\sqrt{7.8^2 + 0.689^2}} = \frac{122.9}{7.83} = 15.7\text{ A}$$

The internal torque T can be calculated from either Equation 7.4.1 or 7.4.8:

$$T = \frac{1}{125.7}(3)(15.7)^2(7.5) = \frac{5{,}546}{125.7} = 44.12\text{ N}\cdot\text{m}$$

From Equation 7.4.2, the internal mechanical power is calculated as

$$P_m = (3)(15.7)^2(7.5)(0.98) = 5{,}435\text{ W}$$

which is also the same as $T\omega_m = T\omega_s(1 - S) = (44.12 \times 125.7 \times 0.98)$. In Example 7.4.1, this value is calculated as 5,460 W; the small discrepancy is due to the approximations.

b. From Equation 7.4.10, it follows that

$$S_{\text{max}\,T} = \frac{0.15}{\sqrt{0.281^2 + 0.689^2}} = \frac{0.15}{0.744} = 0.202$$

or the corresponding speed at T_{max} is $(1 - 0.202)(1{,}200) = 958$ rpm.
  From Equation 7.4.11, the maximum torque can be calculated as

$$T_{\text{max}} = \left(\frac{1}{125.7}\right)\frac{(0.5)(3)(122.9)^2}{0.281 + \sqrt{0.281^2 + 0.689^2}} = \left(\frac{1}{125.7}\right)\frac{(0.5)(3)(122.9)^2}{0.281 + 0.744} = 175.8\text{ N}\cdot\text{m}$$

c. Assuming the rotor-circuit resistance to be constant, with $S = 1$ at starting, from Equations 7.4.7 and 7.4.8, we get the following:

$$I'_{2 \, start} = \frac{122.9}{\sqrt{(0.281 + 0.15)^2 + (0.489 + 0.2)^2}} = \frac{122.9}{0.813} = 151.2 \text{ A}$$

$$T_{start} = \frac{1}{125.7} (3)(151.2)^2 (0.15) = 81.8 \text{ N} \cdot \text{m}$$

For such applications as fans and blowers, a motor needs to develop only moderate starting torque. Some loads, like conveyors, however, require high starting torque to overcome high static torque and load inertia. The motor designer sometimes makes the starting torque equal to the maximum torque by choosing the rotor-circuit resistance at startup to be

$$R'_{2 \, start} = \sqrt{\left(R''_1\right)^2 + \left(X''_1 + X'_{l2}\right)^2} \tag{7.4.13}$$

which can easily be obtained from Equation 7.4.9 with $S_{\max T} = 1$.

The limitations of the equivalent circuit and the conditions under which the circuit parameters are obtained must be borne in mind when investigations involve a wide speed range, as in motor-starting problems. Saturation under the heavy inrush currents associated with starting, or saturation due to large currents corresponding to the maximum torque conditions, can have a significant effect in reducing motor reactances. Moreover, the current distribution in the rotor conductors and the rotor resistance can vary significantly over a wide speed range. By simulating the proposed operating conditions as closely as possible, and determining the equivalent-circuit parameters, designers can keep errors to a minimum in predicting machine performance.

7.5 Speed Control of Polyphase Induction Motors

The induction motor is valuable in so many applications because it combines simplicity and ruggedness. Although a good number of industrial drives run at substantially constant speed, quite a few applications need variable speed. Speed-control capability is essential in such applications as conveyors, hoists, and elevators. Because the induction motor is essentially a constant-speed machine, designers have sought creative ways to easily and efficiently vary its speed continuously over a wide range of operating conditions. We discuss the methods of speed control here only briefly. The interested reader can consult the bibliography at the end of the chapter.

The appropriate equation to be examined, based on Equation 5.3.29, is

$$n = (1 - S)\,n_1 = (1 - S)\,120 f_s / P \tag{7.5.1}$$

where n is the actual speed of the machine in revolutions per minute, S is the per-unit slip, f_s is the supply frequency in hertz, P is the number of poles, and n_1 is the synchronous speed in revolutions per minute. Equation 7.5.1 suggests that the speed of the induction motor can be varied by varying either the slip or the synchronous speed, which in turn can be varied by changing either the number of poles or the supply

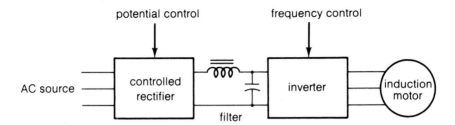

Figure 7.5.1 Solid-state control of induction motor.

frequency. Any method of speed control that depends on variation of slip is inherently inefficient because the efficiency of the induction motor is approximately equal to $(1 - S)$. On the other hand, if the supply frequency is constant, varying the number of poles results only in discrete and stepped variation in motor speed. Indeed, all methods of speed control require some degree of sacrifice in performance, cost, and simplicity; these disadvantages must be weighed carefully against the advantages of speed variability.

Pole-Changing Method.   If a machine is provided with two stator windings arranged for different numbers of poles, and preferably with a squirrel-cage rotor, so that no change of connections is needed on the secondary, two synchronous speeds are available. Special connections can make it possible to have one winding that can be reconnected simply to yield two or even three numbers of poles. Remember that only discrete changes in the motor speed can be obtained by this technique. Recent developers have created many ingenious methods of varying the number of poles by means of pole-amplitude modulation and phase-modulated pole-changing.

   If changes in speed are effected without change in air gap flux density, the motor develops the same maximum torque for any speed setting, in which case the system is said to have a *constant-torque drive*. On the other hand, if the changes in speed are made in such a way that the air-gap flux changes with different stator connections, so that the flux is inversely proportional to the speed setting, the system is said to have a *constant-horsepower drive*.

   Another method, *concatenation,* formerly used for changing the motor speed, employs two or more separate wound-rotor motors, all connected to the same shaft mechanically or through gears. If the stator of one motor, with P_1 poles, is connected to the supply line, and its rotor slip rings are connected to the stator of a second motor, with P_2 poles, whose rotor winding (squirrel-cage or wound-rotor) is short-circuited, the common shaft will be driven at a speed corresponding to $(P_1 + P_2)$ poles. Take a moment to reason out why this is so.

Variable-Frequency Method.   If it is practicable to vary the supply frequency, then the machine's synchronous speed can be varied. However, in order to maintain constant air-gap flux density, the line voltage should also be varied directly with the frequency; in such a case, the maximum torque remains nearly constant. The inherent difficulty in applying this method has been the lack of an effective and economical source of adjustable frequency. One way is to employ a wound-rotor induction machine as a frequency changer. The advent of solid-state devices with relatively large power ratings have made it possible to use solid-state frequency converters.

   One possible combination of power semiconductor converters that would yield the desired variable-frequency, variable-potential excitation from a fixed-frequency, fixed-potential ac supply source is shown in the open-loop system of Figure 7.5.1. For a regenerative system, the single controlled rectifier must be replaced by a dual converter, which is a combination of a rectifier and an inverter that is able to accept the negative current and thereby return power to the supply system.

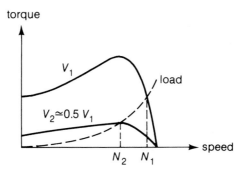

Figure 7.5.2 Speed control by changing the slip by means of line-voltage control.

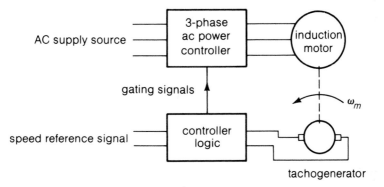

Figure 7.5.3 A closed-loop speed control system.

Variable Line-Voltage Method.   The internal torque developed by an induction motor is proportional to the square of the voltage applied to its stator terminals, as seen from Equation 7.4.8 and as shown in Figure 7.5.2 by the two torque-speed characteristics. If the load has the torque-speed characteristic indicated by the dashed line, the speed is reduced from N_1 to N_2 and the voltage is changed from V_1 to V_2. If the voltage can be varied continuously from V_1 to V_2, the motor speed can be varied continuously from N_1 to N_2 for the given load.

By employing an ac power controller, the fundamental component of the voltage applied to the motor can be controlled from zero to the value of the supply. A simple closed-loop speed control system is shown in Figure 7.5.3.

Variable Rotor-Resistance Method.   The effect of changing rotor-circuit resistance on the torque-slip characteristic of a polyphase wound-rotor induction motor is illustrated in Figure 7.4.5 (page 315). If the load has the torque-speed curve shown, the speeds corresponding to the rotor-circuit resistance values of $(R_2')_A$, $(R_2')_B$, $(R_2')_C$, and $(R_2')_D$, are N_A, N_B, N_C, and N_D, respectively. By continuous variation of the rotor-circuit resistance, continuous variation of speed is possible.

The main disadvantages of both line-voltage and rotor-resistance control are low efficiency at reduced speeds and poor speed regulation with respect to change in load. You may recall that per-unit speed regulation is the drop in speed from no-load to full-load expressed as a ratio of the full-load base (normal, rated, or name plate) speed.

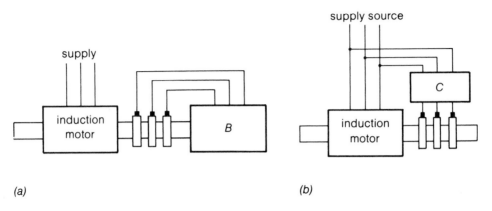

Figure 7.5.4 Basic schemes for induction-motor speed control by auxiliary devices.

Rotor-Slip-Frequency Control. Without sacrificing efficiency at low-speed operation and without affecting the speed by load variation, designers control the induction-motor speed with power semiconductor converters. Equation 7.4.2 may be expressed as

$$T = \frac{1}{\omega_m} m_1 \left(I_2'\right)^2 R_2' \frac{1-S}{S} = m_1 \left(I_2'\right)^2 R_2'/\omega_r \tag{7.5.2}$$

because

$$\omega_r = S\omega_s = \omega_s - \omega_m = \frac{\omega_r}{S} - \omega_m \tag{7.5.3}$$

(where all ω's are expressed in mechanical radians per second) and a rearrangement yields

$$\frac{1-S}{S\omega_m} = \frac{1}{\omega_r} \tag{7.5.4}$$

Equation 7.5.2 shows that, if the rotor frequency ω_r is held constant, the internal torque per rotor ampere is also a constant. Moreover, if ω_r is kept at a lower value, then the torque per rotor ampere is large. Since ω_s is given by $(\omega_r + \omega_m)$, if ω_r in a control system is determined by a constant reference signal Ω_r, and if the required value of the stator frequency ω_s can be computed from this reference signal and a motor-speed command signal Ω_m, then we can obtain an exact value of motor speed ω_m equal to the command signal. Such control systems have been designed to control the speed of induction motors.

Rotor-Slip Energy Recovery Method. Considering the power flow in an induction motor as illustrated in Figure 7.4.2, the fraction S of the power transferred across the air gap is transformed by electromagnetic induction to electric power in the rotor circuits. If the rotor circuits are short-circuited, this slip-frequency electric power is wasted as rotor copper loss; hence, operation at reduced speeds is inherently inefficient. The effect of the external resistors introduced into rotor circuits of a wound-rotor induction motor in the variable rotor-resistance method of speed control is to produce slip-frequency voltages that oppose the emfs induced in the rotor windings. If the energy that would be dissipated in the external resistors in such a rheostatic speed control method could instead be recovered and returned to the ac source, the overall efficiency of the speed control system would be increased.

Numerous schemes have been tried for recovering the slip-frequency electric power and controlling the slip by means of auxiliary devices. Some are rather complicated in their details. However, all of them consist of some means for injecting adjustable voltages of slip frequency into the rotor circuits of a wound-rotor induction motor. The basic schemes are illustrated in Figure 7.5.4. In part a, the

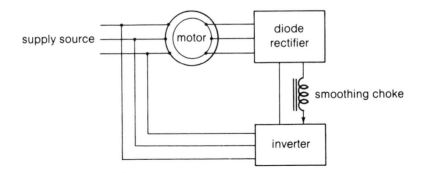

Figure 7.5.5 Rotor-slip energy recovery method of speed control.

slip rings of the induction motor are connected to auxiliary apparatus, represented by the box *B*, in which the slip-frequency electric power is converted to mechanical power and added to the shaft power developed by the main induction motor that is being controlled. In part b, the slip rings of the induction motor are connected to auxiliary frequency-changing devices, represented by the box *C*, in which the slip-frequency electric power is converted to electric power at line frequency and returned to the supply. By adjusting the magnitude and phase of the slip-frequency emfs of the auxiliary devices, the speed and power factor of the main induction motor can be controlled. The auxiliary equipment can consist of rotating machines, adjustable-ratio transformers, or solid-state frequency-converting devices, as shown schematically in Figure 7.5.5, in which an inverter is inserted in the rotor circuit. The diode rectifier connected to the three-phase rotor terminals receives the slip-frequency electric power. The smoothed output of the rectifier is then applied to the dc terminals of an inverter operating at the frequency of the supply source, and the three-phase power from the ac terminals of the inverter is fed back to the induction-motor stator source. Such a system permits closed-loop speed control of the motor with increased overall efficiency.

Other schemes for induction-motor speed control by auxiliary devices include concatenation, the brush-shifting Schrage motor, the Kramer control, and the Scherbius control. For details on these methods, refer to the bibliography at the end of the chapter. In spite of advances made in speed control schemes, researchers are still actively looking for a more satisfactory, economical, and efficient method of speed control for induction motors, particularly for the squirrel-cage machines.

Article 12.3 explores in some detail the solid-state control of induction motors under the Chapter on Power Semiconductor Controlled Drives.

7.6 Starting Methods for Polyphase Induction Motors

When high starting torques are required, a wound-rotor induction motor, with external resistances inserted in its rotor circuits, can be used. The starting current can be reduced and high values of starting torque per ampere of starting current can be obtained with *rotor-resistance starting*. The external resistances are generally cut out in steps as the machine runs up to speed.

For cage-rotor machines, the problem is to keep down the starting current while maintaining adequate starting torque. The input current, for example, can be no more than 6 times the full-load current, while the starting torque may be about 1.5 times the full-load torque. Depending on the capacity of the available supply system, *direct-on-line starting* may be suitable only for relatively small machines, up to 10-hp rating. Other staring methods include *reduced-voltage starting* by means of *wye-delta starting, autotransformer starting,* or *stator-impedance starting.*

For employing the wye-delta starting method, a machine designed for delta operation is connected in wye during the starting period. Because the impedance between line terminals for wye connection is three times that for delta connection for the same line voltage, the line current at standstill for wye connection is reduced to one-third of the value for delta connection. Since the phase voltage is reduced by a factor of $\sqrt{3}$ during starting, it follows that the starting torque will be one-third of the normal. For autotransformer starting, the setting of the autotransformer can be predetermined to limit the starting current to any desired value. An autotransformer, which reduces the voltage applied to the motor to x times the normal voltage, will reduce the starting current in the supply system as well as the starting torque of the motor to x^2 times the normal values.

While employing the wye-delta or autotransformer starting, if all three line switches are opened simultaneously during the changeover from starting to the normal running condition, the air-gap flux and the speed will decrease during this period, and the transient current surge can be high when the switches are reclosed. Thus, such methods of starting should reduce the time duration of the starting current but they do not necessarily reduce its peak value. The current surge during switching can be reduced, however, by introducing transition impedances between the starting and motor terminals to maintain the continuity of the current during changeover.

Stator-impedance starting may be employed if the starting-torque requirement is not severe. Series resistances (or impedances) are inserted in the lines to limit the starting current. These resistances are shorted out when the motor gains speed. This method has the obvious disadvantage of inefficiency caused by the extra losses in the external resistances.

Other methods of starting, such as *part-winding starting,* and *multicircuit starting,* are sometimes used, in which the motor may be connected asymmetrically during the starting period. In some stator-impedance starting methods, a variable impedance is inserted in only one supply line to the machine during the run-up period.

||||||||||||||

EXAMPLE 7.6.1

An induction motor has a starting current that is six times the full-load current and a per-unit full-load slip of 0.04. The machine is to be provided with an autotransformer starter. If the minimum starting torque must be 0.3 times the full-load torque, determine the required tapping on the transformer and the per-unit supply-source line current at starting.

Solution

From Equation 7.4.1, with slip at starting equal to 1, we have

$$\frac{T_s}{T_{fl}} = \left(\frac{I_s}{I_{fl}}\right)^2 S_{fl}$$

where subscripts s and fl correspond to starting and full-load conditions.

Substituting values, $0.3 = (I_s/I_{fl})^2 \times 0.04$, from which

$$I_s/I_{fl} = \sqrt{0.3/0.04} = 2.74$$

or

$$I_s = 2.74 \text{ per unit, with } I_{fl} \text{ taken as one per unit}$$

The applied voltage to the motor must reduce the current from six per unit to 2.74 per unit; i.e., the tapping must be sufficient to reduce the voltage to 2.74/6 = 0.456 of the full value. The supply-source line current will then be 2.74 × 0.456 = 1.25 per unit, where 0.456 is the secondary-to-primary turns ratio.

Deep-Bar and Double-Squirrel-Cage Rotors.   Recall that, at standstill, the rotor frequency is the same as the stator frequency; and that, as the motor accelerates, the rotor frequency decreases to a low value, on the order of 2 or 3 Hz at full load in a typical 60-Hz motor. By using suitable shapes and arrangements for rotor bars, it is possible to design squirrel-cage rotors so that their effective resistance at 60 Hz is several times their resistance at 2 or 3 Hz. This design takes advantage of the inductive effect of the slot-leakage flux on the current distribution in the rotor bars.

Rotor bars embedded in deep slots, whose depth is two or three times greater than the slot width as shown in Figure 7.6.1a, provide a high effective resistance and a large torque at starting. Because of the skin effect, the current has a tendency to concentrate at the top of the bars at starting, when the frequency of the rotor currents is high. Under normal running conditions with low slips, because the frequency of rotor currents is much lower, skin effect is negligible and the current tends to distribute almost uniformly throughout the entire rotor-bar cross section. The rotor resistance thus becomes lower, leading to a higher efficiency.

An alternative to the deep-bar rotor is the double-cage arrangement illustrated in Figure 7.6.1b. This design allows higher starting torque and better running efficiency. The squirrel-cage winding consists of two layers of bars short-circuited by end rings. The inner cage, made of low-resistance bottom bars, is deeply embedded in iron, while the outer cage has relatively high-resistance bars situated close to the inner stator surface. At starting, because of skin effect, the influence of the outer cage dominates, thereby producing a high starting torque. During the normal running period, the current penetrates to full depth into the lower cage because of insignificant skin effect, resulting in an efficient steady-state operation. The inductance of the lower bars is greater than that of the upper ones because of the general nature of the slot-leakage field; the difference in inductance can be made quite large by properly proportioning the slot dimensions, particularly the constriction in the slot between the two layers of bars.

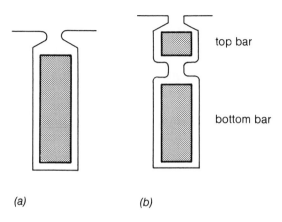

top bar

bottom bar

(a)   (b)

Figure 7.6.1 Forms of a slot for induction-motor rotors with high starting torque.

■■■■■■■■■■■■■■■■■■ ‖‖‖‖‖‖‖‖‖‖‖‖
7.7 Design of a Three-Phase Induction Motor

Let us walk through a three-phase induction motor design project to show the basic elements and design considerations. The motor specifications are listed here:

- The machine is a 15-hp, three-phase, 60-Hz, six-pole, 220-V, squirrel-cage general-purpose induction motor.
- The full-load efficiency and power factor are not less than 87% and 85%, respectively.
- The starting torque is to be not less than 135% of full-load torque for normal voltage.
- The maximum running torque is to be not less than 200% of full-load torque.
- The temperature rise should not exceed 40° C for any part of the motor during continuous full-load operation.
- The synchronous speed n is given by $\frac{120 \times 60}{6} = 1,200$ rpm.

$$\frac{\text{hp}}{n} \times 10^3 = \frac{15}{1,200} \times 1,000 = 12.5$$

- The output constant C is chosen to be 1.004. Empirical data curves are available in the literature for the output constant as a function of $\left(\frac{\text{hp}}{\text{rpm}} \times 10^3 \right)$. For a typical ratio of length to pole-pitch, i.e. l/τ of 0.9, and a typical ratio r (outside diameter to inside diameter of stator) of 1.36, the outside diameter of the armature or stator,

$$D_0 = \sqrt[3]{\frac{C\,(\text{hp})\,pr}{(\text{rpm})\,\pi\,(l/\tau)}}$$

or

$$D_0 = \sqrt[3]{\frac{1.004 \times 15 \times 6 \times 1.36}{1,200 \times \pi \times 0.9}} \simeq 0.33 \text{ m or } 33 \text{ cm}$$

$$l = \frac{C\,(\text{hp})}{D_0^2 n} = \frac{1.004 \times 15}{(0.33)^2\,1200} = 0.115 \text{ m or } 11.5 \text{ cm}$$

The gap diameter (the inside diameter) of the stator, D, is $\frac{33}{1.36} = 24.26$ cm; round it to 24 cm.

So far we have selected the following dimensions: $D = 0.24$ m; $l = 0.115$ m; $\tau = 0.126$ m; $l/\tau = 0.91$.

Radial ventilating ducts in the stator core are not necessary for the diameter and length selected, so the length of the air-gap section is equal to the total length of the stator core:

$$l_g = l = 0.115 \text{ m}$$

The flux per pole for 60-Hz polyphase induction motors can be determined from the following relationship:

$$\frac{\phi}{\sqrt{\text{hp}}} = C_1$$

in which the limits for C_1 are given in the literature. Choosing $C_1 = 1.7 \times 10^{-3}$ the flux per pole or $\phi = 1.7 \times 10^{-3}\sqrt{15} = 0.0066$ Wb.

If we choose the wye-connected stator winding, the phase voltage is

$$E_{ph} = \frac{220}{\sqrt{3}} = 127 \text{ V}$$

Using the voltage equation (5.3.15), we can obtain

$$N_{ph} \cdot k_w = \frac{E_{ph}}{4.44 f \phi} = \frac{127}{4.44 \times 60 \times 0.0066} = 72.2$$

Let us choose two parallel circuits per phase, in which case the winding can be reconnected to one circuit per phase for 440 V. With three slots per pole per phase, i.e., spp= 3, the total number of stator slots is

$$S_s = 3 \times 3 \times 6 = 54$$

The slot or tooth pitch is calculated as

$$\frac{\pi \times 0.24}{54} = 0.014 \text{ m}$$

The winding distribution

$$k_d = \frac{\sin 30°}{3 \sin 10°} = 0.96$$

and the coil-pitch factor for a coil throw, slot 1 to 9, i.e., short-pitched by one slot pitch,

$$k_p = \sin\left(\frac{8}{9} \times 90°\right) = 0.985$$

The number of conductors per slot is then

$$\frac{(72.2 \times 2)\, 2 \times 3}{0.96 \times 0.985 \times 54} = 16.97$$

Since the number of slots must be an even integer, choose 18 conductors per slot. Then the conductors in series per phase are $\frac{18 \times 54}{3 \times 2} = 162$; the number of series turns per phase is $162/2 = 81$; and the flux per pole is

$$\phi = \frac{127}{4.44 \times 60 \times 81 \times 0.96 \times 0.985} = 0.006 \text{ Wb}$$

The air-gap flux density is then

$$B_g = \frac{\phi p}{\pi D l_g f_d} = \frac{0.006 \times 6}{\pi \times 0.24 \times 0.115 \times 0.637} = 0.652 \text{ T}$$

where f_d is the flux-distribution factor, which is the ratio of the average value to the maximum value of the flux-density wave. The distributed stator winding of the induction motor produces an air-gap flux wave that is very nearly sinusoidal. For a sine wave, the flux-distribution factor is 0.637. The stator

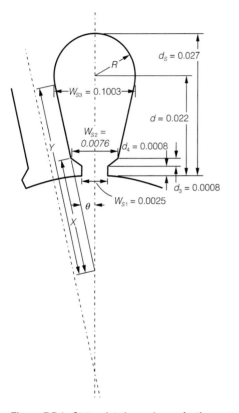

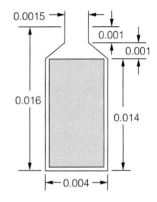

Figure 7.7.1 Stator slot shape chosen for the induction motor design project.

Figure 7.7.2 Rotor slot configuration.

current per phase is $I = \frac{15 \times 746}{3 \times 127 \times 0.87 \times 0.85} = 39.7$ A. For a current density of 465 A/cm², the section area of the stator conductor is calculated as $\frac{39.7}{2 \times 465} = 0.043$ cm².

Suitable insulated wires may be selected. The shape of the slot chosen for this design is shown in Figure 7.7.1. By choosing the width of the stator tooth such that the tooth flux density is about 1.5 T, the stator slot dimensions can be finalized.

The rotor-slot dimensions are then 0.4 cm wide by 1.56 cm deep as shown in Figure 7.7.2. Keeping the current density in the end ring the same as that in the bars, we can design the end-ring section.

The total ampere-turns per pole can be calculated from the flux densities and saturation curves. Various motor characteristics can then be computed and summarized as in the enclosed design sheet, Figure 7.7.3. The student is encouraged to pursue the design and justify the details given here.

7.8 Single-Phase Induction Motors

For reasons of simplicity and cost, a single-phase power supply is universally preferred for fractional-horsepower motors. It is also widely used for motors up to about 5 hp. Single-phase induction motors are usually two-pole or four-pole, rated at 2 hp or less, while slower and larger motors can be manufactured for special purposes. Single-phase induction motors are widely used in domestic appliances and for a

hp.: 15   Syn. rpm: 1,200   Hertz: 60   Poles: 6   Phases: 3   Volts: 220   Amperes per line: 39.3
Amperes per phase: 39.3   Volts per phase: 127

Stator	
Sheet steel . . . . 0.48×10^{-3}, 1% Si	
Outside diameter 0.33	
Gap diameter. 0.24	
Total length 0.115	
Ducts, number and size . . . . None	
Gross iron length 0.115	
Effective length 0.107	

Slots:
  Number 54
  Depth. . . . . 0.027 round bottom
  Width. 0.0076–0.1003
  Opening. 0.0025
Minimum tooth width 0.0066
Conductors:
  Per slot. 18
  Size . . . . 2–14 round in parallel
  Area . . . . 4.2×10^{-6}
  Total section 0.004
  In series per phase 162
Amperes per square meter . . . . 4.77×10^6
Conductors arranged in slot . . . . Random
Insulation allowance:
  Depth 0.006
  Width 0.002
Circuits per phase. 2.0
Connections. . . . . Wye
Coil throw . . . . Slots 1 and 9
Fraction or coil pitch 0.889
Length half mean-turn 0.329
Resistance per phase, 65° C 0.13
Copper weight. . . . .

Ducts, number and size . . . . None
Effective length 0.107
Slots:
  Number 65
  Depth 0.016
  Width 0.004
  Opening. 0.0015
Minimum tooth width 0.006
End-ring:
  Section. . . . . $0.007 \times 0.02 = 0.14 \times 10^{-3}$
  Material. . . . . Copper
Length of bar 0.18
Resistance per phase,

0.12 @ 65° C or 0.104 @ 25° C

Stator and Rotor

Total resistance per phase   0.15 + 0.12 = 0.27
Total reactance per phase   0.39 + 0.42 = 0.81fil
Short-circuit power factor. 0.384
Short-circuit current. 202
Friction and windage loss. 224
No-load stator I^2R. 65.0
Magnetizing current I_m 12.3-31%
Watt component I_c. 1.59
No-load current I_0 . . . . 12.4
No-load power factor. 0.128
Full-load current. 39.3
Full-load slip. 0.037
Full-load speed . . . . . hpm 1157
Full-load torque 69.0
Starting torque 80.0
Maximum torque 137.2

Rotor	
Sheet steel . . . . 0.48×10^{-3} 1% Si	
Gap diameter. 0.2389	
Inside diameter 0.16	
Total length 0.115	

	$\frac{1}{4}$	$\frac{1}{2}$	$\frac{3}{4}$	$\frac{4}{4}$	$\frac{5}{4}$
Efficiency	. . . . . .	. . . . . .	. . . . . .	87	. . . . . .
Power factor	. . . . . .	. . . . . .	. . . . . .	87	. . . . . .

Total flux: 0.057 k_p: 0.985   Flux per pole: 0.006 k_d: 0.96

	Length	Section	Density	Ampere Turns	Weight	Core loss
Stator teeth.	0.0238	0.038	1.5	23.8	6.95	50.5
Stator yoke.	0.08	0.004	1.5	44.8	15.09	108.0
Rotor teeth.	0.016	0.046	1.24	6.8		158.5
Rotor yoke	0.048	0.005	1.14	7.6		$\times 2$
Air gap	$\left(0.55 \times 10^{-3}\right) \times (1.155)$	0.086	0.66	340.0		317.0
				423		

Figure 7.7.3 Typical induction motor design sheet. Note: SI system of units is used. (Adapted from J. H. Kuhlmann, *Design of Electrical Apparatus*, 3/e, New York: John Wiley & Sons, 1950).

Photo 7.8.1 Cross-sectional view of a drip-proof, fractional-horsepower single-phase induction motor. (Photo courtesy of Marathon Electric Manufacturing Company.)

very large number of low-power drives in industry. The single-phase induction machine resembles a small, three-phase, squirrel-cage motor except that, at full speed, only a single winding on the stator is usually excited. See Photo 7.8.1 for a cross-sectional view of a fractional-horsepower, single-phase induction motor.

The single-phase stator winding is distributed in slots so as to produce an approximately sinusoidal space distribution of mmf. As discussed in Section 5.3, such a motor inherently has no starting torque, and as we saw in Section 5.4, it must be started by an auxiliary winding, by being displaced in phase position from the main winding, or by some similar device. Once started by auxiliary means, the motor will continue to run. Thus, nearly all single-phase induction motors are actually two-phase motors, with the main winding in the direct axis adapted to carry most or all of the current in operation, and an auxiliary winding in the quadrature axis with a different number of turns adapted to provide the necessary starting torque.

Since the power input in a single-phase circuit pulsates at twice the line frequency, all single-phase motors have a double-frequency torque component which causes slight oscillations in rotor speed and imparts vibration to the motor supports. The design must provide a means to prevent this vibration from causing objectionable noise.

Exciting the stator with an alternating emf causes the alternating flux along the main axis, as shown in Figure 7.8.1a, designated by ϕ_M. The stator-mmf wave is stationary in space but pulsates in magnitude, the stator-field strength alternating in polarity and varying sinusoidally with time. Currents are induced in the squirrel-cage-rotor conductors by transformer action, these currents being in such a direction as to produce an mmf opposing the stator mmf, as illustrated in Figure 7.8.1a. The axis of the rotor-mmf wave coincides with that of the stator field, and the torque angle is zero; hence, no starting torque is produced. At standstill, the motor is merely a single-phase static transformer with a short-circuited secondary. When the rotor is made to revolve, however, a speed voltage in addition to the transformer voltage is generated in the rotor by virtue of its rotation in the stationary stator field. The direction of rotational voltages in the rotor conductors being as shown in Figure 7.8.1b, a component rotor current is produced along with a component rotor-mmf wave (whose axis is displaced 90 electrical degrees from the stator

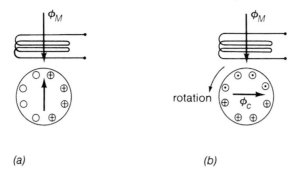

Figure 7.8.1 Single-phase induction motor: Elements of cross-field theory (a) At standstill. (b) During rotation.

axis). The torque angle for this component of the rotor mmf being 90°, a torque is obtained. Further analysis using the *cross-field theory*[1] shows that this torque in fact acts in the direction of rotation and that the necessary conditions for the continued production of torque are satisfied.

The other viewpoint adopted in explaining the operation of the single-phase motor, based on the conditions already established for polyphase motors, is known as the *revolving-field theory.* Each viewpoint has its own advantages. For computational purposes, the revolving-field point of view, already introduced in Section 5.3, is followed hereafter to parallel the analysis we applied to the polyphase induction motor. As stated in Section 5.3 and shown in Section 5.4, a stationary pulsating field can be represented by two counterrotating fields of constant magnitude. The equivalent circuit of a single-phase induction motor, then, consists of the series connection of a forward-rotating field equivalent circuit and a backward-rotating one. Each circuit is similar to that of a three-phase machine, but in the backward-rotating field circuit, the parameter **S** is replaced by $(2 - \mathbf{S})$, as shown in Figure 7.8.2a. The forward and backward torques are calculated from the two parts of the equivalent circuit, and the total torque is given by the algebraic sum of the two. As shown in Figure 7.8.2b, the torque-speed characteristic of a single-phase induction motor is thus obtained as the sum of the two curves, one corresponding to the forward-rotating field and the other to the backward-rotating field.

The slip $\mathbf{S}_f$ of the rotor with respect to the forward-rotating field is given by

$$\mathbf{S}_f = \mathbf{S} = \frac{n_s - n}{n_s} = 1 - \frac{n}{n_s} \tag{7.8.1}$$

where n_s is the synchronous speed and n is the actual rotor speed. The slip $\mathbf{S}_b$ of the rotor with respect to the backward-rotating field is given by

$$\mathbf{S}_b = \frac{n_s - (-n)}{n_s} = 1 + \frac{n}{n_s} = 2 - \mathbf{S} \tag{7.8.2}$$

Since the amplitude of the rotating fields is one-half of the alternating flux, as seen from Equation 5.4.9, the total magnetizing and leakage reactances of the motor can be divided equally so as to correspond to

---

[1] For a discussion of the cross-field theory, see A. F. Puchstein, T. C. Lloyd, and A. G. Conrad, *Alternating-Current Machines,* 3rd ed. (New York: John Wiley & Sons, 1954), chap. 30; and C. G. Veinott, *Fractional- and Subfractional-Horsepower Electric Motors* (New York: McGraw-Hill, 1970), chapter 2.

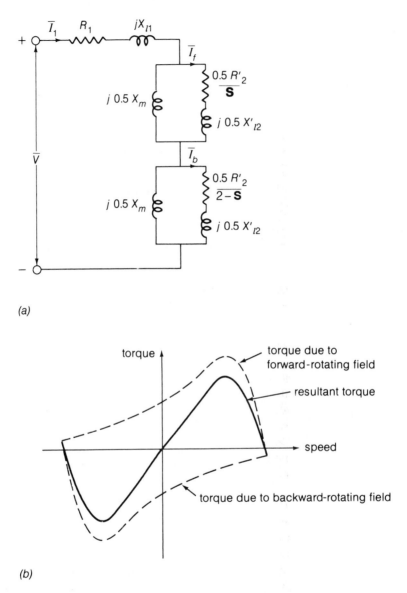

(a)

(b)

Figure 7.8.2 (a) Equivalent circuit of a single-phase induction motor based on the revolving-field theory. (b) Torque-speed characteristics of a single-phase induction motor based on the revolving-field theory.

the forward and backward rotating fields. In the equivalent circuit shown in Figure 7.8.2a, then, R_1 and X_{l1} are respectively the resistance and leakage reactance of the main winding, X_m is the magnetizing reactance, and R'_2 and X'_{l2} are the standstill values of the rotor resistance and leakage reactance referred to the main stator winding by the use of the appropriate turns ratio. The core loss, which is omitted here, can be accounted for later as if it were a rotational loss. The resultant torque of a single-phase induction motor can thus be expressed as

$$T_e = \frac{I_f^2(1-S)}{\omega_m S}R_2' - \frac{I_b^2(1-S)}{\omega_m(2-S)}R_2' \tag{7.8.3}$$

The following example illustrates the usefulness of the equivalent circuit in evaluating the motor performance.

IIIIIIIIIII
EXAMPLE 7.8.1

A $\frac{1}{4}$-hp, 230-V, 60-Hz, four-pole single-phase induction motor has the following parameters and losses:

$$R_1 = 10\ \Omega; \qquad X_{l1} = X_{l2}' = 12.5\ \Omega; \qquad R_2' = 11.5\ \Omega; \qquad X_m = 250\ \Omega$$
$$\text{Core loss at 230 V} = 35\ \text{W}; \qquad \text{Friction and windage loss} = 10\ \text{W}$$

For a slip of 0.05, determine the stator current, power factor, developed power, shaft-output power, speed, torque, and efficiency when the motor is running as a single-phase motor at rated voltage and frequency with its starting winding open.

Solution
From the given data applied to the equivalent circuit of Figure 7.8.2a, we see that

$$\frac{0.5R_2'}{S} = \frac{11.5}{2\times 0.05} = 115\ \Omega; \qquad \frac{0.5R_2'}{2-S} = \frac{11.5}{2(2-0.05)} = 2.95\ \Omega$$

$$j0.5X_m = j125\ \Omega \qquad \text{and} \qquad j0.5X_{l2}' = j6.25\ \Omega$$

For the forward-field circuit, the impedance

$$Z_f = \frac{(115+j6.25)j125}{115+j131.25} = 59 + j57.65 = R_f + jX_f$$

and for the backward-field circuit, the impedance

$$Z_b = \frac{(2.95+j6.25)j125}{2.95+j131.25} = 2.67 + j6.01 = R_b + jX_b$$

The total series impedance Z_e is given by

$$Z_e = Z_1 + Z_f + Z_b = (10+j12.5) + (59+j57.65) + (2.67+j6.01)$$
$$= 71.67 + j76.16 = 104.6\angle 46.74°$$

Input stator current $\bar{I}_1 = \dfrac{230}{104.6\angle 46.74°} = 2.2\angle -46.74°\ \text{A}$

Power factor $= \cos 46.74° = 0.685$ lagging

Developed power $P_d = \left[I_1^2 R_f\right](1-S) + \left[I_1^2 R_b\right][1-(2-S)]$
$$= I_1^2\left(R_f - R_b\right)(1-S) = (2.2)^2(59-2.67)(1-0.05) = 259\ \text{W}$$

Shaft-output power $P_0 = P_d - P_{\text{rot}} - P_{\text{core}} = 259 - 10 - 35 = 214\ \text{W or } 0.287\ \text{hp}$

Speed $= (1-S)(\text{synchronous speed}) = 0.95 \times \frac{120\times 60}{4} = 1{,}710\ \text{rpm or } 179\ \text{rad/s}$

$$\text{Torque} = \frac{214}{179} = 1.2 \text{ N} \cdot \text{m}$$

$$\text{Efficiency} = \frac{\text{Output}}{\text{Input}} = \frac{214}{230 \times 2.2 \times 0.685} = \frac{214}{346.6} = 0.6174 \text{ or } 61.74\%$$

The parameters of the equivalent circuit of a single-phase induction motor may be approximately determined from no-load and blocked-rotor tests similar to those made on a polyphase induction motor. These tests are usually made with the auxiliary winding kept open, except for the capacitor-run motor (see Section 7.9). The no-load test is conducted by running the motor without load at rated voltage and rated frequency. Since the no-load slip is small, the equivalent circuit can be considered to be like the one in Figure 7.8.3a. For the blocked-rotor-test conditions (i.e., for **S = 1**), Figure 7.8.3b shows the approximate equivalent circuit, neglecting the magnetizing current. Thus, on no-load we have

$$Z_0 = \left(R_1 + jX_{l1}\right) + j0.5X_m + \left(0.25R_2' + j0.5X_{l2}'\right) \tag{7.8.4}$$

and under blocked-rotor conditions

$$Z_{bl} = \left(R_1 + jX_{l1}\right) + \left(R_2' + jX_{l2}'\right) \tag{7.8.5}$$

Assuming that $X_{l1} = X_{l2}'$, and measuring the stator resistance R_1, the equivalent-circuit parameters can all be determined.

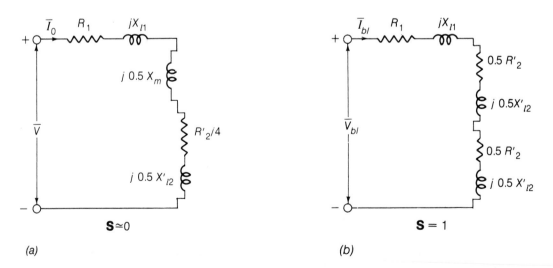

Figure 7.8.3 Equivalent circuit of a single-phase induction motor under (a) No-load conditions. (b) Blocked-rotor conditions.

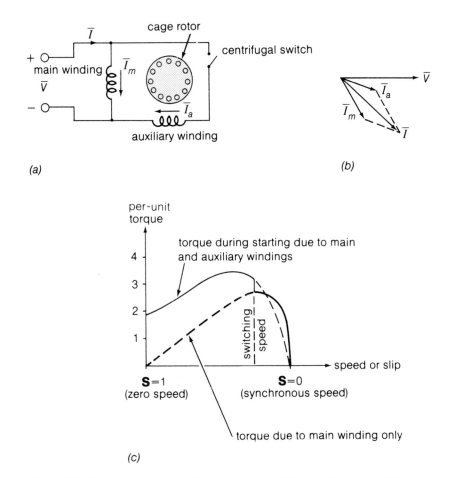

Figure 7.9.1 Split-phase motor. (a) Schematic diagram. (b) Phasor diagram at starting. (c) Typical torque-speed (or slip) characteristic.

7.9 Starting Methods for Single-Phase Induction Motors

The various forms of a single-phase induction motor are grouped into four principal types, depending on how they are started:

- *Split-phase* or *resistance-split-phase motors:* Split-phase motors have two stator windings (a main winding and an auxiliary winding) with their axes displaced 90 electrical degrees in space. As schematically represented in Figure 7.9.1a, the auxiliary winding in this type of motor has a higher resistance-to-reactance ratio than the main winding. The two currents are thus out of phase, as indicated in the phasor diagram of Figure 7.9.1b. The motor is equivalent to an unbalanced two-phase motor. The rotating stator field produced by the unbalanced two-phase winding currents

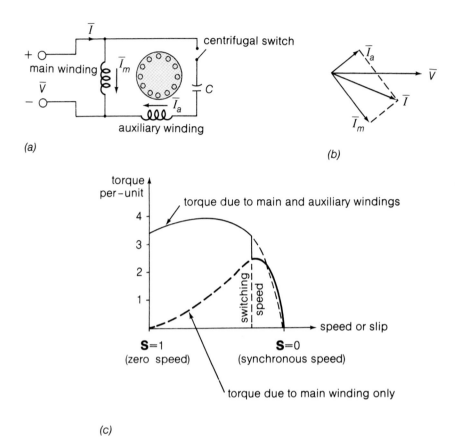

Figure 7.9.2 Capacitor-start motor. (a) Schematic diagram. (b) Phasor diagram at starting. (c) Typical torque-speed characteristic.

causes the motor to start. The auxiliary winding is disconnected by a centrifugal switch or relay when the motor comes up to about 75% of the synchronous speed. The torque-speed characteristic of the split-phase motor is of the form shown in Figure 7.9.1c. A split-phase motor can develop a higher starting torque if a series resistance is inserted in the starting auxiliary winding. A similar effect can be obtained by inserting a series inductive reactance in the main winding; this additional reactance is short-circuited when the motor builds up speed.

- *Capacitor motors:* Capacitor motors have a capacitor in series with the auxiliary winding and come in three varieties: capacitor-start, two-value-capacitor, and permanent-split-capacitor. As their names imply, the first two use a centrifugal switch or relay to open the circuit or reduce the size of the starting capacitor when the motor comes up to speed. A two-value-capacitor motor, with one value for starting and one for running, can be designed for optimum starting and running performance; the starting capacitor is disconnected after the motor starts. The relevant schematic diagrams and torque-speed characteristics are shown in Figures 7.9.2, 7.9.3 and 7.9.4. Motors in

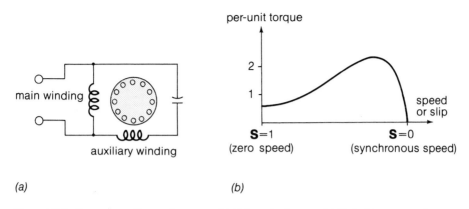

Figure 7.9.3 Permanent-split capacitor motor. (a) Schematic diagram. (b) Typical torque-speed characteristic.

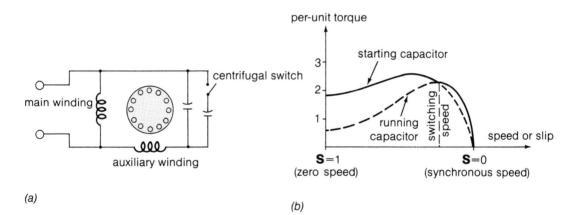

Figure 7.9.4 Two-value-capacitor motor. (a) Schematic diagram. (b) Typical torque-speed characteristic.

which the auxiliary winding and the capacitor are not cut out during the normal running conditions operate, in effect, as unbalanced two-phase induction motors.

- *Shaded-pole motors:* The least expensive of the fractional-horsepower motors, generally rated up to $\frac{1}{20}$ hp, they have salient stator poles, with one-coil-per-pole main windings. The auxiliary winding consists of one (or rarely two) short-circuited copper straps wound on a portion of the pole and displaced from the center of each pole, as shown in Figure 7.9.5a. The shaded-pole motor got its name from these shading bands. Induced currents in the shading coil cause the flux in the shaded portion of the pole to lag the flux in the other portion in time. The result is then like a rotating field moving in the direction from the unshaded to the shaded portion of the pole. A low starting torque is produced; a typical torque-speed characteristic is shown in Figure 7.9.5b. Shaded-pole motors have a rather low efficiency.

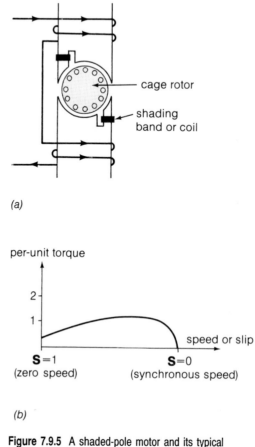

(a)

per-unit torque

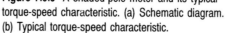

S=1
(zero speed)

S=0
(synchronous speed)

speed or slip

(b)

Figure 7.9.5 A shaded-pole motor and its typical torque-speed characteristic. (a) Schematic diagram. (b) Typical torque-speed characteristic.

- *Repulsion-induction motors.* Before low-cost capacitors became available, repulsion-start and induction-run machines were the most widely used kind of single-phase motor in the range of $\frac{1}{3}$ to 5 hp. This kind of motor has distributed rotor-windings connected to a commutator (like a dc machine) with short-circuited brushes and a distributed single-phase stator winding in the direct axis only. Repulsion-start motors have a centrifugal device that short-circuits all the commutator segments when the motor comes up to speed and lifts the brushes off the commutator. The permanently short-circuited rotor brushes are displaced from the direct axis by an angle of less than 90 electrical degrees so a rotor current is induced by transformer action. The flux produced by this current, which is fixed in position, acts with the main winding flux to create a torque.

    To avoid the centrifugal mechanism, some widely used repulsion-induction motors were designed with an additional deeply buried, low-resistance squirrel cage in the rotor to limit the no-load speed to a little above synchronism. Such machines fall under the general category of ac commutator machines; however, they have been replaced by capacitor motors.

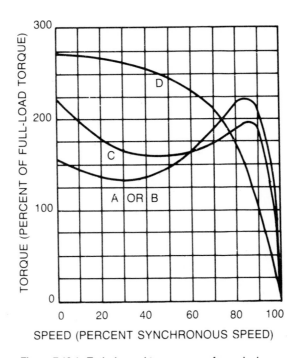

Figure 7.10.1 Typical speed-torque curves for squirrel-cage induction motors with NEMA Design classification A, B, C, and D. (Adapted from NEMA Standards Publication MG10, *Energy Management Guide for Selection and Use of Polyphase Motors,* New York, 1988.)

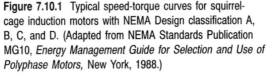

7.10 Applications for Induction Motors

Before specifying a particular motor for a given application, the designer must know the load characteristics, such as horsepower requirement, starting torque, acceleration capability, speed variation, duty cycle, and the environment in which the motor is to operate. Table 7.10.1 lists motors that are readily available and standardized according to generally accepted criteria established by the National Electrical Manufacturer Association (NEMA).[2] Typical speed-torque curves for squirrel-cage induction motors with NEMA Design Classifications A, B, C, and D are shown in Figure 7.10.1. Having selected the appropriate motor for a given application, the next step is to specify a controller for the motor to furnish proper starting, stopping, and reversing without damaging the motor, other connected loads, or the power system.

Typical characteristics and applications of single-phase fractional-horsepower induction motors are provided in Table 7.10.2 (pages 340–341). The ranges of standard power ratings given by NEMA for single-phase motors are listed in Table 7.10.3.

Two-phase induction motors with high rotor resistance are employed as servomotors for control system applications that usually require positive damping over the full speed range.

---

[2] NEMA Standards Publication MG1, *Motors and Generators,* New York, 1987.

7.10.1 Typical characteristics and applications of polyphase 60-hz induction motors

Type classification	hp range	Starting torque (% of normal)	Maximum torque (% of normal)	Starting current (% of normal)
General-purpose, normal torque and starting current, NEMA class A	0.5–200	Poles–Torque 2–150 4–150 6–135 8–125 10–120 12–115 14–110 16–105	Up to 250 but not less than 200	500–1000
General-purpose, normal torque, low starting current, NEMA class B	0.5–200	Same as above or larger	About the same as class A but may be less	About 500–550, less than average of class A
High torque, low starting current, NEMA class C	1–200	200–250	Usually a little less than class A but not less than 200	About same as class B
High torque, medium and high slip, NEMA class D	0.5–150	Medium slip 350 High slip 250–315	Usually same as standstill torque	Medium slip 400–800, high slip 300–500
Low starting torque, either normal starting current, NEMA class E, or low starting current, NEMA class F	40–200	Low, not less than 50	Low, but not less than 150	Normal 500–1000, low 300–500
Wound-rotor	0.5–5000	Up to 300	200–250	Depends upon external rotor resistance but may be as low as 150

Adapted from M. Liwschitz-Garik and C. C. Whipple, *Electric Machinery,* vol. II, D. Van Nostrand Co., Inc., Princeton, N.J., 1946.

Slip (%)	Power factor (%)	Efficiency (%)	Typical applications
Low, 3–5	High, 87–89	High, 87–89	Constant-speed loads where excessive starting torque is not needed and where high starting current is tolerated. Fans, blowers, centrifugal pumps, most machinery tools, wood-working tools, line shafting. Lowest in cost. May require reduced voltage starter. Not to be subjected to sustained overloads, because of heating. Has high maximum torque
3–5	A little lower than class A	87–89	Same as class A — advantage over class A is lower starting current, but power factor slightly less
3–7	Less than class A	82–84	Constant-speed loads requiring fairly high starting torque and lower starting current. Conveyors, compressors, crushers, agitators, reciprocating pumps. Maximum torque at standstill
Medium, 7–11; high, 12–16	Low	Low	Medium slip. Highest starting torque of all squirrel-cage motors. Used for high-inertia loads such as shears, punch presses, die stamping, bulldozers, boilers. Has very high average accelerating torque. High slip used for elevators, hoists, etc., on intermittent loads
1 to $3\frac{1}{2}$	About same as class A or class B	About same as class A or class B	Direct-connected loads of low inertia requiring low starting torque, such as fans and centrifugal pumps. Has high efficiency and low slip.
3–50	High, with rotor shorted same as class A	High, with rotor shorted same as class A, but low when used with rotor resistor for speed control	For high-starting-torque loads where very low starting current is required or where torque must be applied very gradually and where some speed control (50%) is needed. Fans, pumps, conveyors, hoists, cranes, compressors. Motor with speed control more expensive and may require more maintenance.

7.10.2 Typical characteristics and applications of single-phase fractional-horsepower induction motors

Type designation	Starting torque (% of normal)	Approx. comparative price (%)	Breakdown torque (% of normal)	Starting current at 115 V
General-purpose split-phase motor	90–200 Medium	85	185–250 Medium	23 1/4 hp
High-torque split-phase motor	200–275 High	65	Up to 350	32 High 1/4 hp
Permanent-split capacitor motor	60–75 Low	155	Up to 225	Medium
Permanent-split capacitor motor	Up to 200 Normal	155	260	
Capacitor-start general-purpose motor	Up to 435 Very high	100	Up to 400	
Capacitor-start capacitor-run motor	380 High	190	Up to 260	
Shaded-pole motor	50	—	150	

Adapted from M. Liwschitz-Garik and C. C. Whipple, *Electric Machinery,* vol. II, D. Van Nostrand Co., Inc., Princeton, N.J., 1946.

Power factor (%)	Efficiency (%)	Horsepower range	Application and general remarks
56–65	62–67	1/20 to 3/4	Fans, blowers, office appliances, food-preparation machines. Low- or medium-starting torque, low-inertia loads. Continuous-operation loads. May be reversed.
50–62	46–61	1/6 to 1/3	Washing machines, sump pumps, home workshops, oil burners. Medium- to high-starting torque loads. May be reversed.
80–95	55–65	1/20 to 3/4	Direct-connected fans, blowers, centrifugal pumps. Low-starting torque loads. Not for belt drives. May be reversed.
80–95	55–65	1/6 to 3/4	Belt-driven or direct drive fans, blowers, centrifugal pumps, oil burners. Moderate-starting torque loads. May be reversed.
80–95	55–65	1/8 to 3/4	Dual voltage. Compressors. stokers, conveyors, pumps. Belt-driven loads with high static friction. May be reversed.
80–95	55–65	1/8 to 3/4	Compressors, stokers, conveyors, pumps. High-torque loads. High power factor. Speed may be regulated.
30–40	30–40	1/300 to 1/20	Fans, toys, hair dryers, unit heaters. Desk fans. Low-starting torque loads.

Table 7.10.3 Ranges of standard power ratings for single-phase induction motors.

Motor	Power Range
Capacitor start	1 mhp to 10 hp
Resistance start	1 mhp to 10 hp
Two-value capacitor	1 mhp to 10 hp
Permanent-split capacitor	1 mhp to 1.5 hp
Shaded-pole	1 mhp to 1.5 hp

For more information, consult NEMA Standards Publication MG1, *Motors and Generators,* New York, 1987.

■■■■■■■▮▮▮▮
BIBLIOGRAPHY

Alger, P. L. *Induction Machines—Their Behavior and Uses.* 2d ed. New York: Gordon and Breach, 1970.

American National Standards Institute. *ANSI Standard No. C.56.20–1954: Test Code for Induction Motors.* New York: ANSI, 1954.

Cochran, P. L. *Polyphase Induction Motors.* New York: Marcel Dekker, Inc., 1989.

Del Toro, V. *Electromechanical Devices for Energy Conversion and Control Systems.* Englewood Cliffs, N. J.: Prentice-Hall, 1968.

Fitzgerald, A. E.; C. Kingsley, Jr.; and A. Kusko. *Electrical Machinery.* 3d ed. New York: McGraw-Hill, 1971.

Hindmarsh, J. *Electrical Machines and Their Applications.* 2d ed. New York: Pergamon, 1970.

Institute of Electrical and Electronics Engineers. S *tandard No. 112-1991: Test Procedures for Polyphase Induction Motors and Generators.* New York: IEEE, 1991.

Knowlton, A. E., ed. *Standard Handbook for Electrical Engineers.* 8th ed. New York: McGraw-Hill, 1949.

Matsch, L. W. *Electromagnetic and Electromechanical Machines.* New York: Intext, 1972.

Nasar, S. A., and L. E. Unnewehr. *Electromechanics and Electric Machines.* New York: John Wiley & Sons, 1979.

National Electrical Manufacturers Association. *Publication No. MG1-1972: Motors and Generators.* New York: NEMA, 1972.

————. *Publication No. MG2-1951: Standards for Fractional Horsepower Motors.* New York: NEMA, 1951.

Puchstein, A. F.; T. C. Lloyd; and A. G. Conrad. *Alternating-Current Machines.* 3d ed. New York: John Wiley & Sons, 1954.

Slemon, G. R., and A. Straughen *Electric Machines.* Reading, Mass.: Addison-Wesley, 1980.

Veinott, C. G. *Design Analysis of Polyphase Motors (Eng. or SI).* Sarasota, Florida: C. G. Veinott, 1991.

————. *Fractional and Subfractional Horsepower Electric Motors,* 4th ed. New York: McGraw-Hill, 1987.

————. *Fractional- and Subfractional-Horsepower Electric Motors.* New York: McGraw-Hill, 1970.

————. *How to Design a Metric 1-Phase Motor on a Personal Computer.* Sarasota, Florida: C. G. Veinott, 1991.

————. *How to Design a Single-Phase Motor on a Personal Computer.* Sarasota, Florida: C. G. Veinott, 1988.

————. *Multispeed Capacitor Motors (Tapped Wdg.) (Eng. or SI).* Sarasota, Florida: C. G. Veinott, 1991.

————. *Theory and Design of Small Induction Motors.* Sarasota, Florida: C. G. Veinott, 1986.

■■■■■■■■■■■■■■■■▮▮▮▮▮▮▮▮▮▮▮
PROBLEMS

7-1. A balanced three-phase 60-Hz voltage is applied to a three-phase, two-pole induction motor. Corresponding to a per-unit slip of 0.05, determine the following:

   a. The speed of the rotating stator-magnetic field relative to the stator winding.

   b. The speed of the rotor field relative to the rotor winding.

   c. The speed of the rotor field relative to the stator winding.

   d. The speed of the rotor field relative to the stator field.

   e. The frequency of the rotor currents.

   f. Neglecting stator resistance, leakage reactance, and all losses, if the stator-to-rotor turns ratio is 2:1 and the applied voltage is 100 V, find the rotor-induced emf at standstill and at 0.05 slip.

7-2.   The double-layer stator winding of a cage-type, four-pole, three-phase induction motor has 24 turns per phase distributed in 36 slots. The winding is chorded by one slot pitch. Calculate the factor by which the rotor standstill resistance and leakage reactance must be multiplied for referring them to the stator given that the number of rotor bars is 28.

7-3.   No-load and blocked-rotor tests are conducted on a three-phase wye-connected induction motor with the following results: The line-to-line voltage, line current, and total input power for the no-load test are 220 V, 20 A, and 1,000 W; and for the blocked-rotor test, 30 V, 50 A, and 1,500 W. The stator resistance as measured on a dc test is 0.1 Ω per phase.

    a.   Determine the parameters of the equivalent circuit shown in Figure 7.2.5c.

    b.   Compute the no-load rotational losses.

7-4.   A three-phase, 5-hp, 220-V, six-pole, 60-Hz induction motor runs at a slip of 0.025 at full load. Rotational and stray-load losses at full load are 5% of the output power. Calculate the power transferred across the air gap, the rotor copper loss at full load, and the electromagnetic torque at full load in newton-meters.

7-5.   The power transferred across the air gap of a two-pole induction motor is 24 kW. If the electromagnetic power developed is 22 kW, find the slip. Calculate the output torque, if the rotational loss at this slip is 400 W.

7-6.   The stator and rotor of a three-phase, 440-V, 15-hp, 60-Hz, eight-pole wound-rotor induction motor are both connected in wye and have the following parameters per phase: $R_1 = 0.5$ Ω; $R_2 = 0.1$ Ω; $X_{l1} = 1.25$ Ω; $X_{l2} = 0.2$ Ω. The magnetizing impedance is 40 Ω and the core-loss impedance is 360 Ω, both referred to the stator. The ratio of effective stator turns to effective rotor turns is 2.5. The friction and windage losses total 200 W, and the stray-load loss is estimated as 100 W. Using the equivalent circuit of Figure 7.2.5a, calculate the following values for a slip of 0.05, when the motor is operated at rated voltage and frequency applied to the stator, with the rotor slip rings short-circuited: stator input current, power factor at the stator terminals, current in the rotor winding, output power, output torque, and efficiency.

7-7   a.   Starting from Equations 7.4.8 and 7.4.11, show that

$$\frac{T}{T_{\max}} = \frac{1 + \sqrt{Q^2 + 1}}{1 + \frac{1}{2}\sqrt{Q^2 + 1}\left(\frac{s}{s_{\max T}} + \frac{s_{\max T}}{s}\right)} \qquad \text{where} \quad Q = \frac{X_1'' + X_{l2}'}{R_1''}$$

With an infinite value of Q (i.e., $R_1'' = 0$, or negligible stator resistance), show that Equation 7.4.12 results.

    b.   Show that the ratio of the stator-load current I_2' to that at maximum torque $I_{2\,\max T}'$ is given by

$$\frac{I_2'}{I_{2'\,\max T}} = \sqrt{\frac{\left(1 + \sqrt{1 + Q^2}\right)^2 + Q^2}{\left(1 + \frac{s_{\max T}}{s}\sqrt{1 + Q^2}\right)^2 + Q^2}} \qquad \text{where} \quad Q = \frac{X_1'' + X_{l2}'}{R_1''}$$

Obtain the expression for the limiting case when $Q \to \infty$.

7-8.   Considering only the rotor equivalent circuit shown in Figure 7.2.2 or 7.2.3, find

    a.   The R_2 for which the developed torque would be a maximum.

    b.   The slip corresponding to the maximum torque.

    c.   The maximum torque.

    d. R_2 for the maximum starting torque.

7-9.   A three-phase induction motor, operating at its rated voltage and frequency, develops a starting torque of 1.6 times the full-load torque and a maximum torque of 2 times the full-load torque. Neglecting stator resistance and rotational losses, and assuming constant rotor resistance, determine the slip at maximum torque and the slip at full load.

7–10. A three-phase, wye-connected, 400-V, four-pole, 60-Hz induction motor has primary leakage impedance of $(1 + j2)$ Ω and secondary leakage impedance referred to the primary at standstill of $(1 + j2)$ Ω. The magnetizing impedance is $j40$ Ω and the core-loss impedance is 400 Ω. Using the T-equivalent circuit of Figure 7.2.5a,

    a. Calculate the input current and power (i) on the no-load test ($\mathbf{S} \simeq 0$) at rated voltage, and (ii) on a blocked-rotor test ($\mathbf{S} = 1$) at rated voltage.

    b. Corresponding to a slip of 0.05, compute the input current, torque, output power, and efficiency.

    c. Determine the starting torque and current; the maximum torque and the corresponding slip; and the maximum output power and the corresponding slip.

    For the following parts, use the approximate equivalent circuit obtained by transferring the shunt core loss/magnetizing branch to the input terminals:

    d. Find the same values requested in part b.

    e. When the machine is driven as an induction generator with a slip of -0.05, calculate the primary current, torque, mechanical power input, and electrical power output.

    f. Compute the primary current and the braking torque at the instant of plugging (i.e., reversal of the phase sequence), if the slip immediately before plugging is 0.05.

7–11. A 500-hp wye-connected wound-rotor induction motor, when operated at rated voltage and frequency, develops its rated full-load output at a slip of 0.02; maximum torque of 2 times the full-load torque at a slip of 0.06, with a referred rotor current of 3 times that at full load; and 1.2 times the full-load torque at a slip of 0.2, with a referred rotor current of 4 times that at full load. Neglect rotational and stray-load losses. If the rotor-circuit resistance in all phases is increased to 5 times the original resistance, determine the following:

    a. The slip at which the motor will develop the same full-load torque.

    b. The total rotor-circuit copper loss at full-load torque.

    c. The horsepower output at full-load torque.

    d. The slip at maximum torque.

    e. The rotor current at maximum torque.

    f. The starting torque.

    g. The rotor current at starting.

7–12. The per-phase equivalent circuit shown in Figure 7.4.1 of a three-phase, 600-V, 60-Hz, four-pole, wye-connected wound-rotor induction motor has the following parameters: $R_1 = 0.75$ Ω; $R_2' = 0.80$ Ω; $X_{l1} = X_{l2}' = 2.0$ Ω; $X_M = 50$ Ω. Neglect the core losses.

    a. Find the slip at which the maximum developed torque occurs.

    b. Calculate the value of the maximum torque developed.

    c. What is the range of speed for stable operation of the motor?

    d. Determine the starting torque.

    e. Compute the per-phase referred value of the additional resistance that must be inserted in the rotor circuit in order to obtain the maximum torque at starting.

7–13. A three-phase, wye-connected, 220-V, 10-hp, 60-Hz, six-pole induction motor (using Figure 7.4.1 for notation) has the following parameters in ohms per phase referred to the stator: $R_1 = 0.294$; $R_2' = 0.144$; $X_{l1} = 0.503$; $X_{l2}' = 0.209$; $X_M = 13.25$. The total friction, windage, and core losses can be assumed to be constant at 403 W, independent of load. For a slip of 2.00%, compute the speed, output torque and power, stator current, power factor, and efficiency when the motor is operated at rated voltage and frequency. Neglect the impedance of the source.

7-14. A squirrel-cage induction motor operates at a slip of 0.05 at full load. The rotor current at starting is five times the rotor current at full load. Neglecting stator resistance and rotational and stray-load losses, and assuming constant rotor resistance, calculate the starting torque and the maximum torque in per-unit-of-full-load torque, as well as the slip at which the maximum torque occurs.

7-15. Using the approximate equivalent circuit in which the shunt branch is moved to the stator-input terminals, show that the rotor current, torque, and electromagnetic power of a polyphase induction motor vary almost directly as the slip, for small values of slip.

7-16. A three-phase, 50-hp, 440-V, 60-Hz, four-pole wound-rotor induction motor operates at a slip of 0.03 at full load with its slip rings short-circuited. The motor is capable of developing a maximum torque of two times the full-load torque at rated voltage and frequency. The rotor resistance per phase referred to the stator is 0.1 Ω. Neglect the stator resistance and rotational and stray-load losses. Find the rotor copper loss at full load and the speed at maximum torque. Compute the value of per-phase rotor resistance (referred to the stator) that must be added in series to produce a starting torque equal to the maximum torque.

7-17. A three-phase, 220-V, 60-Hz, four-pole, wye-connected induction motor has a per-phase stator resistance of 0.5 Ω. The following no-load and blocked-rotor test data on the motor are given:

> No-load test: Line-to-line voltage = 220 V
> Total input power = 600 W, of which 200 W is the friction and windage loss
> Line current = 3 A

> Blocked-rotor test: Line-to-line voltage = 35 V
> Total input power = 720 W
> Line current = 15 A

    a. Calculate the parameters of the equivalent circuit shown in Figure 7.2.5c.

    b. Find the output power, output torque, and efficiency if the machine runs as a motor with a slip of 0.05.

    c. Determine the slip at which maximum torque is developed, and obtain the value of the maximum torque.

7-18. The synchronous speed of a wound-rotor induction motor is 900 rpm. Under a blocked-rotor condition, the input power to the motor is 45 kW at 193.6 A. The stator resistance per phase is 0.2 Ω, and the ratio of effective stator turns to effective rotor turns is 2. The stator and rotor are both wye-connected. Negelect the effect of the core-loss and magnetizing impedances. Calculate (a) the value in ohms of the rotor resistance per phase and (b) the motor starting torque.

7-19. No-load and blocked-rotor tests on a three-phase, wye-connected induction motor yield these results:

> No-load test: Line-to-line voltage = 400 V   Blocked-rotor test :   Line-to-line voltage = 45 V
> Input power = 1770 W   Input power = 2700 W
> Input current = 18.5 A   Input current = 63 A
> Friction and windage loss = 600 W

Determine the parameters of the equivalent circuit of Figure 7.2.5a, assuming $R_1 = R_2'$ and $X_{l1} = X_{l2}'$.

7-20. A three-phase induction motor has the per-phase circuit parameters shown in Figure P7–20. At what slip is the maximum power developed?

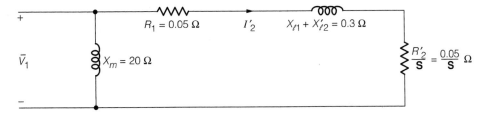

Figure P7–20

7-21. A large induction motor is usually started by applying a reduced voltage across the motor; such a voltage may be obtained from an autotransformer. A motor is to be started on 50% of full-load torque, and the full-voltage starting current is five times the full-load current. The full-load slip is 4%. Determine the percentage reduction in the applied voltage (i.e., the percentage tap on the autotransformer).

7-22. A three-phase, 400-V, wye-connected induction motor takes the full-load current at 45 V with the rotor blocked. The full-load slip is 4%. Calculate the tappings k on a three-phase autotransformer to limit the starting current to four times the full-load current. For such a limitation, determine the ratio of the starting torque to full-load torque.

7-23. A three-phase, 2,200-V, 60-Hz, delta-connected squirrel-cage induction motor, when started at full rated voltage, takes a starting current of 693 A from the line and develops a starting torque of 6,250 N·m.

   a. Neglect the impedance and exciting current of the compensator. Calculate the ratio of a starting compensator (i.e., an autotransformer starter) such that the current supplied by 2,200-V line is 300 A. Compute the starting torque with the starting compensator.

   b. If a wye-delta starting method is employed, find the starting current and the starting torque.

7-24. A three-phase, four-pole, 220-V, 60-Hz induction machine with a per-phase resistance of 0.5 Ω is operating at rated voltage as a generator at a slip of -0.04, delivering 12 A of line current and a total output of 4,000 W. The constant losses from a no-load run as a motor are given to be 220 W, of which 70 W represent friction and windage losses. Calculate the efficiency of the induction generator.

7-25. A 2,200-V, 1,000-hp, three-phase, 60-Hz, 16-pole, wye-connected, wound-rotor induction motor is connected to a 2,200-V, three-phase, 60-Hz bus that is supplied by synchronous generators. The per-phase equivalent circuit of Figure 7.4.1 has the following parameters: $R_1 = 0.1 \ \Omega = R_2'$; $X_{l1} = 0.625 \ \Omega = X_{l2}'$; $X_M = 20 \ \Omega$. If the machine is driven at a speed of 459 rpm to act as a generator of real power, find the rotor current referred to the stator and the real and reactive power outputs of the induction machine.

7-26. A three-phase, 440-V, 60-Hz, four-pole induction motor operates at a slip of 0.025 at full load with its rotor circuit short-circuited. This motor is to be operated on a 50-Hz supply so that the air-gap flux wave has the same amplitude at the same torque as on a 60-Hz supply. Determine the 50-Hz applied voltage and the slip at which the motor will develop a torque equal to its 60-Hz full-load value.

7-27. The rotor of a wound-rotor induction motor is rewound with twice the number of its original turns, with a cross-sectional area of the conductor in each turn of one-half the original value. Determine the ratio of the following in the rewound motor to the corresponding original quantities: (a) full-load current, (b) actual rotor resistance, and (c) rotor resistance referred to the stator. Repeat the problem given that the original rotor is rewound with the same number turns as the original, but with one-half the original cross-section of the conductor. Neglect the changes in the leakage flux.

7-28. A wound-rotor induction machine, driven by a dc motor whose speed can be controlled, is operated as a frequency changer. The three-phase stator winding of the induction machine is excited from a 60-Hz supply, while the variable-frequency three-phase power is taken out of the slip rings. The output frequency range is to be 120 to 420 Hz; the maximum speed is not to exceed 3,000 rpm; and the maximum power output at 420 Hz is to be 70 kW at 0.8 power factor. Assuming that the maximum-speed condition determines the machine size, and neglecting exciting current, losses, and voltage drops in the induction machine, calculate (a) the minimum number of poles for the induction machine, (b) the corresponding minimum and maximum speeds, (c) the kVA rating of the induction-machine stator winding, and (d) the horsepower rating of the dc machine.

7-29. A $\frac{1}{4}$-hp, 110-V, 60-Hz, four-pole capacitor-start single-phase induction motor has the following parameters and losses:

$$R_1 = 2 \ \Omega; \qquad X_{l1} = 2.8 \ \Omega; \qquad X_{l2}' = 2 \ \Omega; \qquad R_2' = 4 \ \Omega; \qquad X_M = 70 \ \Omega$$

$$\text{Core loss at 110 V} = 25 \text{ W}; \qquad \text{friction and windage} = 12 \text{ W}$$

For a slip of 0.05, compute the output current, power factor, power output, speed, torque, and efficiency when the motor is running at rated voltage and rated frequency with its starting winding open.

7-30. The no-load and blocked-rotor tests conducted on a 110-V, single-phase induction motor yield these data:

No-load test: Input voltage = 110 V   Blocked-rotor test: Input voltage = 50 V
Input current = 3.7 A   Input current = 5.6 A
Input power = 50 W

Taking the stator resistance to be 2.0 Ω, friction and windage loss to be 7 W, and assuming $X_{l1} = X'_{l2}$, determine the parameters of the double-revolving-field equivalent circuit.

7-31. The impedance of the main and auxiliary windings of a $\frac{1}{3}$-hp, 120-V, 60-Hz, capacitor-start motor are given as

$$Z_m = (4.6 + j3.8) \ \Omega \quad \text{and} \quad Z_a = (9.6 + j3.6) \ \Omega$$

Determine the value of the starting capacitance that will cause the main and auxiliary winding currents to be in quadrature at starting.

7-32. Prof M. G. Say's design data[1] follow for two slip-ring induction motors. (The slot and core details, including main dimensions, are shown in Figure P7–32.) Pursue the design projects and justify the details furnished.

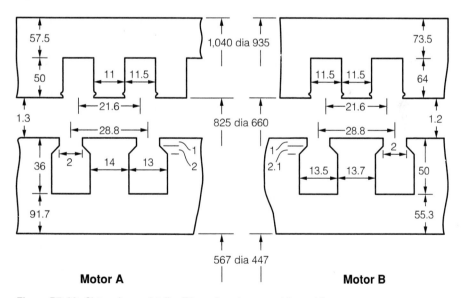

Figure P7–32 Slot and core details. (Dimensions in mm; not to scale).

Motor A: 270-kW, 3.0-kV, three-phase, 50-Hz, 10-pole slip-ring motor to drive a centrifugal water pump, rated with reduced specific magnetic and electric loadings for a 30° C temperature rise in ambient air at 55° C and 100% humidity.

Motor B: 75-kW, 3.0-kV, three-phase, 50-Hz, 8-pole slip-ring motor, flame-proof for driving mine-haulage machinery, rated for a 50° C temperature rise. The restricted ventilation enforces low I^2R loss, especially in the rotor.

[1] M. G. Say, *Alternating Current Machines,* A Halsted Press Book, John Wiley & Sons, New York, 1978.

Data for Problem 7–32

			A	B
Rating	Output	Kw	270	75
	Power factor	p.u.	0.84	0.86
	Efficiency	p.u.	0.93	0.94
	Input	kVA	346	93
	Line/phase current	A	66.6	17.9
Main Dimensions	Magnetic loading	T	0.39	0.225
	Electric loading	kA/m	30.9	14.9
	Stator bore	m	0.825	0.66
	Gross bore length	m	0.40	0.50
	Ducts, no./width	mm	5/10	nil
	Net core length	m	0.35	0.50
	Iron length	m	0.32	0.45
	Radial gap length	mm	1.30	1.20
	Pole pitch	m	0.26	0.26
Stator	Winding/connection		2-layer diamond/wye	
	No-load flux per pole	mWb	40.0	29.1
	Turns per phase	—	200	286
	Number of slots	—	120	96
	Distribution factor	—	0.958	0.958
	Coil-span factor	—	1.0	0.991
	Winding factor	—	0.958	0.950
	Slot pitch	mm	21.6	21.6
	Conductors per slot	—	2×5	2×9
	Conductor size	mm	6.5×3	6.5×2.2
	area	mm^2	19.0	14.0
	Length of mean cond.	m	0.963	1.032
	Resistance	Ω/ph	0.426	0.886
	Current density	A/mm^2	3.50	1.27
	Total f.l. I^2R loss	kW	5.70	0.84
	Copper mass	kg	220	250
Rotor	Winding/connection	—	2-layer bar wave/wye	
	Slip-ring o.c. voltage	V	650	490
	Turns per phase	—	45	48
	Number of slots	—	90	72
	Winding factor	—	0.960	0.960
	Slot pitch	mm	28.8	28.8
	Conductors per slot	—	$2(2 \times 3)$	2×2
	Conductor size	mm	13×3	20×5
	area	mm^2	77	98.5
	Length of mean cond.	m	0.862	0.963
	Resistance	mΩ/ph	21.2	19.7
	Current density	A/mm^2	3.32	2.42
	Total f.l. I^2R loss	kW	4.00	0.49
	Copper mass	kg	175	265

continued

Data for Problem 7–32 *Continued*

			A	B
No-load current	M.M.F. per pole	A-t	1 100	500
	Magnetizing current	A	24.6	6.3
	reactance	Ω	70.5	275
	Core loss	kW	5.7	2.2
	No-load loss	kW	8.6	3.1
	current	A	24.6	6.3
	power factor	p.u.	0.067	0.095
Short-circuit current	Slot reactance	Ω	2.26	9.85
	Overhang reactance	Ω	1.56	3.75
	Zigzag reactance	Ω	1.13	4.40
	Differential reactance	Ω/ph	0.30	0.90
	Total reactance	Ω/ph	5.25	18.9
	Total resistance	Ω/ph	0.85	1.57
	Total impedance	Ω/ph	5.3	19.0
	Short-circuit current		327	91
	power factor	p.u.	0.16	0.083
Performance	Full-load losses:			
	stator I^2R	kW	5.7	0.84
	rotor I^2R	kW	4.0	0.49
	brush	kW	0	0.14
	core	kW	5.7	2.2
	mechanical	kW	2.9	0.9
	load (stray)	kW	0.9	0.33
	total	kW	19.2	4.9
	Output	kW	270.0	75.0
	Input	kW	289.2	79.9
	Efficiency	p.u.	0.93	0.94
	Full-load rotor input	kW	276.9	76.5
	slip	p.u.	0.0145	0.0061
	speed	r/min	591	745
	Pull-out torque	p.u.	2.70	2.87

8

Synchronous Machines

Large ac power networks operating at a constant frequency of 60 Hz in the United States (50 Hz in Europe) rely almost exclusively on synchronous generators to generate electrical energy; they can also have synchronous compensators or condensers at key points for reactive power control. Generators are the largest single-unit electric machines in production, having power ratings in the range of 1,500 MVA; and we can expect machines of several thousand MVA to come into use in future decades. Private, stand-by, and peak-load plants with diesel or gas-turbine prime-movers also have alternators. Nonland-based synchronous plants can be found on oil rigs, on large aircraft with hydraulically driven alternators operating at 400 Hz, and on ships for variable frequency supply to synchronous propeller motors. Synchronous motors provide constant speed industrial drives with the possibility of power-factor correction, although they are not often built in small ratings, for which the induction motor is cheaper.

This chapter develops analytic methods of examining the steady-state performance of synchronous machines. We first consider cylindrical-rotor machines and discuss the effects of salient poles later.

8.1 Constructional Features of Synchronous Machines

The two basic parts of a synchronous machine are the magnetic field structure, carrying a dc-excited winding, and the armature, often having a three-phase winding in which the ac emf is generated. The use of a rotating dc-field system is almost universal, because the ac windings can be placed on the stator, where they are more conveniently braced against electromagnetic forces and insulated for high voltage. Thus almost all modern synchronous machines have stationary armatures and rotating field structures.

The *rotor* may be constructed with *salient poles* or be *cylindrical,* i.e., *round* with no polar projections. The prime-mover speed has a profound influence on the constructional form, and in all large units the limiting feature is the centrifugal force on the stator. Steam-turbine or gas-turbine–driven machines, known as *turbogenerators, turboalternators,* or *turbine generators,* are run at high speeds. They usually

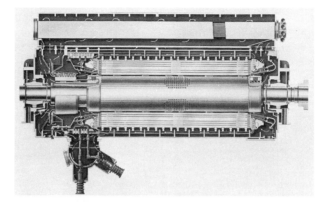

Photo 8.1.1 Cross-sectional view of a large turbogenerator with water-cooled stator and hydrogen-cooled rotor. (Photo courtesy of Westinghouse Electric Corporation.)

have two-pole or four-pole round (cylindrical) rotors of solid forged steel, with a diameter limited to about 1.2 meters and an axial length of several meters. See Photo 8.1.1 for a cross-sectional view of a large turbogenerator. Generators driven by water power, known as *hydrogenerators, hydroelectric generators,* or *water-wheel generators,* are built for a wide range of comparatively low turbine speeds. They are axially short but have a large diameter to accommodate the many salient poles on their rotors. Photo 8.1.2 shows a cutaway view of a hydroelectric generator. *Diesel-electric generators* are also low-speed machines with salient-pole rotors.

The construction of synchronous machines can vary from one kind to another in several aspects. While the water-wheel-driven hydroelectric generators are commonly of the vertical type, with vertically mounted salient-pole rotors, most of the others, including turbine generators, are of the horizontal type. The constructional details also depend greatly on the type of cooling method. The rotor-end windings or overhangs must be properly secured by the use of end-bells or retaining rings. The feasibility of using *superconducting windings*[1] on the rotor is being investigated for large turbogenerators.

The *stator* of a synchronous machine is essentially similar to that of a polyphase induction machine. The stator core consists of punchings (often built up of segmented sectors) of high-quality laminations having slot-embedded double-layer lap windings. Almost all synchronous generators are three-phase wye-connected machines with a 60° phase spread. The end-windings must be securely braced against movement under the impact of short-circuit electromagnetic forces.

Many salient-pole synchronous machines are commonly equipped with *damper (amortisseur) windings,* which usually consist of a set of copper or brass bars set in pole-face slots and connected together at the ends of the machine. (See Figure 5.3.13.) Amortisseur windings have some effect on stability; they serve several useful features like starting synchronous motors as induction motors, damping rotor oscillations, reducing overvoltages under certain short-circuit conditions, and helping to synchronize the machine. The cylindrical rotor of a turbine generator can be considered equivalent to an amortisseur of infinitely many circuits.

---

[1] M. J. Jeffries, et al., "Prospects for Superconductive Generators in the Electric Utility Industry," IEEE Transactions on Power Apparatus and Systems, pp. 1659-69, September/October 1973; Smith, J. L., "Superconductors in Large Synchronous Machines," EPRI Research Project Report, prepared at M.I.T., June 1975.

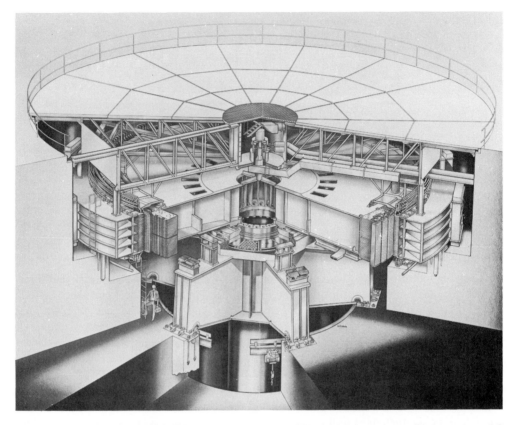

Photo 8.1.2 Cutaway view of a 718/826-MVA, 84-pole, 85.7-rpm, 15-kV hydroelectric generator. (Photo courtesy of General Electric Canada.)

The dc winding on the rotating field structure can be connected to an external dc source through slip rings and brushes. The dc *excitation* can be provided by a self-excited dc generator, known as the *exciter,* mounted on the same shaft as the rotor of the synchronous machine. In slow-speed machines with large ratings, like hydrogenerators, the exciter might not be self-excited; rather, a self-excited or a permanent-magnet type of *pilot exciter* can be employed to activate the exciter. Because of the maintenance problems associated with direct-coupled dc generators beyond a certain rating, an alternative form of excitation is provided, sometimes by silicon diodes and thyristors. The two types of *solid-state excitation* systems are (a) *static systems,* having stationary diodes or thyristors that supply the excitation current to the rotor through brushes and slip rings, and (b) *brushless systems* (Photo 8.1.3), having shaft-mounted rectifiers rotating with the rotor, eliminating the need for brushes and slip rings.

The severe electric and magnetic loadings in a synchronous machine produce heat that must be properly dissipated. The manner in which the active parts of a machine are cooled determines its overall physical size and structure. As stated in Section 5.7, the *cooling* problem is particularly serious in large turboalternators, where economy, mechanical requirements, transportation limitations, and erection and assembly problems demand compactness, especially for the rotor forging. *Direct-conductor cooling* methods, using hydrogen or water as coolant, are employed for both the stator and the rotor of large

DIODE ASSEMBLY

Photo 8.1.3 Brushless exciter assembly of a two-pole turbogenerator. (Photo courtesy of Electric Machinery Manufacturing Company.)

turbogenerators. Such innovative cooling techniques have made it possible to achieve two-pole turbo-generator ratings of 2,000 MVA in a single unit. Superconductive cryoturbogenerators in large sizes with liquid-helium-cooled superconducting field windings should be realizable soon through research and development.[2]

Investigations are continually seeking through better mechanical as well as electromagnetic design for mechanical and electrical stability of the machine. Increased unit ratings bring difficult technological problems, not only in the active part of the generator but also in other parts, particularly the end region. Innovative end-region configurations are being used to overcome the associated problems. New concepts, research, development, testing, and design studies continue to open up possibilities permitting further increases in unit ratings with high reliability at reasonable cost.

8.2 Equivalent Circuit of a Synchronous Machine

A review of the material presented under "Elementary Synchronous Machines" in Section 5.3 is very helpful at this stage to recall the principles of operation for synchronous machines.

To investigate the equivalent circuit of a synchronous machine, for the sake of simplicity, let us begin by considering an unsaturated cylindrical-rotor synchronous machine. Effects of saliency and magnetic saturation are considered later. Since we are concerned for the present with only the steady-state behavior of the machine, circuit parameters of the field and damper windings need not be considered. The effect

---

[2] K. Abegg, "The Growth of Turbogenerators," Philosophical Transactions of the Royal Society of London, vol. 275, no. 1248, pp. 51-68, 1973.

of the field winding, however, is taken care of by the flux produced by the dc field excitation and the ac voltage generated by the field flux in the armature circuit. As for the armature winding, under balanced conditions of operation, we will analyze it on a *per-phase* basis. The armature winding obviously has a resistance; the *effective resistance, R_a*, of the armature per phase, which takes into account the effects of operating temperature and alternating currents (causing skin effect), is about 1.6 times the dc resistance measured at room temperature. The leakage reactance associated with the armature winding is caused by the leakage fluxes (caused in turn by the currents in the conductors) that link the armature conductors but not the field winding. The leakage reactance can, for convenience, be divided into such components as end-connection leakage reactance and slot-leakage reactance. Let the total *leakage reactance* per phase of the armature winding be denoted by X_l, which includes the effects of the leakage across the armature slots and around the coil ends as well as those leakages associated with the space-harmonic fields present in the actual armature-mmf wave.

The resultant air-gap flux, $\bar{\phi}_r$, in the machine can be considered as the space-phasor sum of the component fluxes, $\bar{\phi}_f$ and $\bar{\phi}_{ar}$, produced respectively by the field and armature-reaction mmfs, which are caused by the field and armature currents, respectively. From the viewpoint of the armature windings, these fluxes manifest themselves as generated emfs. Thus, let $\bar{E}_f$ be the ac voltage (known as the *excitation voltage)* generated by the field flux, and $\bar{E}_{ar}$ be the voltage generated by the armature-reaction flux. These voltages are proportional to the field and armature currents respectively, each one lagging the flux generating it by 90°. Note that the armature-reaction flux $\bar{\phi}_{ar}$ is in phase with the armature current $\bar{I}_a$ producing it, and hence the armature-reaction emf $\bar{E}_{ar}$ lags the armature current by 90°. The resultant air-gap voltage $\bar{E}_r$ is then given by the phasor sum of the voltages $\bar{E}_f$ and $\bar{E}_{ar}$. Observe that the voltage $\bar{E}_r$ is then lagging behind the flux $\bar{\phi}_r$ by 90°. These combined space and time phasor relationships are shown in Figure 8.2.1 with rms values of the quantities involved, for a generator and a motor, with armature current lagging behind the resultant air-gap voltage. The effect of armature reaction can be considered to be that of an inductive reactance X_ϕ, accounting for the component voltage generated by the space-fundamental flux, which in turn is created by the armature reaction. This reactance is known as the *magnetizing reactance* or *armature-reaction reactance*. The following may then be written relating the voltage phasors:

$$\bar{E}_f + \bar{E}_{ar} = \bar{E}_f - j\bar{I}_a X_\phi = \bar{E}_r \qquad (8.2.1)$$

The air-gap voltage $\bar{E}_r$ differs from the terminal voltage $\bar{V}_t$ by the voltage drops due to the armature resistance and leakage reactance. Thus, Figure 8.2.2 shows the equivalent circuits, with all per-phase quantities, for a cylindrical-rotor synchronous machine working as either a generator or a motor. The sum of the armature leakage reactance and armature-reaction reactance is known as the *synchronous reactance:*

$$X_s = X_\phi + X_l \qquad (8.2.2)$$

and $(R_a + jX_s)$ is called the *synchronous impedance* Z_s. Thus, the equivalent circuit for an unsaturated cylindrical-rotor synchronous machine under balanced polyphase conditions reduces to that shown in Figure 8.2.2b, in which the machine is represented on a per-phase basis by its excitation voltage $\bar{E}_f$ in series with the synchronous impedance. The excitation voltage takes into account the flux produced by the field current, while the synchronous reactance takes into account all the flux produced by the balanced polyphase armature currents. For an unsaturated cylindrical rotor machine at constant frequency, note that the synchronous reactance is a constant, and the excitation voltage is proportional to the field current. Figure 8.2.3 shows the corresponding combined space and time phasor diagrams of a cylindrical-rotor

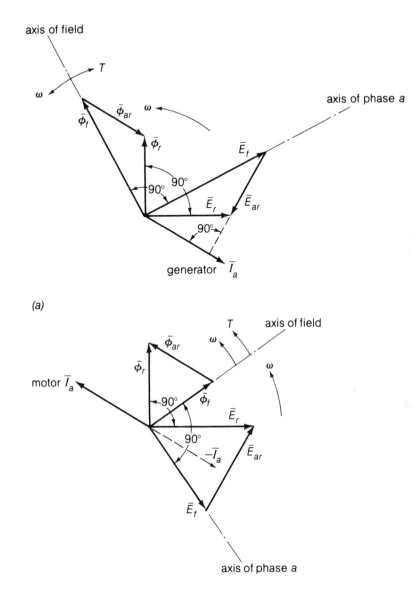

(a)

(b)

Figure 8.2.1 Phasor relationships among the fluxes and the corresponding voltages with the armature current lagging the resultant air-gap voltage. (a) Generator. (Field poles lead the resultant air-gap flux wave. The electromagnetic torque on the rotor acts in opposition to the rotation.) (b) Motor. (Field poles lag the resultant air-gap flux wave. The electromagnetic torque acts in the direction of rotation.)

poles lead the resultant air-gap flux wave. The electromagnetic torque on the rotor acts in opposition to the rotation.) (b) Motor. (Field poles lag the resultant air-gap flux wave. The electromagnetic torque acts in the direction of rotation.)

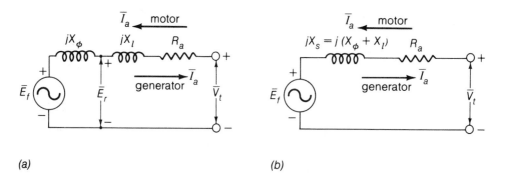

Figure 8.2.2 Per-phase equivalent circuits of a cylindrical-rotor synchronous machine.

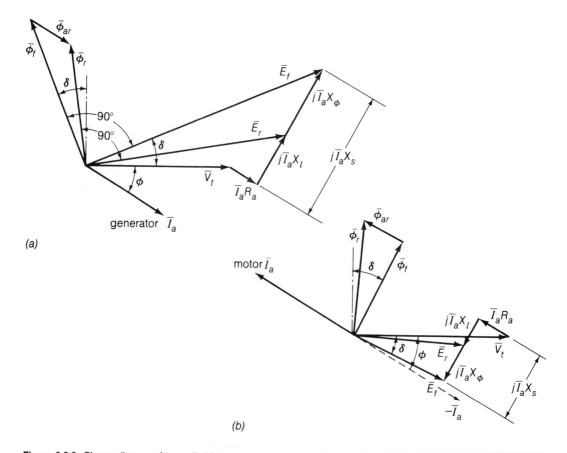

(a)

(b)

Figure 8.2.3 Phasor diagrams for a cylindrical-rotor synchronous machine with a lagging power factor, with the armature current lagging behind the terminal voltage. (a) Generator. (b) Motor.

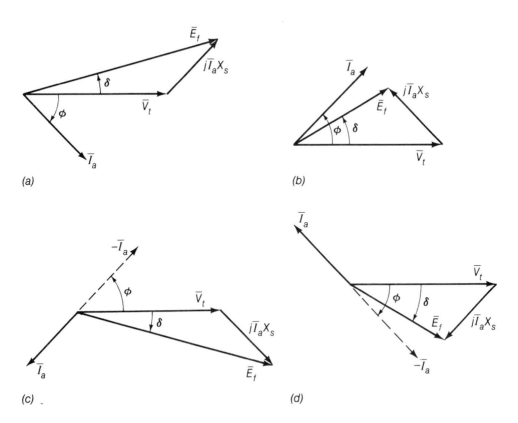

Figure 8.2.4 Four possible cases of operation of a round-rotor synchronous machine, with negligible arma-ture resistance. (a) Overexcited generator (power factor lagging) $P > 0$; $Q > 0$; $\delta > 0$; (b) Underexcited generator (power factor leading) $P > 0$; $Q < 0$; $\delta > 0$; (c) Overexcited motor (power factor leading) $P < 0$; $Q > 0$; $\delta < 0$; (d) Underexcited motor (power factor lagging) $P < 0$; $Q < 0$; $\delta < 0$.

synchronous machine working as either a generator or a motor, for the case of a lagging power factor, with the armature current lagging behind the terminal voltage by an angle ϕ.

In all but small machines, the armature resistance is usually neglected except for its effect on losses (and hence the efficiency) and heating. With this simplification, Figure 8.2.4 shows the four possible cases of operation of a round-rotor synchronous machine, in which the following relation holds good:

$$\bar{V}_t + j\bar{I}_a X_s = \bar{E}_f \tag{8.2.3}$$

Observe that the motor armature current is taken in the direction opposite to that of the generator. The machine is said to be *overexcited* when the magnitude of the excitation voltage exceeds that of the terminal voltage; otherwise, it is said to be *underexcited*. The angle δ between the excitation voltage $\bar{E}_f$ and the terminal voltage $\bar{V}_t$ is known as the *torque angle* or *power angle* of the synchronous machine. The power-angle performance characteristics are discussed in Section 8.4.

The voltage regulation of a synchronous generator at a given load, power factor, and rated speed is defined as follows:

$$\text{Per-unit voltage regulation} = \frac{E_f - V_t}{V_t} \tag{8.2.4}$$

where V_t is the terminal voltage on load, and E_f is the no-load terminal voltage at rated speed when the load is removed without changing the field current.

||||||||||||━━━━━━━━━━━━━━━━━━━━━━━━━━━
EXAMPLE 8.2.1

The per-phase synchronous reactance of a three-phase, wye-connected, 2.5-MVA, 6.6-kV, 60-Hz turboalternator is 10 ohms. Neglect the armature resistance and saturation. Calculate the voltage regulation when the generator is operating at full load (a) with 0.8 power-factor lagging, and (b) with 0.8 power-factor leading.

Solution

Per-phase terminal voltage $V_t = \dfrac{6.6 \times 1,000}{\sqrt{3}} = 3,811$ V

Full-load per-phase armature current $I_a = \dfrac{2.5 \times 10^6}{\sqrt{3} \times 6.6 \times 1,000} = 218.7$ A

a. Referring to Figure 8.2.4a for an overexcited generator operating at 0.8 power-factor lagging, and applying Equation 8.2.3, we have

$$\bar{E}_f = 3,811 + j218.7\,(0.8 - j0.6)\,(10) = 5,414 \angle \tan^{-1} 0.3415$$

Per-unit voltage regulation $= \dfrac{5,414 - 3,811}{3,811} = .042$

b. Referring to Figure 8.2.4b for an underexcited generator operating at 0.8 power-factor leading, and applying Equation 8.2.3, we have

$$\bar{E}_f = 3,811 + j218.7\,(0.8 + j0.6)\,(10) = 3,050 \angle \tan^{-1} 0.7$$

Per-unit voltage regulation $= \dfrac{3,050 - 3,811}{3,811} = -0.2$

As we have just seen, the voltage regulation for a synchronous generator can become negative.

━━━━━━━━━━━━━━━━━━━━━━━━━━━||||||||||||
8.3 Open-Circuit and Short-Circuit Characteristics

To include saturation effects and determine the appropriate machine parameters, let us consider two basic characteristic curves corresponding to open-circuit and short-circuit for a synchronous machine. The *open-circuit characteristic* (OCC or the *saturation curve*) is a plot of the armature terminal voltage (usually line-to-line) on open-circuit as a function of the field-excitation current, when the machine is running at rated synchronous speed. See Figure 8.3.1a for a diagram of connections for the open-circuit test. The curve, indicated as OCC in Figure 8.3.1b, would be linear, as represented by the air-gap line, if it were not for the magnetic saturation of the iron. The OCC is often plotted in per-unit terms, as shown by the alternate set of scales in Figure 8.3.1b, where one per-unit voltage is the rated voltage

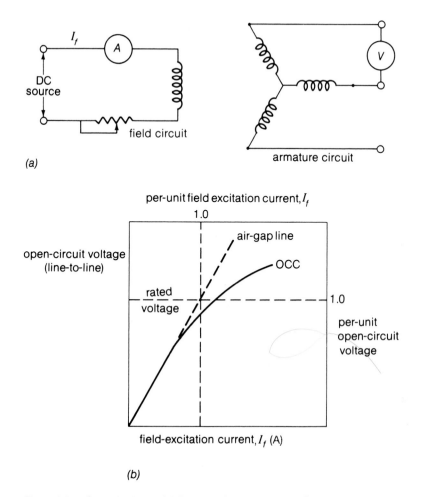

Figure 8.3.1 Open-circuit test. (a) Diagram of connections. (b) Open-circuit characteristic (OCC).

and one per-unit field current is the excitation corresponding to the rated voltage on the air-gap line. On the basis of the present choice, the per-unit representation is such as to make the air-gap lines of all synchronous machines identical. Essentially, the OCC gives the relation between the space-fundamental component of the air-gap flux and the mmf of the magnetic circuit, when the field winding is the only mmf source.

The open-circuit characteristic is usually determined experimentally (as in Figure 8.3.1a) by driving the synchronous machine mechanically at synchronous speed with its armature terminals on open-circuit and reading the terminal voltage corresponding to various values of dc excitation current. The voltage can be stepped down by means of instrument potential transformers for measurement. From the no-load test, we determine the no-load rotational losses (friction, windage, and core losses). The friction and windage losses at synchronous speed are constant, while the core loss is a function of the flux in the machine at no-load, which in turn is proportional to the open-circuit voltage.

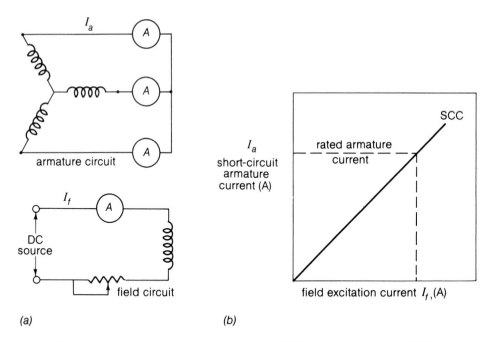

Figure 8.3.2 Short-circuit test. (a) Diagram of connections. (b) Short-circuit characteristic (SCC).

For the short-circuit test, the armature terminals are short-circuited through a current-measuring circuit, which can be an instrument current transformer with an ammeter in its secondary. See Figure 8.3.2a for a diagram of connections; current transformers are not shown. By driving the machine at rated synchronous speed, measurements of armature short-circuit current are made for different values of dc field excitation current; the field current is gradually increased until the armature current is about 1.5 to 2 times the rated current. The plot of short-circuit armature current versus the field current is known as the *short-circuit characteristic,* indicated by SCC in Figure 8.3.2b. It is practically linear because the magnetic field iron is unsaturated. By drawing the phasor diagram (which is left as an exercise for the student) corresponding to the short-circuit conditions, it can be seen that the armature current $\bar{I}_a$ lags the excitation voltage $\bar{E}_f$ by very nearly 90° (because the resistance is much smaller than the synchronous reactance). Consequently the magnetic axes of the armature reaction and field are very nearly in line, but with the field and armature mmfs nearly opposing each other. The resultant air-gap flux is only about 0.15 of its normal-voltage value. Hence, the machine is operating in an unsaturated condition.

By representing the open-circuit and short-circuit characteristics on the same graph, as in Figure 8.3.3, the *unsaturated synchronous impedance* per phase is obtained in this way:

$$Z_{s(ag)} = \frac{ot}{\sqrt{3}\,o'd} \ \Omega \text{ per phase} \tag{8.3.1}$$

where the subscript (ag) indicates air-gap line conditions, ot is the line-to-line terminal voltage (in volts) of the wye-connected synchronous machine, and $o'd$ is the short-circuit armature line current in amperes corresponding to the dc field excitation current oa. By expressing the voltage and current per unit, we can express the synchronous impedance per unit.

For operation at or near rated terminal voltage, it is sometimes assumed that the machine is equivalent to an unsaturated one for which the saturation characteristic is a straight line through the origin and the

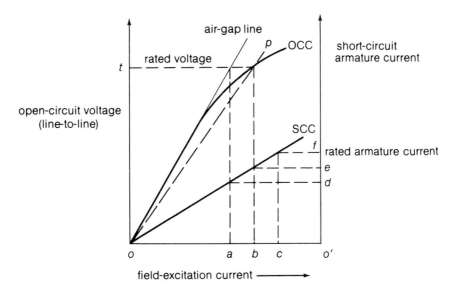

Figure 8.3.3 Open-circuit and short-circuit characteristics of a synchronous machine.

rated voltage point on the OCC, as shown by the dashed line *op* in Figure 8.3.3. In this approximation, the *saturated synchronous impedance* at rated voltage is given by

$$Z_s = \frac{ot}{\sqrt{3}\, o'e} \ \Omega \text{ per phase} \tag{8.3.2}$$

This method of handling the effects of saturation usually gives fairly satisfactory results if great accuracy is not required. Since the armature resistance is usually much smaller than the synchronous reactance, the value of the synchronous impedance is often taken as the synchronous reactance itself.

The *short-circuit ratio* is defined as

$$\text{SCR} = \frac{ob}{oc} \tag{8.3.3}$$

where *ob* is the field current required to generate rated voltage at rated speed on open-circuit, and *oc* is the field current required to produce rated armature current under a sustained three-phase short circuit. It can be shown that the short-circuit ratio is the reciprocal of the per-unit value of the saturated synchronous impedance at rated voltage. Saturation has the effect of increasing the short-circuit ratio.

EXAMPLE 8.3.1

The following data are obtained from the open-circuit and short-circuit characteristics of a three-phase, wye-connected, 135-MVA, 13.8-kV, 60-Hz hydrogen-cooled turbine generator with negligible armature resistance:

From the open-circuit characteristic: Line-to-line voltage 13.8 kV 17.3 kV
　　　　　　　　　　　　　　　　　Field current　　　　550 A　　1,208 A
From the air-gap line: Line-to-line voltage 13.8 kV
　　　　　　　　　　　Field current　　　　480 A

From the short-circuit characteristic: Armature current   3,900 A   5,648 A
                                                                      Field current   550 A   797 A

Compute (a) the unsaturated synchronous reactance, (b) the saturated synchronous reactance at rated voltage, (c) the short-circuit ratio, (d) the estimated field current for rated voltage and rated current at 0.8 power-factor lagging, and (e) the voltage regulation.

Solution

a. The field current of 480 A required for rated line-to-line voltage of 13.8 kV on the air-gap line produces a short-circuit armature line current of $480 \times 3,900/550 = 3,404$ A. Hence, the unsaturated synchronous reactance is given by

$$X_{s(ag)} = Z_{s(ag)} = \frac{13.8 \times 1,000}{\sqrt{3} \times 3,404} = 2.34 \ \Omega \text{ per phase}$$

In per-unit terms, with the machine rating as a base, since the rated armature current is $135 \times 10^6 / (\sqrt{3} \times 13,800) = 5,648$ A and $3,404$ A $= 3,404/5,648 = 0.6$ per unit, then,

$$X_{s(ag)} = Z_{s(ag)} = \frac{1}{0.6} = 1.67 \text{ per unit}$$

b. The field current of 550 A produces rated voltage on the open-circuit characteristic and a short-circuit armature current of 3,900 A, from which the saturated synchronous reactance at rated voltage is given by

$$X_s = Z_s = \frac{13.8 \times 1,000}{\sqrt{3} \times 3,900} = 2.04 \ \Omega \text{ per phase}$$

In per-unit terms, since 3,900 A $= \frac{3,900}{5,648} = 0.69$ per unit, then

$$X_s = Z_s = \frac{1}{0.69} = 1.45 \text{ per unit}$$

c. The short-circuit ratio as given by Equation 8.3.3 is

$$\text{SCR} = \frac{550}{797} = 0.69$$

which is the same as the reciprocal of the per-unit saturated synchronous reactance at rated voltage.

d. We shall assume that the machine is equivalent to an unsaturated one whose saturation characteristic is a straight line through the origin and the rated voltage point on the OCC, as shown by the dashed line *op* in Figure 8.3.3. Applying Equation 8.3.3, we get

$$\bar{E}_f = \frac{13,800}{\sqrt{3}} + j(5,648)(0.8 - j0.6)2.04$$

$$= 7,968 + 6,913 + j9,218 = 14,881 + j9,218 = 17,505\angle\tan^{-1}0.62$$

and the line-to-line magnitude of the induced voltage is

$$\sqrt{3} \times 17,505 = 30,319 \text{ V}$$

The field current required to produce this voltage on the line *op* in Figure 8.3.3 is

$$\frac{30,319 \times 550}{13,800} = 1,208 \text{ A}$$

e. The field current of 1,208 A produces a no-load line-to-line voltage of 17,300 V on the OCC. Hence

$$\text{Per-unit regulation} = \frac{17,300 - 13,800}{13,800} = 0.254$$

8.4 Power-Angle and Other Performance Characteristics

The real and reactive power delivered by a synchronous generator or received by a synchronous motor can be expressed in terms of the terminal voltage V_t, generated voltage E_f, synchronous impedance Z_s, and the power angle or torque angle δ. Referring to Figure 8.2.4, it is convenient to adopt a convention that makes positive the real power P and the reactive power Q delivered by an overexcited generator. Accordingly, the generator action corresponds to positive values of δ, while the motor action corresponds to negative values of δ. This notation is equivalent to omitting the negative sign in Equation 5.3.20, with the understanding, of course, that the electromagnetic torque acts in the direction to bring the interacting fields into alignment. With the adopted notation, it follows that $P > 0$ for generator operation, while $P < 0$ for motor operation. Further, positive Q means delivering inductive vars for a generator action or receiving inductive vars for a motor action; negative Q means delivering capacitive vars for a generator action or receiving capacitive vars for a motor action. It can be observed from Figure 8.2.4 that the power factor is lagging when P and Q have the same sign, and leading when P and Q have opposite signs.

The complex power output of the generator in volt-amperes per phase is given by

$$S = P + jQ = \bar{V}_t \bar{I}_a^* \tag{8.4.1}$$

where $\bar{V}_t$ is the terminal voltage per phase, $\bar{I}_a$ is the armature current per phase, and the (*) indicates a complex conjugate. Referring to Figure 8.2.4a, in which the effect of armature resistance has been neglected, and taking the terminal voltage as reference, we have the terminal voltage,

$$\bar{V}_t = V_t + j0 \tag{8.4.2}$$

the excitation voltage or the generated voltage,

$$\bar{E}_f = E_f (\cos \delta + j \sin \delta) \tag{8.4.3}$$

and the armature current,

$$\bar{I}_a = \frac{\bar{E}_f - \bar{V}_t}{jX_s} = \frac{(E_f \cos \delta - V_t) + jE_f \sin \delta}{jX_s} \tag{8.4.4}$$

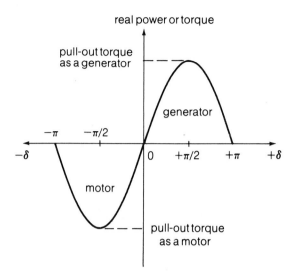

Figure 8.4.1 Steady-state power-angle or torque-angle characteristic of a cylindrical-rotor synchronous machine (with negligible armature resistance).

where X_s is the synchronous reactance per phase.

$$\hat{I}_a^* = \frac{(E_f \cos \delta - V_t) - jE_f \sin \delta}{-jX_s} = \frac{E_f \sin \delta}{X_s} + j\frac{E_f \cos \delta - V_t}{X_s} \qquad (8.4.5)$$

Therefore

$$P = \frac{V_t E_f \sin \delta}{X_s} \qquad (8.4.6)$$

and

$$Q = \frac{V_t E_f \cos \delta - V_t^2}{X_s} \qquad (8.4.7)$$

Equations 8.4.6 and 8.4.7 hold good for a cylindrical-rotor synchronous generator with negligible armature resistance. To obtain the total power for a three-phase generator, Equations 8.4.6 and 8.4.7 should be multiplied by 3 when the voltages are line-to-neutral. If the line-to-line magnitudes are used for the voltages, however, these equations give the total three-phase power.

The power-angle or torque-angle characteristic of a cylindrical-rotor synchronous machine is shown in Figure 8.4.1, neglecting the effect of armature resistance. The maximum real power output per phase of the generator for a given terminal voltage and a given excitation voltage is

$$P_{max} = \frac{V_t E_f}{X_s} \qquad (8.4.8)$$

Any further increase in the prime-mover input to the generator causes the real-power output to decrease; the excess power goes into accelerating the generator, thereby increasing its speed and causing it to pull

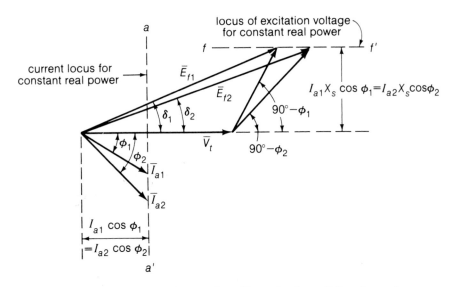

Figure 8.4.2 Phasor diagram showing the effect of increasing the excitation of a synchronous generator, when the real power, frequency, and terminal voltage are constant.

out of synchronism. Hence the *steady-state stability limit* is reached when $\delta = \pi/2$. For normal steady operating conditions, the power angle or torque angle is well under 90°. The maximum torque or *pull-out torque* per phase that a round-rotor synchronous motor can develop for a *gradually applied* load is

$$T_{\max} = \frac{P_{\max}}{\omega_m} = \frac{P_{\max}}{2\pi \left(n_s/60\right)} \tag{8.4.9}$$

where n_s is the synchronous speed in rpm. The pull-out torque of a synchronous motor, according to the American Standard Definitions of Electrical Terms, is the maximum sustained torque that the motor will develop at synchronous speed for one minute, with rated voltage applied at rated frequency and with normal excitation.

In the steady-state theory of the synchronous machine, with known terminal bus voltage V_t and a given synchronous reactance X_s, the six operating variables are given by P, Q, δ, ϕ, I_a, and E_f. We can write only four independent equations relating these six variables; hence the synchronous machine is said to have two degrees of freedom. The selection of any two, such as ϕ and I_a, P and Q, or δ and E_f, determines the operating point and establishes the other four quantities.

The principal steady-state operating characteristics are the interrelations among the terminal voltage, field current, armature current, real power, reactive power, torque angle, power factor, and efficiency. These characteristics can be computed for application studies by means of phasor diagrams, such as those shown in Figure 8.2.4, corresponding to various conditions of operation. Let us consider, for example, the loci of the excitation voltage and armature current for constant real power and variable excitation of a synchronous generator operating at constant terminal voltage and frequency. Neglecting armature resistance and for a given synchronous reactance, Figure 8.4.2 shows the phasor diagram corresponding to two different values of excitation when the real power, frequency, and terminal voltage are constant. For the real-power output of the generator, $P = V_t I_a \cos \phi$, to be a constant, and with constant terminal voltage V_t, the term $\left(I_a \cos \phi\right)$ must be a constant. Thus the locus of the current for constant real power

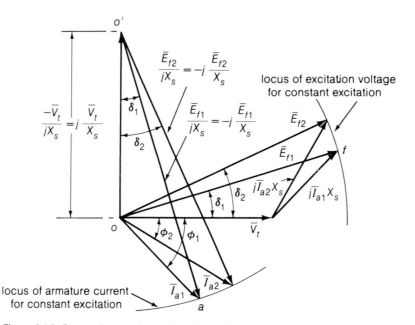

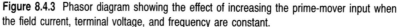

Figure 8.4.3 Phasor diagram showing the effect of increasing the prime-mover input when the field current, terminal voltage, and frequency are constant.

is given by the vertical line aa', while the locus of the excitation voltage is given by the horizontal line ff'. Observe that the variation in the power-factor angle ϕ is quite significant, whereas the variation in the torque angle δ is only slight.

Next let us consider the loci of the excitation voltage and armature current for constant excitation and variable real-power output of a synchronous generator operating at constant terminal voltage and frequency. Figure 8.4.3 shows the phasor diagram corresponding to two different values of the prime-mover input when the field current, terminal voltage, and frequency are constant. The locus of excitation voltage for constant excitation is the circle with o as center and the line-segment of as radius, while the locus of the armature current is given by the circle with o' as center and the line-segment $o'a$ as radius, because, neglecting armature resistance,

$$\bar{V}_t + j\bar{I}_a X_s = \bar{E}_f \tag{8.4.10}$$

or

$$\bar{I}_a = \frac{\bar{E}_f}{jX_s} - \frac{\bar{V}_t}{jX_s} = j\frac{\bar{V}_t}{X_s} - j\frac{\bar{E}_f}{X_s} \tag{8.4.11}$$

in which V_t and X_s are constant, and E_f is also a constant for constant excitation. Observe that an increase in the real-power output $(V_t I_a \cos\phi)$ obtained by an increase in the prime-mover input is associated with a decrease in the reactive-power output $(V_t I_a \sin\phi)$.

A *generator-compounding curve* is a curve showing the field current required to maintain rated terminal voltage as the constant-power-factor load is varied. Three compounding curves at constant power factors of unity, 0.8 power-factor lagging, and 0.8 power-factor leading are shown in Figure 8.4.4.

For synchronous-motor operation, the power factor as well as the armature current can be controlled by adjusting the field excitation. A *synchronous-motor V-curve* is a curve showing the relation between

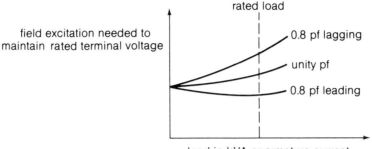

Figure 8.4.4 Generator-compounding curves.

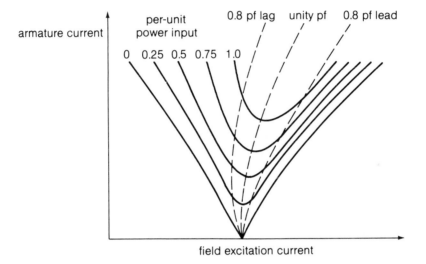

Figure 8.4.5 Synchronous motor V-curves.

the armature current and field current at a constant terminal voltage and with a constant shaft load. Figure 8.4.5 shows a family of V-curves, named because of their characteristic V-shape. Observe that, for constant power output, the armature current is a minimum at unity power factor and increases as power factor decreases. The loci of constant power factors are shown by the dashed lines, known as *synchronous-motor compounding curves*. When the effect of armature resistance is neglected, the motor and generator compounding curves are identical except that the lagging and leading-power-factor curves are interchanged.

A *synchronous condenser* is a synchronous motor running idle with an overexcited field, which can be used in a power system solely for power-factor correction or for control of reactive power flow. Such a machine in a larger size can be more economical than static capacitors. The V-curve for zero mechanical power in Figure 8.4.5 approximately represents the synchronous-condenser operation. When rotational losses and armature resistance are neglected, the real power of a synchronous condenser is zero with an associated zero torque angle, and the reactive-power output is given by Equation 8.4.7 with cos δ equal to unity.

An electrical system supplied by only one synchronous generator is known as an *isolated system*. For this system, the concept of an infinite bus having constant voltage and constant frequency does not

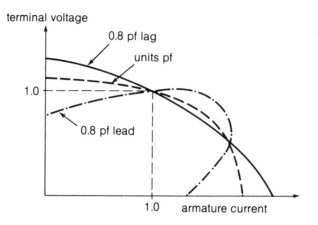

Figure 8.4.6 Synchronous-generator constant-field-current volt-ampere characteristic curves.

apply because there are no other parallel synchronous generators to compensate for changes in field excitation and in prime-mover output in order to maintain constant terminal voltage and frequency. If the generator is driven at a constant speed, thereby maintaining constant frequency, and the field current is increased, the terminal voltage increases; this effort is associated in general with an increase in the real and reactive-power output to an isolated system. Similarly, an increase in prime-mover output at constant field excitation generally produces an increase in frequency, terminal voltage, and real as well as reactive power. Figure 8.4.6 shows the *generator constant-field-current volt-ampere characteristics* for three constant power factors and three values of constant field current. In each case, the field current chosen is the value required to yield rated terminal voltage at rated armature current, as given by the compounding curves of Figure 8.4.4.

Synchronous generators are usually rated in terms of maximum kVA or MVA load at a specific voltage and power factor, such as 0.8, 0.85, or 0.9 lagging, that they can deliver continuously without overheating. Let us now take a look at the generator capabilities within which the machine can be operated, and the effects of field excitation on the individual machine as well as on the system of which that machine is a part. As mentioned earlier, a synchronous machine can be operated in any quadrant of Figure 8.4.7 in which point A is plotted corresponding to the name-plate-rated generator conditions for an assumed typical 0.85 power-factor lagging. Note that in Figure 8.4.7 the positive kilovars ($+Q$) are taken along the positive y-axis; hence the lagging-power-factor condition corresponding to an overexcited generator is shown in the upper right quadrant. We will limit our discussion primarily to generator operation; therefore we must restrict our attention to the right or positive-kilowatt side ($+P$) of Figure 8.4.7. Because it is usual for generators to be suitable for delivering rated kilovolt-amperes at unity power factor (corresponding to point B in Figure 8.4.7), we can draw the arc AB with its center at zero and its radius equal to the rated armature current.

Starting with rated terminal voltage $\bar{V}_t$ and rated armature current $\bar{I}_a$, Figure 8.4.8 gives a typical phasor diagram corresponding to a lagging power-factor angle of ϕ for an overexcited cylindrical-rotor generator with negligible armature resistance. $\bar{E}_f$ is then the internal (generated) voltage corresponding to the rated terminal conditions. It would also be equal to the terminal voltage if full load were removed without making any change in field current.

Dividing the three sides of the phasor triangle by X_s and reorienting the diagram to suit the kilowatt-kilovar coordinates of Figure 8.4.9, we can construct triangle OAC, in which OA represents the rated

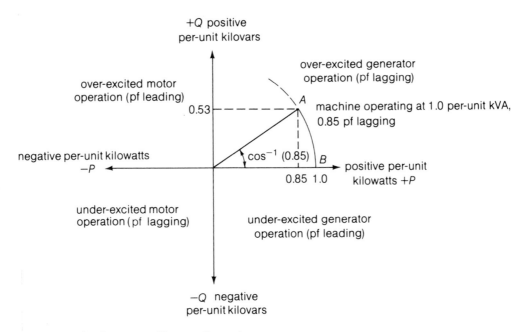

Figure 8.4.7 Synchronous machine operating modes.

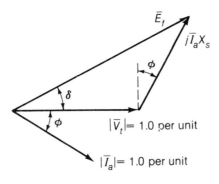

Figure 8.4.8 Typical phasor diagram of an overexcited round-rotor generator with negligible armature resistance.

armature current, OC corresponds to $(1/X_s)$ or the short-circuit ratio, and CA represents the rated full-load field current E_f/X_s. Then, with C as center and CA as radius, the arc AD is drawn, representing the locus of the rated field current, thereby closing off the top of the area within which the machine may be operated. Thus we see that the output of the synchronous generator is limited by *field heating* from D to A, and by *armature heating* from A to B. The vertical intercept OD gives the maximum permissible per-unit kilovars to scale, which is a measure of the capability as a synchronous condenser.

Corresponding to the underexcited generator operation in the fourth quadrant (lower left), because this is a region of low field current, we might be tempted to extend the armature current arc AB all the way around to E. But such an extension could be wrong from the viewpoints of system stability and

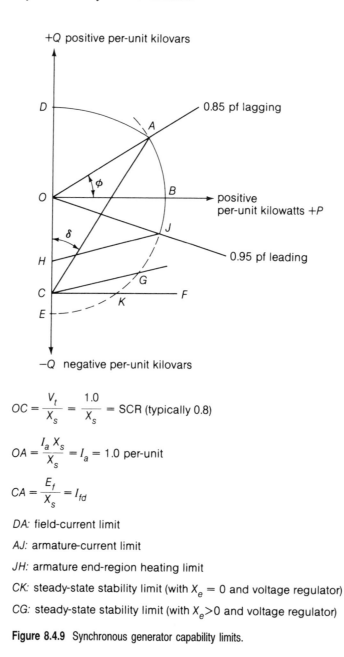

$$OC = \frac{V_t}{X_s} = \frac{1.0}{X_s} = \text{SCR (typically 0.8)}$$

$$OA = \frac{I_a X_s}{X_s} = I_a = 1.0 \text{ per-unit}$$

$$CA = \frac{E_f}{X_s} = I_{fd}$$

DA: field-current limit

AJ: armature-current limit

JH: armature end-region heating limit

CK: steady-state stability limit (with $X_e = 0$ and voltage regulator)

CG: steady-state stability limit (with $X_e > 0$ and voltage regulator)

Figure 8.4.9 Synchronous generator capability limits.

localized heating in the machine iron. Modern generators can be operated successfully in the underexcited region down to a line like *HJ* in Figure 8.4.9, where the point *H* is typically at 60% rated kVA with zero power-factor leading and the point *J* is typically at rated kVA with 0.95 power-factor leading. The line *HJ* then represents the *end-region heating limit,* which applies to steam-turbine generators in particular. Note that it is not a limitation of water-wheel generators because of their generally different construction. The problem of end-region heating stems from the armature reaction end-leakage flux at both ends of the stator.

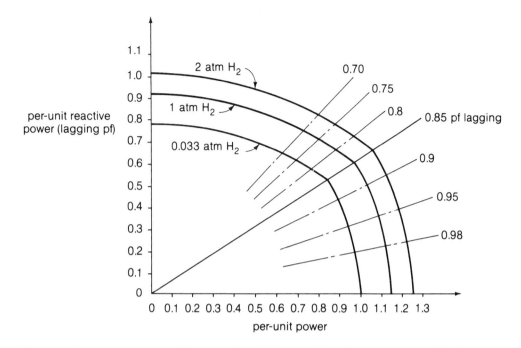

Figure 8.4.10 Reactive-power capability curves for a hydrogen-cooled turbine generator 0.85 pf lagging, 0.8 SCR, with rated kVA at 0.04 atm hydrogen as base kVA.

If the machine under consideration is connected to an infinitely large system through negligibly small impedance, then the *stability limit* may be represented by a straight line CF corresponding to a δ of 90°. Operation along the arc EK in Figure 8.4.9 is impossible because the machine will not remain in synchronism with the system, even if it did not exceed any heating limitation. Since the machine operates in practice through impedance representing transformers, lines, and the paralleled value of the impedances of all the other machines on the system, the resultant external impedance is typically about 0.2 to 0.4 per unit, based on the individual machine rating. The effect of this external impedance X_e is to bend the line CF upward to some position like CG. Thus we can establish the part of the boundary dictated by the steady-state stability. With no voltage regulator, the curve CG of Figure 8.4.9 will be bent even further upward.

The permissible generator-operating region is bounded by limits determined by the field heating, the armature end-region heating, and the steady-state stability. It can be seen from Figure 8.4.9 that an increase of the generator load, without corresponding adjustment of the field current, pushes the power factor of the machine toward the leading (underexcited) region, and an increase of excitation causes the generator power factor to lag more. The more the initial load power factor lags, the greater total load the machine can carry before running into one of the limits. Figure 8.4.10 shows a typical set of reactive-power capability curves for a large hydrogen-cooled turbine-generator, along with the effect of increased hydrogen pressure on allowable machine loadings.

The *efficiency* of a synchronous generator at a specified power output and power factor is determined by the ratio of the output to the input; the input power is given by adding the machine losses to the power output. The efficiency is conventionally computed in accordance with a set of rules agreed upon by the ANSI. Six losses are included in the computation:

- *Armature winding copper loss* for all the phases, calculated after correcting the dc resistance of each phase for an appropriate allowable temperature rise depending on the class of insulation used.
- *Field copper loss,* based on the field current and measured field-winding dc resistance, corrected for temperature in the same way armature resistance is corrected. Note that the losses in the field rheostats that are used to adjust the generated voltage are not charged to the synchronous machine.
- *Core loss,* which is read from the open-circuit core-loss curve at a voltage equal to the internal voltage behind the resistance of the machine.
- *Friction and windage loss.*
- *Stray load losses,* which account for the fact that the effective ac resistance of the armature is greater than the dc resistance because of the skin effect, and for the losses caused by the armature leakage flux.
- *Exciter loss,* but only if the exciter is an integral component of the alternator, i.e., shares a common shaft or is permanently coupled. Losses from a nonintegral exciter are not charged to the alternator.

EXAMPLE 8.4.1

A synchronous generator with negligible resistance and a synchronous reactance of 1.0 per unit is connected to an infinite bus of 1.0 per-unit voltage through two parallel three-phase transmission circuits, each having negligible resistance and a reactance of 0.6 per unit, including step-up and step-down transformers at each end, as shown in Figure 8.4.11. The generator rating is chosen as the base for expressing the per-unit values.

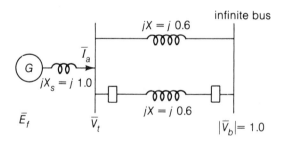

Figure 8.4.11

a. Under steady-state operation, compute the generator terminal voltage, the excitation voltage, the real-power output, and the reactive power delivered to the infinite bus, when the machine is delivering rated current at unity power factor at its terminals.

b. Let the throttle of the prime mover be adjusted so that no real power is transferred between the generator and the infinite bus. Determine the generator terminal and excitation voltages when the generator-field current is adjusted for delivering 0.5 per unit lagging reactive kVA to the infinite bus.

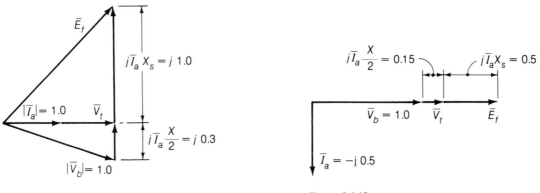

Figure 8.4.12   **Figure 8.4.13**

c. Let the system be returned to the operating conditions of part a, and let one of the two parallel transmission circuits be disconnected by tripping the circuit breakers at its ends, while the generator excitation is kept constant, as in part a. Calculate the maximum power transfer possible under these conditions, and check whether the generator will remain in synchronism.

Solution

a. With unity power factor at the generator terminals and a generator current I_a of 1.0 per unit, the phasor diagram is sketched in Figure 8.4.12.

   Note that the equivalent reactance of the two transmission circuits in parallel is 0.3 per unit.

   Generator terminal voltage $V_t = \sqrt{(1.0)^2 - (0.3)^2} = \sqrt{0.91} = 0.954$ per unit

   Generator excitation voltage $E_f = \sqrt{(0.954)^2 + (1.0)^2} = \sqrt{1.91} = 1.382$ per unit

   Real power output $= P = V_t I_a \cos \phi = 0.954$ per unit

Reactive power at infinite bus is 0.3 per unit, with the current $\tilde{I}_a$ leading the infinite-bus voltage $\overline{V}_b$.

b. The phasor diagram corresponding to the conditions stated in the problem is given in Figure 8.4.13.

c. When one of the transmission circuits is disconnected, the transmission-circuit reactance is 0.6 per unit, and the total reactance including the synchronous reactance is then 1.6 per unit. With E_f of 1.382 as in part a, the maximum power transfer possible is

$$P_{max} = \frac{E_f V_b}{(X + X_s)}$$

$$= \frac{(1.382)(1.0)}{1.6} = 0.864 \text{ per unit}$$

which is less than the original power of 0.954 in part a. The generator runs over the normal speed and loses synchronism unless a quick-response excitation system and governor action come into play.

|||||||||||

EXAMPLE 8.4.2

A 1,000-hp, 2,300-V, wye-connected, three-phase, 60-Hz, 20-pole synchronous motor, for which cylindrical-rotor theory can be used and all losses can be neglected, has a synchronous reactance of 5.00 ohms per phase.

a. The motor is operated from an infinite bus supplying rated voltage and rated frequency, and its field excitation is adjusted so that the power factor is unity when the shaft load is such as to require an input of 750 kW. Compute the maximum torque that the motor can deliver, given that the shaft load is slowly increased with the field excitation held constant.

b. Instead of the infinite bus as in part a, let the power to the motor be supplied by a 1,000-kVA, 2,300-V, wye-connected, three-phase, 60-Hz synchronous generator whose synchronous reactance is also 5.00 ohms per phase. The generator is driven at rated speed, and the field excitations of the generator and motor are adjusted so that the motor absorbs 750 kW at unity power factor and rated terminal voltage. If the field excitations of both machines are then held constant and the mechanical load on the synchronous motor is gradually increased, compute the maximum motor torque under the conditions. Also determine the armature current, terminal voltage, and power factor at the terminals corresponding to this maximum load.

c. Calculate the maximum motor torque if, instead of remaining constant as in part b, the field currents of the generator and motor are gradually increased so as to always maintain rated terminal voltage and unity power factor while the shaft load is increased.

Solution
The solution neglects reluctance torque because cylindrical-rotor theory is applied.

a. The equivalent circuit and the corresponding phasor diagram for the given conditions are shown in Figure 8.4.14, with subscript m attached to the motor quantities.

$$\text{Rated voltage per phase } V_t = \frac{2,300}{\sqrt{3}} = 1,328 \text{ V line-to-neutral}$$

$$\text{Current per phase } I_{am} = \frac{750 \times 10^3}{\sqrt{3} \times 2,300 \times 1.0} = 188.3 \text{ A}$$

$$I_{am}X_{sm} = 188.3 \times 5 = 941.5 \text{ V}$$

$$E_{fm} = \sqrt{(1,328)^2 + (941.5)^2} = 1,628 \text{ V}$$

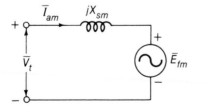
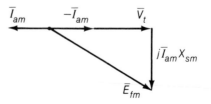

Figure 8.4.14

$$P_{max} = \frac{E_{fm}V_t}{X_{sm}} = \frac{(1,628)(1,328)}{5}$$

$$= 432,397 \text{ W per phase or } 1,297,191 \text{ W for three phases or } \frac{1,297.2}{1,000 \times 0.746} = 1.74 \text{ per unit}$$

$$\text{Synchronous speed} = \frac{120 \times 60}{20} = 360 \text{ rpm or 6 rps}$$

$$\omega_s = 2\pi \times 6 = 37.7 \text{ rad/s}$$

$$T_{max} = \frac{1,297.2 \times 10^3}{37.7} = 34,408 \text{ N·m}$$

b. With the synchronous generator as the power source, the equivalent circuit and the corresponding phasor diagram for the given conditions are shown in Figure 8.4.15, with subscript g attached to the generator quantities:

$$E_{fg} = E_{fm} = 1,628 \text{ V}$$

$$P_{max} = \frac{E_{fg}E_{fm}}{X_{sg} + X_{sm}} = \frac{(1,628)(1,628)}{10}$$

$$= 265 \text{ kW per phase or } 795 \text{ kW for three phases or } \frac{795}{746} = 1.07 \text{ per unit}$$

$$T_{max} = \frac{795 \times 10^3}{37.7} = 21,088 \text{ N.m}$$

If a load torque greater than this amount were applied to the motor shaft, synchronism would be lost; the motor would stall, the generator would tend to overspeed, and the circuit would be opened by circuit-breaker action.

Corresponding to the maximum load, the angle between $\bar{E}_{fg}$ and $\bar{E}_{fm}$ is 90°; from the phasor diagram, it follows that

$$V_t = \frac{E_{fg}}{\sqrt{2}} = \frac{1,628}{\sqrt{2}} = 1,151.3 \text{ V line-to-neutral or } 1,994 \text{ V line-to-line}$$

$$I_{ag}X_{sg} = 1,151.3 \quad \text{or} \quad I_{ag} = \frac{1,151.3}{5} = 230 \text{ A}$$

The power factor is unity at the terminals.

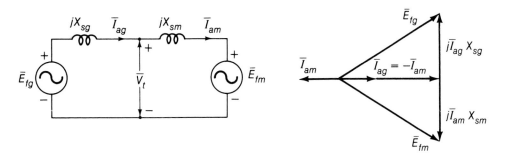

Figure 8.4.15

c. $V_t = 1,328$ V, and the angle between $\bar{E}_{fg}$ and $\bar{E}_{fm}$ is 90°. Hence it follows that

$$E_{fg} = E_{fm} = (1,328)\sqrt{2} = 1,878 \text{ V}$$

$$P_{max} = \frac{E_{fg}E_{fm}}{X_{sg} + X_{sm}} = \frac{(1,878)(1,878)}{10} = 352.7 \text{ kW per phase or } 1,058 \text{ kW for three phases}$$

$$T_{max} = \frac{1,058 \times 10^3}{37.7} = 28,064 \text{ N.m}$$

EXAMPLE 8.4.3

A three-phase, 350-kVA, 3,300-V, 60-Hz, six-pole, wye-connected synchronous machine, for which cylindrical rotor theory can be applied and all losses can be neglected, has a synchronous reactance of 10.0 ohms per phase.

a. Given that it operates as a generator delivering full load at rated voltage and with a lagging power factor of 0.9, construct the circle diagram and hence determine the excitation voltage and the load angle.
b. Compute the current and power factor when the machine is operating as a motor at maximum torque.
c. Determine the current and torque when the machine is operating as a motor at 0.6 power-factor leading.
d. With the excitation voltage adjusted to the same value as the terminal voltage, find the power factor when the machine is running as a generator delivering 80 A at rated voltage.

Solution

a. Full-load current $= \dfrac{350 \times 10^3}{\sqrt{3} \times 3,300} = 61.24$ A

$\dfrac{\text{voltage per phase}}{X_s} = \dfrac{3,300}{\sqrt{3}} \times \dfrac{1}{10} = 190.5$ A

$\cos^{-1} 0.9 = 25.9°$

In the diagram shown in Figure 8.4.16, following the construction of Figure 8.4.9, OC represents 190.5 A, while OA represents 61.24 A at an angle of 25.9° from the horizontal. The excitation voltage is proportional to the closing phasor CA:

$$226 \times 10 \times \sqrt{3} = 3,914.3 \text{ V}$$

The load angle δ is given by the angle OCA: 14°.

b. As seen from Figure 8.4.7, the motoring operation occurs in the second and third quadrants. The maximum torque corresponds to a torque angle of 90°. With C as center and CA as radius, draw a circle and let angle OCD be 90°, in which case OD gives the corresponding current of 293 A and the *lagging* power-factor angle is given by the angle $O'OD$ of 40°. The power factor is then 0.766 lagging.

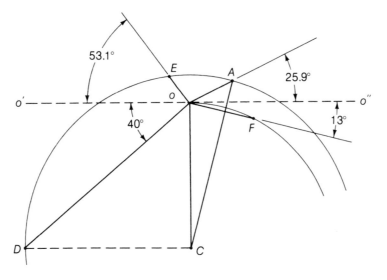

Figure 8.4.16

c. Draw OE at an angle of $\cos^{-1} 0.6 = 53.1°$ from the horizontal, as shown. The corresponding current is given by OE: 45 A. The power is then given by $\sqrt{3} \times 3{,}300 \times 45 \times 0.6 = 154{,}321$ watts. With six poles at 60 Hz, the synchronous speed is 1,200 rpm or 20 rps. Hence

$$T = \frac{154{,}321}{2\pi \times 20} = 1{,}227.55 \text{ N·m}$$

d. With C as center and CO as radius, construct a circle and draw OF equal to 80 A. The machine is operating as a generator with a *leading* power-factor angle $O'OF$ equal to 13°. The power factor is then 0.974 leading.

All the above answers can be calculated analytically; note that the graphical solution is quicker, although less accurate. The student is encouraged to solve the problem analytically and compare the results.

8.5 Effects of Saliency: Two-Reactance Theory of Salient-Pole Synchronous Machines

Because of saliency, the reactance measured at the terminals of a salient-pole synchronous machine as opposed to a cylindrical-rotor machine varies as a function of the rotor position. The effects of saliency are taken into account by the two-reactance theory proposed by Blondel and extended by Doherty, Nickle, and Park[3] and others. The armature current $\bar{I}_a$ is resolved into two components: $\bar{I}_d$ in

[3] A. Blondel, "The two-reaction method for study of oscillatory phenomena in coupled alternators," *Revue génerale de l'électricité,* vol. 13, pp. 235–51, 515–31, February/March 1923; Doherty, R. E., and C. A. Nickle, "Synchronous Machines I: An extension of Blondel's two-reaction theory," *AIEE Transactions,* vol. 45, pp. 927–42, 1926; Park, R. H., "Two-reaction Theory of Synchronous Machines–Part I," *AIEE Transactions,* vol 48, pp. 716–27, 1929.

time quadrature with, and $\bar{I}_q$ in time phase with the excitation voltage $\bar{E}_f$, as shown in Figure 8.5.1a for an unsaturated salient-pole generator operating at a lagging power factor. The component $\bar{I}_d$ of the armature current produces a space-fundamental component armature-reaction flux $\bar{\phi}_{ad}$ along the axes of the field poles, while the component $\bar{I}_q$ produces a space fundamental component armature-reaction flux $\bar{\phi}_{aq}$ in space quadrature with the field poles. Direct-axis mmfs act on the main magnetic circuit with their magnetic effect centered on the axes of the field poles, whereas the quadrature-axis quantities have their magnetic effect centered on the interpolar axes. The armature-reaction flux $\bar{\phi}_{ar}$, for the case of an unsaturated machine, is given by the space-phasor sum of the components $\bar{\phi}_{ad}$ and $\bar{\phi}_{aq}$. The resultant flux $\bar{\phi}_r$ in the machine is given by the space-phasor sum of $\bar{\phi}_{ar}$ plus the main field flux $\bar{\phi}_f$, as in Figure 8.5.1a. The angle ψ between $\bar{E}_f$ and $\bar{I}_a$ is known as the *internal power-factor angle,* and β is the space angle between the fundamentals of the ϕ_f and ϕ_r waves.

Similar to the magnetizing reactance X_ϕ of the cylindrical-rotor theory, the inductive effects of the direct-axis and quadrature-axis armature-reaction flux waves can be accounted for by the direct-axis and quadrature-axis magnetizing reactances $X_{\phi d}$ and $X_{\phi q}$, respectively. The *direct-axis and quadrature-axis synchronous reactances* are then given by

$$X_d = X_l + X_{\phi d} \tag{8.5.1}$$

$$X_q = X_l + X_{\phi q} \tag{8.5.2}$$

where X_l is the armature leakage reactance, assumed to be the same for both the direct-axis and quadrature-axis currents. Following the analysis presented in Section 8.2, we know that the excitation voltage $\bar{E}_f$ is equal to the phasor sum of the terminal voltage $\bar{V}_t$, the armature-resistance drop $\bar{I}_a R_a$, and the synchronous-reactance drops $\left(j\bar{I}_d X_d + j\bar{I}_q X_q \right)$:

$$\bar{E}_f = \bar{V}_t + \bar{I}_a R_a + j\bar{I}_d X_d + j\bar{I}_q X_q \tag{8.5.3}$$

We can see this relationship in Figure 8.5.1b, in which δ is the torque angle or the power angle, ϕ is the power-factor angle, and $(\phi + \delta)$ is the internal power factor angle ψ.

Because of the greater reluctance of the air gap in the quadrature axis, X_q is less than X_d, usually on the order of 0.6 to 0.7 times X_d. The general range of X_d is anywhere from 0.6 to 2.2 per unit, while X_q typically has the range of 0.4 to 1.4 per unit, with the machine kVA rating as the base. Even though turbogenerators are essentially cylindrical-rotor machines, a small salient-pole effect is present because the rotor slots have an effect on the quadrature-axis reluctance.

Once the internal power-factor angle ψ or $(\phi + \delta)$ is known, it is straightforward to draw Figure 8.5.1b by resolving the armature current into its d-axis and q-axis components. However, it is the external power-factor angle ϕ at the machine terminals that is usually explicitly known, rather than the internal power-factor angle. Neglecting the armature resistance, it can be shown that

$$\tan \delta = \frac{I_a X_q \cos \phi}{V_t + I_a X_q \sin \phi} \tag{8.5.4}$$

Figure 8.5.1c shows the construction of the diagram for finding $\bar{E}_f$ from the given values of V_t, I_a, ϕ, R_a, X_d, and X_q, while satisfying Equation 8.5.3. The phasor sum $\left(\bar{V}_t + \bar{I}_a R_a + j\bar{I}_a X_q \right)$ given by OD determines the angular position of the q-axis. Hence, the d-axis, perpendicular to the q-axis, can be

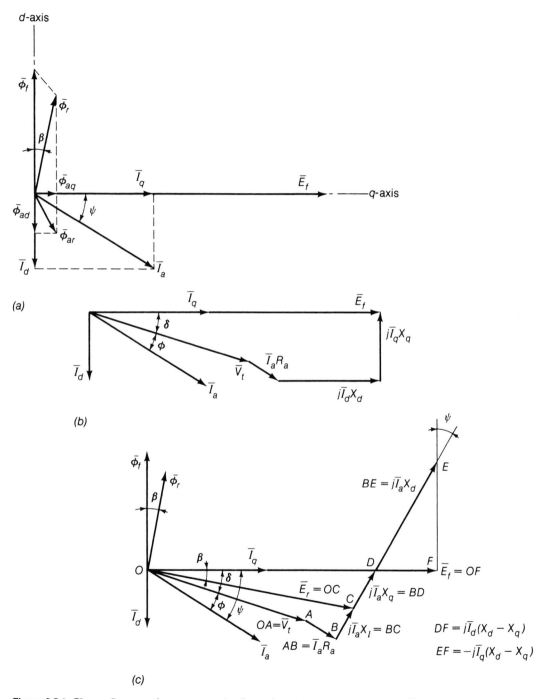

Figure 8.5.1 Phasor diagrams of an unsaturated salient-pole synchronous generator operating at a lagging power factor.

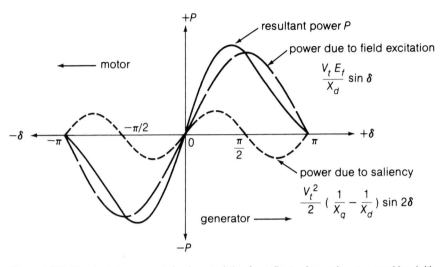

Figure 8.5.2 Steady-state power angle characteristic of a salient-pole synchronous machine (with negligible armature resistance).

located. By adding DF equal to $j\bar{I}_d\left(X_d - X_q\right)$ to OD, we can get the excitation voltage $\bar{E}_f$ for the chosen terminal conditions. Note that in Figure 8.5.1c, DE is an extension of BD, DF is an extension of OD, and EF is perpendicular to OF. Triangle DEF will vanish if $X_d = X_q$, which is the case for a cylindrical-rotor machine.

Let us next consider how the saliency affects the power-angle characteristic of a synchronous machine. Neglecting the armature resistance, the real power output or the developed power P is given by $V_t I_a \cos \phi$. From Figure 8.5.1b, it follows that

$$I_q X_q = V_t \sin \delta \tag{8.5.5}$$

$$I_d X_d = E_f - V_t \cos \delta \tag{8.5.6}$$

$$I_d = I_a \sin (\phi + \delta) \tag{8.5.7}$$

$$I_d = I_a \cos (\phi + \delta) \tag{8.5.8}$$

Solving for $\left(I_a \cos \phi\right)$, we obtain

$$I_a \cos \phi = \frac{E_f}{X_d} \sin \delta + \frac{V_t}{2X_q} \sin 2\delta - \frac{V_t}{2X_d} \sin 2\delta \tag{8.5.9}$$

and

$$P = V_t I_a \cos \phi = \frac{V_t E_f}{X_d} \sin \delta + \frac{1}{2} V_t^2 \left(\frac{1}{X_q} - \frac{1}{X_d} \right) \sin 2\delta \tag{8.5.10}$$

Figure 8.5.2 shows the power-angle characteristic of a salient-pole synchronous machine given by Equation 8.5.10. The resulting power has two terms: one due to field excitation, given by the first term of Equation 8.5.10, and the other due to saliency, given by the second term of Equation 8.5.10. If $X_d = X_q$, Equation 8.5.10 reduces to Equation 8.4.6. Because of the reluctance torque, for equal voltages and equal values of X_d, a salient-pole machine develops a given torque at a smaller value of δ, and the maximum torque that can be developed is somewhat greater. Hence a salient-pole machine is said to be *stiffer* than one with a cylindrical rotor. Equation 8.5.10 is applicable both for generator and motor operation, with δ being positive for the generator and negative for the motor.

EXAMPLE 8.5.1

The reactances X_d and X_q of a salient-pole synchronous generator are 1.00 and 0.60 per unit, respectively. The armature resistance is negligible. Compute the per-unit excitation voltage when the generator delivers rated kilovolt-amperes at 0.80 lagging power factor and rated terminal voltage.

Solution
First, the phase of $\bar{E}_f$ must be found so that $\bar{I}_a$ can be resolved into its direct-axis and quadrature-axis components. The phasor diagram is shown in Figure 8.5.3.

$$\bar{I}_a = 0.8 - j0.60 = 1.00\angle{-36.9°}$$

$$j\bar{I}_aX_q = j(0.80 - j0.60)(0.60) = 0.36 + j0.48$$

$$\bar{V}_t = \text{reference phasor} = 1.00 + j0$$

$$\bar{V}_t + j\bar{I}_aX_q = \bar{E}' = 1.36 + j0.48 = 1.44\angle{19.4°}$$

The angle $\delta = 19.4°$, and the phase angle between $\bar{E}_f$ and $\bar{I}_a$ is $36.9° + 19.4° = 56.3°$.

The armature current can now be resolved into its direct-axis and quadrature-axis components. Their magnitudes are

$$I_d = 1.00\sin 56.3° = 0.832 \; ; \qquad I_q = 1.00\cos 56.3° = 0.555$$

As phasors,

$$\bar{I}_d = 0.832\angle{-90°} + 19.4° = 0.832\angle{-70.6°}$$

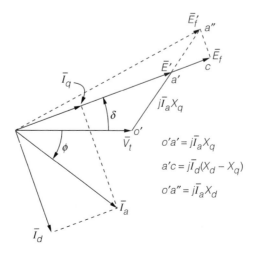

Figure 8.5.3 Generator phasor diagram

and

$$\bar{I}_q = 0.555\angle 19.4°$$

We can now find E_f by adding the length

$$a'c = I_d (X_d - X_q)$$

numerically to the magnitude of $\bar{E}'$; thus, the magnitude of the excitation voltage is the algebraic sum

$$E_f = E' + I_d (X_d - X_q) = 1.44 + 0.832\,(0.40) = 1.77 \text{ per unit}$$

As a phasor, $\bar{E}_f = 1.77\angle 19.4°$.

8.6 Determining Reactances by Test Data

Referring to Figure 8.5.1b, the terminal current $\bar{I}_a$ may be expressed as follows, while taking the terminal voltage $\bar{V}_t$ as reference:

$$\bar{I}_a = \frac{E_f - V_t \cos \delta}{X_d} e^{j[\delta-(\pi/2)]} + \frac{V_t \sin \delta}{X_q} e^{j\delta} \tag{8.6.1}$$

which may be rewritten as

$$\bar{I}_a = V_t \frac{X_d + X_q}{2X_d X_q} e^{j\pi/2} + V_t \frac{X_d - X_q}{2X_d X_q} e^{j(2\delta - \pi/2)} + \frac{E_f}{X_d} e^{j(\delta - \pi/2)} \tag{8.6.2}$$

Thus the salient-pole synchronous machine delivers three components of current to the bus. The second term on the right side of Equation 8.6.2 vanishes for a round-rotor synchronous machine, because X_d equals X_q. Even for no excitation (i.e., $E_f = 0$), as for a salient-pole synchronous machine, it can be seen that the current has an active component. With sufficient excitation, the current can be made equal to zero. For example, at no-load, for $\delta = 0$ and $E_f = V_t, I_a$ becomes zero. Further, it can be seen that, for $E_f = 0$,

$$I_a = I_{max} = \frac{V_t}{X_q} \quad \text{for } \delta = \frac{\pi}{2} \tag{8.6.3}$$

and

$$I_a = I_{min} = \frac{V_t}{X_d} \quad \text{for } \delta = 0 \tag{8.6.4}$$

The current wave is an amplitude-modulated wave, provided δ is varied slowly.

The foregoing analysis suggests a method of determining the direct-axis and quadrature-axis steady-state (or synchronous) reactances of a synchronous machine by a test known as the *slip test*. The machine is unexcited, and balanced voltages are applied at the armature terminals. The rotor is driven at a speed differing slightly from synchronous speed (which is easily calculated from the frequency of the applied voltages and the number of poles of the machine). The armature currents are then modulated at slip frequency by the machine, having maximum amplitude when the quadrature axis is in line with the mmf wave and minimum amplitude when the direct axis aligns with the mmf wave. The armature voltages are also usually modulated at slip frequency because of impedances in the supply lines, the amplitude

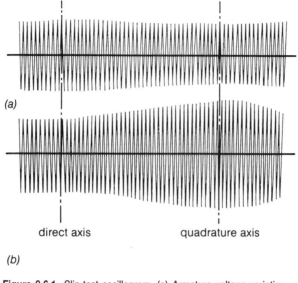

(a)

direct axis   quadrature axis

(b)

Figure 8.6.1 Slip-test oscillogram. (a) Armature voltage variation.
(b) Armature current variation.

being greatest when the current is smallest, and vice versa. Such variations of voltage and current are illustrated in the oscillograms of Figure 8.6.1. The maximum and minimum values of the voltage and current can also be read on a voltmeter and an ammeter, provided the slip is small. The field winding should be kept open in the slip test so that the slip-frequency current is not induced in it.

The direct-axis synchronous reactance X_d can now be calculated from the ratio of maximum voltage to minimum current, i.e., the ratio of applied volts per phase to the armature amperes per phase for the direct-axis position. The quadrature-axis synchronous reactance X_q is given by the ratio of minimum voltage to maximum current, i.e., the ratio of applied volts per phase to the armature amperes per phase for the quadrature-axis position. For the test to be successful, the slip must be sufficiently small.

The open-circuit and short-circuit tests discussed in Section 8.3 can also be used to determine the direct-axis armature reactance X_d from measurements of the steady-state armature open-circuit voltage and short-circuit current.

▬▬▬▬▬▬▬▬▬▬▬▬▬▬▬ ‖‖‖‖‖‖‖‖‖‖‖‖‖
8.7 Effect of Saturation on Reactances and Regulation

While both the synchronous reactance X_s of a round-rotor machine and X_d of a salient-pole machine are affected by saturation, the reactance X_q is usually assumed to be unaffected by saturation. Equations 8.3.1 and 8.3.2 yield the unsaturated and saturated values of the synchronous reactance, respectively, based on open-circuit and short-circuit characteristics, as explained in Section 8.3. Closer approximations of the saturated synchronous reactance can be obtained, however, for round-rotor as well as for salient-pole machines, from their zero-power-factor and open-circuit characteristics.

The *zero-power-factor rated-current saturation curve* is a graph of armature terminal voltage versus field current, with the armature current held constant at its rated full-load and the lagging power-factor angle held constant at or near 90°. Figure 8.7.1 shows the no-load and zero-power-factor rated-current saturation curves and the *Potier triangle* (named after its inventor) ABC or $A'B'C'$. With the assumptions

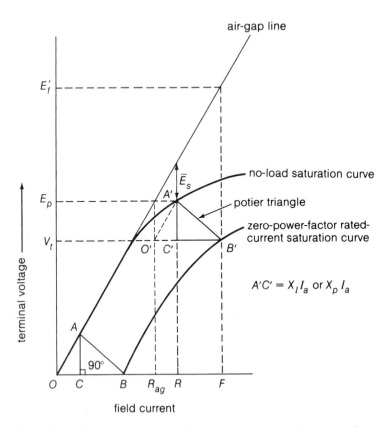

Figure 8.7.1 No-load and zero-power-factor rated-current saturation curves of a synchronous machine and the Potier triangle. (Note: $\bar{E}_p$ in general is a phasor combination of $\bar{V}_t$ and $jX_p\dot{I}_a$, as shown in Figure 8.7.2).

that the leakage reactance is a constant, the armature resistance is negligibly small, and the open-circuit characteristic is also the load saturation curve, the zero power-factor characteristic is a curve of exactly the same shape as the no-load saturation curve shifted vertically downward by an amount equal to the leakage-reactance voltage drop and horizontally to the right by an amount equal to the armature-reactance mmf. $O'A'$ in Figure 8.7.1 is parallel to OA, which is a part of the air-gap line.

The *Potier reactance* X_p, as determined from the vertical side $\left(AC \text{ or } A'C' = X_pI_a\right)$ of the Potier triangle replaces the leakage reactance X_l, and the voltage behind the Potier reactance $\bar{E}_p$ (obtained by the phasor combination of $\bar{V}_t$ and $jX_p\dot{I}_a$, as shown in Figure 8.7.2) is the air-gap voltage, of which saturation is now assumed to be a function. E_f of Figure 8.5.1 is proportional to the field current of an unsaturated machine. $\bar{E}_s$, the correction for saturation, is a function of $\bar{E}_p$, as shown in Figure 8.7.1, and would be parallel to $\bar{E}_p$ in the phasor diagram. $\bar{E}_f'$ is a phasor summation of the voltages $\bar{E}_f$ and $\bar{E}_s$. For convenience, however, $\bar{E}_s$ is sometimes taken to be parallel to $\bar{E}_f$ itself, as shown in Figure 8.7.2, rather than to $\bar{E}_p$; the theoretical justification is possibly greater with a salient-pole machine than with a nonsalient-pole machine. E_f', a fictitious armature voltage, is now proportional to the field current of a saturated machine. Figure 8.7.2 shows the phasor diagram of a saturated salient-pole synchronous generator in the steady state, while that of a nonsalient-pole synchronous generator is left to the reader's imagination. The per-unit

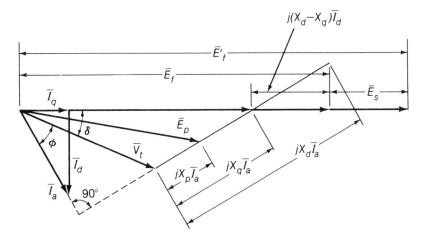

Figure 8.7.2 Steady-state phasor diagram of a salient-pole synchronous generator, while taking the approximate effect of saturation into account.

Potier reactance X_p is on the order of 0.15 for the case of turboalternators and of 0.32 for water-wheel generators as well as synchronous condensers. While X_p is assumed equal to X_l in a number of cases, the ratio (X_p/X_l) is about 1.3 for round-rotor machines and can be as much as 3 for some salient-pole machines with long slender poles and sizable field-leakage flux.

Referring to Figure 8.7.1, a *saturation factor* is defined by the ratio

$$k = R/R_{ag} \tag{8.7.1}$$

and the saturated synchronous reactance X_{ds} is obtained in terms of the unsaturated synchronous reactance X_{du}, as follows:

$$X_{ds} = X_p + \frac{X_{du} - X_p}{k} \tag{8.7.2}$$

The calculation of voltage regulation with the use of saturated synchronous reactance is illustrated in Example 8.3.1.

Since the almost universal practice now is to operate synchronous generators with automatic voltage regulators to hold the voltage constant under varying loads, a knowledge of the voltage regulation is of less importance than a knowledge of the change in the field excitation that is required to maintain constant voltage under varying loads. Both involve the same factors for consideration, however.

8.8 Parallel Operation of Interconnected Synchronous Generators

In order to assure continuity of the power supply within prescribed limits of frequency and voltage at all the load points scattered over the service area, it becomes necessary in any modern power system to operate several alternators in parallel, interconnected by various transmission lines, in a well-coordinated and optimized manner for the most economical operation. A generator can be paralleled with an infinite bus (or with another generator running at rated voltage and frequency supplying the load) by driving it at synchronous speed corresponding to the system frequency and adjusting its field excitation so that its terminal voltage equals that of the bus. If the frequency of the incoming machine is not exactly equal to that of the system, the phase relation between its voltage and bus voltage will vary at a frequency equal to the difference between the frequencies of the machine and bus voltages. In normal practice, this difference can usually be made quite small, to a fraction of a hertz; in polyphase systems, it is essential that the same phase sequence be maintained on either side of the synchronizing switch. Thus *synchronizing* requires the following conditions of the incoming machine:

- Correct phase sequence
- Phase voltages in phase with those of the system
- Frequency almost exactly equal to that of the system
- Machine terminal voltage approximately equal to the system voltage

A *synchroscope* (also called a *bright-lamp* or *dark-lamp synchronizing method*) [4] is used for indicating the appropriate moment for synchronization. After the machine has been synchronized and is part of the system, it can be made to take its share of the active and reactive power by appropriate adjustments of its prime-mover throttle and field rheostat. The system frequency and the division of active power among the generators are controlled by means of prime-mover throttles regulated by governors and automatic frequency regulators, whereas the terminal voltage and the reactive volt-ampere division among the generators are controlled by voltage regulators acting on the generator-field circuits and by transformers with automatic tap-changing devices.

||||||||||||

EXAMPLE 8.8.1

Two three-phase, 6.6-kV, wye-connected synchronous generators, operating in parallel, supply a load of 3,000 kW at 0.8 power-factor lagging. The synchronous impedance per phase of machine A is $(0.5 + j10)$ Ω and of machine B is $(0.4 + j12)$ Ω. The excitation of machine A is adjusted so that it delivers 150 A at a lagging power factor, and the governors are so set that the load is shared equally between the two machines. Determine the armature current, power factor, excitation voltage, and power angle of each machine.

Solution
One phase of each generator and one phase of the equivalent wye of the load are shown in Figure 8.8.1. The load current $\bar{I}_L$ is calculated as follows:

$$\bar{I}_L = \frac{3,000}{\sqrt{3} \times 6.6 \times 0.8} \angle - \cos^{-1} 0.8 = 328\,(0.8 - j0.6) = (262.4 - j196.8)\ \text{A}$$

For machine A: $\cos \phi_A = \dfrac{1,500}{\sqrt{3} \times 6.6 \times 150} = 0.875$ lagging; $\phi_A = 29°$; $\sin \phi_A = 0.485$

$$\bar{I}_A = 150\,(0.874 - j0.485) = (131.1 - j72.75)\ \text{A}$$

For machine B: $\bar{I}_B = \bar{I}_L - \bar{I}_A = (131.3 - j124) = 180.6\angle - \cos^{-1}\left(\dfrac{131.3}{180.6}\right)$

$$\cos \phi_B = \frac{131.3}{180.6} = 0.726\ \text{lagging}$$

---

[4] For details, see R. R. Lawrence and H. E. Richards, *Principles of Alternating-Current Machinery*, 4th ed. (New York: McGraw-Hill, 1953), chap. 29; A. S. Langsdorf, *Theory of Alternating-Current Machinery*, 2d. (New York: McGraw-Hill, 1955), chap. 12.

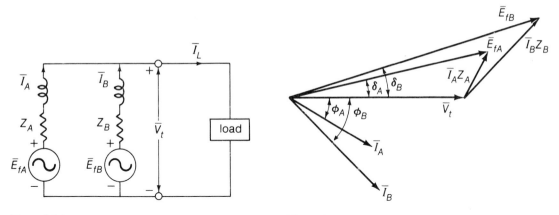

Figure 8.8.1   **Figure 8.8.2**

With the terminal voltage $\bar{V}_t$ as reference, we have

$$\bar{E}_{fA} = \bar{V}_t + \bar{I}_A Z_A = \left(6.6/\sqrt{3}\right) + (131.1 - j72.75)(0.5 + j10) \times 10^{-3}$$
$$= (4.6 + j1.27) \text{ kV per phase}$$

Power angle $\delta_A = \tan^{-1}(1.27/4.6) = 15.4°$

Line-to-line excitation voltage for machine $A = \sqrt{3}\sqrt{(4.6)^2 + (1.27)^2} = 8.26$ kV

$$\bar{E}_{fB} = \bar{V}_t + \bar{I}_B Z_B = \left(6.6/\sqrt{3}\right) + (131.3 - j124)(0.4 + j12) \times 10^{-3}$$
$$= (5.35 + j1.52) \text{ kV per phase}$$

Power angle $\delta_B = \tan^{-1}(1.52/5.35) = 15.9°$

Line-to-line excitation voltage for machine $B = \sqrt{3}\sqrt{(5.35)^2 + (1.52)^2} = 9.6$ kV

The corresponding phasor diagram is sketched in Figure 8.8.2.

▬▬▬▬▬▮▮▮▮▮▮
8.9 Steady-State Stability

The property of a power system that ensures it will remain in equilibrium under both normal and abnormal conditions is known as *power-system stability*. *Steady-state stability* is concerned with slow and gradual changes, while *transient stability* is concerned with severe disturbances like sudden changes in load or fault conditions. The largest possible flow of power through a particular point without the loss of stability is known as the *steady-state stability limit* when the power is increased gradually, and as the *transient-stability limit* when a sudden disturbance occurs.

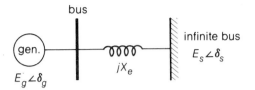

Figure 8.9.1 Generator connected to an infinite bus.

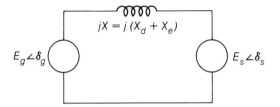

Figure 8.9.2 Positive-sequence reactance diagram corresponding to the system shown in Figure 8.9.1.

When the power system is subjected to disturbances small and gradual enough that the system can be regarded electrically as being in steady state, the steady-state equations of the machines can be used as an approximation for the steady-state stability analysis, while neglecting the effect of the excitation system. Such an approach is justified only for generators equipped with noncontinuously acting automatic voltage regulators that cannot act in time to keep the machine stable because of inherent dead-bands and time delays. The ability of interconnected synchronous machines to remain in synchronism is analyzed under small and gradual disturbances affecting the system, without representing the excitation-system behavior.

The effect of resistance is usually neglected. When a machine is connected through appreciable impedance to other machines in the system, which is most often the case, the saliency effects can be neglected for all practical calculations. In terms of steady-state stability, however, the effect of saturation in machines is usually more significant than saliency effects.

For a generator connected to a system very large compared to its own size, the system in Figure 8.9.1 can be used; the corresponding positive-sequence reactance diagram is shown in Figure 8.9.2. Recall that, for a single machine, the load angle δ is the angle between the terminal voltage and the q-axis. However, for a system in which several machines are interconnected, it is more convenient to measure an angle between the quadrature axis of each machine and a common reference axis that rotates at synchronous speed at all times. Let δ_g and δ_s, associated respectively with voltages $\bar{E}_g$ and $\bar{E}_s$ of Figure 8.9.1, be such angles, measured from a common reference axis. The power-angle equation for the system under consideration becomes

$$P = \frac{E_g E_s}{X} \sin \delta_{gs} \tag{8.9.1}$$

where δ_{gs} is the angular difference between δ_g and δ_s. The power-angle characteristic given by Equation 8.9.1 is plotted in Figure 8.9.3. Let us now examine the operating conditions at three points, a, b, and c, shown in Figure 8.9.3.

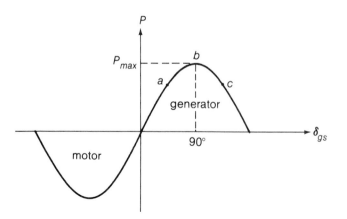

Figure 8.9.3 Power-angle curve of Equation 8.9.1.

At operating point a, the slope of the curve $dP/d\delta_{gs}$ is positive. With a small increase in the angle δ_{gs}, the electrical power can be seen to increase and will exceed the mechanical power input, given that the mechanical power input corresponding to the initial power input at the operating point a remains constant. This change results in deceleration, which will make the angle approach the initial operating point a. Then the machine is said to be stable. The peak of the power-angle curve given by P_{max} is known as the *steady-state power limit,* representing the maximum power that can theoretically be transmitted in a stable manner. A zero slope corresponds to such an operating condition, as shown at point b in Figure 8.9.3. A machine is usually operated at less than the power limit, thereby leaving a steady-state margin, as otherwise even a slight increase in the angle δ_{gs} would lead to instability. The operation at point c shown in Figure 8.9.3 is clearly unstable; note the negative slope of the curve. A slight increase in the angle δ_{gs} leads to a decrease in electrical power, which becomes smaller than the mechanical power. This decrease in turn would accelerate the machine, thereby making the angle larger and eventually losing synchronism.

Steady-state stability analysis neglecting saturation leads to unduly pessimistic estimates regarding the ability of a machine in a system to remain in synchronism given small and gradual disturbances. The effect of saturation is usually taken into account by suitably modifying the synchronous reactance X_d, as discussed in Section 8.7. The saturated value usually lies within the range of 0.4 to 0.8 of the unsaturated synchronous reactance, smaller values being attributed to the newer and larger turbogenerators. It is also a common practice in overexcited operation to choose the saturated value of $0.8X_d$ for light loads and $0.6X_d$ for full loads.

The slope $\left(dP/d\delta_{gs}\right)$ of the power-angle curve gives a measure of the steady-state stability of a system; it is most frequently used to check the steady-state stability of synchronous machines in a system. The test is made between pairs of machines that are electrically most remote from each other, since this situation is the one most likely to give rise to instability. If the slopes of both machines are positive, the system is stable; if either slope is negative, the system is unstable. This criterion leads to conservative estimates, particularly when one of the tested machines is much larger than the other. Refined methods, presented in texts on stability, can then be used.[5]

---

[5] See, for example, E. W. Kimbark, *Power-System Stability,* vols. 1, 2, and 3 (New York: John Wiley & Sons, 1948).

The steady-state stability limits of a system can be increased by an increase in the excitation of the generator or by a reduction in the reactance of the network, as seen from Equation 8.9.1. Installing parallel transmission lines contributes to an increase in the stability limit and the dependability of the system. Series capacitors are sometimes used in lines to improve the voltage regulation or to raise the stability limit by effectively decreasing the line reactance. Continuously acting automatic voltage regulators can raise the steady-state stability limits considerably. Such regulators maybe represented mathematically by differential equations, and the overall performance—including the regulator, excitation system, generator, and system transient characteristics—can be evaluated by solving the equations for small changes.

8.10 Excitation Systems

The source of field current for the excitation of a synchronous machine is an excitation system that includes the exciter, regulator, and manual control. The modern excitation control system of a large synchronous machine is a *feedback control system,* the essential elements of which are illustrated in Figure 8.10.1. Figure 8.10.2 shows the general layout of the components included in an excitation system. The use of solid-state components and the invention of new excitation systems prompted some new definitions.[6]

The main *exciter* can be one of the following kinds:

- *DC generator-commutator exciter,* whose energy is derived from a dc generator with its commutator and brushes. The exciter can be driven by a motor, a prime mover, or the shaft of the synchronous machine.

- *Alternator-rectifier exciter,* whose energy is derived from an alternator and converted to dc by rectifiers. The power rectifiers can be controlled or noncontrolled; they can be stationary or rotating with the alternator shaft. The alternator can be driven by a motor, a prime mover, or the shaft of the synchronous machine.

- *Compound-rectifier exciter,* whose energy is derived from the currents and potentials of the ac terminals of the synchronous machine and converted to dc by rectifiers. The exciter includes the power transformers (current and potential), power reactors, and power rectifiers, which can be either noncontrolled or controlled, including the gate circuitry.

- *Potential-source-rectifier exciter,* whose energy is derived from a stationary ac potential source and converted to dc by rectifiers. The exciter includes the power potential transformers, if used, and power rectifiers, which can be controlled or noncontrolled, including the gate circuitry.

A synchronous-machine *regulator* couples the output variables of the synchronous machine to the input of the exciter through feedback and forward controlling elements for the purpose of regulating the

---

[6] IEEE Committee Report, "Proposed Excitation System Definitions for Synchronous Machines," *IEEE Transactions on Power Apparatus and Systems,* vol. PAS–88, no. 8 (August 1969), pp. 1248–58.

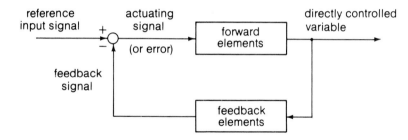

Figure 8.10.1 Essential elements of an automatic feedback control system. (Adapted from IEEE Comittee Report, "Proposed Excitation Systems for Synchronous Machines," *IEEE Transactions on Power Apparatus and Systems,* vol. PAS–88, no. 8. Copyright ©1969 IEEE.)

synchronous machine output variables. The regulator generally consists of an error detector, a preamplifier, stabilizers, auxiliary inputs, and limiters. A regulator can be continuously or noncontinuously acting; the former type indicates a corrective action for a sustained infinitesimal change in the controlled variable, while the latter requires a sustained finite change in the controlled variable to initiate corrective action. Regulators in which the regulating function is accomplished by mechanically varying a resistance are known as *rheostatic-type* regulators.

Power-system stabilizers provide additional input to the regulator to improve power-system dynamic performance. Quantities such as shaft speed, frequency, and synchronous machine power can be used as input to the power-system stabilizer. Other signal modifiers include *limiters* and *compensators*. Limiters act to limit a variable by modifying or replacing the function of the primary detector element when predetermined conditions have been reached. Example conditions include maximum excitation, minimum excitation, and maximum volts per hertz. Compensators act to compensate for the effect of a variable by modifying the function of the primary detecting element. Examples include reactive-current compensators, active-current compensators, and line-drop compensators. Generator-voltage regulators are equipped with reactive compensators in order to share reactive current among generators operating in parallel. Line-drop compensators introduce within the regulator-input circuit a voltage equal to the impedance drop to thereby modify the generator voltage. The voltage drops of the resistive and reactive portions of the impedance are obtained respectively by active and reactive compensators.

The *excitation-control-system stabilizer* modifies the forward signal by series or feedback compensation to improve the dynamic performance of the excitation-control system; it eliminates undesired oscillations and overshoot of the regulated voltage.

The *exciter-control system* provides the proper field voltage to maintain a desired system voltage. An important characteristic is this control system's ability to respond rapidly to voltage deviations during normal and abnormal or emergency system operations. While many types of exciter-control systems are employed in practice, the basic principle of operation is the same: The voltage deviation signal is amplified to produce the signal required to change the exciter field current; that change in turn produces a change in exciter-output voltage, thereby resulting in a new level of excitation for the synchronous generator.

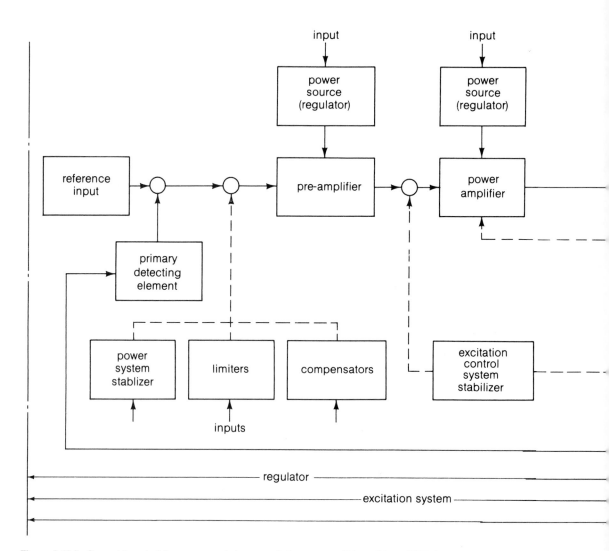

Figure 8.10.2 General layout of the components in an excitation system. (Adapted from IEEE Comittee Report, "Proposed Excitation System Definitions for Synchronous Machines," *IEEE Transactions on Power Apparatus and Systems,* vol. PAS–88, no. 8. Copyright ©1969 IEEE.)

8.11 Synchronous Machine Design

Let us work through a three-phase synchronous motor design project to show the basic elements and some considerations of the design aspect. Here are the specifications for the motor:

A 200-hp, 440-V, three-phase, 60-Hz, 900-rpm, synchronous motor is to be self-starting for direct connection to a centrifugal pump and is to operate at unity power factor at full load without field rheostat. The temperature rise for continuous full-load operation must not exceed 50° C, and the full-load efficiency must not be less than 93.3%.

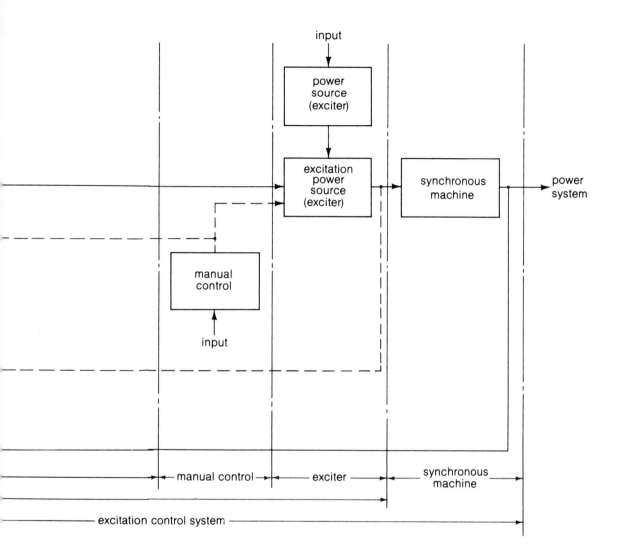

$$\text{The output kVA} = \frac{\text{hp} \times 0.746}{\text{power factor}} = \frac{200 \times 0.746}{1} = 149$$

$$\text{Number of poles} = \frac{120 \times 60}{900} = 8$$

$$\frac{\text{kVA}}{\text{rpm}} = \frac{149}{900} = 0.166$$

Based on empirical curves available in the literature, choose the output coefficient as $2.2 \times 10^4 \text{J/m}^3$. From Equation 5.6.1, it follows

$$D^2L = \frac{W}{C\omega_s} = \frac{149 \times 10^3}{(2.2 \times 10^4)\left(\frac{900}{60} \times 2\pi\right)} = 0.072 \text{ m}^3$$

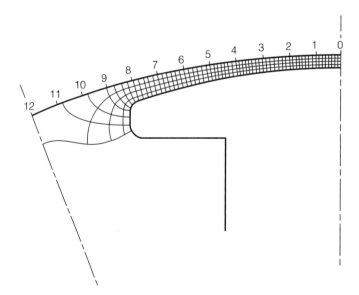

Figure 8.11.1 Pole-shoe shape and the flux plot.

For

$$L/\tau = 0.7, \quad L = \frac{0.7 \times \pi \times D}{p} = \frac{0.7 \times \pi \times D}{8} = 0.275D$$

$$D^2L = 0.275D^3 = 0.072 \text{ m}^3 \quad \text{or} \quad D = \sqrt[3]{0.262} = 0.64 \text{ m}$$

and

$$L = 0.275 \times 0.64 = 0.176 \text{ m}$$

or choose

$$D = 63.5 \text{ cm}; \qquad L = 18 \text{ cm}$$

The pole pitch τ is then

$$\tau = \pi D/p = \pi (63.5)/8 \simeq 25 \text{ cm}$$

For 70% pole embrace, the pole arc is 0.7 x 25 = 17.5 cm. We can use the method of shaping the pole shoe suggested by Wieseman.[7]

From empirical curves available in the literature, based on kVA and rpm, the minimum air-gap length is selected as 0.48 cm.

$$g_{min} = 0.48 \text{ cm}; \qquad \frac{g_{max}}{g_{min}} = 1.75; \qquad \frac{g_{min}}{\tau} = 0.0192; \qquad \frac{\text{pole arc}}{\text{pole pitch}} = 0.7$$

The shape of the pole shoe along with the flux plot is shown in Figure 8.11.1.

From the air-gap flux distribution curve, we can make calculations for the fundamental, third, fifth, and seventh harmonics. The flux-distribution factor f_d, which is the ratio of the average ordinate of the

---

[7] R. W. Wieseman, "Graphical Determination of Magnetic Fields: Practical Applications to Salient Pole Synchronous Machine Design," *AIEE Journal,* vol. 46, May 1927, p.433.

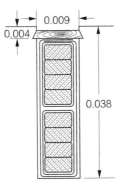

Figure 8.11.2 Stator
slot configuration for the
synchronous motor.

flux wave to the maximum, is computed as 0.62, while the form factor f_b, which is the ratio of the root-mean-square ordinate to the average, is 1.14.

Choosing the air-gap flux density of 0.67 T, we can calculate the total flux:

$$\phi_t = \pi DLB_g = \pi \times 0.635 \times 0.18 \times 0.67 = 0.24 \text{ Wb}$$

The winding constant $C_w = f_d f_b k_d = 0.62 \times 1.14 \times 0.956 = 0.676$, where k_d is the stator-winding distribution factor (discussed in Section 6.5). The number of conductors in series per phase of a wye-connected winding with full-pitch coils is then

$$\frac{E \times 60}{\phi_t \, (\text{rpm}) \, k_p C_w} = \frac{\left(440/\sqrt{3}\right) 60}{0.24 \times 900 \times 1 \times 0.676} = 104$$

For one circuit per phase, the total number of conductors $= 104 \times 3 = 312$. Choosing $3\frac{1}{2}$ slots per pole per phase, the number of armature slots $= \frac{7}{2} \times 3 \times 8 = 84$. The tooth pitch at the armature surface, $t_1 = \frac{\pi \times 63.5}{84} = 2.376$ cm. Since the number of conductors per slot must be an even integer, let us select 4 conductors per slot so that the total number of conductors is 4 x 84 = 336. Choosing the coil span to be two-thirds full-pitch, the coil throw will be slot 1 to slot 8, and the pitch factor (Section 6.5) will be

$$k_p = \sin\left(\frac{7}{10.5}90°\right) = 0.866$$

The final value of the total flux is then

$$\phi_t = \frac{\left(440/\sqrt{3}\right) 60}{(336/3) \, 900 \times 0.866 \times 0.676} = 0.258 \text{ Wb}$$

The armature current per phase is

$$I = \frac{(149/0.933) \, 10^3}{440\sqrt{3}} = 210 \text{ A}$$

From empirical curves available in the literature, based on kVA and rpm, and choosing a current density of 530 A/cm² in the armature copper, the section area of armature conductor is

$$s_a = \frac{210}{530} = 0.4 \text{ cm}^2$$

Suitable conductors can be selected from the tables available in the literature. Two conductors that are in parallel and arranged in the slot are shown in Figure 8.11.2. Slot dimensions are then finalized.

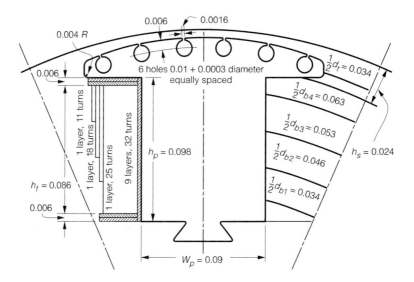

Figure 8.11.3 Salient-pole configuration for the synchronous motor.

The flux per pole is

$$\phi = \frac{\phi_t f_d}{p}$$

$$= \frac{0.258 \times 0.62}{8} = 0.02 \text{ Wb}$$

Choosing the leakage constant of 1.14, the section area of the pole body is

$$s_p = \frac{0.02 \times 1.14}{1.3}$$

$$= 0.0175 \text{ m}^2 \quad \text{or} \quad 175 \text{ cm}^2$$

where the flux density of 1.3 T is assumed in the pole body. Since the length of the pole parallel to the shaft is equal to that of the armature, the width of the pole body is

$$w_p = \frac{175}{18} = 9.7 \text{ cm}; \quad \text{choose 9 cm}$$

The pole configuration for our design project is shown in Figure 8.11.3.

While we will not go through the detailed calculations related to the open-circuit saturation curve, reactances, damper windings, and so forth here, the design sheet for the synchronous motor is given in Figure 8.11.4. The student is encouraged to pursue the design and justify the details furnished here.

Hp. 200	Volts 440	Phase 3	Amperes 209	Hertz 60	Poles 8
	Rpm. 900	Kva./rpm. 0.166	Output constant 2.2×10^4		

Armature

Sheet steel	0.5×10^{-3} Electrical grade
Outside diameter	0.826
Gap diameter	0.635
Total length	0.18
Ducts, number and size	Two 0.013
Gross length	0.15
Effective length	0.14
Slots per pole per phase	$3\frac{1}{2}$
Total number of slots	84
Type of winding	Wye
Circuits per phase	One
Coil throw	Slots 1 and 8
Per cent pitch	$66\frac{2}{3}$

Conductors:

Per slot	4
Dimensions	0.0038×0.007
Area	$2 \times (21 \times 10^{-6})$
In series per phase	112
Total	336
Current density	5×10^6
Length, one-half mean-turn	0.466
Resistance per phase, 25°C	0.022
Resistance per phase, 75°C	0.0262
Resistance drop, volts	6.87
Resistance drop, per cent	2.71
Reactance drop, volts	24.8
Reactance drop, per cent	9.75
Impedance drop, volts	25.7
Impedance drop, per cent	10.1
Armature reaction, AT per pole	3290
Armature reaction factor	0.86
eq. fld. AT	2830
Ampere conductors, per m	30.63×10^3
Short-circuit ratio	1.25
$\frac{D^2 Ln}{Kva}$	0.437
ATP_{100}	5220
Square meter per watt	0.45×10^{-3}

Field

Total air gap length	2×0.0056
Rotor diameter	0.6238
Peripheral speed	29.4
Pole pitch	0.25
Pole arc	0.175
Material spider	Hot-rolled steel
Damper slot pitch	0.03
Damper bars per pole	6
Size of bar	0.01 m diam.
Material of bar	Copper
Section end-ring	$0.005 \times 0.03 = 0.00015$
Material end-ring	Copper
Air gap coefficient	1.11
Effective length of gap	0.006
Leakage constant	1.184

f_d, 0.62, f_b, 1.14, k_p, 0.956, C_w, 0.676, k_p, 0.866
Total flux, 0.259.   Flux per pole, 0.0201

	Section	Den-sity	Length	Amp.-turns
Air gap	0.355	0.73	1.11×0.0056	3600
Teeth	0.185	1.4	0.038	41
Armature yoke	0.016	1.24	0.15	65
Poles	0.016	1.5	0.13	275
Field yoke	0.034	0.62	0.07	23
Total ampere turns per pole	......	....	....	4004

Size of conductor	No. 11 square
Turns per pole*	342
Amperes no load	11.7 Maximum, 16.2
Length of mean-turn	0.645
Resistance, 25°C	6.21
Resistance, 75°C	7.4
IR no-load	69.6 Maximum, 120
I^2R no-load	815 Maximum, 1940
Square meter per watt maximum I	0.63×10^{-3}
Kva	159
Power factor	1.00
Amperes	15.26
IR 75°C	113
I^2R 75°C	1830
Square meter per watt	0.66×10^{-3}
Exciter voltage	125
Exciter capacity	2.11

Full-Load Losses

Friction and windage	1200
Core	2150
Stray load	870
Armature copper	3460
Field copper	2020
Total losses	9700

Weights

Armature copper	58.6
Filed copper	78.6
Armature teeth	55.9
Armature yoke	151.8
Filed poles	142.7

Remarks: *9 layers—32 turns $X_d = 0,89$, $X_q = 0.468$
1 layer—25 turns $\delta_e = 26.2°$
1 layer—18 turns
1 layer—11 turns

Figure 8.11.4 Synchronous motor design sheet. Note: All dimensions are in SI system of units. (Adapted from J. H. Kuhlmann, *Designing Electrical Apparatus*, 3d ed. New York: John Wiley & Sons, 1954.)

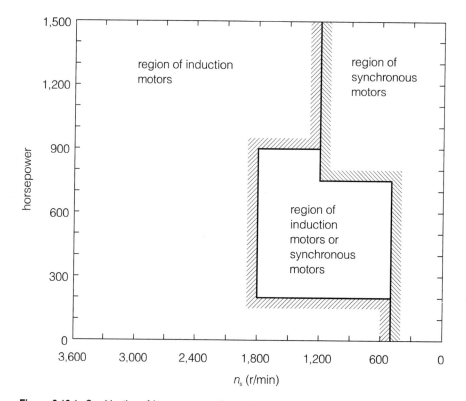

Figure 8.12.1 Combination of horsepower and speed regions for application of synchronous and induction motors. (Courtesy of Del Toro, Vincent, *Electromechanical Devices for Energy Conversion and Control Systems,* ©1968, p. 398. Reprinted by permission of Prentice-Hall, Englewood Cliffs, N.J.)

8.12 Applications for Synchronous Motors

With constant-speed operation, power-factor control, and high operating efficiency, three-phase synchronous motors are employed in a wide range of applications. An overexcited synchronous motor, known as a synchronous condenser, is used to improve the system power factor. Synchronous motors are used as prime movers of dc generators and variable-frequency ac generators. Indicating the general areas of application of synchronous and induction motors, Figure 8.12.1 gives an idea of an optimum combination of horsepower and speed. Table 8.12.1 shows typical characteristics and applications of synchronous motors.

Single-phase reluctance motors find application in such devices as clocks, electric shavers, electric clippers, vibrators, sandpapering machines, and engraving tools. A hysteresis motor is employed for driving high-quality record players and tape recorders, electric clocks, and other timing devices. The horsepower range for hysteresis motors is up to 1 horsepower, and up to 100 horsepower for reluctance motors.

Table 8.12.1 Typical Characteristics and Applications of Synchronous Motors

	Synchronous Speed	
	$n_s > 500$ rpm	$n_s < 500$ rpm
Starting Torque (% of normal)	25 up to several thousand	Usually above 25 up to several thousand
Pull in torque (% of Normal)	Up to 120	Low 40
Pull-out Torque	Up to 200	Up to 180
Starting Current	800–700	200–350
Slip	0	0
Power Factor	High, but varies with load and excitation.	High, but varies with excitation
Efficiency	92–96%; the highest of all motors	92–96%; the highest of all motors
Typical Applications	power-factor correction, constant-speed generations, frequency changing, fans, blowers, dc generators, line shafts, centrifugal pumps and compressors, reciprocating pumps and compressors	power-factor control, constant-speed generation, flywheel used for pulsating loads, low-speed, direct-connected loads such as reciprocating compressors when started unloaded, dc generators, rolling mills, band mills, ball mills, pumps, flour-mill line shafts, rubber mills, mixers, crushers, pulp grinders, refiners in papermaking industry.

Adapted from M. LiwSchitz-Garik and C.C. Whipple, *Electric Machinery,* vol II. Van Nostrand Co., Inc., Princeton, N.J., 1946.

BIBLIOGRAPHY

Del Toro, V. *Electromechanical Devices for Energy Conversion and Control Systems.* Englewood Cliffs, N.J.: Prentice-Hall, 1968.

Fitzgerald, A. E., C. Kingsley Jr., and A. Kusko. *Electric Machinery.* 3d ed. New York: McGraw-Hill, 1971.

Hindmarsh, J. *Electrical Machines and Their Applications.* 2d ed. New York: Pergamon, 1970.

Institute of Electrical and Electronics Engineers. *Standard No. 115–1965, Code SH00554, Test Procedures for Synchronous Machines.* New York, 1965.

Institute of Electrical and Electronics Engineers Committee Report. "Proposed Excitation System Definitions for Synchronous Machines." *IEEE Transactions on Power Apparatus and Systems,* vol. PAS–88, no. 8 (August 1969), pp. 1248–58.

Kimbark, E. W. *Power System Stability: Synchronous Machines.* New York: Dover, 1968.

Knowlton, A. E., ed. *Standard Handbook for Electrical Engineers.* 8th ed. New York: McGraw-Hill, 1949.

Matsch, L. W. *Electromagnetic and Electromechanical Machines.* New York: Intext, 1972.

Nasar, S. A., and L. E. Unnewehr. *Electromechanics and Electric Machines,* New York: John Wiley & Sons, 1979.

Puchstein, A. F., T. C. Lloyd, and A. G. Conrad. *Alternating-Current Machines.* 3d ed. New York: John Wiley & Sons, 1954.

Sarma, M. S. *Synchronous Machines (Their Theory, Stability and Excitation Systems).* New York: Gordon and Breach, 1979.

Say, M. G. *Alternating-Current Machines.* New York: John Wiley & Sons, Halsted Press, 1976.

Slemon, G. R., and A. Straughen. *Electric Machines.* Reading, Mass.: Addison-Wesley, 1980.

Westinghouse Electric Corporation Central Station Engineers. *Electrical Transmission and Distribution Reference Book.* 4th ed. East Pittsburg, Pa.: Westinghouse Electric Corporation, 1950.

▬▬▬▬▬▬▬▬ ||||||||||||

PROBLEMS

8-1. A three-phase, wye-connected, cylindrical-rotor synchronous generator rated at 10 kVA and 230 V has a synchronous reactance of 1.5 Ω per phase and an armature resistance of 0.5 Ω per phase.

    a. Determine the voltage regulation at full load
       (i)   with 0.8 lagging power factor.   (ii)   with 0.8 leading power factor
    b. Calculate the power factor for which the voltage regulation becomes zero on full load.

8-2. The following readings are taken from the results of open-circuit and short-circuit tests on a 10-MVA, three-phase, wye-connected, 13.8-kV, two-pole turbine generator driven at synchronous speed:

Field current, A:	170	200
Armature current, short-circuit test, A:	418	460
Line-to-line voltage, open-circuit test, V	13,000	13,800
Line-to-line voltage, air-gap line, V	15,500	17,500

Neglect the armature resistance.

    a. Determine the unsaturated value of the synchronous reactance in ohms per phase and per unit.
    b. Compute the saturated value of the synchronous reactance in ohms per phase and per unit.
    c. Find the short-circuit ratio.
    d. Calculate the field current required at rated voltage, rated kVA, and 0.8 lagging power factor, while accounting for saturation under load.

8-3. The loss data for the synchronous generator of Problem 8–2 are given below:

    Open-circuit core loss at 13.8 kV is 70 kW.   Field-winding resistance at 75°C is 0.3 Ω.

    Short-circuit load loss at 418 A, 75°C is 50 kW.   Stray-load loss at full load is 20 kW.

    Friction and windage loss is 80 kW.

Determine the efficiency of the generator at rated load, rated voltage, and 0.8 power factor lagging.

8-4. The following data are obtained from the open-circuit and short-circuit characteristics of a three-phase, wye-connected, four-pole, 150-MW, 0.85-pf, 12.6-kV, 60-Hz, hydrogen-cooled turbine generator with negligible armature resistance:

    Open-circuit characteristic:

Field current, A:	200	300	400	500	600	700	800	900
Line-to-line terminal voltage, kV:	3.8	5.8	7.8	9.8	11.3	12.6	13.5	14.2

    Short-circuit characteristic:

Armature current, A:	4,043	8,086
Field current, A:	350	700

Determine (a) the unsaturated synchronous reactance, (b) the saturated synchronous reactance at rated voltage, (c) the short-circuit ratio, (d) the estimated field current and regulation for rated voltage, rated current at 0.85 power-factor lagging, and (e) the estimated field current and regulation for rated voltage, rated current at 0.85 power-factor leading.

8–5. The following data are taken from the open-circuit and short-circuit characterictics of a 45-kVA, three-phase, wye-connected, 220-V, six-pole, 60-Hz synchronous machine.

From the open-circuit characteristic: Line-to-line voltage = 220 V;   Field current = 2.84 A

From the short-circuit characteristic: Armature current, A: 118   152
                                                                          Field current, A: 2.20   2.84

From the air-gap line: Field current = 2.20 A;   Line-to-line voltage = 202 V

Compute the unsaturated value of the synchronous reactance, its saturated value at rated voltage, and the short-circuit ratio. Express the synchronous reactance in ohms per phase and in per unit on the machine rating as a base.

8–6. For a 45-kVA, three-phase, wye-connected, 220-V synchronous machine at rated armature current, the short-circuit load loss (total for three phases) is 1.80 kW at a temperature of 25°C. The dc resistance of the armature at this temperature is 0.0335 Ω per phase. Compute the effective armature resistance in per unit and in ohms per phase at 25° C.

8–7. A 4,000-V, 5,000-hp, 60-Hz, 12-pole synchronous motor with a synchronous reactance of 4 ohms per phase (based on cylindrical-rotor theory) is excited to produce unity power factor at rated load. Neglect all losses.

a. Find the rated and maximum torques.

b. What is the armature current corresponding to the maximum torque?

8–8. A synchronous motor with $X_d = 1.0$ per unit has a rated power factor of 0.8 leading. Neglecting the effect of saliency and losses, determine the ratio between the pull-out torque when the field is excited for 0.8 leading power factor at rated mechanical load and the pull-out torque when the motor delivers its rated load at unity power factor.

8–9. A three-phase, wye-connected, four-pole, 400-V, 60-Hz, 15-hp synchronous motor has a synchronous reactance of 3 Ω per phase and negligible armature resistance. The data for its no-load magnetization curve follow:

Field current, A:	2	3.5	4.4	6	8	10	12
Line-to-neutral voltage, V:	100	175	200	232	260	280	295

a. When the motor operates at full load at 0.8 leading power factor, determine the power angle and the field current. Neglect all losses.

b. Compute the minimum line current for the motor operating at full load and the corresponding field current.

c. When the motor runs with an excitation of 10 A while taking an armature current of 25 A, calculate the power developed and the power factor.

d. If the excitation is adjusted such that the magnitudes of the excitation voltage and terminal voltage are equal, and if the motor is taking 20 A, find the torque developed.

8–10. Equation 8.4.6 is derived by neglecting the effect of armature resistance. Including the effect of armature resistance R_a, show that the power developed per phase by a round-rotor synchronous machine is

$$P = -\frac{E_f V_t}{Z_s} \cos(\delta + \theta) + E_f^2 \frac{R_a}{Z_s^2}$$

where V_t is the terminal voltage per phase, E_f is the excitation voltage per phase, Z_s is the synchronous impedance per phase, θ is the impedance angle, and δ is the power angle.

8–11. Consider the per-phase equivalent circuit shown in Figure P8–11 of a three-phase synchronous generator supplying a synchronous motor through reactors. The motor drives a mechanical load, with a torque such that the mechanical power is 3.2 kW at rated speed.

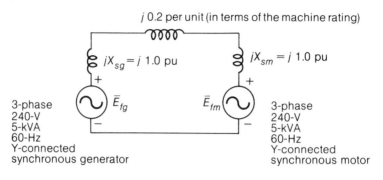

Figure P8–11

a. If the field currents of the two machines are adjusted so that rated voltage is applied to the motor terminals at rated frequency with a motor power factor of 0.8 lagging, determine the internally generated emfs E_{fg} and E_{fm}.

b. With the generator field current the same as in part a, what is the minimum value of the motor-excitation voltage for which the machines will remain in synchronism?

8–12. A turbogenerator rated at 200 MVA, 11 kV, and 0.9 lagging power factor has a synchronous reactance of 1.7 per unit and a field current of 300 A for rated load. Draw an operating chart (circle diagram) for the machine when it is running on 11-kV infinite bus bars. Then find from the diagram the field current required when the machine delivers 100 MW at 0.8 lagging power factor. Also determine the load angle and the armature current.

8–13. A three-phase, wye-connected, 2,300-V, 60-Hz round-rotor synchronous motor has a synchronous reactance of 2 Ω per phase and negligible armature resistance.

a. If the motor takes a line current of 350 A operating at 0.8 power factor leading, calculate the excitation voltage and the power angle.

b. If the motor is operating on load with a power angle of $-20°$, and the excitation is adjusted so that the excitation voltage is equal in magnitude to the terminal voltage, determine the armature current and the power factor of the motor.

8–14. A 2300-V, three-phase, wye-connected, round-rotor synchronous motor has a synchronous reactance of 3 Ω per phase and an armature resistance of 0.25 Ω per phase. The motor operates on load with a power angle of $-15°$, and the excitation is adjusted so that the internally induced voltage is equal in magnitude to the terminal voltage. Determine (a) the armature current and (b) the power factor of the motor. Neglect the effect of armature resistance.

8–15. An induction motor takes 350 kW at 0.8 power factor lagging while driving a load. When an overexcited synchronous motor taking 150 kW is connected in parallel with the induction motor, the overall power factor is improved to 0.95 lagging. Determine the kVA rating of the synchronous motor.

8–16. An industrial plant consumes 500 kW at a lagging power factor of 0.6.

a. Find the required kVA rating of a synchronous capacitor to improve the power factor to 0.9.

b. If a 500-hp, 90% efficient synchronous motor operating at full load and 0.8 leading power factor is added instead of the capacitor in part b, calculate the resulting power factor.

8–17. A three-phase, wye-connected, cylindrical-rotor synchronous motor with negligible armature resistance and a synchronous reactance of 1.27 Ω per phase is connected in parallel with a three-phase, wye-connected load taking 50 A at 0.707 lagging power factor from a three-phase, 220-V, 60-Hz source. If the power developed by the motor is 33 kW at a power angle of 30°, determine (a) the overall power factor of the motor and the load, and (b) the reactive power of the motor.

8–18. A three-phase salient-pole synchronous generator has the reactances of $X_d = 1.0$ per unit and $X_q = 0.7$ per unit. When the machine is delivering full-load current at rated terminal voltage of 3.3 kV and a power factor of 0.8 lagging, determine the excitation voltage (internally generated emf). If the excitation is removed, what is the maximum per-unit power that can be generated?

8–19. A salient-pole synchronous generator with $X_d = 1.0$ per unit and $X_q = 0.6$ per unit is connected to an infinite bus.

    a. If the field circuit is open, determine the maximum real power per unit that this generator can deliver without losing synchronism.

    b. Calculate the reactive power per unit, current per unit, and the power factor. Is the power factor leading or lagging?

8–20. A three-phase, wye-connected, salient-pole synchronous generator rated at 20 kVA, 220 V has negligible armature resistance and the reactances per phase of $X_d = 4\ \Omega$, $X_q = 2\ \Omega$. When the machine is supplying rated load at 0.8 lagging power factor, determine (a) the power angle, (b) the voltage regulation, and (c) the power developed due to saliency.

8–21. A salient-pole synchronous generator is operating under balanced steady-state conditions at rated voltage, delivering rated current at a lagging power-factor angle of 15°. The reactances in per-unit are given to be $X_d = 1.2$ and $X_q = 0.8$.

    a. Compute the per-unit excitation voltage and the power angle.

    b. Using the excitation voltage found in part a, to what value could the real power output be increased without loss of synchronism?

8–22. Equation 8.5.10 gives the steady-state power-angle equation for a salient-pole synchronous machine with negligible armature resistance and fixed field excitation. Show that the condition for maximum power is

$$\cos \delta = -\frac{X_q E_f}{4\left(X_d - X_q\right)V_t} + \sqrt{\left[\frac{X_q E_f}{4\left(X_d - X_q\right)V_t}\right]^2 + \frac{1}{2}}$$

8–23. Starting from Equation 8.6.2, and taking the terminal bus voltage $\bar{V}_t$ as reference, draw the steady-state phasor diagram of a synchronous machine, showing the three components of current making up the armature current $\bar{I}_a$. With reference to your diagram, try to identify the four possible cases of operation of a synchronous machine.

8–24. The following readings are obtained in a slip test on a 5-kVA, 240-V, 60-Hz synchronous machine:

    Line-to-line, V: 200 maximum   180 minimum
    Line current, A: 12 maximum   8 minimum

Calculate X_d and X_q in ohms and per unit, assuming the armature to be wye-connected. What are the per-phase and per-unit values if the armature is delta-connected?

8–25. A four-pole, 60-Hz, three-phase, wye-connected, 13.2-kV, 93.75-MVA turbogenerator is delivering rated load at 0.8 power-factor lagging. The Potier reactance of the generator is given as 0.248 Ω per phase; effective armature resistance is 0.004 Ω per phase; unsaturated synchronous reactance is 2.13 Ω per phase; E_s, the correction for saturation, corresponding to E_p of 8,280 V per phase (see Figure 8.7.1), is 1,155 V per phase; a field current of 425 A corresponds to the rated voltage on the air-gap line of the open-circuit saturation curve of the machine. Compute the field current required for the given operating condition of the turboalternator.

8–26. Compute the field current required for a power factor of 0.80-leading current when a 45-kVA, three-phase, wye-connected, 220-V, six-pole, 60-Hz synchronous machine is run as a synchronous motor at a terminal voltage of 230 V, with a power input to its armature of 45 kW. It is given that the unsaturated value of the synchronous reactance, based on cylindrical-rotor theory, is 0.92 per unit, Potier reactance is 0.227 per unit, effective armature resistance is 0.04 per unit, and E_s, the correction for saturation, corresponding to

E_p of 1.2 per unit (see Figure 8.7.1), is 0.45 per unit. Further, it is given that 1.00 per-unit excitation is 2.40 field amperes, when unit excitation is defined as the value corresponding to unit voltage on the air-gap line of the open-circuit characteristic of the machine.

8–27. Two identical three-phase, 33-kV, wye-connected synchronous generators operating in parallel share equally a total load of 12 MW at 0.8 lagging power factor. The synchronous reactance of each machine is 8 Ω per phase and the armature resistance is negligible.

　　a. If one of the machines has its field excitation adjusted such that it delivers 125 A lagging current, determine the armature current, power factor, excitation voltage, and power angle of each machine.

　　b. If the power factor of one of the machines is 0.9 lagging, find the power factor and current of the other.

8–28. Consider two identical three-phase, 13.8-kV, 100-MVA, 60-Hz turbine generators, each with a synchronous reactance of 1.10 per unit, operating in parallel supplying a total load of 150 MVA at 0.8 lagging power factor, rated voltage, and rated frequency. Calculate for each generator (i) the real power, (ii) the reactive power, (iii) the armature current, (iv) the power factor, (v) the excitation voltage, and (vi) the torque angle, under these conditions:

　　a. If the prime movers are adjusted so that the machines share the real power equally, and the field excitations are such that the reactive power output of each of the two machines is the same.

　　b. If the machines deliver equal real power, but with the field excitation of one of the generators (say, generator 1) increased by 20% above its value in part a, while the field current of the other generator is adjusted so that the terminal voltage remains constant at rated value.

　　c. If the excitation of generator 1 is kept constant at the value in part b, while the input to its prime mover is increased by 20%, and adjustments are made in the field current of generator 2 and in its prime mover such that the frequency and terminal voltage remain at their rated values.

8–29. Repeat Problem 8–28 for two generators with the same voltage and frequency ratings as those in Problem 8–28, but with generator 1 rated at 80 MVA with a synchronous reactance of 1.0 per unit and generator 2 rated at 120 MVA with a synchronous reactance of 1.2 per unit.

8–30. The project is to design a 2500-kVA, three-phase, 60-Hz, 2400-V, 225-rpm salient-pole synchronous generator of the vertical, water-wheel type. The efficiency at full load, unity power factor, rated speed, and voltage is to be not less than 95.6%, and in no part of the machine will the temperature rise exceed 50°C when operating continuously at rated load, voltage, and speed. Prof. John H. Kuhlmann's[8] synchronous machine design sheet in the American Customary System of Units[9] is given in Figure P8–30. Pursue the design project and justify the details furnished.

8–31. Prof. M. G. Say's design data[10] are given on pages 406 and 407 for three salient-pole synchronous machines along with their essential stator and rotor dimensions. Pursue the design projects and justify the details furnished here.

　　Generator A: 1,250 kVA, power factor 0.8 lagging, 6.6 kV, three-phase, 50 Hz, 20-pole, 300 rpm for engine drive

　　Generator B: 3,750 kVA, power factor 0.8 lagging, 10 kV, three-phase, 50 Hz, 10-pole, 600 rpm for hydraulic drive, requiring an inertia of 3,400 kg-m^2

　　Compensator C: 5 MVA leading and 2.5 MVA lagging, 11 kV, three-phase, 50 Hz, 6 pole, 1,000 rpm machine with a full-load loss of 125 kW (0.025 per unit). To reduce load (stray) loss, the stator core ends are stepped and the tooth stiffeners are insulated from the core plates.

[8] J. H. Kuhlmann, *Design of Electrical Apparatus,* 3d ed., N.Y., John Wiley & Sons, 1950.

[9] Wildi, T., *Units and Conversion Factors.,* IEEE Press, N. Y., 1991.

[10] M. G. Say, *Alternating Current Machines,* A Halstead Press Book, John Wiley & Sons, N. Y., 1978.

Generator

| Kva. 2500 | Volts 2400 | Phase 3 | Amperes 600 | Hertz 60 | Poles 32 |

R.p.m., 225   Kva./r.p.m., 11.1   Output constant, 1.58×10^4

Armature

Sheet steel . . . . 0.019-Dynamo grade
Outside diameter 112.0
Gap diameter 100.0
Total length 17.5
Ducts, number and size . . . . $5 - \frac{1}{2}$
Gross length 15.0
Effective length 14.0
Slots per pole per phase . . . . $3\frac{1}{2}$
Total number of slots 336
Type of winding . . . . 2 layer Wye
Circuits per phase 2
Coil throw . . . . Slots 1 and 10
Per cent pitch 85.7
Conductors:
    Per slot 4
    Dimensions . . . . $0.149 \times 0.276 - 4$
    Area . . . . 4×0.0325
    In series per phase 224
    Total 1326
Current density 2310
Length, one-half mean-turn 33.78
Resistance per phase, 25°C 0.0202
Resistance per phase, 75°C 0.024
Resistance drop, volts 18.7
Resistance drop, per cent 1.35
Reactance drop, volts 195
Reactance drop, per cent 14
Impedance drop, volts 196
Impedance drop, per cent 14.1
Armature reaction, AT per pole 5280
Armature reaction factor 0.852
Armature reaction, eq. fld. AT 4500
Square inches per watt 0.82
Ampere conductors, per inch 1,200
Short-circuit ratio 1.22
$\dfrac{D^2 Ln}{\text{Kva}}$ . . . . 1.58×10^4
ATP_{100} 8620
ATP_{80} 11,180
Regulation, 80% Pf 27.0%
Regulation, 100% Pf 16.5%

Field

Total air gap length . . . . 2 × 0.406
Rotor diameter 99.188
Peripheral speed 5840
Pole pitch 9.82
Pole arc 6.75
Material spider . . . . Hot-rolled steel
Damper bars per pole . . . .
Size of bar . . . .
Material of bar . . . .
Section end-ring . . . .
Material end-ring . . . .
Air gap coefficient 1.07
Effective length of gap 0.435
Leakage constant 1.35
f_d,   0.666, f_b,   1.14, k_p,   0.956, C_w,   0.725, k_p,   0.975
Total flux, 234,000 K.L. Flux per pole, 4879 K.L.

	Section	Den-sity	Length	Amp.-turns
Air gap	5310	44.1	1.07×0.406	5980
Teeth	2470	95	2.83	142
Armature yoke	88.5	55	5.34	19
Poles	70	93.6	7.00	315
Field yoke . . .	101	65.0	4.05	65
Total ampere turns per pole	. . . . . .	. . .	. . . .	6521

Size of conductor . . . . 0.109×1.25
Turns per pole 48
Amperes no load . . . . 136 Maximum, 253
Length of mean-turn 50.1
Resistance, 25°C 0.398
Resistance, 75°C 0.474
IR no-load . . . . 53.8 Maximum, 120
I^2R no-load . . . . 7360 Maximum, 30,400
Square inch per watt maximum I 0.74
Kva 2500
Power factor 80%
Amperes 233.0
IR 75°C 110.0
I^2R 75°C 25,700
Square inch per watt 0.875
Exciter voltage 125.0
Exciter capacity 33.0

Full-Load Losses, 100% P.F.

Friction and windage 15,000
Core 13,840
Stray load 7,800
Armature copper 25,900
Field copper 15,350
Total losses 77,890
Flywheel effect WR^2 . . . . 159,800 lb-ft.²

Weights

Armature copper 1890
Filed copper 3310.0
Armature teeth 1970
Armature yoke 4260.0
Filed poles 4300.0

Remarks: Vertical Water-wheel Type $X_d = 0.892$ $X_q = 0.54$
                                          δ_e f.l. −0.80 p.f.= 18.05

Figure P8–30 Synchronous Machine Design Sheet. Note: American Customary System of Units.

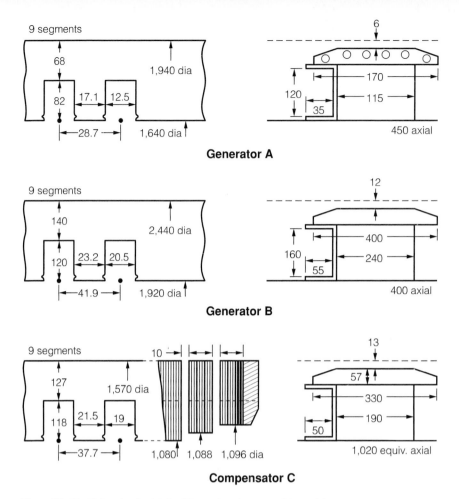

Generator A

Generator B

Compensator C

Figure P8–31 Slot and pole details. (Dimensions in mm; not to scale)

Data for Problem 8–31

			A	**B**	**C**
Rating	Output	kVA	1250	3750	5000
		Kw	1000	3000	0
	Line/phase voltage	kV	6.6/3.8	10/5.8	11/6.35
	Line-phase current	A	109	217	263
Main Dimensions	Magnetic loading	T	0.51	0.48	0.46
	Electric loading	kA/m	38.2	51.5	55
	Stator bore	m	1.64	1.92	1.08
	Gross core length	m	0.45	0.40	1.00
	Ducts,no./width	mm	6/10	5/10	14/10
	Net core length	m	0.39	0.35	0.86
	Iron length	m	0.351	0.315	0.775
	Radial gap length	mm	6	12	13
	Pole pitch	m	0.26	0.60	0.57

continued

Data for Problem 8–31 *Continued*

			A	B	C
Stator	Winding/connection	—		double-layer diamond/wye	
	No-load flux per pole	mWb	59	116	261
	Turns per phase	—	300	240	120
	Number of slots	—	180	144	90
	Conductors per slot	—	2×5	2×5	2×4
	Coil pitch	—	9	12	12
	Conductor size	mm	5×5	$2(10 \times 4)$	$2(8.5 \times 4.5)$
	area	mm^2	25	80	77
	Length of mean cond.	m	1.11	1.38	2.07
	Resistance	mΩ/ph	560	174	136
	Current density	A/mm^2	4.4	2.7	3.4
	Total f.l. I^{2}R loss	kW	20	24.5	28
	Copper mass	kg	475	1520	1060
Rotor	Type of pole	—	laminated	laminated	solid
	fixing	—	bolted	dove-tailed	screwed
	Damped bars/pole	—	8	0	0
	Turns/pole	—	$71\frac{1}{2}$	$99\frac{1}{2}$	$99\frac{1}{2}$
	Conductor size	mm	35×1.4	55×1.3	50×1.75
	area	mm^2	49	71	87
	Full-load current	A	135	207	270
	Current density	A/mm^2	2.75	2.9	3.1
	Length of mean turn	m	1.49	1.41	2.52
	Resistance	Ω	0.89	0.42	0.37
	Volt drop	V	120	86.5	100
	I^{2}R loss	kW	16.1	18	27
	Copper mass	kg	950	630	1300
	Peripheral speed	m/s	25.5	60.5	11.2
	Overspeed	p.u.	0	0.8	0
Magnetization	Flux density, core	T	1.24	1.31	1.33
	pole	T	1.20	1.23	1.40
	teeth ($\frac{1}{3}$)	T	1.61	1.67	1.74
	gap	T	0.77	0.725	0.79
	M.M.F./pole, no-load	kA-t	7.5	8.95	10.35
	full-load	kA-t	9.65	20.6	26.9
	Armature m.m.f./pole	kA-t	4.91	15.1	16.0
Performance	Full-load losses:				
	stator I^{2}R	kW	20.0	24.5	28.0
	rotor I^{2}R	kW	16.1	18.0	27.0
	core	kW	20.0	35.0	38.0
	mechanical	kW	3.5	14.5	23.0
	load (stray)	kW	7.4	12.0	5.0
	total	kW	67.0	104.0	121.0
	Output	kW	1000	3000	0
	Input	kW	1067	3104	121
	Efficiency	p.u.	0.937	0.967	0

9

Direct-Current Machines

Generally speaking, conventional dc generators are becoming obsolete and increasingly often are being replaced by solid-state rectifiers in most applications for which an ac supply is available. The same is not true with dc motors however. The torque-speed characteristics of dc motors are what make them extremely valuable in many industrial applications. The significant features of the direct-current drives include adjustable motor speed over wide ranges, constant mechanical power output or constant torque, rapid acceleration or deceleration, and responsiveness to feed-back signals.

The dc commutator machines are built in a wide range of sizes, from small control devices with a one-watt power rating up to the enormous motors of 10,000 hp or more used in rolling mill applications. The dc machines today are principally applied as industrial drive motors, particularly when high degrees of flexibility in controlling speed and torque are demanded. Such motors are used in steel and aluminum rolling mills, traction motors, overhead cranes, forklift trucks, electric trains, and golf carts. Commutator machines are also used in portable tools supplied from batteries, in automobiles as starter motors, in blower motors, and in control applications as actuators and speed-sensing or position-sensing devices.

In this chapter, we examine the steady-state performance characteristics of dc machines to help us understand their applications. Study of the flux, mmf conditions, and switching action at the commutator brings out conditions limiting machine capability and suggests methods for combating such conditions. The analysis, restricted in this chapter to the steady state, illustrates the versatility of the dc machine.

9.1 Constructional Features of DC Machines

The dc machines differ from ac machines in having a commutator along with the armature on the rotor. They also have salient poles on the stator, and, except for a few small machines, what are known as *commutating poles* between the main poles. Medium and large machines are provided with large heat-dissipating surfaces and effectively placed ventilating ducts for passage of cooling air in order to prevent hot spots. Rotor punchings for small machines are mounted solidly on the shaft, whereas, on large machines, it is customary to employ a spider, consisting of a hub and projecting arm, so that the annular

Photo 9.1.1 Cross-sectional view of a typical dc machine. (Photo courtesy of Westinghouse Electric Corporation.)

laminations can be rigidly fastened to the shaft, thereby allowing free flowing air to keep the armature ventilated and cooled. See Photo 9.1.1 for a cross-sectional view of a typical dc machine.

The armature-core punchings, slotted to receive the insulated armature winding, are usually of high-permeability electrical sheet steel, 0.4 to 0.65 mm thick, with an insulating film between them. Small and medium units use doughnut-shaped circular punchings, but large units (above 1 m in diameter) use segmental punchings along with appropriately placed ventilating ducts. The main and commutating-pole punchings are generally thicker than rotor punchings because only the pole faces are subjected to high-frequency flux changes. The pole punchings range from 1.5 to 3 mm thick and are normally riveted. The frame yoke is usually made from rolled mild steel plate, although, on high-demand large generators for rapidly changing loads, laminations can be used. The electromagnets are bolted to the cylindrical yoke or frame. End bells with their bearings and brush-rigging become part of the stator when the machine is assembled. The yoke may have a base with feet or a supporting bracket upon which the entire structure rests. See Photo 9.1.2 for a view of a fully assembled dc machine constructed on a pedestal.

The commutator is truly the heart of the dc machine. It is made up of hard copper bars, drawn accurately in a wedge shape, separated from each other by mica plate segments 0.5 to 1.25 mm thick, depending on the size of the machine and on the maximum voltage than can be expected between bars during operation. The mica segments and bars are clamped between two metal V-rings and insulated from them by cones of mica. On very high-speed commutators, shrink rings of steel are used to hold the bars. Mica is used under the rings. The carbon brushes with rounded contact surfaces ride on the commutator bars and carry the load current from the rotor coils to the external circuit. The brush holders hold the brushes against the commutator surface by springs to maintain a fairly constant pressure and smooth riding. See Photos 9.1.3 and 9.1.4 for a cross-sectional view of a V-ring commutator and a completely wound armature and commutator.

A dc generator or motor may have as many as four field windings, depending on the type and size of the machine and the kind of service it will have. These field windings consist of two normal exciting fields, the *shunt* and *series* windings, and two fields that act in a corrective capacity to combat the detrimental effects of armature reaction, called the *commutating* (*compole* or *interpole*) and *compensating*

Photo 9.1.2 View of a fully assembled dc machine with pedestal-type construction. (Photo courtesy of Westinghouse Electric Corporation.)

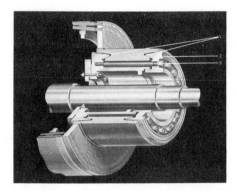

Photo 9.1.3 Cross-sectional view of a typical V-ring commutator of a dc machine. (Photo courtesy of Westinghouse Electric Corporation.)

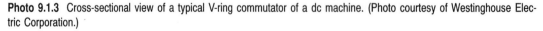

windings, which are connected in series with the armature. The type of machine, whether shunt, series, or compound, is determined solely by the normal exciting-field circuit connections, as shown in Figure 5.3.14. Figure 9.1.1 illustrates how various field windings are arranged with respect to one another in part of a cross-section of a dc machine, while Figure 9.1.2 shows the schematic connection diagram of a dc machine. The commutating and compensating windings, their purpose as well as circuit connections, are discussed in detail in the sections to follow. See Photo 9.1.5 for a view of commutating poles and compensating windings of an assembled dc machine.

Photo 9.1.4 A completely wound armature along with the commutator of a dc machine. (Photo courtesy of Westinghouse Electric Corporation.)

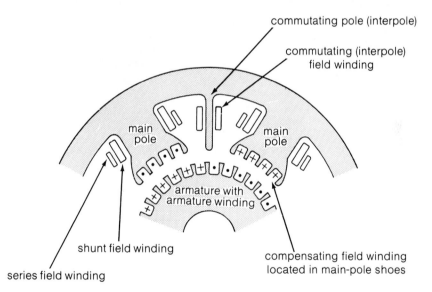

Figure 9.1.1 Section of a dc machine illustrating the arrangement of various field windings.

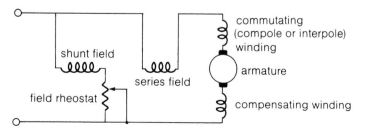

Figure 9.1.2 Schematic connection diagram of a dc machine.

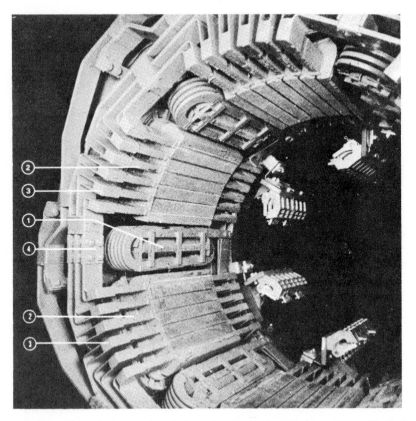

Photo 9.1.5 Commutating poles and compensating windings of a dc machine. (Photo courtesy of Westinghouse Electric Corporation.)

9.2 Equivalent Circuit of a DC Machine

A review of the material presented under "Elementary Direct-Current-Machines" in Section 5.3 can be helpful at this stage to recall the principles of operation for dc machines.

We discussed various arrangements of armature windings in Chapter 6, where we noted that the commutator brush is normally situated on the center line of a main pole, although it is connected to a coil in the interpolar gap. It is a common practice to use a schematic representation of a dc machine in which the brushes are shown in the position of the coil to which they are connected. The physical arrangements together with the schematic diagram and circuit representations are shown in Figure 9.2.1. Under *steady-state conditions*, the interrelationships between the voltage and current are given by

$$V_f = I_f R_f \tag{9.2.1}$$

and

$$V_t = E_a \pm I_a R_a \tag{9.2.2}$$

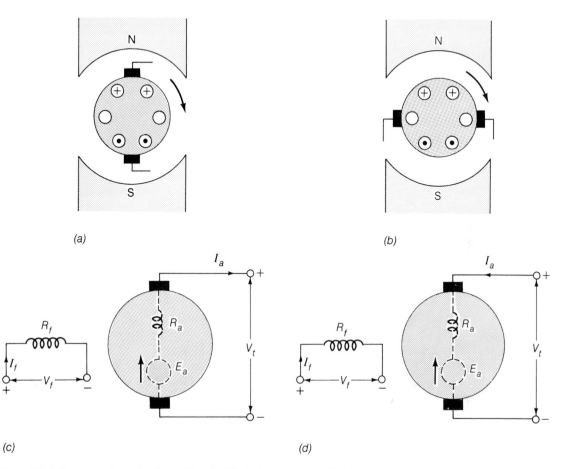

Figure 9.2.1 Representations of a dc machine. (a) Physical arrangement. (b) Schematic diagram. (c) Circuit representation of a dc generator under steady-state conditions $V_f = R_f I_f$; $V_t = E_a - I_a R_a$. (d) Circuit representation of a dc motor under steady-state conditions $V_f = I_f R_f$, $V_t = E_a + I_a R_a$.

where the plus sign signifies a motor and the minus sign a generator. V_f is the voltage applied to the field circuit, I_f is the field current, and R_f is the field-winding resistance. V_t is the terminal voltage, E_a is the generated emf, I_a is the armature current, and R_a is the armature resistance. The generated emf E_a is given by Equation 5.3.24 as

$$E_a = K_a \phi \omega_m \qquad (9.2.3)$$

which is the speed (motional) voltage induced in the armature circuit due to the flux of the stator-field current. The electromagnetic torque T_e is given by Equation 5.3.26 as

$$T_e = K_a \phi I_a \qquad (9.2.4)$$

where K_a is the design constant. The product $E_a I_a$, known as the electromagnetic power being converted, is related to the electromagnetic torque by the relation

$$P_{em} = E_a I_a = T_e \omega_m \qquad (9.2.5)$$

For a motor, the terminal voltage is always greater than the generated emf, and the electromagnetic torque produces rotation against a load. For a generator, the terminal voltage is less than the generated emf, and the electromagnetic torque opposes that applied to the shaft by the prime mover. If the magnetic circuit of the machine is not saturated, note that the flux ϕ in Equations 9.2.3 and 9.2.4 is proportional to the field current I_f producing the flux.

▮▮▮▮ ‖‖‖‖
9.3 Commutator Action

In the discussion on dc armature in Section 6.3, we saw why a rotating commutator winding is called a pseudostationary winding that produces a stationary flux when carrying a direct current. As a consequence of the arrangement of the commutator and brushes, the currents in all conductors under the north pole are in one direction and the currents in all conductors under the south pole are in the opposite direction. Thus, the magnetic field of the armature currents is stationary in space in spite of the rotation of the armature. As seen from Figures 6.3.2 and 6.3.3, during the time when the brushes are simultaneously in contact with two adjacent commutator segments, the coils connected to these segments are short-circuited by the brushes and are temporarily removed from the main circuits through the windings; the directions of the currents in the coils are then reversed. The coils are then said to be undergoing *commutation* during this interval of time, and the process of reversal of currents in the coil is known as commutation. If the width of the brush is infinitesimally small, then the current reversal is instantaneous. In practice, however, the current reversal takes place during some finite time Δt, depending on the speed of the rotor and the width of the brushes. The current changes from $+I$ to $-I$ in time Δt. Ideally, the current in the coils being commutated should reverse linearly with time, as shown in Figure 9.3.1. Serious departure from *linear commutation* results in sparking at the brushes. Means for achieving sparkless commutation are discussed in Section 9.5. As shown in Figure 9.3.1, with linear commutation, the waveform of the current in any coil as a function of time is trapezoidal.

The air-gap flux distribution produced by the field windings is symmetrical about the center line of the field poles, known as the *field axis* (also called the *direct* or *d*-axis). The axis of the armature-mmf wave is 90 electrical degrees away from the axis of the field poles, i.e., in the *interpolar axis (also called the quadrature or q-axis),* as shown in Figure 5.3.5. The brushes are located so that commutation occurs when the coil sides are in the *neutral zone,* midway between the field poles. Because of the position of the coils to which the brushes are connected, the brushes are shown in the quadrature axis in Figure 5.3.5, and the armature-mmf wave is then directed along the brush axis as shown. The geometrical (physical) position of the brushes in an actual machine, however, is approximately 90 electrical degrees

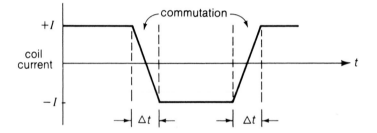

Figure 9.3.1 Waveform of current in an armature coil undergoing linear commutation.

away from their position in the schematic diagram, i.e., along the center lines of the main poles, because of the shape of the end connections to the commutator, as can be seen from Figures 6.3.2 and 6.3.3. As stated earlier, the magnetic torque and the speed voltage appearing at the brushes are independent of the spatial waveform of the flux distribution.

9.4 Armature Reaction

We showed in Section 6.7 and by Figure 6.7.4 that the armature-mmf wave can be closely approximated by a triangle corresponding to the wave produced by a finely distributed armature winding or current sheet. Figure 9.4.1a shows the flux-density distribution due to the main field alone, while Figure 9.4.1b shows the armature-mmf distribution for a machine with brushes in the geometric neutral axes as well as the flux-density distribution due to the armature mmf alone. Note that the armature-reaction flux is appreciably decreased in the interpolar space. Observe also that the axis of the armature mmf is fixed at 90 electrical degrees from the main-field axis by the brush position. Hence, armature reaction of this type is known as the *cross-magnetizing armature reaction*, which apparently causes a decrease in the resultant air-gap flux density distribution under one half of the pole and an increase under the other half. The resultant air-gap flux-density distribution, when the armature and field windings are both excited, is shown in Figure 9.4.1c. We can also see from Figure 9.4.1c that the effect of the armature reaction is to distort the flux wave and shift the position of the magnetic neutral axis in the direction of rotation for a generator and against the direction of rotation for a motor.

In general, because of the nonlinearity of the iron-magnetic circuit, the resultant flux-density distribution is not the algebraic sum of the flux distributions due to main field and armature mmf. Because of the saturation of iron, the flux density is decreased by a greater amount under one pole tip than it is increased under the other. Hence, the resultant flux per pole is lower than the one that would be produced by the field winding alone. This phenomenon is known as the *demagnetizing effect of the cross-magnetizing armature reaction*, caused by saturation. At the flux densities usually employed for normal machine operation, the effect is significant, especially at heavy loads.

The distortion of the flux distribution caused by the cross-magnetizing armature reaction can have a detrimental influence on commutation and can limit the short-time overload of a dc machine. Its effect can be limited, however, during the design and construction of the machine by increasing the reluctance of the cross-flux path; such techniques as using chamfered or eccentric pole faces, which increase the air gaps at the pole tips, can be employed. The best yet most expensive corrective measure is to compensate the armature mmf by means of a distributed *compensating winding,* embedded in slots in the main-pole faces. Such a design is presented in Section 9.5.

9.5 Interpoles and Compensating Windings

When the commutator bar that is connected to a particular armature coil rotates past a fixed commutator brush, the coil current must be reversed. In normal practice, a brush spans more than one commutator segment so that, during commutation, several coils in series are short-circuited. The reversal of the coil current in a short time induces a voltage of self-induction in the commutated coil, but the voltage of mutual induction from other coils (particularly from those in the same slots, undergoing commutation at the same time) opposes changes in current in the commutated coil. The sum of these two voltages (self-induction and mutual induction) is often referred to as the reactance voltage. While the ideal process of

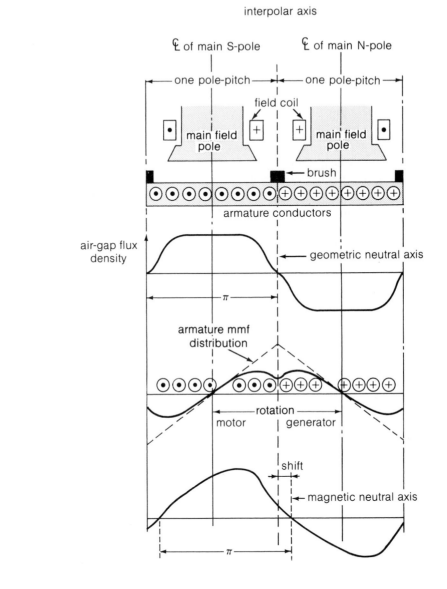

Figure 9.4.1 Main field, armature, and resultant flux-density distributions with brushes on the geometric neutral axes. (a) Flux-density distribution due to main field alone, (b) flux-density distribution due to armature mmf alone, (c) resultant flux-density distribution.

commutation is linear, as shown in Figure 9.3.1, the reactance voltage causes the condition known as *undercommutation* (or *delayed commutation,*) as illustrated in Figure 9.5.1.

The effect of a given reactance voltage in delaying commutation is minimized by the resistive brush-contact voltage drop, which is why carbon brushes with appreciable contact drop are usually employed. When good commutation is achieved by virtue of resistive drops, the process is known as the *resistance commutation*. It is used these days only in fractional horsepower machines.

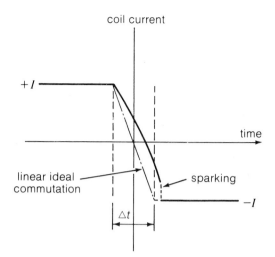

Figure 9.5.1 Undercommutation (delayed commutation).

It can be seen from Figure 9.4.1 that one of the reasons for poor commutation is the effect of armature-reaction flux in shifting the position of the magnetic neutral axis. The cross-magnetizing armature reaction causes distortion in the resultant air-gap flux distribution and creates flux in the interpolar region. The rotational voltage induced in the short-circuited coil becomes another important factor in the commutation process. The direction of the induced voltage being the same as that of the current under the immediately preceding pole face, this induced voltage then encourages the continuance of the current in the old direction and, like the reactance voltage, opposes the reversal of the current in the commutated coil. The general principle of producing a rotational voltage that approximately compensates for the reactance voltage in the coil undergoing commutation is known as *voltage commutation*. Voltage commutation is used in almost all modern commutating machines. The appropriate flux density is introduced in the commutating zone by means of small, narrow auxiliary poles located between the main poles and centered on the interpolar gap; these auxiliary poles are known as *interpoles, commutating poles,* or simply as *compoles*. They produce the so-called *commutation voltage* to counteract the adverse effects of flux distortion and the reactance voltage on commutation. For small machines, such a voltage can be introduced by shifting the brush position. *Brush shift* in the direction of rotation in a generator or against rotation in a motor produces a direct demagnetizing mmf that can result in unstable operation of a motor or excessive drop in the voltage of a generator.

The most generally used method for aiding commutation is by providing the machine with interpoles. The commutating (interpole) winding must be connected in series with the armature, since both the armature mmf and the reactance voltage are proportional to the armature current. The interpole mmf must be sufficient not only to neutralize the cross-magnetizing armature mmf in the interpolar region but also to furnish the flux density required for the rotational voltage in the short-circuited armature coil to counteract the reactance voltage. Since the compole must produce flux proportional to armature current (to preserve the desired linearity), it must operate with an unsaturated magnetic circuit; hence, the air gap between the armature surface and the compole is normally greater than that between the armature and the main pole. The arrangement of a two-pole generator with interpoles is shown in Figure 9.5.2; the compole has the same polarity as the main pole ahead in the direction of rotation. For a motor, the arrangement is shown in Figure 9.5.3; the interpole has the same polarity as the main pole behind,

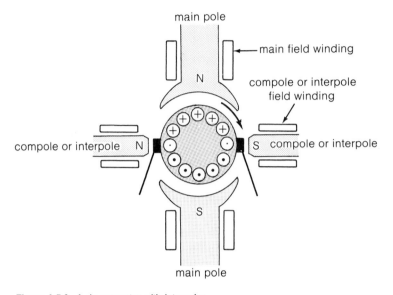

Figure 9.5.2 A dc generator with interpoles.

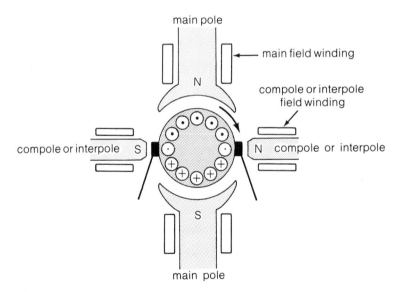

Figure 9.5.3 A dc motor with interpoles.

relative to the direction of rotation. Under these circumstances, with appropriate design of interpoles, a dc machine can be operated as either a motor or a generator with either direction of rotation.

Even with interpoles, when a machine is subjected to heavy overloads, rapidly changing loads, or operation with a weak main field, the armature reaction under the poles can badly distort the flux wave and cause induced coil voltage high enough to result in flashover or arcing between commutator segments. This process can quickly spread around the entire commutator, leading to a direct short circuit on the line, besides its destructive effects on the commutator itself. The demagnetizing effect of the armature mmf under pole faces can be compensated for or neutralized by providing a *compensating (pole-face) winding,* arranged in slots in the pole face, having a polarity opposite to that of the adjoining armature winding. With the proper number of turns and connected in series with the armature in order to carry a proportional current, the compensating winding has the same axis as that of the armature and will almost completely neutralize the armature reaction of the armature inductors under the pole faces. The net effect of the main field, armature, commutating winding, and compensating winding on the air-gap flux distribution is that, except for the commutation zone in the interpolar region, the resultant flux-density distribution is substantially the same as that produced by the main field alone, as sketched in Figure 9.5.4.

Furthermore, with the addition of a compensating winding, the armature-circuit time constant is reduced and hence the speed of response is improved. Because they are costly, pole-face windings are usually employed only in machines designed for heavy overload or rapidly changing loads, such as steel-mill motors subjected to reverse duty cycles or in motors intended to operate over wide speed ranges by shunt-field control. The schematic diagram of Figure 9.1.2 shows the relative positions of various windings, indicating that the commutating and compensating fields act along the armature axis (i.e., the quadrature axis), and the shunt as well as series fields act along the axis of the main poles (i.e., the direct axis). It is thus possible to achieve rather complete control of the air-gap flux around the entire armature periphery, along with smooth sparkless commutation.

EXAMPLE 9.5.1

A six-pole, 600-Kw dc generator has a simplex lap winding with 720 armature conductors and is equipped with six commutating poles.

a. If the mmf of the commutating poles is 1.2 times that of the armature, calculate the number of turns in the interpole winding.
b. If the generator is also provided with a compensating winding, and the pole-face covers 70% of the pole-span, compute the number of conductors for the compensating winding in each pole-face.
c. When the pole-face winding of part b is in the circuit, what should be the number of turns per pole in the interpole winding?

Solution

a. The number of armature conductors being 720, the number of turns is $720/2 = 360$; the number of armature turns per pole is $360/6 = 60$; and the number of parallel paths in the armature is 6, the same as the number of poles (given that the winding is simplex lap).

Because the interpole winding is in series with the armature carrying the armature current I_a, the mmf per interpole must be $1.2 \times 60 I_a/6$, and the number of turns in the interpole winding per pole is given by 12.

b. The armature conductors per pole being $720/6 = 120$, then the number of armature conductors under each pole face is given by $120 \times 0.7 = 84$. Since the compensating winding carries the entire armature current, i.e., 6 times that of the armature conductors, the number of conductors per pole for the compensating winding is then $84/6 = 14$.

c. Because good commutation requires 12 turns per pole, of which $14/2 = 7$ are provided by the compensating winding, the interpole winding should make up the difference, i.e., $12 - 7 = 5$ turns per pole.

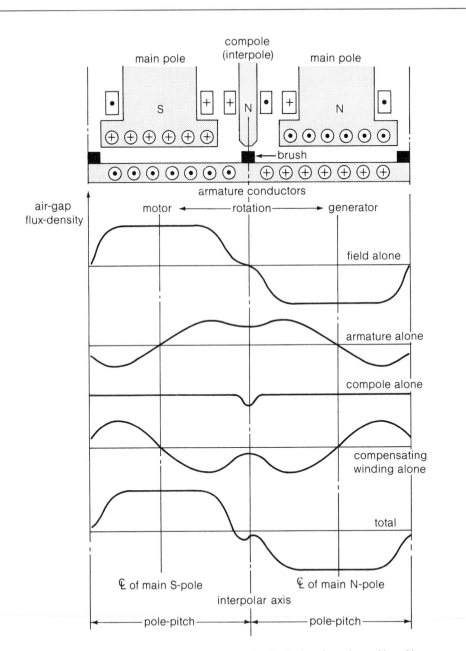

Figure 9.5.4 Component and total air-gap flux-density distributions for a dc machine with interpoles and compensating winding.

■■■■■■■■■■■■■**IIIIIIIII**
9.6 DC Generator Characteristics

Figure 5.3.14 shows schematic diagrams of field-circuit connections for dc machines without including commutating-pole or compensating windings. Shunt generators can be either *separately excited,* or *self-excited,* as shown in Figures 5.3.14a and b, respectively. Compound machines can be connected *long shunt*, as in Figure 5.3.14d, or *short shunt*, in which the shunt-field circuit is connected directly across the armature, without including the series field.

In general, three characteristics specify the *steady-state* performance of a dc generator:

- The *open-circuit characteristic* (abbreviated as OCC, and also known as the *no-load magnetization curve*), which gives the relationship between the generated emf and field current at constant speed
- The *external characteristic*, which gives the relationship between the terminal voltage and load current at constant speed
- The *load characteristic*, which gives the relationship between the terminal voltage and field current, with constant armature current and speed

All other characteristics depend on the form of the open-circuit characteristic, the load, and the method of field connection. Under steady-state conditions, the currents being constant or, at most, slowly varying, voltage drops due to inductive effects are negligible.

As stated earlier and shown in Figure 9.2.1, the terminal voltage V_t of a dc generator is related to the armature current I_a and the generated emf E_a by

$$V_t = E_a - I_a R_a \tag{9.6.1}$$

where R_a is the total internal armature resistance, including resistance of interpole and compensating windings as well as that of the brushes. The value of the generated emf E_a, given by Equation 9.2.3, is governed by the direct-axis field flux (which is a function of the field current and armature reaction) and the angular velocity ω_m of the rotor.

The open-circuit and load characteristics of a separately excited dc generator, along with its schematic diagram of connections, are shown in Figure 9.6.1. It can be seen from the form of the external volt-ampere characteristic, shown in Figure 9.6.1c, that the terminal voltage falls slightly as the load current increases. *Voltage regulation* is defined as the percentage change in terminal voltage when full load is removed, so that, from Figure 9.6.1c, it follows that

$$\text{Voltage regulation} = \frac{E_a - V_t}{V_t} \times 100\% \tag{9.6.2}$$

Because the separately excited generator requires a separate dc field supply, its use is limited to applications in which a wide range of controlled voltage is essential.

Let us next consider the case of a self-excited dc shunt generator for which the schematic diagram of connections and the open-circuit characteristic are given in Figures 9.6.2a and b, respectively. With a field resistance, given by a line *OR* of slope *OA/OB*, shown in Figure 9.6.2b, the open-circuit voltage will build up to a value of *OA*. It follows that a *critical-field resistance* must occur when the corresponding critical-field resistance line OR_c is coincident with the linear part of the characteristic. It is the maximum permissible value of the field resistance if the armature voltage is to build up. Reasons for failure of voltage buildup can be summarized as follows:

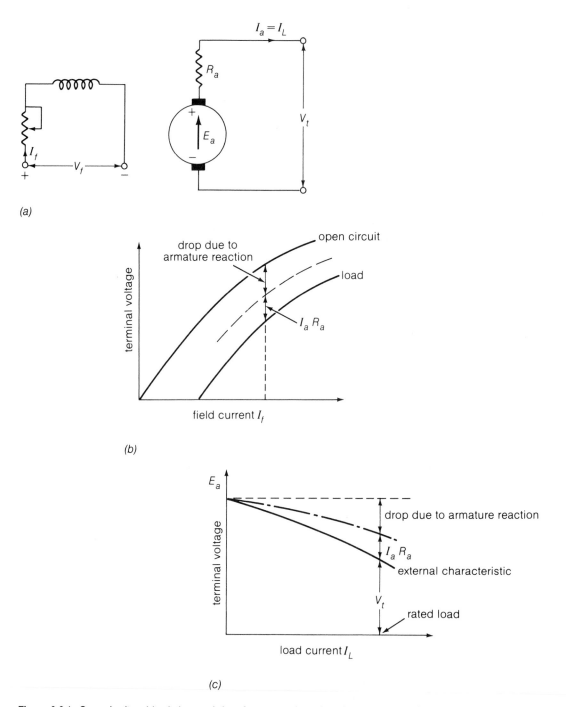

Figure 9.6.1 Open-circuit and load characteristics of a separately excited dc generator. (a) Schematic diagram of connections. (b) Terminal voltage vs. field current relationships. (c) External volt-ampere characteristic.

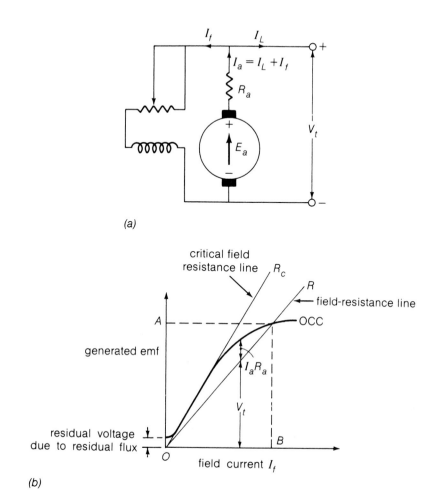

Figure 9.6.2 Self-excited dc shunt generator. (a) Schematic diagram of connections. (b) Open-circuit characteristic.

- Insufficient residual flux
- Incorrect polarity of the field winding, causing the field current to run in a direction that will reduce the residual flux
- Field resistance above critical value (if the speed is normal)
- Speed below critical value (if the resistance of the field circuit is normal)

When the form of the open-circuit characteristic, the field resistance and turns, the armature resistance, and the demagnetizing ampere-turns of armature reaction are all known, various graphical techniques can be used to determine the variation of terminal voltage with armature current. One such technique is illustrated in Figure 9.6.3. For a value of armature current $O'L$, when we know the effective reduction in total ampere-turns caused by demagnetization, the equivalent reduction in field current OA is given by the demagnetization ampere-turns divided by the field turns. The armature resistance drop

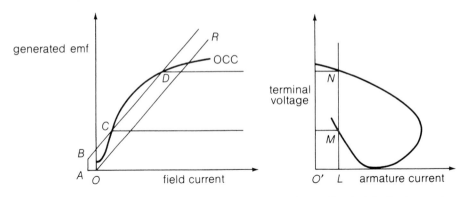

Figure 9.6.3 Graphical technique to compute the $V_t - I_a$ curve for a self-excited dc shunt generator.

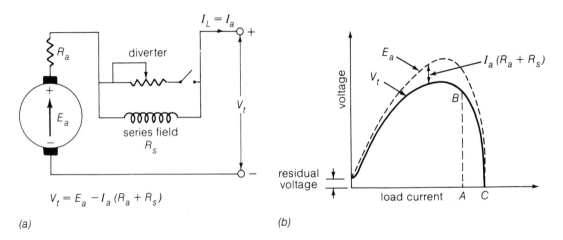

$$V_t = E_a - I_a(R_a + R_s)$$

(a)

(b)

Figure 9.6.4 A dc series generator. (a) Schematic diagram of connections. (b) Volt-ampere characteristic at constant speed.

for the same armature current is given by AB. A line through B parallel to the field-resistance line OR is now drawn to meet the OCC at the points C and D. Horizontal lines CM and DN are then drawn to meet the vertical from L at points M and N, respectively. M and N are then the points on the terminal-voltage versus armature-current characteristic; by repeating this process for several values of armature current, the complete $V_t - I_a$ characteristic can be obtained, as in Figure 9.6.3. By subtracting the field current I_f from the armature current I_a for each point on the $V_t - I_a$ characteristic, the external $V_t - I_L$ characteristic can be obtained. We can observe from Figure 9.6.3 that a maximum value of armature current exists beyond which a decrease in load resistance will cause a decrease in armature current. Under short-circuit conditions, the armature current is given by the ratio of the residual voltage to the internal armature resistance. If the demagnetizing effect is neglected, the distance OA becomes zero and the construction for the $V_t - I_a$ or $V_t - I_L$ characteristic is simplified.

A shunt generator maintains approximately constant voltage on load; it finds wide application as an exciter for the field circuit of large ac generators. The shunt generator is also sometimes used as a tachogenerator when a signal proportional to a motor speed is required for control or display purposes.

The schematic diagram and the volt-ampere characteristic of a dc series generator at constant speed are shown in Figure 9.6.4. The resistance of the series-field winding must be low for efficiency as well

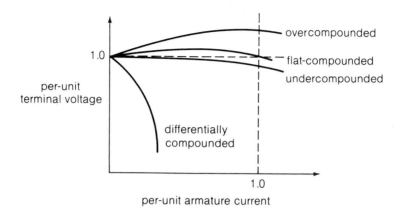

Figure 9.6.5 Volt-ampere characteristics of dc compound generators at constant speed.

as for low voltage drop. The series generator was used in early constant-current systems by operating in the range *BC*, where the terminal voltage fell off very rapidly with increasing current.

The volt-ampere characteristics of dc compound generators at constant speed are shown in Figure 9.6.5. *Cumulatively compound* generators, in which the series and shunt-field-winding mmfs are aiding, may be *overcompounded, flat-compounded,* or *undercompounded,* depending on the strength of the series field. Overcompounding can be used to counteract the effect of a decrease in the prime-mover speed with increasing load or to compensate for the line drop when the load is at a considerable distance from the generator. *Differentially compounded* generators, in which the series-winding mmf opposes that of the shunt-field winding, are used in applications in which a wide variation in load voltage can be tolerated and when the generator might be exposed to load conditions approaching short circuit.

▌▏▎▍▎▏▎▍▌▏▎▍▐

EXAMPLE 9.6.1

The open-circuit and short-circuit test data of a four-pole dc generator rated at 200 V, 100 A, and 1,000 rpm are given in Table 9.6.1. The armature is wave-wound with 704 conductors. The armature resistance is 0.15 Ω; the series-field resistance is 0.01 ohm; and the separately excited field resistance is 100 Ω with 2,500 turns per pole. The brush drop is 2 V.

Table 9.6.1

Open-Circuit Characteristic								
field current, A	0.1	0.4	0.6	1.0	1.3	1.5	2.1	2.3
terminal voltage, V	17	80	120	170	190	200	217	220

Short-Circuit Characteristic			
field current, A	0.1	0.2	0.3	0.4
armature current, V	50	80	100	120

Calculate the following values:

a. The demagnetizing mmf due to full-load armature reaction, expressed as a percentage of full-load armature mmf.
b. The no-load voltage when the separately excited field current is adjusted to give 200 V on full load.
c. The new field current and the no-load voltage when a series winding having 25 turns per pole is cumulatively connected in circuit, and the voltage is adjusted to 200 V on full load.
d. The value of the series-diverter resistance to give a flat-compounded curve at 200 V.

Solution

a. Full-load armature mmf $F_a = [(100/2)/4] \times 704/2 = 4{,}400$ ampere-turns (At) per pole. (Note that the number of parallel paths for a wave winding is 2.) Total excitation for 100 A on short circuit $= 0.3$ A $= F_f/2{,}500$. The internal voltage drop at full load $= (100 \times 0.15) + 2 = 17$ V. Excitation for 17 V from the OCC $= 0.1$ A $= F_r/2{,}500$.

   Armature-demagnetizing mmf in terms of the equivalent field amperes is $0.3 - 0.1 = 0.2$ A $= -F'_a/2{,}500$. (Note that the effective or resultant mmf is $F_r = F_f + F'_a$, where F'_a is usually negative because of the demagnetizing effect.) As a percentage, $F'_a/F_a = (0.2 \times 2{,}500/4{,}400) \times 100 = 11.36\%$.

b. The full-load voltage plus the voltage drop at full load $= 200 + 17 = 217$ V. The required number of resultant ampere-turns, F_r, as read from the OCC corresponding to 217 V, is $2.1 \times 2{,}500 = 5{,}250$. The demagnetizing mmf due to full-load armature reaction $-F'_a = 0.2 \times 2{,}500 = 500$. Because F'_a is negative, the total excitation required, F_f, is 5,750 At; the total field current required is then $5{,}750/2{,}500 = 2.3$ A, yielding 220 V on no load from the OCC.

c. Neglecting the small additional series-field resistance drop, the total excitation must still be 5,750 At; the series-field winding contributes $25 \times 100 = 2{,}500$ At; hence, the separate-field winding must provide $5{,}750 - 2{,}500 = 3{,}250$ At or a field current of $3{,}250/2{,}500 = 1.3$ A, corresponding to which the no-load voltage from the OCC is 190 V.

d. For 200 V on no load, the separate-field excitation is 1.5 (as read from the OCC) $\times 2{,}500 = 3{,}750$ At; for the total excitation at full load to be $5{,}750$ At, the series-field contribution must be reduced to $5{,}750 - 3{,}750 = 2{,}000$ At; and the corresponding series-field current must be $2{,}000/25 = 80$ A. Therefore, the required series-diverter resistance is $(80/20) \times 0.01 = 0.04$ Ω.

▮▮▮▮▮▮▮▮▮▮▮▮|||||||
9.7 DC Motor Characteristics

We can gain an understanding of the speed-torque characteristics of a dc motor from Equations 9.2.2 through 9.2.4. In shunt motors, the field current can be simply controlled by the use of a variable resistance in series with the field winding; the load current influences the flux only through armature reaction, and its effect is therefore relatively small. In series motors, the flux is largely determined by the armature current, which is also the field current; it is somewhat difficult to control the armature and field currents independently. In the compound motor, the effect of the armature current on the flux depends on the degree of compounding. Most motors are designed to develop a given horsepower at a specified speed, and it follows from Equations 9.2.2 and 9.2.3 that the angular velocity ω_m can be expressed as

$$\omega_m = \frac{V_t - I_a R_a}{K_a \phi} \qquad (9.7.1)$$

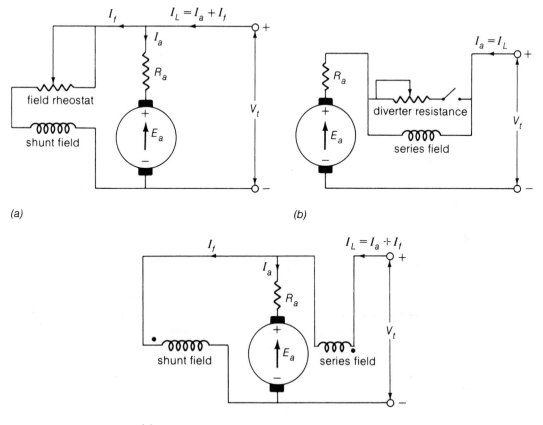

(a)

(b)

(c)

Figure 9.7.1 Schematic diagrams of dc motors. (a) Shunt motor. (b) Series motor. (c) Cumulatively compounded motor.

Thus the speed of a dc motor depends on the values of the applied voltage V_t, the armature current I_a, the resistance R_a and the field flux per pole ϕ.

The schematic arrangement of a shunt motor is shown in Figure 9.7.1a. For a given applied voltage and field current, Equations 9.2.4 and 9.2.3 can be rewritten as

$$T_e = (K_a\phi)I_a = K_m I_a \tag{9.7.2}$$

$$E_a = (K_a\phi)\omega_m = K_m \omega_m \tag{9.7.3}$$

Because

$$V_t = E_a + I_a R_a, \text{ or } I_a = (V_t - E_a)/R_a$$

it follows that

$$T_e = \frac{K_m V_t}{R_a} - \frac{K_m^2 \omega_m}{R_a} \tag{9.7.4}$$

$$P_{em} = E_a I_a = T_e \omega_m = \frac{V_t K_m \omega_m}{R_a} - \frac{K_m^2 \omega_m^2}{R_a} \tag{9.7.5}$$

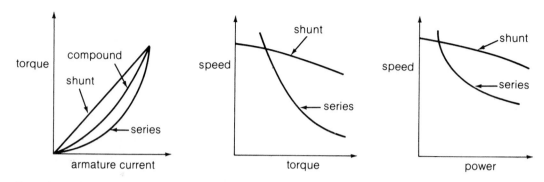

Figure 9.7.2 Characteristic curves for dc motors.

The forms of the torque-armature current, speed-torque, and speed-power characteristics for a shunt-connected dc motor are illustrated in Figure 9.7.2.

The shunt motor is essentially a constant speed machine with a low speed regulation. As seen from Equation 9.7.1, the speed is inversely proportional to the field flux, and thus it can be varied by controlling the field flux. When the motor operates at very low values of field flux, however, the speed will be high, and if the field becomes open-circuited, the speed will rise rapidly beyond the permissible limit governed by the mechanical structure. In order to limit the speed to a safe value, when a shunt motor is to be designed to operate with a low value of shunt-field flux, it is usually fitted with a small cumulative series winding, known as a *stabilizing winding*.

The schematic diagram of a series motor is shown in Figure 9.7.1b. The field flux is directly determined by the armature current so that

$$T_e = K_a \phi I_a = K I_a^2 \tag{9.7.6}$$

and with negligible armature resistance,

$$V_t = E_a = K_a \phi \omega_m = K I_a \omega_m \tag{9.7.7}$$

$$T_e = \frac{V_t^2}{K \omega_m^2} \tag{9.7.8}$$

$$P_{em} = \omega_m T_e = \frac{V_t^2}{K \omega_m} \tag{9.7.9}$$

and the speed-power curve is a rectangular hyperbola. The forms of the torque-armature current, speed-torque, and speed-power characteristics for a series-connected dc motor are illustrated in Figure 9.7.2. Note that the no-load speed is very high; care must be taken to ensure that the machine always operates on load. In practice, however, the series machine normally has a small shunt-field winding to limit the no-load speed. The assumption that the flux is proportional to the armature current is valid only on light load in the linear region of magnetization; in general, performance characteristics of the series motor must be obtained by using the magnetization curve. The series motor is ideally suited to traction, when large torques are required at low speeds and relatively low torques are needed at high speeds.

A schematic diagram for a cumulatively compounded dc motor is shown in Figure 9.7.1c. The operating characteristics of such a machine lie between those of the shunt and series motors, as shown in Figure 9.7.2. (The differentially compounded motor has little application, since it is inherently unstable,

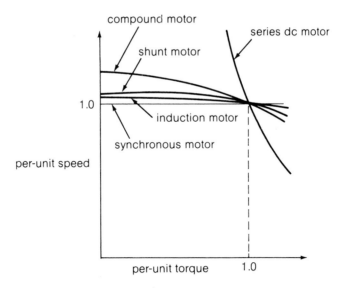

Figure 9.7.3 Typical speed-torque characteristics of various electric motors.

particularly at high loads.) Figure 9.7.3 shows a comparison of the speed-torque characteristics for various types of electric motors; 1.0 per unit represents rated values.

EXAMPLE 9.7.1

A 200-V dc shunt motor has a field resistance of 200 Ω and an armature resistance of 0.5 Ω. On no-load, the machine operates with full field flux at a speed of 1,000 rpm with an armature current of 4 A. Neglect magnetic saturation and armature reaction.

a. If the motor drives a load requiring a torque of 100 N·m, find the armature current and speed of the motor.

b. If the motor is required to develop 10 hp at 1,200 rpm, compute the required value of external series resistance in the field circuit.

Solution

a. Full field current $I_f = 200/200 = 1$ A. On no load, $E_a = V_t - I_a R_a = 200 - (4 \times 0.5) = 198$ V. Since $E_a = k_1 I_f \omega_m$, where k_1 is a constant,

$$k_1 = \frac{198}{1 \left(\frac{2\pi}{60} \times 1,000 \right)} = 1.89$$

On load, $T_e = k_1 I_f I_a$, or $100 = 1.89 \times 1.0 \times I_a$; therefore the armature current $I_a = 100/1.89 = 52.9$ A. Now, $V_t = E_a + I_a R_a$, or $E_a = 200 - (52.9 \times 0.5) = 173.55$ V. Since $E_a = k_1 I_f \omega_m$, it follows that

$$\omega_m = \frac{173.55}{1.89 \times 1.0} = 91.8 \text{ rad/s}$$

That is, the load speed = $(91.8 \times 60)/2\pi = 876$ rpm.

b. For 10 hp at 1,200 rpm,

$$T_e = \frac{10 \times 746}{\frac{2\pi}{60} \times 1,200} = 59.34 \text{ N} \cdot \text{m}$$

Then $59.34 = 1.89 I_f I_a$, or $I_f I_a = 31.4$. Since $V_t = E_a + I_a R_a$, it follows that

$$200 = 1.89 \left(\frac{2\pi}{60} \times 1,200 \right) I_f + 0.5 I_a$$

or

$$200 = 237.6 I_f + 0.5 I_a = 237.6 I_f + \frac{0.5 \times 31.4}{I_f}$$

$I_f = 0.754$ A or 0.088 A; and $I_a = 31.4/I_f = 41.6$ A or 356.8 A. Since the value of $I_f = 0.088$ A will produce very high armature currents, it will not be considered. Thus, with $I_f = 0.754$ A,

$$R_f = 200/0.754 = 265.25 \ \Omega$$

$$\text{External resistance required} = 265.25 - 200 = 65.25 \ \Omega$$

■■■■■■■■|||||||
9.8 Control of DC Motors

Equation 9.7.1 showed that the speed of a dc motor can be varied by control of the field flux, the armature resistance, and the armature applied voltage. The three most common speed-control methods are *shunt-field-rheostat control, armature-circuit-resistance control,* and *armature-terminal-voltage control.* The *base speed* of the machine is defined as the speed with rated armature voltage and normal armature resistance and field flux. Speed control above the base value can be obtained by varying the field flux. By inserting a series resistance in the shunt-field circuit of a dc shunt motor (or a compound motor), we can achieve speed control over a wide range above the base speed. It is important to note, however, that a reduction in the field flux causes a corresponding increase in speed, so that the generated emf does not change appreciably while the speed is increased, but the machine torque is reduced as the field flux is reduced. The dc motor with shunt-field-rheostat speed control is accordingly referred to as a *constant-horsepower drive.* This method of speed control is suited to applications in which the load torque falls as the speed increases. For a machine with a series field, speed control above the base value can be achieved by placing a diverter-resistance in parallel with the series winding, so that the field current is less than the armature current.

When speed control below the base speed is required, the effective armature resistance can be increased by inserting external resistance in series with the armature; this method can be applied to shunt, series, or compound motors. It has the disadvantage, however, that the series resistance, carrying full armature current, will cause significant power loss with an associated reduction in overall efficiency. The speed of the machine is governed by the value of the voltage drop in the series resistor and is therefore a function of the load on the machine. The application of this method of control is thus limited; because of its low initial cost, however, the series-resistance method or a variation of it is often attractive economically for short-time or intermittent slowdowns.

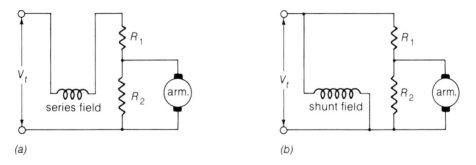

Figure 9.8.1 Shunted-armature method of speed control. (a) As applied to a series motor. (b) As applied to a shunt motor.

Unlike shunt-field control, armature-resistance control offers a *constant-torque drive* because both flux and, to a first approximation, allowable armature current remain constant as speed varies. The *shunted-armature method* is a variation of this control scheme, as shown in Figure 9.8.1a as applied to a series motor and Figure 9.8.1b as applied to a shunt motor. Resistors R_1 and R_2 act as a voltage divider applying a reduced voltage to the armature. They offer greater flexibility in their adjustment to provide the desired performance.

The *Ward-Leonard system* (or any of its several variations) is a conventional scheme that uses the armature-terminal-voltage control. While this system is one of the most versatile methods of speed control, the recent development of solid-state controlled rectifiers capable of handling many kilowatts has opened up a whole new field of solid-state dc motor drives with precise control of motor speed. These are not presented here, however, and we discuss only the basic elements of the Ward-Leonard method of speed control. Since the available power is usually of constant-voltage alternating current, auxiliary equipment in the form of a rectifier or a motor-generator set is required to provide the controlled armature voltage for the motor whose speed is to be controlled.

The schematic diagram of a basic Ward-Leonard system is shown in Figure 9.8.2. An individual motor-generator set supplies power to the armature of the main motor *M* whose speed is to be controlled; the field-rheostat adjustment in the separately excited generator *G* controls the armature voltage of the main dc motor *M*, thereby permitting close control of speed over a wide range. The motor field control (effected by means of the rheostat in the field of motor *M*) combined with the control of generator voltage makes it possible to achieve the widest possible speed range. With such dual control, and with base speed being defined as the full-field speed of the motor at the normal armature voltage, speeds above base speed are obtained by motor-field control at approximately constant horsepower, and speeds below base speed are obtained by armature-voltage control at approximately constant torque. The overall output limitations are as shown in Figure 9.8.3; continuous speed control over very wide ranges is available, with the major advantage that large torques are available at low speeds.

The Ward-Leonard system also has the advantage that, when the armature voltage of the main motor *M* in Figure 9.8.2 is reduced to produce a decrease in speed, the motor *M* will momentarily operate as a generator and the generator *G* will operate as a motor, driving the ac motor *A* as a generator. Thus, the change in kinetic energy of the main motor during retardation is converted to electrical energy and returned to the power system, in a process known as *regeneration*. The obvious disadvantage of the Ward-Leonard method of speed control is the requirement of a separate motor-generator set and its associated initial investment.

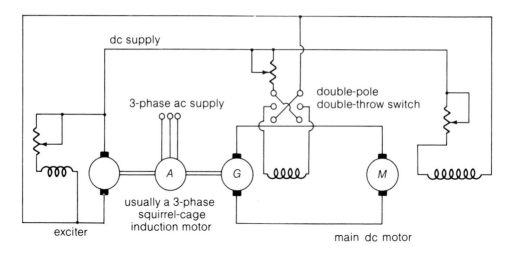

Figure 9.8.2 Schematic diagram of the Ward-Leonard method of speed control.

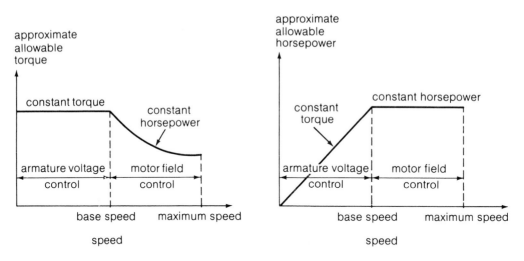

Figure 9.8.3 Output limitations of the Ward-Leonard system combining the armature-voltage and field-rheostat methods of speed-control.

Next let us consider *dc motor starting*. When voltage is applied to the armature of a dc motor with the rotor stationary, no emf is generated and the armature current is limited only by the internal armature resistance of the machine. So, to limit the starting current to the value that the motor can commutate successfully, all except very small dc motors are started with variable external resistance in series with their armatures. This starting resistance is cut out manually or automatically as the motor comes up to speed.

Typical forms of manual starters for dc motors are shown in Figure 9.8.4. The *three-point starter* shown in Figure 9.8.4a has the holding coil connected in series with the field current circuit so that the

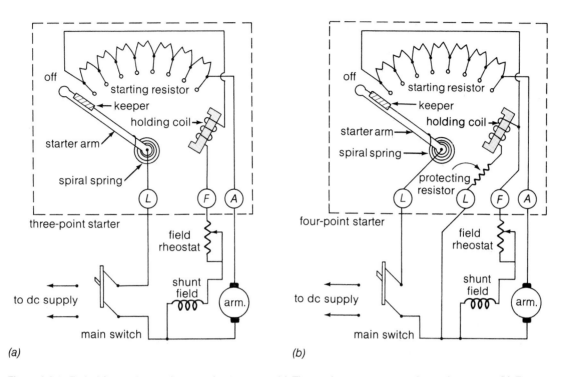

Figure 9.8.4 Typical forms of manual starters for dc motors. (a) Three-point starter connected to a shunt motor, (b) Four-point starter connected to a shunt motor.

starter arm, which is spring loaded, drops back to the *off* position if the field circuit is open-circuited. With sufficient resistance in the field circuit, the holding-coil current might no longer be able to create an electromagnetic pull strong enough to overcome the spring tension, and the starter arm falls back to the *off* position. To overcome this objection, in the *four-point starter* shown in Figure 9.8.4b, the holding coil is connected directly across the supply along with a current-limiting resistor in series; with this arrangement in parallel with the armature and shunt field, the holding-coil current is independent of any field-rheostat changes. The starter arm drops back to the *off* position if the voltage supply to the machine is removed. In both cases, the field circuit is completed when the moving arm is connected to the first stud of the variable resistor; it is common to include a relay in the armature circuit to give overload protection. The starting series resistance is cut out in steps, with values of resistance between steps sufficient to keep the initial current at each step within the proper limits as the motor accelerates. Automatic starters are generally used for larger motors, with relays to short out the resistance in series with the armature in successive steps.

Facilities for changing the field current that are incorporated within the starter are known as a *controller*. It is common to have mechanical interlocks in the controller to ensure that the machine can be started only with full field flux. The *drum controller* for a series motor, with the external series resistor rated to carry full-load current continuously provides facilities for starting, speed control, and reversal of the direction of rotation.

The solid-state control of dc motors is presented in some detail in Section 12.2 as a part of the chapter on Power Semiconductor-Controlled Drives.

||||||||||||

EXAMPLE 9.8.1

A 10-hp, 230-V, 500-rpm shunt motor having a full-load armature current of 37 A is started with a four-point starter. The resistance of the armature circuit, including the interpole winding, is 0.39 Ω, and the resistances of the steps in the starting resistor are 1.56, 0.78, and 0.39 Ω, in the order in which they are successively cut out. When the armature current has dropped to its rated value, the starting box is switched to the next point, thus eliminating a step at a time in the starting resistance. Neglecting field-current changes, armature reaction, and armature inductance, find the initial and final values of the armature current and speed corresponding to each step.

Solution

First step: At this point, the entire resistance of the starting resistor is in series with the armature circuit; thus,

$$R_{T1} = 1.56 + 0.78 + 0.39 + 0.39 = 3.12 \ \Omega$$

At starting, the counter emf is zero; the armature starting current is then

$$I_{st} = \frac{V_t}{R_{T1}} = \frac{230}{3.12} = 73.7 \ \text{A}$$

By the time the armature current drops to its rated value of 37 A, the counter emf is

$$E = 230 - (3.12)(37) = 230 - 115.44 = 114.56 \ \text{V}$$

The counter emf, when the motor is delivering rated load at rated speed of 500 rpm with series-starting resistor completely cut out, is

$$230 - (0.39)(37) = 230 - 14.43 = 215.57 \ \text{V}$$

The speed corresponding to the counter emf of 114.56 V is then given by

$$N = \frac{114.56}{215.57} \times 500 = 265.7 \ \text{rpm}$$

Second step: The 1.56-Ω step is cut out, leaving a total resistance of $3.12 - 1.56 = 1.56 \ \Omega$ in the armature circuit. The initial motor speed at this step is 265.7 rpm, which means that the counter emf is still 114.56 V, if the effect of armature reaction is neglected. Accordingly, the resistance drop in the armature circuit is still 115.44 V, so that

$$I_a R_{T2} = 115.44 \quad \text{or} \quad I_a = \frac{115.44}{1.56} = 74 \ \text{A}$$

That is, if the inductance of the armature is neglected, the initial current is 74 A. With the final current at 37 A, the counter emf is

$$E = 230 - (37)(1.56) = 230 - 57.72 = 172.28 \ \text{V}$$

corresponding to which the speed is

$$N = \frac{172.28}{215.57} \times 500 = 399.6 \ \text{rpm}$$

Third step: The total resistance included in the armature circuit is 0.78 Ω at the beginning of this step. The initial motor speed is 399.6 rpm, and the counter emf is 172.28 V. The initial armature current is then $(230 - 172.28)/0.78 = 57.72/0.78 = 74$ A. With the final current at 37 A, the counter emf is

$$E = 230 - (37)(0.78) = 230 - 28.86 = 201.14 \text{ V}$$

corresponding to which the speed is

$$N = \frac{201.14}{215.57} \times 500 = 466.5 \text{ rpm}$$

Thus we have the results shown in Table 9.8.1.

Table 9.8.1

Step Number	Current, A		Speed, rpm	
	Initial	Final	Initial	Final
1	74	37	0	266
2	74	37	266	400
3	74	37	400	467
4	74	37	467	500

9.9 Testing and Efficiency

As is true for any other machine, the efficiency of a dc machine can be expressed as

$$\text{Efficiency} = \frac{\text{Output}}{\text{Input}} = \frac{\text{Input} - \text{Losses}}{\text{Input}} = 1 - \frac{\text{Losses}}{\text{Input}} \tag{9.9.1}$$

The losses are made up of rotational losses (3 to 15%), armature-circuit copper losses (3 to 6%), and shunt-field copper loss (1 to 5%). Figure 9.9.1 shows the schematic diagram of a dc machine along with the power division in a generator and a motor. The resistance voltage drop, also known as the *arc drop*, between the brushes and commutator is generally assumed constant at 2 V, and the brush-contact loss is therefore calculated as $2I_a$; in such a case, the resistance of the armature circuit should not include the resistance between brushes and commutator.

EXAMPLE 9.9.1

The following data apply to a 100-kW, 250-V, six-pole, 1,000-rpm long-shunt compound generator:

No-load rotational losses = 4,000 W   Interpole field resistance at 75° C = 0.005 Ω

Armature resistance at 75° C= 0.015 Ω Shunt-field current = 2.5 A

Series-field resistance at 75° C =0.005 Ω

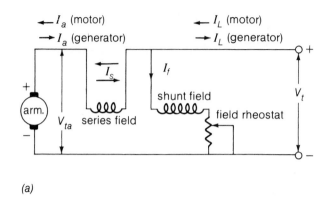

(a)

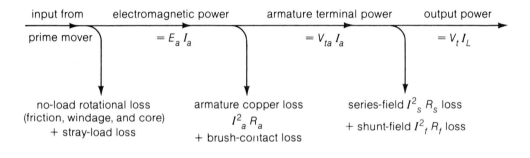

(b)

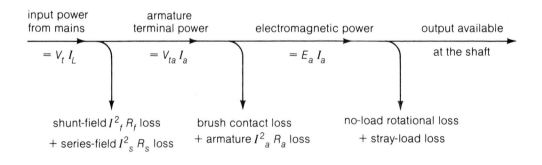

(c)

Figure 9.9.1 (a) Schematic diagram of a dc machine. (b) Power division in a dc generator. (c) Power division in a dc motor.

Assuming a stray-load loss of 1% of the output and a brush-contact resistance drop of 2 V, compute the rated-load efficiency.

Solution

The total armature-circuit resistance (not including that of brushes) is

$$R_a = 0.015 + 0.005 + 0.005 = 0.025 \ \Omega$$

$$I_a = I_L + I_f = \frac{100,000}{250} + 2.5 = 402.5 \ \text{A}$$

The losses are then computed as follows:

No-load rotational loss	4,000
Armature-circuit copper loss $= (402.5)^2 (0.025)$	4,050
Brush-contact loss $= 2I_a = 2 \times 402.5$	805
Shunt-field circuit copper loss $= 250 \times 2.5$	625
Stray-load loss $= 0.01 \times 100,000$	1,000
Total losses	10,480

The efficiency at rated load is then given by

$$\eta = 1 - \frac{10,480}{100,000 + 10,480} = 1 - 0.095 = 0.905 \quad \text{or} \quad 90.5\%$$

It is usual to determine the efficiency by some method based on the measurement of losses. See the bibliography at the end of the chapter for test codes and standards.

In the method known as *Swinburne's Test,* the machine is run on no load at rated speed with rated applied voltage. The armature resistance R_a is measured and the armature copper loss is calculated for an assumed armature current I_a. The remainder of the losses is assumed to be constant and independent of the armature current. For a no-load applied voltage V, if the no-load armature current is I_O and the field current is I_f, the efficiency for a shunt motor is given by

$$\eta_M = 1 - \frac{V(I_f + I_O) + I_a^2 R_a}{V(I_f + I_a)} \tag{9.9.2}$$

The corresponding efficiency as a generator is given by

$$\eta_G = 1 - \frac{V(I_f + I_0) + I_a^2 R_a}{V(I_a + I_0) + I_a^2 R_a} \tag{9.9.3}$$

The value obtained for efficiency by this method is usually greater than the actual efficiency.

A more accurate method of predicting the efficiency is the *Kapp-Hopkinson Test,* in which two similar machines are mechanically coupled and electrically connected *back-to-back,* as shown in Figure 9.9.2.

Machine A is run up to normal speed as a motor with machine B unexcited. Excitation is then applied to machine B and increased until its open-circuit voltage is equal to the system voltage. Machine B is then connected to the system; by increasing the excitation of machine B while decreasing that of machine A so that the speed is unchanged, machine A can be made to motor while machine B

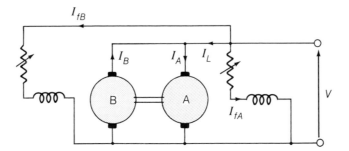

Figure 9.9.2 Schematic diagram of connections for the Kapp-Hopkinson test.

generates. In this situation, the supply system provides total losses. Assuming η to be the efficiency of each machine, it follows that

$$\text{Generator input} = \text{Motor output} = \eta\left(\text{Motor input}\right) = \eta VI_A \qquad (9.9.4)$$

$$\text{Generator input} = VI_B = \eta\left(\text{Generator input}\right) = \eta^2 VI_A \qquad (9.9.5)$$

Then

$$\eta = \sqrt{I_B/I_A} \qquad (9.9.6)$$

When the efficiencies of the two machines are not the same, the following analysis applies. The total fixed loss of the system, as seen from Figure 9.9.2, is given by

$$P = VI_L - \left(I_A^2 R_A + I_B^2 R_B\right) \qquad (9.9.7)$$

so we can assume the fixed loss of each machine to be $(P/2)$.

With the input power of the motor being $VI_A + VI_{fA}$ and the output power from the generator being VI_B, the efficiency is given by

$$\text{Motor efficiency} = 1 - \frac{(P/2) + I_A^2 R_A + VI_{fA}}{V\left(I_A + I_{fA}\right)} \qquad (9.9.8)$$

and

$$\text{Generator efficiency} = 1 - \frac{(P/2) + I_B^2 R_B + VI_{fB}}{V\left(I_B + I_{fB}\right) + (P/2) + I_B^2 R_B} \qquad (9.9.9)$$

The values of efficiency found in this test are close to the actual efficiency.

9.10 DC Machine Design

Let us carry on with a DC motor design project to show the basic elements and considerations of the design aspect. The specifications are given below:

20-hp, 230-V, 1,150-rpm, constant-speed, commutating-pole, dc motor. The motor is to be of the general-purpose class with continuous-duty rating. The full-load temperature rise is not to exceed 40° C in any part of the machine. The efficiency at rated load, voltage, and speed should not be less than 88.5%.

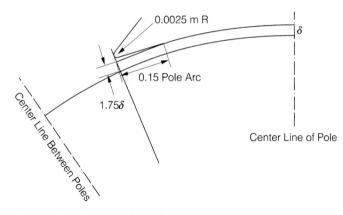

Figure 9.10.1 Pole shoe design for the dc motor.

The motor output is $20 \times 0.746 = 14.92$ KW

$$\frac{\text{KW}}{\text{rpm}} = \frac{14.92}{1,150} = 0.013$$

Based on empirical curves available in literature, one may choose the output coefficient 1.5×10^4 J/m^3. From Equation 5.6.1, it follows that

$$D^2L = W/C\,\omega_s = 14,920/(1.5 \times 10^4 \times 1,150 \times 2\pi/60) = 8.26 \times 10^{-3} \text{ m}^3$$

Selecting four poles for the design based on speed and KW-output for $L/\tau = 0.6$,

$$L = \frac{0.6 \times \pi \times D}{p} = \frac{0.6 \times \pi \times D}{4} = 0.47D$$

$$D^2L = 0.47D^3 \quad \text{or} \quad D = \sqrt[3]{8.26 \times 10^{-3}/0.47} = 0.26 \text{ m}$$

$$L = 0.47 \times 0.26 = 0.12 \text{ m}$$

Choose $D = 26$ cm and $L = 13$ cm. The pole pitch τ is then

$$\tau = \pi D/p = \pi(26)/4 = 20.4 \text{ cm} \quad \text{and} \quad L/\tau = 13/20.4 = 0.64$$

The peripheral speed is

$$v = \pi D \frac{(\text{rpm})}{60} = \pi \times 0.26 \times \frac{1,150}{60} \approx 15.7 \text{ m/s}$$

The frequency of flux reversals in the armature core is

$$f = \frac{p(\text{rpm})}{2 \times 60} = \frac{4 \times 1,150}{120} = 38.3 \text{ Hz.}$$

For 66% pole embrace, the pole arc is $0.66 \times 20.4 = 13.46$ cm. From empirical curves available in literature, based on the armature diameter, the approximate minimum length of airgap is selected as 2.5 mm. Since the shape of the field form depends upon the shape of the pole and the percent pole-embrace, the pole shoe is designed as indicated in Figure 9.10.1. The approximate flux plot for this pole shoe is shown in Figure 9.10.2. The air-gap flux distribution factor f_d, which is the ratio of the average ordinate of the flux wave to the maximum, is computed as 0.67. From the empirical curves available in the literature, based on kilowatts and revolutions per minute, the air-gap flux density is about 0.66 tesla (T). The total flux is then

$$\phi_t = \pi D L B_g = \pi \times 0.26 \times 0.13 \times 0.66 = 0.07 \text{ Wb} \tag{9.10.1}$$

The induced voltage per conductor is

$$e_i = \frac{\phi_t f_d \times \text{rpm}}{60} = \frac{0.07 \times 0.67 \times 1200}{60} = 0.938 \text{ V} \tag{9.10.2}$$

This value being quite low, a simplex wave winding can be used to advantage, and the total number of conductors is calculated as

$$N = \frac{Ea}{e_i} = \frac{230 \times 2}{0.938} = 490 \tag{9.10.3}$$

where a is the number of parallel paths, which is always 2 for a wave winding. (See Section 6.3).

Because the minimum number of armature slots should be 9 per pole, let us choose 41 armature slots with $10\frac{1}{4}$ slots per pole. With 12 conductors per slot, the total number of armature conductors is

$$N = 12 \times 41 = 492 \tag{9.10.4}$$

and

$$\phi_t = \frac{230 \times 2 \times 60}{492 \times 1200 \times 0.67} = 0.07 \text{ Wb} \tag{9.10.5}$$

The armature is then simplex-wave–wound with 41 slots, 12 conductors per slot, and two turns per coil. The number of commutator bars is

$$K = 3 \times 41 = 123 \tag{9.10.6}$$

According to Equation 6.3.2, the commutator pitch y_c is given by

$$y_c = y_b + y_f = \frac{123 \pm 1}{2} = 61 \quad \text{or} \quad 62 \tag{9.10.7}$$

The armature winding can be developed as in Figure 6.3.3. The full-load terminal current

$$I = \frac{20 \times 746}{230 \times 0.885} = 73.3 \text{ A} \tag{9.10.8}$$

Taking the shunt-field current to be approximately 1% of the armature terminal current, i.e., 0.7 A, in our case, then

$$I_a = 73.3 - 0.7 = 72.6 \text{ A} \tag{9.10.9}$$

The ampere conductors per centimeter of armature circumference is then

$$Q = \frac{I_a N}{\pi D a} = \frac{72.6 \times 492}{\pi \times 26 \times 2} = 218.6 \tag{9.10.10}$$

Choosing the current density in the armature conductor to be 450 A/cm^2, the section area of the armature conductor is

$$s_a = \frac{72.6}{450 \times 2} = 0.08 \text{ cm}^2 \tag{9.10.11}$$

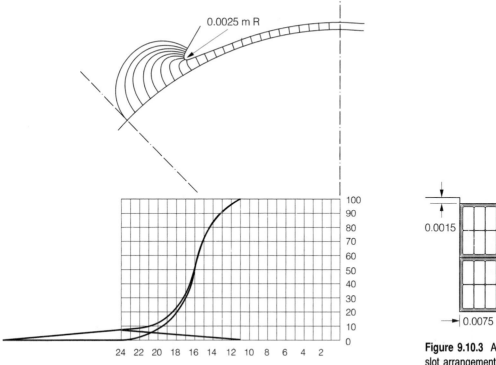

Figure 9.10.2   Pole shoe along with its approximate flux plot.

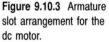

Figure 9.10.3 Armature slot arrangement for the dc motor.

Suitable conductors can be selected from the tables available in the literature. The conductors are arranged in the slot as shown in Figure 9.10.3. Slot dimensions are then finalized.

Without the detailed calculations related to the open-circuit saturation curve, field winding, and so forth, the design sheet for a dc motor is given in Figure 9.10.4. The interested student is encouraged to pursue the design and justify the details furnished here.

9.11 Applications for DC Machines

DC motors find wide applications in which control of speed, voltage, or current is essential. Table 9.11.1 gives typical characteristics and applications for dc motors. Universal motors, operating with either dc or ac excitation, are employed in vacuum cleaners, food processors, hand tools, and several other household applications. They are available in sizes of fractional horsepower up to and well beyond 1 hp, in speed ranges between 2,000 and 12,000 rpm. The dc shunt generators are often used as exciters to provide dc supply. The series generator is employed as a voltage booster and also as a constant source in welding machines. In applications for which a constant dc voltage is essential, the cumulative-compound generator finds its use. The differential-compound generator is used in applications like arc welding, where a large voltage drop is desirable when the current increases.

Hp.: 20	Volts: 230	Amps: 72.5	Rpm: 1150	Poles: 4
Watts/rpm: 13		Output constant: 1.5×10^4		Type: Shunt-wound Commutating pole

Outside diameter 0.26
Inside diameter 0.1
Total length 0.13
Ducts—number size .... none
Length—gross 0.13; .... effective 0.117
Number of slots 41
Slots per pole .... $10\frac{1}{4}$
Type of winding .... Simplex wave
Number of coils 123
Turns per coil 2
Conductors total 492
Conductors per slot 12
Total flux 0.07
Distribution constant 0.671
Flux per pole 0.0117
Conductor dimensions .... 0.002×0.0055
Conductor section .... 8×10^{-6}
Current density .... 4.5×10^6
Equa.-conn. number, size .... none
Coils in slots .... 1 and 11
Coils in bar .... 1 and 63
One-half mean-turn 0.4
Resistance, 25°C 0.107
Resistance, 75°C 0.128
Square meter per watt 0.001
Cal. temperature rise 37.9
Ampere conductors per meter .... 22.2×10^3
Ampere-turns per pole 2240

Commutator and Brush

Diameter 0.184
Length 0.076
Peripheral speed 11.08
Number of bars 123
Bar pitch 0.005
Thickness of mica 0.0008
Volts bar average .... 7.48; max. 14.5
Number of arms 4
Amperes per arm 35.9
Brushes per arm 3
Size of brush .... 0.01×0.01
Current density .... 80.3×10^3
Square meter per watt 0.0005

Weights

Armature copper 14.14
Shunt field copper 16.73
Series field copper 2.64
Commutating field copper 11.55
Armature teeth 9.55
Armature yoke 21.73

Full-Load Losses

Friction and windage 200
Brush friction 49
Core 372
Armature copper 676
Shunt field copper+rheo. 167
Series field copper 29.6
Comm. field copper 106
Brush contact I^2R 145.4
Total losses 1705

Full-Load Resistance Drop

Armature 9.3–4.04%
Series field 0.4–0.18
Commutating field 1.45–0.63
Brushes 2.00–0.87
Total 13.15–5.7
Shunt field IR 230

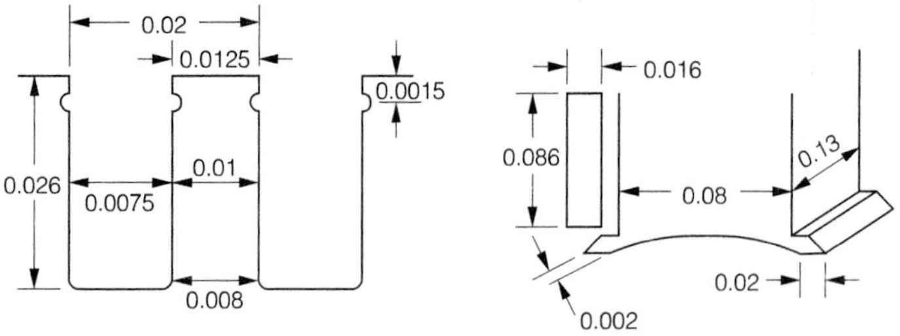

Poles and Yoke	Main	Comm.
Material	Sheet steel	Sheet steel
Body, length and width ...	0.13×0.08	0.13×0.032
Shoe, length and width....	0.13×0.132	0.13×0.032
Pole pitch................	0.2	
Percentage of pole embrace	66	
Pole arc	0.132	
Total air gap length.......	2×0.0025	2×0.0025

Material yoke .... Hot rolled steel
Outside diameter of yoke 0.53
Inside diameter of yoke 0.47
Length of yoke 0.254
Magnetic section .... $0.064 \times 0.254 = 0.016$

Field Winding	Shunt	Series	Comm.
Size of conductor....	No. 21 / No. 22	0.005×0.009	$2(0.004 \times 0.006)$
Conductor section ...	0.4×10^{-6} / 0.3×10^6	33.4×10^{-6}	$2 \times 18.5 \times 10^{-6}$
Amperes...........	0.728	71.8	71.8
Current density......	1.8×10^6 / 2.2×10^6	2.15×10^6	1.97×10^{-6}
Turns per pole	848 / 1888	4	42
Length of mean turn.	0.5	0.53	0.42
Resistance, 25°C	265	0.0047	0.168
Resistance, 75°C	316	0.0056	0.20
Watts loss...........	167	29.6	106
Square meter per watt	0.0024		0.0018

Field leakage constant..1.20

Magnetic Circuit

Volts 230				R.P.M. 1200
	Section	Density	Length	AT
Gap ...	0.1	0.69	1.18×0.0025	1635
Teeth ..	0.044	1.57	0.026	67
A. Yoke	0.012	1	0.06	16
Pole ...	0.01	1.45	0.105	187
F. Yoke	0.016	0.87	0.2	99
Total ..				2004

Shunt AT per pole full-load 1990
Series AT per pole full-load 287
Commutating AT per pole full-load 3020

Commutation

Bars covered by brush.. 2.03
Commutating zone at armature surface 0.034
Commutating zone, per cent neutral zone 50.5
Reactance volts... 0.77
Density in commutating pole air gap... 0.2

Remarks: Armature laminations 0.36×10^{-3} m open-hearth sheet steel

Figure 9.10.4 Design sheet for dc motor. Note: All dimensions are in the SI system of units. (Adapted from J. H. Kulhmann, *Design of Electrical Apparatus*, 3d ed. New York: John Wiley & Sons, 1950.)

Table 9.11.1 Typical characteristics and applications of dc motor

Type	Starting torque (%)	Max. running torque, momentary (%)	Speed-regulation or characteristic (%)	Speed control (%)	Typical application and general remarks
Shunt, constant speed	Medium—usually limited to less than 250 by a starting resistor but may be increased	Usually limited to about 200 by commutation	5–10	Increase up to 200 by field control; decrease by armature-voltage control	Essentially for constant-speed applications requiring medium starting torque. May be used for adjustable speed not greater than 2:1 range. For centrifugal pumps, fans, blowers, conveyors, woodworking machines, machine tools, printing presses, positioning devices
Shunt, adjustable speed	Same as above	Same as above	10–15	6:1 range by field control, lowered below base speed by armature-voltage control	Same as above, for applications requiring adjustable speed control, either constant torque or constant output. For rolling mills, papermills, centrifugal pumps, fans, conveyors, woodworking machines, machine tools
Compound	High—up to 450, depending on degree of compounding	Higher than shunt—up to 350	Varying, depending on degree of compounding—up to 25–30	Not usually used but may be up to 125 by field control	For drives requiring high starting torque and only fairly constant speed; pulsating loads with flywheel action. For plunger pumps, shears, conveyors, crushers, bending rolls, punch presses, hoists
Series	Very high—up to 500	Up to 400	Widely variable, high at no-load	By series rheostat	For drives requiring very high starting torque and where adjustable, varying speed is satisfactory. This motor is sometimes called the traction motor. Loads must be positively connected, not belted. For electric locomotives, hoists, cranes, bridges, car dumpers. To prevent overspeed, lightest load should not be much less than 15 to 20 per cent of full-load torque

Adapted from M. Liwschitz-Garik and C. C. Whipple, *Electric Machinery*, vol. I, Van Nostrand Co., Inc., Princeton, N.J., 1946.

■■■■■■■■■■IIII

BIBLIOGRAPHY

American National Standards Institute. *ANSI Std. No. C.50.4–1965, Rotating Electrical Machinery.* New York: ANSI, 1965.

Clayton, A. E. *Performance and Design of DC Machines.* London: Pitman, 1938.

Daniels, A. R. *Introduction to Electric Machines.* London: Macmillan, 1976.

Del Toro, V. *Electromechanical Devices for Energy Conversion and Control Systems.* Englewood Cliffs, N.J.: Prentice-Hall, 1968.

Fitzgerald, A. E., et al. *Electric Machinery.* 3d ed. New York: McGraw-Hill, 1971.

Hindmarsh, J. *Electrical Machines and Their Applications.* 2d ed. New York: Pergamon, 1970.

Institute of Electrical and Electronics Engineers. *Standard No. 113–1962, Test Code for Direct-Current Machines.* New York: IEEE, 1962.

Kloeffler, R. G., et al. *Direct-Current Machinery.* New York: Macmillan, 1950.

Knowlton, A. E., ed. *Standard Handbook for Electrical Engineers.* 8th ed. New York: McGraw-Hill, 1949.

Langsdorf, A. S. *Principles of Direct-Current Machines.* New York: McGraw-Hill, 1959.

Matsch, L. W. *Electromagnetic and Electromechanical Machines.* New York: Intext, 1972.

Say, M. G., and E. O. Taylor. *Direct-Current Machinery.* New York: McGraw-Hill, 1952.

Siskind, C. S. *Direct-Current Machinery.* New York: McGraw-Hill, 1952.

Thaler, G. J., and M. L. Wilcox. *Electric Machines: Dynamic and Steady-State.* New York: John Wiley & Sons, 1966.

■■■■■■■■■■■■■■■■■IIIIIIIIIII

PROBLEMS

9–1. A 100-kW, 250-V shunt generator has an armature-circuit resistance of 0.05 Ω and field-circuit resistance of 60 Ω. With the generator operating at rated voltage, determine the induced voltage at (a) full load, and (b) half-full load. Neglect brush-contact drop.

9–2. A 100-kW, 230-V shunt generator has $R_a = 0.05$ Ω and $R_f = 57.5$ Ω. If the generator operates at rated voltage, calculate the induced voltage at (a) full-load and (b) half full-load. Neglect brush-contact drop.

9–3. A 10-hp, 250-V shunt motor has an armature-circuit resistance of 0.5 Ω and a field resistance of 200 Ω. At no load, rated voltage, and 1,200 rpm, the armature current is 3 A. At full load and rated voltage, the line current is 40 A, and the flux is 5% less than its no-load value because of armature reaction. Compute the full-load speed.

9–4. When delivering rated load a 10-kW, 230-V self-excited shunt generator, has an armature-circuit voltage drop that is 6% of the terminal voltage and a shunt-field current equal to 4% of the rated load current. Calculate the resistance of the armature circuit and the field circuit.

9–5. A 20-hp, 250-V shunt motor has a total armature-circuit resistance of 0.25 Ω and a field-circuit resistance of 200 Ω. At no load and rated voltage, the speed is 1,200 rpm and the line current is 4.5 A. At full load and rated voltage, the line current is 65 A. Assume the field flux to be reduced by 6% from its value at no load due to the demagnetizing effect of armature reaction. Compute the full-load speed.

9–6. A dc series motor is connected to a load. The torque varies as the square of the speed. With the diverter-circuit open, the motor takes 20 A and runs at 500 rpm. Determine the motor current and speed when the diverter-circuit resistance is made equal to the series-field resistance. Neglect saturation and the voltage drops across the series-field resistance, as well as the armature resistance.

9-7. A 50-kW, 230-V compound generator has the following data:

Armature-circuit resistance = 0.05 Ω

Series-field resistance = 0.05 Ω

Shunt-field circuit resistance = 125 Ω

Assuming the total brush-contact drop to be 2 V, find the induced armature voltage at rated load and rated terminal voltage for (a) short-shunt, and (b) long shunt compound connection.

9-8. A 50-kW, 250-V short-shunt compound generator has the following data: R_a = 0.06 Ω, R_S = 0.04 Ω, and R_f = 125 Ω. Calculate the induced armature voltage at rated load and terminal voltage. Take 2 V as the total brush-contact drop.

9-9. Repeat the calculations of Problem 9–8, for a machine that is a long-shunt compound generator.

9-10. A 20-hp, 440-V, 500-rpm dc shunt motor has rotational losses of 800 W at rated speed. The armature-circuit resistance is 0.5 ohm and the field resistance is 220 Ω.

a. Calculate the armature current when the motor is delivering rated load at rated voltage and rated rpm.

b. Determine the electromagnetic torque, shaft-output torque, and efficiency.

c. Let the load on the shaft be removed. Find the no-load speed of the motor for the given field current and terminal voltage, neglecting saturation.

9-11. A 10-kW, 230-V shunt generator, with an armature-circuit resistance of 0.1 Ω and a field-circuit resistance of 230 Ω, delivers full load at rated voltage and 1,000 rpm. If the machine is run as a motor while absorbing 10 kW from 230-V mains, find the speed of the motor. Neglect the brush-contact drop.

9-12. The magnetization curve taken at 1,000 rpm on a 200-V dc series motor has the following data:

Field current, A:	5	10	15	20	25	30
Voltage, V:		80	160	202	222	36	244

The armature-circuit resistance is 0.25 ohm and the series-field resistance is 0.25 Ω. Calculate the speed of the motor (a) when the armature current is 25 A, and (b) when the electromagnetic torque is 36 N·m. Neglect armature reaction.

9-13. The open-circuit characteristic data of a shunt generator at 1,200 rpm are given below:

Field current, A:	1.0	1.5	2.0	2.5	3.0	3.5	4.0	4.5	5.0	6.0
Terminal voltage, V:	67	100	134	160	180	200	210	220	230	242

a. Determine the critical field resistance for self-excitation at 1,200 rpm.

b. Find the total field-circuit resistance if the induced voltage is 230 V.

9-14. A dc series motor operates at 750 rpm with a line current of 100 A from the 250-V mains. Its armature-circuit resistance is 0.15 Ω and its series-field resistance is 0.1 Ω. Assuming that the flux corresponding to a current of 25 A is 40% of that corresponding to a current of 100 A, determine the motor speed at a line current of 25 A at 250 V.

9-15. The data for the magnetization curve of a 250-V dc series motor taken at 300 rpm are given as follows:

Field current, A:	20	30	40	50
Terminal voltage, V:	162	215	250	274

The armature-circuit resistance is 0.3 Ω and the series-field resistance is 0.1 Ω. Determine the speed-torque curves when operating on a 250-V supply for the following cases:

a. Without any external resistance.

b. With an external series resistance of 2 Ω.

c. With a series-winding diverter of 0.1 Ω.

d. With an external series resistance of 2 Ω and an armature diverter of 10 Ω connected across the armature.

9-16. A 7.5-hp, 250-V, 1,800-rpm shunt motor having a full-load line current of 26 A is started with a four-point starter. The resistance of the armature circuit including the interpole winding is 0.48 Ω and the resistance of the shunt-field circuit including the field rheostat is 350 Ω. The resistances of the steps in the starting resistor are 2.24, 1.47, 0.95, 0.62, 0.40, and 0.26 Ω, in the order in which they are successively cut out. When the armature current is dropped to its rated value, the starting box is switched to the next point, thus eliminating a step in the starting resistance. Neglecting field-current changes, armature reaction, and armature inductance, find the initial and final values of the armature current and speed corresponding to each step.

9-17. Two shunt generators operate in parallel to supply a total load current of 3,000 A. Each machine has an armature resistance of 0.05 Ω and a field resistance of 100 Ω. If the generated emfs are 200 V and 210 V respectively, determine the load voltage and the armature current of each machine.

9-18. Three dc generators are operating in parallel with excitations such that their external characteristics are almost straight lines over the working range with the following pairs of data points:

Load Current (A)	Terminal Voltage (V)		
	Generator I	Generator II	Generator III
0	492.5	510	525
2,000	482.5	470	475

Compute the terminal voltage and current of each generator

a. When the total load current is 4,350 A.

b. When the load is completely removed without change of excitation currents.

9-19. The external-characteristics data of two shunt generators in parallel are given as follows:

Load current (A)	0	5	10	15	20	25	30
Terminal voltage (V) I	270	263	254	240	222	200	175
Terminal voltage (V) II	280	277	270	263	253	243	228

Calculate the load current and terminal voltage of each machine

a. When the generators supply a load resistance of 6 Ω.

b. When the generators supply a battery of emf 300 V and resistance 1.5 Ω.

9-20. Two similar shunt machines connected to a 240-V dc supply are tested by the Hopkinson method. The generator is delivering an output of 42 kW, and the current taken from the supply is 40 A. The shunt-field circuit resistances of the motor and generator are 300 Ω and 240 Ω, respectively. The armature-circuit resistance of each machine is 0.03 Ω. Calculate the efficiency of the generator.

9-21. A 25-hp, 500-V, 500-rpm dc shunt motor has a field circuit resistance of 500 Ω and an armature-circuit resistance of 0.5 Ω. The full-load armature current is 42 A. The magnetization-characteristic data taken at 400 rpm are given as follows:

> Field current, A: 0.4   0.6   0.8   1.0   1.2
> Generated emf, V: 236   300   356   400   432

The difference between the electromagnetic torque and the useful-output torque can be assumed to vary directly in proportion with speed. Calculate the following:

a. The field current required to operate at full-load torque when running at rated speed. With this same field current, also find the no-load speed.

b. The speed, with this same field current, at which the machine must be driven in order to regenerate with full-load armature current.

c. The extra field-circuit resistance required to run at 600 rpm on no load and on full-load torque.

d. The external armature-circuit resistance needed to operate at 300 rpm with full-load torque.

9–22. The data for the magnetization curve of a 150-hp, 250-V, 500-A dc series motor taken at 1,000 rpm are given as follows:

Field current, A: 100   200   300   400   500   600

Voltage, V: 93   163   194   212   220   227

The resistance of the armature-circuit (including that of the interpole winding and brushes) is 0.015 Ω and that of the series-field winding is 0.01 Ω. The series-field winding has 10 turns per pole. The demagnetizing mmf of armature reaction is 250 At per pole at rated current and can be assumed to vary linearly with current.

a. Determine (i) the speed, (ii) the electromagnetic power, and (iii) the electromagnetic torque when the armature current is 500 A.

b. Repeat part a for an armature current of 250 A.

9–23. The field of the series motor of Problem 9–22 is replaced by a shunt field of 600 turns and a series field of four turns per pole, connected cumulatively, for a long-shunt compound motor operation at 250 V. The armature winding is unchanged; the resistance of the shunt field winding is 50 Ω and that of the series-field winding is 0.005 Ω. Calculate (a) the no-load speed and (b) the following values given an armature current of 500 A: (i) the speed, (ii) the electromagnetic power, and (iii) the electromagnetic torque.

9–24. A 100-kW, 250-V long-shunt cumulatively compound generator has an armature-circuit resistance of 0.025 Ω and a series-field resistance of 0.005 Ω. The open-circuit test data taken at 1,200 rpm are given as follows:

Shunt-field current, A: 0   1   2   3   4   5   6   7   8   9

Generated voltage, V: 6   57   114   171   218   250   275   293   308   320

The generator has 1,000 shunt-field turns per pole and four series-field turns per pole. Determine the terminal voltage at rated-current output when the shunt-field current is 5 A and the speed is 1,000 rpm. Neglect armature reaction.

9–25. The magnetization curve, including armature-reaction effects, at an armature-current of 400 A at 1,200 rpm is given for the generator of Problem 9–24 as follows:

Shunt-field current, A: 0   1   2   3   4   5   6   7   8   9

Generated voltage, V: 6   57   114   171   210   240   262   282   299   313

If the generator is to be flat compounded so that the full-load voltage at 1,200 rpm is 250 V when the shunt-field rheostat is adjusted to yield a no-load voltage of 250 V at 1,200 rpm, calculate the diverter resistance to be placed across the series field.

9–26. A 100-hp, 250-V dc shunt motor with 1,000 shunt-field turns per pole has an armature-circuit resistance of 0.025 Ω. The data at 1,200 rpm for the no-load saturation curve and for the magnetization curve including armature-reaction effects at an armature current of 400 A are the same as given in Problems 9–24 and 9–25. The field rheostat is adjusted for a no-load speed of 1,100 rpm.

a. Determine the speed corresponding to an armature current of 400 A.

b. If a stabilizing winding consisting of two cumulative series turns per pole is added, and the resistance of this winding is negligible, find the speed corresponding to an armature current of 400 A.

9–27. The open-circuit-characteristic data of a 500-kW, 500-V, six-pole, dc shunt generator obtained at 1,000 rpm are given as follows:

Shunt-field current, A: 0   1   2   3   4   5   6   6.5   7   7.5   8

Generated voltage, V: 10   100   200   300   390   450   490   500   510   520   525

The armature-circuit resistance including that of brushless is 0.02 ohm.

a. Calculate the resistance of the shunt-field circuit when the no-load voltage is the rated voltage at a speed of 1,000 rpm.

b. With the resistance of the shunt-field circuit as in part a, and neglecting the effect of armature reaction, determine (by graphical means) the terminal voltage when the armature current is 1,000 A and the speed is 1,000 rpm.

c. If the demagnetizing mmf of armature reaction corresponds to a field current of 0.3 A, how would the answer in part b be affected?

9–28.   A 230-V dc shunt motor has the no-load magnetization characteristic at 1,800 rpm shown in Figure P9–28. The full-load armature current is 100 A; the armature-circuit resistance (including brushes and interpoles) is 0.12 Ω.

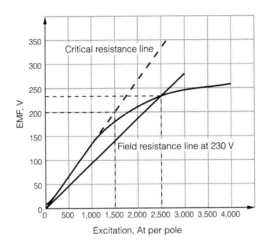

Figure P9–28

a. If the motor runs at 1800 rpm at both full load and no load, determine the demagnetizing effect of armature reaction at full load, in ampere-turns per pole.

b. If a long-shunt, cumulative, series-field winding having eight turns per pole and a resistance of 0.08 Ω, is added to the motor, find the speed at full-load current and rated voltage. (Assume magnetic linearity in the operating range.)

9–29.   The shunt-field winding of the 500-kW, 500-V generator of Problem 9–27 has 2,000 turns per pole. The armature characteristic or field-compounding curve of the shunt generator obtained at a speed of 1,000 rpm is almost a straight line, with the following two data points:

Shunt-field current, A: 0   1,000
Generated voltage, V: 6.5   7.8

Neglect the resistance of the series-field winding. Determine the number of turns per pole of a series-field winding for flat-compounding at 1,000 rpm. Check whether a diverter resistance is necessary if the number of turns in the series-field winding is to be an integer.

9–30.   A separately excited dc generator with an armature-circuit resistance R_a is operating at a terminal voltage V_t while delivering an armature current I_a and has a constant loss P_c. Find the value of I_a for which the generator efficiency is a maximum.

9–31. A 100-kW, 230-V, dc shunt generator, with $R_a = 0.05\ \Omega$ and $R_f = 57.5\ \Omega$, has no-load rotational loss (friction, windage, and core loss) of 1.8 kW. Compute (a) the generator efficiency at full-load and (b) the horsepower output from the prime mover to drive the generator at this load.

9–32. The open-circuit-characteristic data of a 25-hp, 250-V, 85-A shunt motor taken at 1,800 rpm are given as follows:

Shunt-field current, A: 0.6   0.8   1.0   1.2   1.3   1.5   1.7   2.0   2.4

Terminal voltage, V: 150   185   214   237   245   260   270   280   290

The resistance of the shunt-field circuit, including the field rheostat, is 185 Ω, and the resistance of the armature circuit (including that of the interpole winding and brushes) is 0.08 Ω. The field winding has 3,000 turns per pole. The demagnetizing mmf of armature reaction, at rated armature current, is 0.08 A in terms of the shunt-field current. The friction, windage, and core losses (known as the no-load rotational losses) are 1,500 W. Assume the stray-load loss is 1% percent of the output and that the field-circuit resistance is constant at 185 ohms. Calculate the following when the motor draws 85 A from the line: (a) speed of the motor, (b) electromechanical power, (c) mechanical power output, (d) output torque, and (e) efficiency.

9–33. A dc series motor with a design constant $K_a = 40$ and flux per pole of 46.15 mWb operates at 200 V while taking a current of 325 A. The total series-field and armature-circuit resistance are 25 mΩ and 50 mΩ, respectively. The core loss is 220 W; friction and windage loss is 40 W. Determine (a) electromagnetic torque developed, (b) motor speed, (c) mechanical power output, and (d) motor efficiency.

9–34. A 230-V dc shunt motor delivers 30 hp at the shaft at 1,120 rpm. If the motor has an efficiency of 87% at this load, find (a) the total input power and (b) the line current. (c) If the torque lost due to friction and windage is 7% of the shaft torque, calculate the developed torque.

9–35. A 10-kW, 250-V dc shunt generator, having an armature resistance of 0.1 Ω and a field resistance of 250 Ω, delivers full load at rated voltage and 800 rpm. The machine is now run as a motor while taking 10 kW at 250 V. Neglect brush-contact drop. Determine the speed of the motor.

9–36. A 10-hp, 230-V dc shunt motor takes a full-load line current of 40 A. The armature and field resistances are 0.25 Ω, and 230 Ω, respectively. The total brush-contact drop is 2 V, and the core and rotational losses are 380 W. Assume that stray-load loss is 1% of output. Compute the efficiency of the motor.

9–37. The project is to design a 300-kW, 900-rpm, 230-V, compound-wound, dc generator to be a part of a synchronous motor-generator set. The machine is to have commutating poles and be overcompounded to give a full-load voltage of 250 V. The efficiency at full load and normal voltage should be not less than 92%. The temperature rise of no part of the generator should exceed 50° C when operating on continuous full load. Prof. John H. Kuhlmann's[1] dc generator design sheet in the *American Customary System of Units*[2] is given in Figure P9–37 (page 450). Pursue the design project and justify the details furnished.

9–38 Prof. M. G. Say's[3] design data are given below for two dc machines:

Machine A: Shunt generator, 350 kW, 440 V, 600 rpm, natural cooling

Machine B: Shunt motor, 110 kW, 500 V, 450 rpm, fan cooled, with a series stabilizing winding to reduce the field-weakening effect of armature reaction and to provide a slightly drooping speed/load characteristic

The main dimensions are given in Figure P9–38. Pursue the design projects and justify the details furnished on pages 451–454.

[1] J. H. Kuhlmann, *Design of Electrical Apparatus*, 3d ed. New York: John Wiley & Sons, 1950.

[2] T. Wildi, *Units and Conversion Factors*. IEEE Press, New York, 1991.

[3] M. G. Say, and E. O. Taylor, *Direct Current Machines,* 2nd Ed., Longman Group, U.K., 1986.

Kw. 300	Volts: 230-250	Amps: 1200	Rpm: 900	Poles: 6
Watts/rpm: 333	Output constant: 1.97×10^4		Type: Compound-wound Commutating pole	

Outside diameter.............................25
Inside diameter..............................13.5
Total length.................................10.5
Ducts—number, size ....3 × .375
Length—gross 9.75; .... effective 8.62
Number of slots 81
Slots per pole...............................13.5
Type of winding ....Simplex lap
Number of coils 162
Turns per coil...............................1
Conductors total 324
Conductors per slot 4
Total flux................................48,300 K.L.
Distribution constant 0.665
Flux per pole.............................5350 K.L.
Conductor dimensions...............0.109 × 0.625
Conductor section...........................0.0656
Current density 3070
Equa.-conn. no., size .... 27 − 0.078 × 0.5
Coils in slots ....1 and 14
Coils in bar............................1 and 2
One-half mean-turn...........................27.1
Resistance, 25°C............................0.00258
Resistance, 75°C............................0.00307
Square inch per watt 1.375
Cal. temperature rise ....40°C
Ampere conductors per inch 830
Ampere-turns per pole 5440

Commutator and Brush

Diameter 16.0
Length 11.0
Peripheral speed 3770
Number of bars..............................162
Bar pitch 0.31
Thickness of mica...........................0.03
Volts bar average...................9.63 max. 18.8
Number of arms 6
Amperes per arm 402
Brushes per arm 6
Size of brush ....0.75 × 1.50
Current density.............................5.96
Square inch per watt 0.444

Weights

Armature copper.............................185
Shunt field copper 213
Series field copper 92
Commutating field copper 241
Armature teeth 182
Armature yoke 559

Full-Load Losses

Friction and windage........................3000
Brush friction 1220
Core..6060
Stray load..................................3000
Armature copper.............................4460
Shunt field.................................1265
Series field................................820
Comm. field 1080
Brush contact I^2R 2414
Total losses................................23,330

Full-Load Resistance Drop

Armature 3.71-1.48%
Series field................................0.68-0.27
Commutating field...........................0.89-0.36
Brushes.....................................2.00-0.80
Total 7.28-2.91
Shunt field IR 179

Remarks: Armature laminations 0.014 open-hearth sheet steel

Poles and Yoke	Main	Comm.
Material Sheet steel	0.023	0.023
Body, length and width	$10\frac{1}{2} \times 6\frac{1}{2}$	7 × 1.75
Shoe, length and width	$10\frac{1}{2} \times 8\frac{3}{4}$	
Pole pitch	13.1	
Percentage of pole embrace	65.8	
Pole arc............................	$8\frac{3}{4}$	
Total air gap length................	2 × 0.20	2 × 0.25

Material yoke .... Cast Steel
Outside diameter of yoke...45.0
Inside diameter of yoke..38.5
Length of yoke 15.5
Magnetic section .... 6.5 × 15.5 = 100.7

Field Winding	Shunt	Series	Comm.
Size of conductor	No. 13 sq.	3 − 0.219 × 1.0	3 − 0.25 × 1.25
Conductor section......	0.00465	3 × 0.216	3 × 0.305
Amperes	7.06	1207	1207
Current density	1518	1860	1320
Turns per pole........	620	1.5	5.5
Length of mean-turn ...	38.4	49.2	24.8
Resistance, 25°C.......	21.3	0.000473	0.00062
Resistance, 75°C.......	25.4	0.000564	0.00074
Watts loss	1265		
Square inch per watt ...	2.21		

Field leakage constant...1.20

Magnetic Circuit

Volts 260   Rpm 900

	Section	Density	Length	AT
Gap ...	825	58.5	1.067 × 0.2	3900
Teeth ..	404	120	1.56	540
A. Yoke	72.3	74	4.6	42
Pole ...	68.2	94.1	6.55	301
F. Yoke	100.7	63.7	10.9	207
Total ..				4990

Shunt AT per pole full-load 4370
Series AT per pole full-load 1810
Commutating AT per pole full-load 6640

Commutation

Bars covered by brush..2.42
Commutating zone at arm. surface.....................................2.14
Commutating zone per cent neutral zone...............................48.1
Reactance, volts 2.01
Density in commutating pole air gap 12,200

Figure P9–37 Design sheet for a dc generator. Note: American Customary system of units.

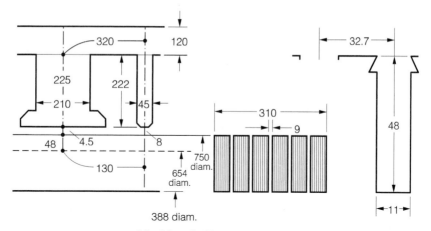

Machine A: Shunt generator

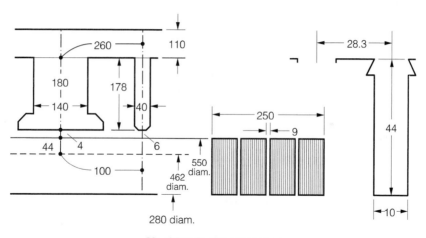

Machine B: Shunt motor

Figure P9–38 Main dimensions (in mm), not to scale. (From M. G. Say and E. O. Taylor, *Direct Current Machines*, 2nd ed., Longman Group, U.K., 1986.)

Data sheet for Problem 9–38

			A	B
Rating	Output	kW	350	110
	Voltage	V	440	500
	Speed	r/min	600	450
	Efficiency	p.u.	0.95	0.92
	Input	kW	368	120
	Terminal current	A	795	240
	Shunt field current	A	5	2
	Armature current	A	800	238
	Number of poles	—	6	6

continued

Data sheet for Problem 9–38, Continued

			A	B
Main Dimensions	Specific magnetic loading	T	0.64	0.60
	Specific electric loading	kA/m	32.6	34.2
	Armature diameter	m	0.75	0.55
	Core length (gross)	m	0.31	0.25
	Peripheral speed	m/s	23.6	13.0
	Armature frequency	Hz	30	22.5
	Pole-pitch	m	0.393	0.288
	Pole-arc	m	0.275	0.202
	Gap flux per pole	mWb	78.4	43.0
Armature Winding	Type	—	lap	wave
	Number of 1-turn coils	—	288	244
	Number of slots/conductor per slot	—	72/8	61/8
	Current per conductor	A	133	119
	Current density	A/mm^2	4.35	4.96
	Conductor size	mm^2	18 × 1.7	16 × 1.5
	Length of turn	m	1.78	1.40
	Resistance between brushes (hot)	mΩ	9.7	73
	M.M.F. per pole	A-t	6400	4950
Magnetic Circuit	Core: number of ducts/width	mm	5/9	3/9
	net length	m	0.265	0.223
	iron length	m	0.236	0.198
	path area	m^2	0.032	0.018
	flux density	T	1.22	1.19
	path length	m	0.13	0.10
	Teeth: number pe pole-arc	—	8.4	7.1
	effective area (at 1/3)	m^2	0.0375	0.0215
	flux density	T	2.09	2.00
	path length	m	0.048	0.044
	Pole: leakage factor	—	1.2	1.2
	effective area	mWb	94	51.6
	area	m^2	0.060	0.033
	flux density	T	1.67	1.56
	path length	m	0.225	0.180
	Yoke: area	m^2	0.036	0.020
	flux	mWb	47	25.8
	flux density	T	1.30	1.29
	path length	m	0.32	0.26
	Airgap: length	mm	4.5	4.0
	opening/gap coefficient	—	1.13 × 1.05	1.13 × 1.04
	effective gap length	mm	5.31	4.70
	flux density	T	0.920	0.851

continued

Data Sheet for Problem 9–38, Continued

			A	B
Excitation per Pole	Core	A-t	50	30
	Teeth	A-t	1250	1100
	Pole	A-t	890	360
	Yoke	A-t	130	100
	Gap	A-t	3950	3200
	Armature reaction (15%)	A-t	960	740
	Total f.l. m.m.f./pole	A-t	7230	5530
	Shunt field m.m.f.	A-t	7230	4800
	Series field m.m.f.	A-t	0	730
Shunt Field Coil	Voltage	V	64	83.3
	Current	A	5.30	2.17
	Number of turns	—	1360	2240
	Conductor diameter	mm	1.90	1.20
	Conductor area	mm^2	2.83	1.13
	Mean length of turn	m	1.20	0.94
	Resistance (hot)	Ω	12.1	38.3
	Cooling surface	m^2	3.06	1.80
Series Field Coil	Current	A	—	238
	Number of turns	—	—	3
	Conductor area	mm^2	—	110
	Mean length of turn	m	—	0.95
	Resistance (hot)	mΩ	—	0.55
Commutator	Diameter	m	0.55	0.40
	Peripheral speed	m/s	17.3	9.4
	Number of sectors	—	288	244
	Sector pitch	mm	6.00	5.15
	Volts between sectors (mean/max)	V	9.2/12.9	12.3/17.5
Brushgear	Current per brush-arm	A	267	79
	Number of brushes per arm	—	8	4
	Contact area per brush	mm^2	15.9 × 31.7	12.7 × 31.7
	Contact area per arm	mm^2	4040	1610
	Current density	mA/mm^2	67	49
	Brush pressure	mN/mm^2	12	14
Compole	Axial length	m	0.16	0.14
	Circumferential width	m	0.045	0.040
	Flux density	T	0.35	0.35
	Airgap	mm	8	6

continued

Data Sheet for Problem 9–38, Continued

			A	B
Compole,	Airgap coefficient	—	1.10	1.10
continued	Net m.m.f	A-t	2440	1850
	Armature m.m.f.	A-t	6400	4950
	Compole m.m.f.	A-t	8840	6800
	Current	A	800	238
	Conductor area	mm^2	350	110
	Number of turn	—	11	28
	Mean length of turns	m	0.50	0.42
	Resistance (hot)	mΩ	0.33	2.24
Core Loss	Frequency	Hz	30	22.5
	Armature core: flux density	T	1.22	1.19
	specific loss	W/kg	2.4	1.8
	mass	kg	410	172
	loss	W	980	310
	Teeth: flux density	T	2.09	2.09
	specific loss	W/kg	6.8	5.0
	mass	kg	120	63
	loss	W	820	320
	Pole-face pulsation loss	W	720	250
	Total core loss	kW	2.52	0.88
I^2R *Loss*	Armature	kW	6.21	4.14
	Shunt field	kW	2.33	1.09
	Series field	kW	0	0.19
	Compole field	kW	1.27	0.76
	Total I^2R loss	kW	9.81	6.18
Other Losses	Brush contact	kW	1.60	0.48
	Brush friction	kW	0.54	0.27
	Bearing friction and windage	kW	2.80	1.60
	Stray load	kW	1.66	0.65
	Total other loss	kW	6.60	3.00
Performance (f.l.)	Rotational e.m.f.	V	451	477
	Output	kW	350	110
	Loss summation	kW	18.9	10.1
	Input	kW	368.9	120.1
	Efficiency	p.u.	0.95	0.915

Source: M. G. Say and E. O. Taylor, *Direct Current Machines,* 2nd ed., Longman Group, U.K., 1986.

3

Transients and Dynamics

10

Transients and Dynamics of AC Machines

10.1 Mathematical Description of a Three-Phase Synchronous Machine
10.2 Synchronous-Machine Transient Reactances and Time Constants
10.3 Synchronous-Machine Transient Parameters from Sudden Three-Phase Short-Circuit Test Data
10.4 Synchronous-Machine Dynamics
10.5 Mathematical Description of a Three-Phase Induction Machine
10.6 Induction-Machine Electrical Transients
10.7 Induction-Machine Dynamics

Energy storage in magnetic and electric fields is the key to energy conversion by electromagnetic methods. As pointed out in Chapter 5, however, almost all industrial electric machines are magnetic-field devices. Changes in stored magnetic energy in response to changes in operating conditions cannot occur instantaneously; a period of accommodation intervenes between the initial and final operating conditions. Both the electric transients and the mechanical transients need to be considered.

An ac machine constitutes a multitude of windings characterized by time-varying self-inductaces and mutual inductances. Also, saturation and magnetic nonlinearity tend to complicate matters considerably. Unless one is very careful, the physical behavior of machines can become obscured by the mathematical analysis. The main objective of this chapter is to present a unified development and an adequate background for the fundamental coupled-circuit theory of the transient performance of ac machines. Out of the results, steady-state theory can be deduced as a special case. Emphasis is placed on a more or less rigorous mathematical development that is sound enough from a practical engineering point of view and on presenting a fundamental physical understanding of the machine so that the reader is well equipped with the concepts required to extend the theory as needed.

10.1 Mathematical Description of a Three-Phase Synchronous Machine

We base the mathematical model of a salient two-pole machine and the present analysis on the following assumptions:

- The stator windings are sinusoidally distributed around the periphery along the air gap, as far as all mutual effects with the rotor are concerned.
- The effect of stator slots on the variation of any of the rotor inductances with changes of rotor angle is neglected.
- Saturation is neglected, at least for the present.

Justification for any assumption comes from comparing performance calculated on the assumed basis to actual performance obtained by test. Judging by such comparisons, the first two assumptions are quite reasonable, while the third one is not because saturation does significantly affect certain aspects of machine performance.

The electrical performance equations of a synchronous machine will now be developed from a coupled-circuit viewpoint. First we adopt the generator convention for polarities so that positive current corresponds to generator action. Then we can write the voltage relations for the armature (stator) circuits as follows:

$$e_a = p\lambda_a - ri_a \tag{10.1.1}$$
$$e_b = p\lambda_b - ri_b \tag{10.1.2}$$
$$e_c = p\lambda_c - ri_c \tag{10.1.3}$$

where e_a is the terminal voltage of phase a, λ is the total flux linkage of phase a, and i_a is the current in phase a; a, b, and c are the three phases; r is the resistance of each armature winding, assumed to be the same for each of the three phases; and p is the derivative operator d/dt (where t denotes time).

As for the rotor field circuit, whenever the flux linkage is subject to change, a voltage drop in the direction of positive current occurs in the coil, and the resistance of the field circuit also contributes to a voltage drop in the same direction, so that the voltage equation can be written as

$$e_{fd} = p\lambda_{fd} + r_{fd}i_{fd} \tag{10.1.4}$$

The symbols e, λ, and i have the same meanings as for Equations 10.1.1 through 10.1.3, and the subscript fd denotes the circuit in question and further denotes that the field axis aligns itself with the pole (direct) axis.

From the physical description of a synchronous machine, the field winding has its axis in line with the pole axis. Assuming that the rotor magnetic paths and electric circuits are symmetrical about the pole and interpolar axes for a salient-pole machine, We can conveniently choose the pole axis as the direct axis and the interpolar axis as the quadrature axis, the angle between the two axes being 90 electrical degrees. The quadrature axis is located 90 electrical degrees ahead of the direct axis in the direction of rotor rotation. Figure 10.1.1 shows the schematic diagram of phase windings and field winding, along with the notations adopted for voltages and currents, as well as the locations of the direct and quadrature axes. A *salient two-pole three-phase machine* has been chosen for simplicity and convenience.

A flattened layout of damper circuits is given in Figure 10.1.2, showing the direct-axis circuits (numbered $1d$ and $2d$) and quadrature-axis circuits (numbered $1q$ and $2q$). The symmetrical choice of the rotor circuits has the virtue of making all mutual inductances and resistances between direct-axis and quadrature-axis rotor circuits equal to zero. Considering only one circuit in each axis, the voltage equations of the direct-axis and quadrature-axis amortisseur circuits are given by

$$e_{1d} = 0 = p\lambda_{1d} + r_{1d}i_{1d} \tag{10.1.5}$$
$$e_{1q} = 0 = p\lambda_{1q} + r_{1q}i_{1q} \tag{10.1.6}$$

Inductances

The self-inductance of any stator winding varies periodically from a maximum (when the direct axis coincides with the phase axis) to a minimum (when the quadrature axis is in line with the phase axis).

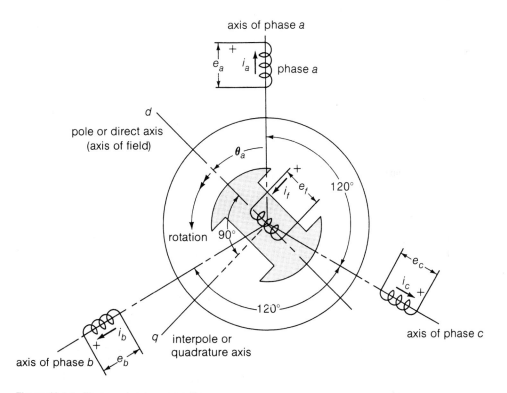

Figure 10.1.1 Diagram showing conventions.

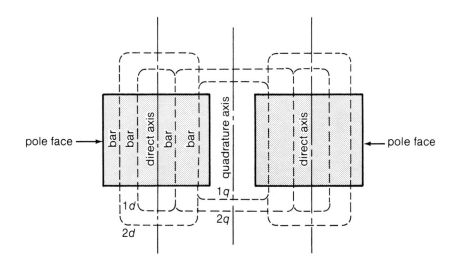

Figure 10.1.2 Flattened layout of damper circuits.

The inductance, which has a period of 180 electrical degrees, is expressible by a series of cosines of even harmonics of angle, because of the rotor symmetry. The variation can be represented approximately as

$$l_{aa} = L_{aa0} + L_{aa2} \cos 2\theta \tag{10.1.7}$$

$$l_{bb} = L_{aa0} + L_{aa2} \cos 2(\theta - 120°) \tag{10.1.8}$$

$$l_{cc} = L_{aa0} + L_{aa2} \cos 2(\theta + 120°) \tag{10.1.9}$$

where $\theta = \theta_a$, the angle between the direct axis and the axis of phase a, as shown in Figure 10.1.1.

The mutual inductances between any two stator phases are also periodic functions of rotor angular position because of the rotor saliency. We can conclude from symmetry considerations that the mutual inductance between phases a and b should have a negative maximum when the pole axis is lined up 30° behind phase a or 30° ahead of phase b, and a negative minimum when it is midway between the two phases. It can be shown[1] that the variable part of the mutual inductance is of exactly the same magnitude as that of the variable part of the self-inductance, and that the constant part of the mutual inductance has a magnitude of very nearly half that of the constant part of the self-inductance. Thus, the variations of stator mutual inductances can be represented as follows:

$$l_{ab} = l_{ba} = -\left[L_{ab0} + L_{aa2} \cos 2\left(\theta + 30°\right)\right] \tag{10.1.10}$$

$$l_{bc} = l_{cb} = -\left[L_{ab0} + L_{aa2} \cos 2\left(\theta - 90°\right)\right] \tag{10.1.11}$$

$$l_{ca} = l_{ac} = -\left[L_{ab0} + L_{aa2} \cos 2\left(\theta + 150°\right)\right] \tag{10.1.12}$$

All the rotor self-inductances, l_{fdfd}, l_{1d1d}, and l_{1q1q}, are constants since the effects of stator slots and saturation are neglected. The mutual inductance between any rotor direct-axis circuit and quadrature-axis circuit vanishes, i.e., $l_{fd1q} = l_{1qfd} = 0$ and $l_{1d1q} = l_{1q1d} = 0$. The mutual inductance between any two circuits both in direct axis (or both in quadrature axis) is constant, i.e., $l_{fd1d} = l_{1dfd} = $ constant.

Next let us consider the mutual inductances between stator and rotor circuits, which are periodic functions of rotor angular position. Because only the space-fundamental component of the produced flux links the sinusoidally distributed stator, all stator-rotor mutual inductances vary sinusoidally, reaching a maximum when the two coils in question align. Thus their variations can be written as follows:

$$l_{afd} = l_{fda} = L_{afd} \cos \theta \tag{10.1.13}$$

$$l_{bfd} = l_{fdb} = L_{afd} \cos (\theta - 120°) \tag{10.1.14}$$

$$l_{cfd} = l_{fdc} = L_{afd} \cos (\theta + 120°) \tag{10.1.15}$$

$$l_{a1d} = l_{1da} = L_{a1d} \cos \theta \tag{10.1.16}$$

$$l_{b1d} = l_{1db} = L_{a1d} \cos (\theta - 120°) \tag{10.1.17}$$

$$l_{c1d} = l_{1dc} = L_{a1d} \cos (\theta + 120°) \tag{10.1.18}$$

$$l_{a1q} = l_{1qa} = L_{a1q} \cos (\theta + 90°) = -L_{a1q} \sin \theta \tag{10.1.19}$$

$$l_{b1q} = l_{1qb} = -L_{a1q} \sin (\theta - 120°) \tag{10.1.20}$$

$$l_{c1q} = l_{1qc} = -L_{a1q} \sin (\theta + 120°) \tag{10.1.21}$$

---

[1] See C. Concordia, *Synchronous Machines* (New York: John Wiley & Sons, 1951), chap. 2.

Flux-Linkage Relations

The rotor flux linkage equations are written as follows:

$$\lambda_{fd} = l_{fdfd}i_{fd} + l_{fd1d}i_{1d} + l_{fd1q}i_{1q} - l_{fda}i_a - l_{fdb}i_b - l_{fdc}i_c \tag{10.1.22}$$

$$\lambda_{1d} = l_{1dfd}i_{fd} + l_{1d1d}i_{1d} + l_{1d1q}i_{1q} - l_{1da}i_a - l_{1db}i_b - l_{1dc}i_c \tag{10.1.23}$$

$$\lambda_{1q} = l_{1qfd}i_{fd} + l_{1q1d}i_{1d} + l_{1q1q}i_{1q} - l_{1qa}i_a - l_{1qb}i_b - l_{1qc}i_c \tag{10.1.24}$$

Equation 10.1.24 can be rewritten as follows after substituting the inductance variations we obtained earlier:

$$\lambda_{fd} = -L_{afd}[i_a \cos\theta + i_b \cos(\theta - 120°) + i_c \cos(\theta + 120°)] + L_{fdfd}i_{fd} + L_{fd1d}i_{1d} \tag{10.1.25}$$

$$\lambda_{1d} = -L_{a1d}[i_a \cos\theta + i_b \cos(\theta - 120°) + i_c \cos(\theta + 120°)] + L_{1dfd}i_{fd} + L_{1d1d}i_{1d} \tag{10.1.26}$$

$$\lambda_{1q} = L_{a1q}[i_a \sin\theta + i_b \sin(\theta - 120°) + i_c \sin(\theta + 120°)] + L_{1q1q}i_{1q} \tag{10.1.27}$$

The form of Equations 10.1.25 through 10.1.27 suggests the introduction of new variables for simplification:

$$i_d = K[i_a \cos\theta + i_b \cos(\theta - 120°) + i_c \cos(\theta + 120°)] \tag{10.1.28}$$

$$i_q = -K[i_a \sin\theta + i_b \cos(\theta - 120°) + i_c \sin(\theta + 120°)] \tag{10.1.29}$$

where K is a convenient constant. The current i_d and i_q are proportional to the components of mmf in the direct and quadrature axes, respectively, produced by the resultant of all three armature currents, i_a, i_b, and i_c. For balanced phase currents of a given maximum magnitude, the maximum values of i_d and i_q can be of the same magnitude. Under balanced conditions, the maximum magnitude of any one of the phase currents is then given by $\sqrt{i_d^2 + i_q^2}$. To achieve this relationship, a value of $\frac{2}{3}$ is assigned to the constant K in equations 10.1.28 and 10.1.29.

The next logical step is to eliminate the old variables, i_a, i_b, and i_c in favor of the newly introduced variables. If three currents are to be eliminated, three substitute variables are required. Hence, we need to introduce another new variable, i_0, which is the conventional zero-phase sequence current of the symmetrical component theory:

$$i_0 = \frac{1}{3}\left(i_a + i_b + i_c\right) \tag{10.1.30}$$

We have now established the *transformation* from the "*abc*" phase variables to the "*dq0*" Blondel variables. The transformation is expressed as follows in matrix notation:

$$\begin{bmatrix} i_d \\ i_q \\ i_0 \end{bmatrix} = \begin{bmatrix} \frac{2}{3}\cos\theta & \frac{2}{3}\cos(\theta - 120°) & \frac{2}{3}\cos(\theta + 120°) \\ -\frac{2}{3}\sin\theta & -\frac{2}{3}\sin(\theta - 120°) & -\frac{2}{3}\sin(\theta + 120°) \\ \frac{1}{3} & \frac{1}{3} & \frac{1}{3} \end{bmatrix} \begin{bmatrix} i_a \\ i_b \\ i_c \end{bmatrix} \tag{10.1.31}$$

Equation 10.1.31 can be rewritten as follows for simplicity:

$$[i_B] = [A][i_p] \tag{10.1.32}$$

where $[i_B]$ is the *Blondel-component column matrix*, $[i_p]$ is the *phase-component column matrix*, and $[A]$ is the *Blondel-transformation matrix*. The same transformation matrix can be applied to the flux linkages as well as the voltages.

One can find the phase components from the Blondel components through inverse Blondel transformation, as follows:

$$[i_p] = [A]^{-1}[i_B] \tag{10.1.33}$$

where $[A]^{-1}$ can be seen to be

$$[A]^{-1} = \begin{bmatrix} \cos\theta & -\sin\theta & 1 \\ \cos(\theta - 120°) & -\sin(\theta - 120°) & 1 \\ \cos(\theta + 120°) & -\sin(\theta + 120°) & 1 \end{bmatrix} \tag{10.1.34}$$

Going back to the rotor flux-linkage equations 10.1.25 through 10.1.27 and substituting the new variable currents, we obtain:

$$\lambda_{fd} = -\frac{3}{2}L_{afd}i_d + L_{fdfd}i_{fd} + L_{fd1d} \tag{10.1.35}$$

$$\lambda_{1d} = -\frac{3}{2}L_{a1d}i_d + L_{1dfd}i_{fd} + L_{1d1d}i_d \tag{10.1.36}$$

$$\lambda_{1q} = -\frac{3}{2}L_{a1q}i_q + L_{1q1q}i_q \tag{10.1.37}$$

The Equations 10.1.35 through 10.1.37 contain inductances that are all *constant*, as indicated by the capital Ls.

Next, the armature flux-linkage relations can be written as follows:

$$\lambda_a = -l_{aa}i_a - l_{ab}i_b - l_{ac}i_c + l_{afd}i_{fd} + l_{a1d}i_{1d} + 1_{a1q}i_{1q} \tag{10.1.38}$$
$$\lambda_b = -l_{ba}i_a - l_{bb}i_b - l_{bc}i_c + l_{bfd}i_{fd} + l_{b1d}i_{1d} + l_{b1q}i_{1q} \tag{10.1.39}$$
$$\lambda_c = -l_{ca}i_a - l_{cb}i_b - l_{cc}i_c + l_{cfd}i_{fd} + l_{c1d}i_{1d} + l_{c1q}i_{1q} \tag{10.1.40}$$

The inductance variations that have been obtained earlier are now substituted in the last three equations. The new variables of flux linkages (λ_d, λ_q, and λ_0) are now introduced in terms of the phase variables (λ_a, λ_b, and λ_c), and the relatively simple relations given below are obtained after considerable simplification, the details of which are left to the student as an instructive exercise:

$$\lambda_d = -\left(L_{aa0} + L_{ab0} + \frac{3}{2}L_{aa2}\right)i_d + L_{afd}i_{fd} + L_{a1d}i_{1d} = -L_d i_d + L_{afd}i_{fd} + L_{a1d}i_{1d} \tag{10.1.41}$$

$$\lambda_q = -\left(L_{aa0} + L_{ab0} - \frac{3}{2}L_{aa2}\right)i_q + L_{a1q}i_{1q} = -L_q i_q + L_{a1q}i_{1q} \tag{10.1.42}$$

$$\lambda_0 = -\left(L_{aa0} - 2L_{ab0}\right)i_0 = -L_0 i_0 \tag{10.1.43}$$

λ_d and λ_q can be interpreted as the flux linkages in coils moving with the rotor and centered over the direct and quadrature axes, respectively. The equivalent direct-axis moving armature circuit is seen to have the self-inductance of

$$L_d = L_{aa0} + L_{ab0} + \frac{3}{2}L_{aa2} \tag{10.1.44}$$

which is known as the *direct-axis synchronous inductance*. The equivalent quadrature-axis moving armature circuit has the self-inductance of

$$L_q = L_{aa0} + L_{ab0} - \frac{3}{2}L_{aa2} \tag{10.1.45}$$

which is known as the *quadrature-axis synchronous inductance*. Also, the zero-sequence axis, has an equivalent coil, which is completely separated magnetically from all the other coils and which has the self-inductance of

$$L_0 = L_{aa0} - 2L_{ab0} \tag{10.1.46}$$

known as the *zero-sequence inductance*.

Armature Voltage Equations

The new voltage e_d, e_q, and e_0 are now defined in the same manner as the currents and flux linkages through Blondel transformation. The equations 10.1.1 through 10.1.3 are then substituted for e_a, e_b, and e_c; the new currents i_d, i_q, and i_0 are introduced in order to eliminate the phase quantities i_a, i_b, and i_c. Then the armature voltage can be rewritten as follows:

$$e_d = p\lambda_d - \lambda_q p\theta - ri_d \tag{10.1.47}$$
$$e_q = p\lambda_q + \lambda_d p\theta - ri_q \tag{10.1.48}$$
$$e_0 = p\lambda_0 - ri_0 \tag{10.1.49}$$

From a physical standpoint, the manipulations through the transformation correspond to the specification of the armature quantities along axes fixed to the rotor and rotating with speed $p\theta$ with respect to the stator. We can therefore naturally expect to find the generated (speed) voltage as well as the induced voltages produced by the rotating flux linkages. The representation of a synchronous machine in terms of the direct-axis, quadrature-axis, and zero-sequence windings is given in Figure 10.1.3.

We have now obtained the complete set of machine-performance equations, consisting of flux-linkage relations and circuit-voltage equations. Known as *Park's equations,* they are linear differential equations with constant coefficients if the rotor speed is constant.

Looking at the flux-linkage equations given by Equations 10.1.35 through 10.1.37, and 10.1.41 through 10.1.43, the *reciprocity* of mutual-inductance coefficients, which is an essential condition for the existence of a static equivalent circuit, is not fulfilled. This difficulty, which occurs because we need the same transformation for both currents and flux linkages with the choice of $K = 2/3$, could have been avoided by other choices of transformation. The difficulty is circumvented, however, by suitably defining a per-unit system, presented later in this section.

To rewrite the armature flux-linkage relations in a more suitable form, we substitute the rotor flux-linkage relations into the rotor-circuit voltage equations, solve for the rotor currents in terms of the field voltage e_{fd} and the armature currents i_d, i_q, and plug the resultant relations into the armature flux-linkage equations. During these manipulations, we can treat the derivative operator p $(= d/dt)$ algebraically, which is not only easier but also legitimate for many problems because all the flux-linkage equations and rotor-circuit voltage relations are linear. As a result, we arrive at the equations of the following form:

$$\lambda_d = G(p)e_{fd} - L_d(p)i_d \tag{10.1.50}$$
$$\lambda_q = -L_q(p)i_q \tag{10.1.51}$$
$$\lambda_0 = -L_0 i_0 \tag{10.1.52}$$

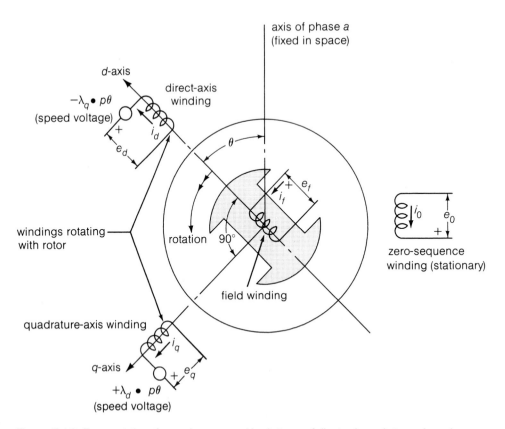

Figure 10.1.3 Representation of a synchronous machine in terms of direct-axis, quadrature-axis, and zero-sequence windings.

where $G(p)$, $L_d(p)$, and $L_q(p)$ are operators expressed as functions of the derivative operator p. $L_d(p)$, $L_q(p)$, and L_0 are known as the *operational inductances*.

Power Output

The instantaneous power output of a three-phase synchronous generator is given by

$$P = \left[e_a i_a + e_b i_b + e_c i_c \right] \tag{10.1.53}$$

In terms of the *dvo*-components, it can be shown to be

$$P = \frac{3}{2} \left[e_d i_d + e_q i_q + 2 e_0 i_0 \right] \tag{10.1.54}$$

For balanced operation, under which zero-sequence quantities vanish, the power output is expressed

$$P = \frac{3}{2} \left[e_d i_d + e_q i_q \right] \tag{10.1.55}$$

Equation 10.1.54 can be rewritten by substituting the armature voltage relations given by Equations 10.1.47 through 10.1.49 for e_d, e_q, and e_0:

$$P = \frac{3}{2}\left[i_d p\lambda_d + i_q p\lambda_q + 2i_0 p\lambda_0\right] + \frac{3}{2}\left[i_q\lambda_d - i_d\lambda_q\right]p\theta - \frac{3}{2}r\left[i_d^2 + i_q^2 + 2i_0^2\right] \qquad (10.1.56)$$

The expression for the power output in Equation 10.1.56 can be interpreted further: The first term on the right side shows the rate of decrease of armature magnetic energy; the second term indicates the power transferred electromagnetically across the air gap; and the third term is the armature resistance loss.

Electromagnetic Torque

The electromagnetic torque is obtained by dividing the power transferred across the air gap by the rotor speed, $p\theta$. Thus we have

$$T = \frac{3}{2}\left[\lambda_d i_q - \lambda_q i_d\right] \qquad (10.1.57)$$

So far we have developed the volt-ampere relations, flux-linkage equations, and expressions for power output and electromagnetic torque produced for the case of a two-pole synchronous machine. For the general case of a P-pole machine, the expression for torque in Equation 10.1.57 is to be multiplied by $P/2$.

L_{ad}-Base Per-Unit Representation

While there are many per-unit systems that were used in the past for analyzing synchronous machines by various researchers, the L_{ad}-base per-unit system, which is now widely accepted by the IEEE and other engineering institutions, is presented here. The rated volt-amperes of the machine and the rated frequency are chosen as the stator three-phase base volt-amperes (or power) and base frequency, respectively. The stator base voltage is selected as the peak value of the rated line-to-neutral voltage, and the stator base current is the peak value of the rated phase current, as follows:

$$e_{s_{base}} = \sqrt{2}V_{ph}; \qquad i_{s_{base}} = \sqrt{2}I_{ph} \qquad (10.1.58)$$

where V_{ph} is the rated rms line-to-neutral voltage, and I_{ph} is the rated rms phase current given by

$$I_{ph} = \frac{\text{Rated volt-amperes}}{3V_{ph}} \qquad (10.1.59)$$

The stator base impedance, base inductance, and base flux linkage are given by the following equations:

$$Z_{s_{base}} = \frac{e_{s_{base}}}{i_{s_{base}}} \qquad (10.1.60)$$

$$L_{s_{base}} = \frac{Z_{s_{base}}}{\omega_{base}} = \frac{Z_{s_{base}}}{2\pi f_{base}} \qquad (10.1.61)$$

$$\lambda_{s_{base}} = \frac{e_{s_{base}}}{\omega_{base}} \qquad (10.1.62)$$

where f is the frequency in hertz and ω is the radian frequency in radians/second.

Furthermore, the volt-ampere rating of the machine is expressed as

$$\text{Stator three-phase base volt-amperes (or power)} = \frac{3}{2} e_{s_{base}} i_{s_{base}} \qquad (10.1.63)$$

The time t (seconds) can be made dimensionless by multiplying it by ω_{base}, and the rotor speed $p\theta$ for steady-state synchronous-speed operation becomes unity in per-unit notation. The per-unit reactance can be seen to be the same as the per-unit inductance, and as such the terms can be used interchangeably.

Having chosen the stator-base quantities, we can now express the armature voltage equations in per-unit notation. It turns out that Equations 10.1.47 through 10.1.49 are unchanged when all quantities involved are expressed in the per-unit system just described.

Next we must specify the rotor-base quantities. To maintian the *reciprocity* of mutual-inductance coefficients in the flux-linkage equations expressed in the per-unit system (i.e., to have $L_{afd} = L_{fda}$; $L_{a1d} = L_{1da}; L_{a1q} = L_{1qa}$ in the per-unit system) it can be shown[2] that we must choose the volt-ampere base of each rotor circuit so that it is equal to the three-phase stator volt-ampere base:

$$e_{fd_{base}} i_{fd_{base}} = e_{1d_{base}} i_{1d_{base}} = e_{1q_{base}} i_{1q_{base}} = \text{Rated three-phase volt-amperes} \qquad (10.1.64)$$

It is desirable to make all per-unit mutual inductances equal between the stator and rotor circuits in each axis. Toward this end, let us consider each of the inductances L_d and L_q to be made up of two parts:

$$L_d = L_{ad} + L_l; \qquad L_q = L_{aq} + L_l \qquad (10.1.65)$$

where L_{ad} is the mutual (magnetizing) inductance between the stator and rotor in the d-axis; L_{aq} is the mutual (magnetizing) inductance between the stator and rotor in the q-axis; and L_l is the leakage inductance (assumed to be the same in both axes) due to the flux (such as air-gap leakage, slot leakage, and end-turn leakage) that does not link any rotor circuit. The per-unit L_{ad}, L_{aq}, and L_l are obtained by dividing the respective inductances (in henries) by the $L_{s_{base}}$ (in henries).

In order to have $L_{ad} = L_{afd} = L_{a1d}$, and $L_{aq} = L_{a1q}$ in per-unit representation, base quantities must be chosen in the following way:

$$i_{fd_{base}} = \frac{L_{ad}}{L_{afd}} i_{s_{base}} \qquad (10.1.66)$$

$$i_{1d_{base}} = \frac{L_{ad}}{L_{a1d}} i_{s_{base}} \qquad (10.1.67)$$

$$i_{1q_{base}} = \frac{L_{aq}}{L_{a1q}} i_{s_{base}} \qquad (10.1.68)$$

From the base voltage and base current, the base impedance and the base inductance can be calculated for each of the rotor circuits. It can also be shown that

$$L_{fd1d_{base}} = \left(L_{fd_{base}} i_{fd_{base}} \right) / i_{1d_{base}} \qquad (10.1.69)$$

---

[2] See M. S. Sarma, *Synchronous Machines (Their Theory, Stability, and Excitation Systems)* (New York: Gordon and Breach, 1979), chap. 1.

The performance equations of a synchronous machine (known as *Park's equations*) in per-unit notation according to the adopted per-unit system (which is also known as the L_{ad}-*base system*), are then given by the following equations:

$$e_d = p\lambda_d - \lambda_q p\theta - ri_d \tag{10.1.70}$$

$$e_q = p\lambda_q + \lambda_d p\theta - ri_q \tag{10.1.71}$$

$$e_0 = p\lambda_0 - ri_0 \tag{10.1.72}$$

$$e_{fd} = p\lambda_{fd} + r_{fd}i_{fd} \tag{10.1.73}$$

$$e_{1d} = 0 = p\lambda_{1d} + r_{1d}i_{1d} \tag{10.1.74}$$

$$e_{1q} = 0 = p\lambda_{1q} + r_{1q}i_{1q} \tag{10.1.75}$$

where

$$\lambda_d = -L_d i_d + L_{ad}i_{fd} + L_{ad}i_{1d} = L_{ad}(-i_d + i_{fd} + i_{1d}) - L_l i_d$$
$$= -L_d(p)i_d + G(p)e_{fd} + L_{ad}i_{1d} \tag{10.1.76}$$

$$\lambda_q = -L_q(p)i_q + L_{aq}i_{1q} \tag{10.1.77}$$

$$\lambda_0 = -L_0 i_0 \tag{10.1.78}$$

$$\lambda_{fd} = -L_{ad}i_d + L_{fdfd}i_{fd} + L_{fd1d}i_{1d} \tag{10.1.79}$$

$$\lambda_{1d} = -L_{ad}i_d + L_{1dfd}i_{fd} + L_{1d1d}i_{1d} \tag{10.1.80}$$

$$\lambda_{1q} = -L_{aq}i_q + L_{1q1q}i_{1q} \tag{10.1.81}$$

Note that the *Heaviside operator p* can be replaced by the *Laplace operator s*.

The L_{ad}-base per-unit system assumes a sinusoidal mutual flux-density distribution in the air gap. The base field current is that which establishes the same space fundamental air-gap flux as the unit-peak three-phase armature currents. The significant features of the L_{ad}-base system are that it permits replacement of the performance equations by simple equivalent circuits in the main axes, and that the equivalent-circuit reactances correspond to those normally calculated by the machine designer. In the *per-unit* notation, the power output and the electromagnetic torque are given by

$$p = e_d i_d + e_q i_q + 2e_0 i_0 \tag{10.1.82}$$

and

$$T = \lambda_d i_q - \lambda_q i_d \tag{10.1.83}$$

For quantities expressed in their normal physical units and in the per-unit system, we do not adopt two distinctly different notations because the only difference is the factor of 3/2 occurring in the rotor flux-linkage equations and the expressions for power output and electromagnetic torque. We will use only the per-unit performance equations hereafter.

We introduce the following new inductances expressed in the per-unit notation, in order to develop simplified equivalent circuits in the direct axis and the quadrature axis:

- The field leakage inductance, $L_{fd} = L_{fdfd} - L_{ad}$
- The direct-axis amortisseur-leakage inductance, $L_{1d} = L_{1d1d} - L_{ad}$
- The quadrature-axis amortisseur–leakage inductance, $L_{1q} = L_{1q1q} - L_{aq}$

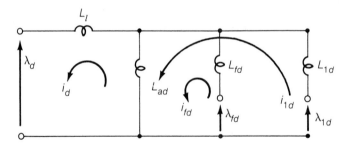

Figure 10.1.4 Direct-axis equivalent circuit.

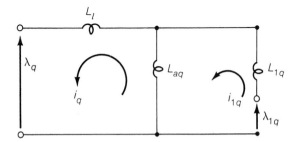

Figure 10.1.5 Quadrature-axis equivalent circuit.

Neglecting the peripheral leakage ($L_{fd1d} - L_{ad}$), (i.e., $L_{fd1d} \simeq L_{ad}$), the equivalent direct-axis and quadrature-axis circuits based on the per-unit flux-linkage equations are shown in Figures 10.1.4 and 10.1.5.

10.2 Synchronous-Machine Transient Reactances and Time Constants

During transients, we are interested in the changes that occur in the values of variables just after ($t = 0_+$) and just before ($t = 0_-$) a disturbance. By defining such differences to be

$$\Delta\lambda_d = \lambda_d|_{t=0_+} - \lambda_d|_{t=0_-}; \qquad \Delta i_d = i_d|_{t=0_+} - i_d|_{t=0_-}; \qquad \text{etc.}$$

we can verify that the flux-linkage relations in terms of the change variables are of the same form as the original equations, and hence the equivalent circuits of the same form as before can be drawn for the change variables, as shown in Figures 10.2.1 and 10.2.2.

From the *theorem of constant flux linkage,* the flux linkage of any closed circuit of finite resistance and emf cannot change instantly, and the flux linkage of any closed circuit having no resistance or emf remains constant. If the changes occur in a time that is short compared to the time constants of

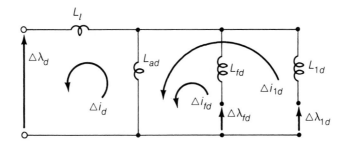

Figure 10.2.1 Direct-axis equivalent circuit.

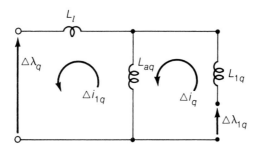

Figure 10.2.2 Quadrature-axis equivalent circuit.

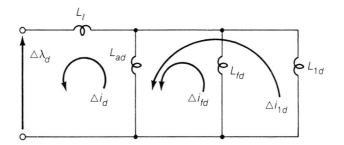

Figure 10.2.3 Direct-axis equivalent circuit for the subtransient period.

the circuit, the flux linkages will remain substantially constant during the change. Then the equivalent cirucits, which hold good for the short period of disturbance, can be redrawn as in Figures 10.2.3 and 10.2.4.

When the damper circuits are considered along with the main field winding, the armature flux linkage per armature ampere is defined as the *direct-axis subtransient inductance* L_d''. The decrement of the currents in the damper circuits is quite rapid compared to that of the field current, so all rotor currents except the field current become negligibly small after a few cycles. The effective armature

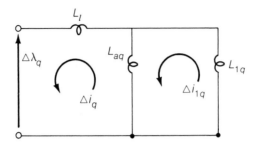

Figure 10.2.4 Quadrature-axis equivalent circuit for the subtransient period.

reactance will then increase from the *subtransient* to the *transient* value. (*Subtransient* is the term used for a transient of very short duration.)

The *subtransient direct-axis reactance* and the *subtransient quadrature-axis reactance* are then defined in per-unit notation as follows:

$$X''_d = \frac{\Delta \lambda_d}{\Delta i_d} \tag{10.2.1}$$

$$X''_q = \frac{\Delta \lambda_q}{\Delta i_q} \tag{10.2.2}$$

It follows from the equivalent circuits of Figures 10.2.3 and 10.2.4 that

$$X''_d = L_l + \frac{1}{\frac{1}{L_{ad}} + \frac{1}{L_{fd}} + \frac{1}{L_{1d}}} \tag{10.2.3}$$

$$X''_q = L_l + \frac{1}{\frac{1}{L_{aq}} + \frac{1}{L_{1q}}} \tag{10.2.4}$$

Neglecting the amortisseur winding circuits, we can go through a similar procedure and define the *transient direct-axis* and *quadrature-axis* reactances:

$$X'_d = L_l + \frac{1}{\frac{1}{L_{ad}} + \frac{1}{L_{fd}}} \tag{10.2.5}$$

$$X'_q = L_l + \frac{1}{\frac{1}{L_{aq}}} = L_l + L_{aq} = X_q \tag{10.2.6}$$

Negative Sequence Reactance X_2

If the unexcited field structure is rotated forward at synchronous speed with all rotor circuits closed and the negative-sequence currents are applied to the armature terminals, a backward mmf rotating at synchronous speed with respect to the armature is set up. This mmf, rotating backward at twice the synchronous speed with respect to the rotor, induces currents of twice the rated frequency in the rotor circuits, while keeping the flux linkages of those circuits almost constantly at zero. The armature flux linkage per armature ampere under these conditions is defined as the negative-sequence inductance L_2. The value of X_2 lies between X_d'' and X_q'', and is usually taken as the average value given by

$$X_2 = \frac{X_d'' + X_q''}{2} \tag{10.2.7}$$

Synchronous-Machine Resistances

The positive-sequence resistance of a synchronous machine is its armature ac resistance, which is somewhat greater than the dc armature resistance. The positive-sequence resistance is mostly neglected in power-system studies under normal operation and in system-stability studies. The negative-sequence resistance of a synchronous machine is greater than the positive-sequence resistance, since currents are induced in all the rotor circuits under negative-sequence conditions. Its value depends greatly on the resistance of the damper windings. The zero-sequence resistance of a synchronous machine is usually neglected.

Synchronous-Machine Time Constants

The *direct-axis transient open-circuit time constant* T_{d0}', also known as the field open-circuit time constant, is given in the per unit by

$$T_{d0}' = \frac{X_{fdfd}}{r_{fd}} = \frac{1}{r_{fd}}\left(X_{ad} + X_{fd}\right) \tag{10.2.8}$$

When the armature is open-circuited and there is no amortisseur winding, the change of field current in response to the sudden application, removal, or change of emf in the field circuit is governed by this time constant. The open-circuit ac terminal voltage of an unsaturated machine is directly proportional to the field currentt, and therefore changes with the same time constant. This time constant, which is on the order of 750 to 4,000 radians (2 to 11 seconds), is greater than any of the other time constants discussed later in this section.

The *direct-axis transient short-circuit time constant* T_d' is related to T_{d0}' as follows:

$$T_d' = \frac{X_d'}{X_d}T_{d0}' \tag{10.2.9}$$

because the ratio of the short-circuit inductance of the field winding (when the armature is short-circuited) to its open-circuit inductance (when the armature is open-circuited) is X_d'/X_d, and its resistance is unchanged. This time constant is about one-fourth the size of the open-circuit time constant.

For a machine with damper windings, the *direct-axis subtransient open-circuit time constant* T''_{d0} and the *short-circuit time constant* T''_d can be introduced. The subtransient time constant is shorter than the transient time constant, and the short-circuit value is related to the open-circuit value as follows:

$$T''_d = \frac{X''_d}{X'_d} T''_{d0} \tag{10.2.10}$$

The open-circuit time constant with one amortisseur circuit is given by

$$T''_{d0} = \frac{1}{r_{1d}} \left[X_{1d1d} - \frac{X^2_{fd1d}}{X_{fdfd}} \right] = \frac{1}{r_{1d}} \left[\frac{X_{ad}X_{fd}}{X_{ad} + X_{fd}} + X_{1d} \right] \tag{10.2.11}$$

Note that the amortisseur winding and field winding are coupled and that a mutual reactance exists between the two.

The *quadrature-axis subtransient open-circuit time constant* T''_{q0} can be introduced for the machine with one additional rotor circuit as the ratio given by

$$T''_{q0} = \frac{X_{1q1q}}{r_{1q}} = \frac{X_{aq} + X_{1q}}{r_{1q}} \tag{10.2.12}$$

and the *short-circuit time constant* T''_q is related to T''_{q0} as follows:

$$T''_q = \frac{X''_q}{X_q} T''_{q0} \tag{10.2.13}$$

Usually, T''_q is very nearly equal to T''_d.

Next, the *quadrature-axis transient open-circuit time constant* T'_{q0} and the *short-circuit time constant* T'_q may be introduced. For the case of a salient-pole machine, these time constants are meaningless; however, for a machine with a solid round rotor, these constants might be applicable; the value of T'_q is about one-half that of T'_d.

The *armature short-circuit time constant* T_a applies to the direct current in the armature windings and to the induced alternating currents in the field and damper windings (when both are closed). It is equal to the ratio of the average armature short-circuit inductance to the armature resistance under the stated conditions. It is approximately given by

$$T_a = \frac{2X''_d X''_q}{r \left(X''_d + X''_q \right)} \tag{10.2.14}$$

The armature time constant is also sometimes taken to be

$$T_a = \frac{X_2}{r} = \frac{X''_d + X''_q}{2r} \tag{10.2.15}$$

for simplicity and convenience.

Table 10.2.1 Typical Average Values of Synchronous-Machine Constants

Machine constant	Turbogenerator (solid rotor)	Waterwheel generator (with dampers)	Synchronous condenser	Synchronous motor
X_d	1.1	1.15	1.80	1.20
X_q	1.08	0.75	1.15	0.90
X_d'	0.23	0.37	0.40	0.35
X_q'	0.23	0.75	1.15	0.90
X_d''	0.12	0.24	0.25	0.30
X_q''	0.15	0.34	0.30	0.40
X_2	0.13	0.29	0.27	0.35
X_0	0.05	0.11	0.09	0.16
r (dc)	0.003	0.012	0.008	0.01
r (ac)	0.005	0.012	0.008	0.01
r_2	0.035	0.10	0.05	0.06
T_{d0}'	5.6×377	5.6×377	9.0×377	6.0×377
T_d'	1.1×377	1.8×377	2.0×377	1.4×377
$T_d'' = T_q''$	0.035×377	0.035×377	0.035×377	0.035×377
T_a	0.16×377	0.15×377	0.17×377	0.15×377

Adapted from E. W. Kimbark, *Power-System Stability: Synchronous Machines* (New York: Dover Publications, 1956 and 1968), chap. 12.

The currents and voltages of a synchronous machine under transient conditions have components whose magnitudes change in accordance with one or more of the time constants discussed in Section 10.2. The next section contains an investigation of these components. A proper understanding of the nature of short-circuit currents of synchronous machines (particularly large ones) is necessary to evaluate short-circuit forces, stresses, and torques, and to design proper protective relaying and switch gear. In practice, of course, the designer needs to consider the completely integrated power system, of which the synchronous machine is just one of the components.

Table 10.2.1 gives typical average values of synchronous-machine constants for various types of machines in per-unit notation; Table 10.2.2 gives a summary of reactances and time constants.

Table 10.2.2 Summary of Reactances and Time Constants

Reactances		
Synchronous	d-axis	$X_d = X_{ad} + X_l$
	q-axis	$X_q = X_{aq} + X_l$
Transient	d-axis	$X'_d = \dfrac{X_{ad}X_{fd}}{X_{ad} + X_{fd}} + X_l$
Subtransient	d-axis	$X''_d = \dfrac{X_{ad}X_{fd}X_{1d}}{X_{ad}X_{fd} + X_{fd}X_{1d} + X_{1d}X_{ad}} + X_l$
	q-axis	$X''_q = \dfrac{X_{aq}X_{1q}}{X_{aq} + X_{1q}} + X_l$
Time Constants		
Open-circuit transient	d-axis	$T'_{d0} = \dfrac{1}{r_{fd}}[X_{ad} + X_{fd}]$
Open-circuit subtransient	d-axis	$T''_{d0} = \dfrac{1}{r_{1d}}\left[\dfrac{X_{ad}X_{fd}}{X_{ad} + X_{fd}} + X_{1d}\right]$
	q-axis	$T''_{q0} = \dfrac{1}{r_{1q}}[X_{aq} + X_{1q}]$
Short-circuit transient	d-axis	$T'_d = \dfrac{1}{r_{fd}}\left[\dfrac{X_{ad}X_l}{X_{ad} + X_l} + X_{fd}\right] = \dfrac{X'_d}{X_d}T'_{d0}$
Short-circuit subtransient	d-axis	$T''_d = \dfrac{1}{r_{1d}}\left[\dfrac{X_{ad}X_{fd}X_l}{X_{ad}X_{fd} + X_{fd}X_l + X_{ad}X_l} + X_{ld}\right] = \dfrac{X''_d}{X'_d}T''_{d0}$
	q-axis	$T''_q = \dfrac{1}{r_{1q}}\left[\dfrac{X_{aq}X_l}{X_{aq} + X_l} + X_{1q}\right] = \dfrac{X''_q}{X_q}T''_{q0}$
Short-circuit armature (dc)		$T_a = \dfrac{1}{r}\left[\dfrac{2X''_dX''_q}{X''_d + X''_q}\right]$

▐▐▐▐▐▐▐▐▐▐▐▐▐▐▐▐▐

10.3 Synchronous-Machine Transient Parameters from Sudden Three-Phase Short-Circuit Test Data

The reactances and time constants of a synchronous machine are of great assistance for predicting short-circuit currents. Conversely, a short-circuit oscillogram can be used to evaluate some of the reactances and time constants. Before we go into details of computing the constants from test data, let us try to get a physical picture of what happens under short-circuit conditions.

Consider a three-phase, initially unloaded, synchronous generator operating at synchronous speed with constant excitation. Let a three-phase short-circuit be suddenly applied at the armature terminals. We will explore the nature of the three armature-phase currents and the field current.

The flux that is produced by the field circuit links the armature circuits. When the three-phase short-circuit fault occurs at $t = 0$, the trapped armature-flux linkages are given by

$$\lambda_a \propto \cos \alpha \tag{10.3.1}$$

$$\lambda_b \propto \cos (\alpha - 2\pi/3) \tag{10.3.2}$$

$$\lambda_c \propto \cos (\alpha + 2\pi/3) \tag{10.3.3}$$

where α is the angle at $t = 0$ between the phase-a axis and the d-axis. Thus the flux linking each phase is different. As the field moves away after $t = 0$, the flux cannot change immediately, so the dc current of appropriate magnitude appears in each phase to preserve the flux. Because the flux is different for each phase depending on the angle α, the dc currents that appear are also of different magnitudes depending on α. These currents damp out eventually with the armature time constant T_a.

If the magnitudes of each of the dc currents appearing in the armature phases were to be the same, no net flux would result. But, because they happen to be all unequal, a flux results. It produces a damped fundamental current in the field circuit because of the relative motion between the field and armature circuits, the damping being dictated by the armature time constant T_a.

To an observer on the rotor, the fundamental damped current component produces a uniaxial pulsating flux, which can be resolved into two rotating flux waves of the same magnitude traveling in opposite directions at synchronous speed. The fluxwave traveling in the positive direction with respect to the rotor travels at twice the synchronous speed with respect to the stator; the other one is stationary with respect to the stator. So the latter fluxwave does not induce anything, but the former produces double-frequency component currents of the same magnitude in the armature phases. These 2nd-harmonic currents give rise to zero net resultant flux. Hence the reflection across the air gap ceases. These 2nd-harmonic currents in the armature phases damp out eventually with the armature time constant T_a.

Thus we see that the unequal dc currents in the armature phases give rise to the fundamental component in the field circuit, which in turn is responsible for producing equal 2nd-harmonic components in the armature phases. Because the process starts in the armature, all these components damp out with the armature time constant T_a.

The constant flux from the sustained dc current in the field circuit induces sustained fundamental three-phase armature currents. These currents produce an mmf that is rotating forward at synchronous speed with respect to the stator but stationary with respect to the field and centered on the direct axis of the field. This armature mmf opposes the field mmf and tends to reduce the field flux linkages and damper-winding flux linkages. In order to prevent such flux linkage changes, increased field current as well as amortisseur currents are induced. Thus the transient and subtransient dc components exist in the rotor windings, damped by the direct-axis transient short-circuit time constant T_d' and the direct-axis subtransient short-cirucit time constant T_d'', respectively. Correspondingly, the fundamental ac components appear in the armature windings; the transient and subtransient fundamental components damp out with time constants T_d' and T_d'', respectively.

Thus we have the following components in the field current: *(i)* sustained dc, *(ii)* damped dc with time constant T_d', *(iii)* damped dc with time constant T_d'', and *(iv)* damped fundamental ac component with time constant T_a.

The armature phase windings have the following component currents: *(i)* sustained fundamental ac, *(ii)* damped fundamental ac with time constant T_d', *(iii)* damped fundamental ac with time constant T_d'', *(iv)* damped dc components with time constant T_a (which depend on the instant the fault occurs), and *(v)* damped 2nd-harmonic ac with time constant T_a. Thus each current wave in general consists

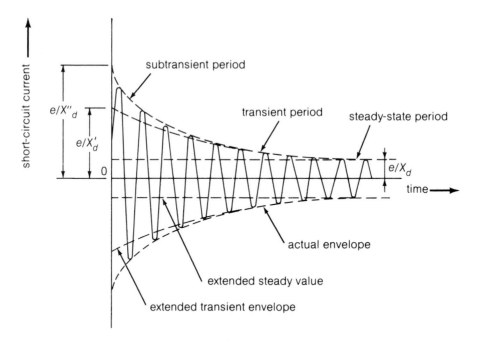

Figure 10.3.1 Alternating component of a symmetrical short-circuit armature current of a synchronous machine.

alternating-current components, which are equal in all the three phases, and direct-current components, which are dependent on the particular point in the cycle at which the short circuit occurs.

Note that the short-circuit currents would not decay if the flux linkages were to remain absolutely constant. Actually, they have decrements controlled by different time constants of the synchronous machine, as explained previously. The induced 2nd-harmonic armature currents are rather small if the machine has damper windings.

Figure 10.3.1 shows an enlarged view of the alternating components of a symmetrical short-circuit armature current of a synchronous machine. The envelope of the current for the subtransient, transient, and steady-state periods can be seen. The subtransient period lasts only for the first few cycles, during which the decrement of the current is very rapid. The transient period covers a relatively longer time, during which the decrement of the current is more moderate. Finally the steady state is attained, during which the current has a sustained value. From the open-circuit prefault armature voltage and initial values of different armature-current components, the direct-axis reactances X_d, X_d', and X_d'' can be computed. The direct-axis short-circuit time constants T_d' and T_d'' can be evaluated from semilog plots of the transient and subtransient components.

For a prolonged short circuit, the armature current finally attains a sustained value (during the steady-state period), the magnitude of which is e/X_d, where e is open-circuit voltage of an unsaturated machine, or the voltage read from the air-gap line of a saturated machine. The initial value of the alternating component of the armature current at the beginning of the transient period is e/X_d', which is seen by extrapolating the transient envelope in Figure 10.3.1. The initial value of the alternating component of the armature current at the commencement of the subtransient period is given by e/X_d''.

Figure 10.3.2 shows the envelope of the symmetrical short-circuit current. The difference $\Delta i'$ indicated in Figure 10.3.2 between the transient envelope (including the extended part) and the steady-state

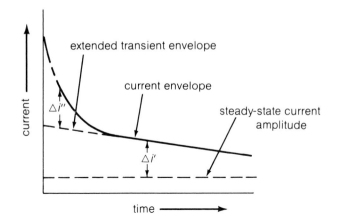

Figure 10.3.2 Envelope of the symmetrical short-circuit current.

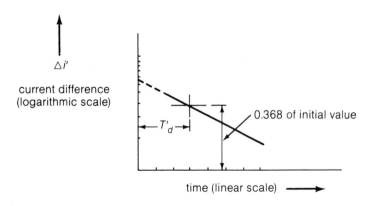

Figure 10.3.3 Semilog plot of $\Delta i'$ as a function of time.

amplitude is plotted to a logarithmic scale as a function of time in Figure 10.3.3. Similiarly, the difference $\Delta i''$ between the subtransient and extrapolated transient envelope is plotted to a logarithmic scale as a function of time in Figure 10.3.4. Both the plots are approximately straight lines, which illustrates the essentially exponential nature of the current decrement. From Figures 10.3.3 and 10.3.4, we can evaluate the time constants T'_d and T''_d by reading the times corresponding to 0.368 of their initial current values.

A typical short-circuit oscillogram in Figure 10.3.5 shows the three-phase armature current waves as well as the field current. The traces of the armature-phase currents are not symmetrical about the zero-current axis, and they definitely exhibit the dc components responsible for offset waves. A symmetrical wave, like that of Figure 10.3.1, is easily obtained by replotting the offset wave with the dc component subtracted. The armature time constant T_a, which controls the decay of the dc component, is equal to the time required for the dc component to decay to 0.368 of its initial value. Semilog plots of the dc components of the armature currents, shown in Figure 10.3.5, and the ac component of the field current can be used to determe the time constant T_a. From the semilog plot of the excess dc component of the field current over sustained value, time constants T'_d and T''_d can also be evaluated.

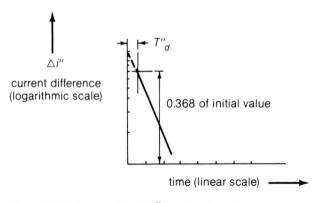

Figure 10.3.4 Semilog plot of $\Delta i''$ as a function of time.

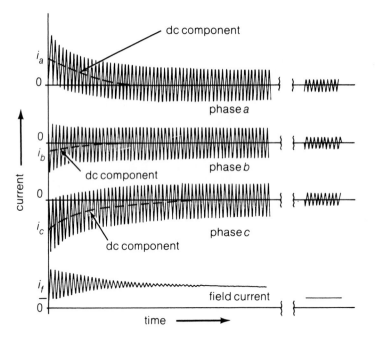

Figure 10.3.5 Short-circuit three-phase armature current and field current waves. (Adapted, by permission, from E.W. Kimbark, *Power System Stability: Synchronous Machines.* New York: Dover Publications, 1968, p. 41.)

For the experimental determination of the quadrature-axis transient and subtransient reactances, the negative-sequence resistance, and the reactance, the reader is referred to the other published literature.[3]

[3] Institute of Electrical and Electronics Engineers, *Standard No. 115–1965, Code SH0054, Test Procedures for Synchronous Machines* (New York: IEEE, 1965); C. F. Wagner, and R. D. Evans, *Symmetrical Components* (New York: McGraw-Hill, 1933); S. H. Wright, "Determination of Synchronous Machine Constants by Test," *AIEE Transactions,* 50 (December 1931): 1331–50.

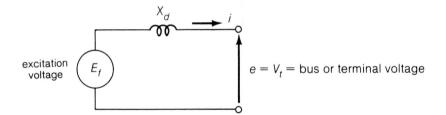

Figure 10.4.1 Simple model of a synchronous machine for steady-state stability analysis.

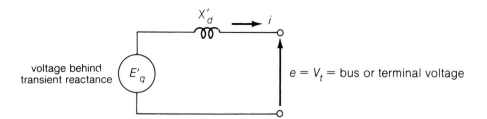

Figure 10.4.2 Simple model of a synchronous machine for transient-stability studies.

10.4 Synchronous-Machine Dynamics

Transient stability is concerned with the operation of an interconnected power system subjected to severe disturbances like faults, circuit switching, and sudden changes in load. Both electrical and mechanical systems will obviously be in the transient state. Transient stability analysis involves electrical and mechanical properties of the machines of the system because the machines must adjust the relative angles of their rotors after every disturbance to meet the conditions imposed be power transfer. The duration of the fault depends on the relay and circuit-breaker characteristics; it is usually on the order of 10 cycles or less for systems with high-speed fault clearing.

The simplest model that can be used for a synchronous machine for steady-state stability studies is given in Figure 10.4.1; it neglects the armature resistance and saliency. For transient stability studies, we neglect transient saliency and armature resistance and assume *constant field-flux linkages;* the simple equivalent circuit of Figure 10.4.2 is often used. Neglecting damper-circuit currents, per-unit notation gives us

$$\lambda_d = -X_d i_d + X_{ad} i_{fd} \tag{10.4.1}$$

and

$$\lambda_{fd} = -X_{ad} i_d + X_{fdfd} i_{fd} \tag{10.4.2}$$

Eliminating the field current i_{fd}, we obtain

$$\lambda_{fd} = \frac{X_{fdfd}}{X_{ad}} \left[\lambda_d + \left(X_d - \frac{X_{ad}^2}{X_{fdfd}} \right) i_d \right] \tag{10.4.3}$$

The quantity

$$X_d - \left(X_{ad}^2/X_{fdfd}\right)$$

is a short-circuit reactance of the armature direct-axis circuit, indicating the demagnetizing effect of the armature current; it can be measured at the direct-axis armature terminals with zero field resistance. Let us define the following for convenience:

$$\left(X_d - \frac{X_{ad}^2}{X_{fdfd}}\right) = X_d' \tag{10.4.4}$$

which is called the *transient reactance,* and

$$\frac{X_{ad}}{X_{fdfd}}\lambda_{fd} = E_q' \tag{10.4.5}$$

which is the *voltage behind transient reactance;* it is a quantity proportional to the field-flux linkage. Using these new definitions, we get the following relationship from Equation 10.4.3:

$$E_q' = \lambda_d + X_d' i_d \tag{10.4.6}$$

and since

$$e_q = \lambda_d - ri_q \tag{10.4.7}$$

it follows then that

$$e_q = E_q' - X_d' i_d - ri_q \tag{10.4.8}$$

From Equations 10.4.5, 10.4.2, and 10.4.4, we obtain

$$E_q' = -\frac{X_{ad}^2}{X_{fdfd}} i_d + E_f = E_f - \left(X_d - X_d'\right) i_d \tag{10.4.9}$$

where

$$E_f = X_{ad} i_{fd}$$

is the excitation voltage.

A phasor diagram similar to that of Figure 8.5.1 can now be developed as shown in Figure 10.4.3 for the case of a salient-pole synchronous generator (with $X_q' = X_q$), operating at a lagging power factor in the transient state. A quantity corresponding to the field-flux linkage, E_q' or E' (neglecting transient saliency), can be identified on the phasor diagram of Figure 10.4.3.

Assuming the field-flux linkage remains constant, we can derive a *transient power-angle character- istic* in terms of the voltage E_q' in a manner similar to finding the steady-state power-angle characteristic of a salient-pole synchronous machine as we developed in Section 8.5. The procedure for derivation is the same as before, except that E_q' replaces E_f, and X_d' replaces X_d. We obtain the following equation with negligible armature resistance similar to that of Equation 8.5.10:

$$P = \frac{V_t E_q'}{X_d'} \sin\delta + \frac{1}{2}V_t^2\left(\frac{1}{X_q} - \frac{1}{X_d'}\right)\sin 2\delta \tag{10.4.10}$$

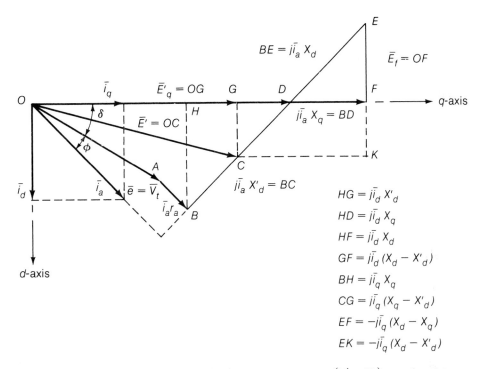

Figure 10.4.3 Phasor diagram of a salient-pole synchronous generator $\left(X'_q = X_q \right)$ operating at a lagging power factor in the transient state.

Noting that $\left(X'_d - X_d \right)$ is negative, the 2nd-harmonic term, which represents the *transient-reluctance power*, reaches its positive maximum at $\delta = 135°$, corresponding to generator action, so that the maximum transient power occurs at an angle between $\pi/2$ and $3\pi/4$. Equation 10.4.10 is sketched in Figure 10.4.4.

The usual assumptions regarding the representation of a synchronous machine for transient-stability analysis are as follows: (a) The effects of speed variation are neglected and the speed is treated as a constant by setting $p\theta$ equal to unity in per-unit notation. (b) The electromagnetic torque T_e is taken to be equal to the electric power developed or the air-gap power P_e for a generator, and the mechanical-torque input is assumed to be a constant. (c) Only the fundamental frequency voltages and currents are considered.

The effects of saturation are either neglected or taken care of approximately. Amortisseur effects are either neglected or approximately considered in transient-stability studies. The transient saliency is usually neglected. Assuming the field-flux linkage to remain constant, the transient power-angle or torque-angle relationship is given in per-unit notation as follows:

$$P_e = T_e = \frac{V_t E'}{X'_d} \sin \delta \tag{10.4.11}$$

in which transient saliency is neglected, V_t is the terminal voltage, and δ is now the angle between $\bar{E}'$ and $\bar{V}_t$.

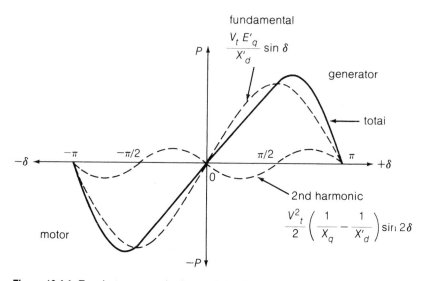

Figure 10.4.4 Transient power-angle characteristic of a salient-pole synchronous machine (with negligible armature resistance).

Variation of field-flux linkage, excitation response, speed changes, and saturation effects can be adequately considered in the step-by-step solution procedure[4] and other digital-computer methods[5] for transient-stability studies.

The Swing Equation

The per-unit mechanical acceleration is given by

$$p^2\theta = \frac{1}{H'}\left(T_m - T_e\right) \tag{10.4.12}$$

where p is the operator d/dt, with t in radians; θ is the angle in electrical radians between the d-axis and the centerline of the phase-a axis; H' is the per-unit inertia constant in (kW-rad)/kVA; T_m represents the per-unit mechanical-torque input; and T_e stands for the per-unit electromagnetic torque developed. Since θ is continuously changing with time, it is more convenient to measure the angular position with respect to a synchronously rotating reference axis. θ is then expressed as

$$\theta = t + \delta \tag{10.4.13}$$

where δ is the angle between the reference axis and the quadrature axis of the machine. Equation 10.4.12 is then rewritten as

$$p^2\delta = \frac{1}{H'}\left(T_m - T_e\right) \tag{10.4.14}$$

[4] W. D. Stevenson, Jr., *Elements of Power System Analysis*, 3d ed. (New York: McGraw-Hill, 1975), chap. 14.

[5] G. W. Stagg, and A. H. El-Abiad, *Computer Methods in Power System Analysis* (New York: McGraw-Hill, 1968), chaps. 7–10.

As it is more common to express the angle in degrees and time in seconds, the acceleration equation is also given by

$$p^2\delta = \frac{180f_b}{H}(T_m - T_e) \tag{10.4.15}$$

where p is the operator d/dt, with t in seconds; δ is in electrical degrees; f_b is the base-rated frequency; H is the per-unit inertia constant in (kW-s)/kVA; and T_m as well as T_e are the per-unit values. The torque caused by friction, windage, and core loss in machine is usually discarded. Since the per-unit electrical torque is equal to the per-unit air-gap power developed, we can rewrite Equation 10.4.15 as

$$p^2\delta = \frac{180f_b}{H}(P_m - P_e) \tag{10.4.16}$$

where P_m is the per-unit shaft-power input (minus rotational losses, if they are to be considered); and P_e represents the per-unit electrical power developed (electrical power output of the machine plus the armature copper losses). While the equations have been developed here with a generator in mind, they can easily be modified appropriately for a motor. Quite often in the literature, the quantities involved are expressed in their natural units without the use of per-unit notation, in which case Equation 10.4.16 becomes

$$p^2\delta = \frac{1}{M}(P_m - P_e) \tag{10.4.17}$$

where p is the operator d/dt, with t in seconds; δ is in electrical degrees; M is the angular momentum given by $(GH/180 f_b)$ kilojoule-seconds per electrical degree; G is the rating of the machine in kilovolt-amperes; H is in (kW-s)/kVA; f_b is the base-rated frequency in hertz; P_m and P_e are expressed in kilowatts. Equation 10.4.16 or 10.4.17 is known as the *swing equation*. Note that the inertia constant, defined as the angular momentum at synchronous speed, is truly a constant, while the angular momentum of a machine is not a constant if the angular velocity is changing. However, M is treated as a constant since the speed of the machine does not differ much from the synchronous speed unless the stability limit is exceeded. The shaft-power input P_m is usually taken as a constant for simplifying the calculations, although it is possible in digital-computer programs to take account of the governor action and hence the change in input from the prime mover. The electrical power developed P_e is of the form given by Equation 10.4.11. The solution of Equation 10.4.17 yields the *swing curve,* which is a graph of δ as a function of time t. The swing equation in the form of Equation 10.4.17 is used for further discussion in this section.

Formal analytical solution of the swing equations is almost impossible for a multimachine system. Elliptic integrals result even for the simple case of one machine connected to an infinite bus, when the resistance is neglected and P_m is taken to be zero. Transient stability studies are usually carried out with the aid of computers using numerical integration schemes, and even then a point-by-point solution is attempted. If the swing curve indicates that the angle δ starts to decrease after reaching a maximum value, it is usually assumed that the system will not lose stability and that the oscillations of δ around the equilibrium point will become successively smaller and eventually damp out. It is possible in certain cases to make use of the equal-area criterion of stability in order to gain an understanding of the transient-stability conditions without formally solving the swing equation. The equal-area criterion cannot be used directly in systems where three or more machines exist and the assumption of an infinite bus is not valid.

Equal-Area Criterion for Transient Stability

For each of the two interconnected machines, the acceleration equations are given by

$$p^2\delta_1 = \frac{P_{a1}}{M_1}; \qquad p^2\delta_2 = \frac{P_{a2}}{M_2} \tag{10.4.18}$$

in which subscripts 1 and 2 refer to the interconnected machines, and P_a is the accelerating power given by $(P_m - P_e)$. Equation 10.4.18 is rearranged as

$$p^2 \delta_{12} = p^2 (\delta_1 - \delta_2) = \frac{P_{a1}}{M_1} - \frac{P_{a2}}{M_2} \qquad (10.4.19)$$

Multiplying by $2p\delta_{12}$ and integrating both sides, we obtain

$$(p\delta_{12})^2 = 2 \int \left(\frac{P_{a1}}{M_1} - \frac{P_{a2}}{M_2} \right) d\delta_{12} \qquad (10.4.20)$$

The relative speed between the two machines becomes zero when $p\delta_{12}$ equals zero, which then forms the basis of the equal-area criterion. The machines will not remain at rest with respect to each other the first time $p\delta_{12}$ becomes zero; but the fact that δ_{12} has momentarily stopped changing can be taken to indicate stability, which is equivalent to the assumption that the swing curve indicates stability when the angle δ_{12} reaches a maximum and starts to decrease.

For the case of one machine connected to an infinite bus, the equal-area criterion is given by

$$\int_{\delta_0}^{\delta_s} \frac{2P_{a1}}{M_1} d\delta_{12} = 0 \quad \text{or} \quad \int_{\delta_0}^{\delta_s} P_{a1} d\delta_{12} = 0 \qquad (10.4.21)$$

in which δ_0 is the angle prior to the disturbance when the machine is operating at synchronous speed, and δ_s is the angle after the disturbance when the machine is again operating at synchronous speed and the angle ceases to change. Note that M_2 corresponding to the infinite bus is infinite in view of the infinite inertia of that system. Figure 10.4.5 shows a generator connected to an infinite bus, as well as the equivalent circuit of the system for transient stability study. The accelerating power P_{a1} is given by

$$P_{a1} = P_m - P_e \qquad (10.4.22)$$

where P_m is usually assumed to be a constant equal to the value of P_{e0} prior to the disturbance, and P_e is given by the corresponding power-angle equation as

$$P_e = \frac{E'_g E_s}{(X'_d + X_e)} \sin \delta_{12} = P_{max} \sin \delta_{12} \qquad (10.4.23)$$

which is shown in Figure 10.4.6.

Let us now consider a solid three-phase fault at the generator terminals occuring at $t = 0$, with the fault cleared at a time $t = t_s$ without any alteration in the reactance X_e. The generator is isolated from the system during the fault period. The shaded area A_1 of Figure 10.4.6 represents the energy that tends to increase the rotor speed and cause loss of synchronism between the generator and the infinite bus, while the shaded area A_2 represents the energy that tends to stabilize and restore synchronism. The algebraic sum of the areas producing stability is given by

$$A_2 - A_1 = \int P_{a1} d\delta_{12} \qquad (10.4.24)$$

When the area A_2 is greater than the area A_1, the machine is transiently stable; when A_2 is less than A_1, the machine is transiently unstable. The limiting case between being stable or unstable arises when A_2

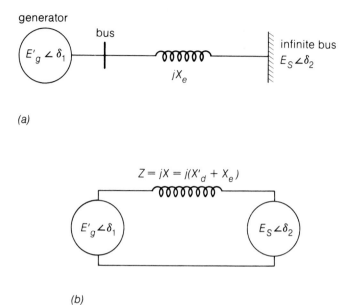

(a)

(b)

Figure 10.4.5 (a) Generator connected to an infinite bus. (b) Equivalent circuit of the system for transient stability study. (Note: The angles δ_1 and δ_2 are measured from a common reference axis. E'_g is the voltage behind transient reactance of the generator. The resistance is neglected here.)

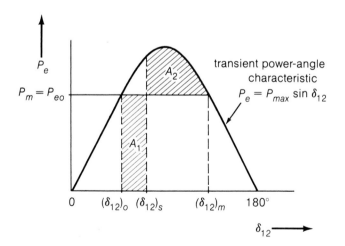

Figure 10.4.6 Equal-area criterion for transient stability of a generator connected to an infinite bus, for critical clearing time.

is exactly equal to A_1, in which case the corresponding $(\delta_{12})_s = \delta_c$ is the *critical-clearing angle* and the corresponding clearing time is the *critical-clearing time*.

The equal-area criterion is also easily applied to find the value to which the input power could be suddenly increased without loss of synchronism, for the case of a synchronous generator, connected to

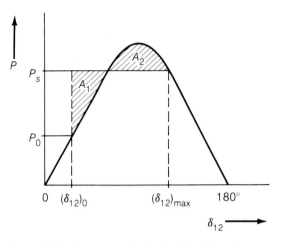

Figure 10.4.7 Equal-area criterion for determining the power limit.

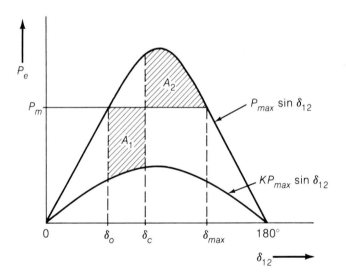

Figure 10.4.8 Equal-area criterion applied to fault-clearing, when power is transmitted during the fault (Note: Areas A_1 and A_2 are equal.)

an infinite bus, and initially operating under steady-state conditions while delivering power P_0, shown in Figure 10.4.7. When areas A_1 and A_2 are equal in Figure 10.4.7, P_s represents such a value.

It is possible to transmit some power even during the fault, to an extent that depends on the nature of the fault. Such a case is easily analyzed by means of the equal-area criterion, as shown in Figure 10.4.8. Let P_m be the mechanical input from the prime mover; $(P_{max} \sin \delta_{12})$ be the electric power that can be transmitted by the generator before the fault and also after the fault is cleared; and $(KP_{max} \sin \delta_{12})$ be the power that can be transmitted during the fault, where K is a constant depending on the nature of the fault. The largest possible value of δ_{12} for clearing to occur without exceeding the transient stability limit is called the *critical-clearing* (or *switching*) *angle;* it is given by δ_c when the areas A_1 and A_2 are equal. The critical-clearing time corresponding to the critical-clearing angle is the time taken for the machine to swing from its original position to its critical-clearing angle. If the fault clearing occurs at a value of δ_{12} greater that δ_c, the area A_2 is less than the area A_1 and δ_{12} continues to increase beyond δ_{max}; the speed increases further, and the generator cannot regain synchronism. The severity of the fault affects the value of K and hence the critical-clearing angle. An increase in the severity of the fault results in a decrease of the value of K. In order of decreasing severity, the fault conditions are given below:

1. Three-phase fault
2. Double line-to-ground fault
3. Line-to-line fault
4. Single line-to-ground fault

The effect of ground impedance in system-neutrals is to decrease the severity. While the three-phase fault is the most severe one, it is the least likely to occur. For complete reliability, a system should be designed for transient stability for three-phase faults at the worst locations. Reliability is sometimes sacrificed for economic reasons, however.

Factors Affecting Transient Stability

It is important for the power-system designer and operating engineer to understand the causes of instability in a system so that appropriate means can be reliably employed to prevent instability from both economic and technical viewpoints. Some of the stability considerations are more remote plant siting, larger plant ratings, larger unit sizes, fewer transmission circuits, more heavily loaded transmission, increased pool interchanges, increased emphasis on reliability, and increased severity of disturbance criteria as well as contingencies. Most of these factors are manifestations of environmental considerations or economic pressures, or both.

This list gives some methods for improving transient stability:

- Increasing the system voltage
- Reducing the overall reactance by appropriately choosing individual system components, providing series compensation, and adding more parallel transmission lines
- Providing a strong transmission network with various interconnections and additional transmission lines
- Controlling the high-response and optimally stabilized excitation systems
- Using high-speed circuit breakers, including reclosing breakers
- Introducing a breaking resistor to guard against a sudden loss of a large load
- Using independent-pole switching of the circuit breakers
- Using single-pole switching to trip only the faulted phase
- Shedding load during the periods of insufficient generation
- Shedding generation in response to the line faults
- Applying fast turbine-valve control
- Providing a dc-transmission line

Countermeasures taken against sudden, undesired changes in the network should match the disturbance with respect to the location, amount, duration, and dependence on angular difference. A proper combination of measures accomplished rapidly under suitable control can improve system stability to a large extent. Combined means for improving power-system stability should be considered in terms of a coordinated system, not individual measures applied locally. Methods such as the resistor breaking, load shedding, generation shedding, and switching of series capacitors on the electrical sides should be coordinated with fast valving and bypassing of steam or water on the mechanical side. Coordination and cooperation among interconnected power system owners are required for economical design and reliable operation. Central control by a microprocessor can be expensive, yet much can be accomplished for less cost by local analog or digital devices.

10.5 Mathematical Description of a Three-Phase Induction Machine

A three-phase induction machine has three distributed stator windings and three distributed rotor windings, as shown schematically in Figure 10.5.1. If we assume the machine to be cylindrical, the self-inductances and mutual inductances between stator phases or rotor phases are constant. Mutual inductances between stator and rotor coils are functions of the rotor position, however.

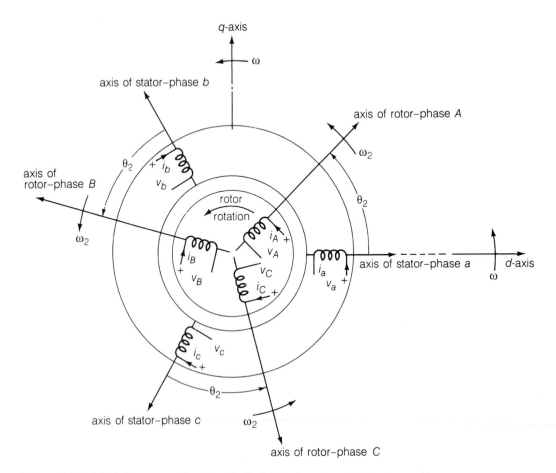

Figure 10.5.1 Schematic representation of an idealized three-phase two-pole induction machine.

Space mmf and flux waves are considered to be sinusoidally distributed, thereby neglecting the effects of teeth and slots. The machine is regarded as a group of linear coupled circuits, permitting superpositions to be applied, while neglecting saturation, hysteresis, and eddy currents. The stator and rotor phases are balanced and can be connected in wye or delta. The rotor can be a squirrel-cage structure short-circuited on itself by end rings, or a wound rotor with the terminals brought to slip rings on the shaft for connecting to an external circuit. With the rotor terminals short-circuited, the rotor-terminal voltages V_A, V_B, and V_C become zero; however, these voltages are shown in Figure 10.5.1 for our study.

The magnetic axes of each of the stator and rotor phases are shown in Figure 10.5.1, in which the axis of a rotor phase is displaced by an angle θ_2 from the axis of the corresponding stator phase. The angle θ_2 varies with time, as the rotor turns; with a constant rotor angular velocity ω_2, or for a constant slip **S,**

$$\theta_2 = \omega_2 t = (1 - S)\,\omega t \tag{10.5.1}$$

where ω is the stator electrical angular velocity.

The voltage relations for the stator and rotor windings from a coupled-circuit viewpoint are given by the following equations:

$$v_1 = p\lambda_1 + r_1 i_1 \tag{10.5.2}$$
$$v_2 = p\lambda_2 + r_2 i_2 \tag{10.5.3}$$

where the subscript 1 denotes a stator phase a, b, or c; subscript 2 denotes a rotor phase A, B, or C; r is the resistance per phase; λ is the flux linkage; and p is the differential operator d/dt. As shown in Figure 10.5.1, the motor conventions are used here.

Let the maximum value of mutual inductance between any stator phase and any rotor phase be M. Let the inductance per phase and mutual inductance between phases be L_1 and M_1 for the stator winding, and L_2 and M_2 for the rotor winding. The flux-linkage equations are then given by the following equations:

Stator: $\lambda_a = L_1 i_a + M_1 \left(i_b + i_c\right) + M\left[i_A \cos\theta_2 + i_B \cos\left(\theta_2 + 120°\right)\right] + i_C \cos\left(\theta_2 - 120°\right)$ (10.5.4)

$\lambda_b = L_1 i_b + M_1 \left(i_a + i_c\right) + M[i_A \cos\left(\theta_2 - 120°\right) + i_B \cos\theta_2 + i_C \cos\left(\theta_2 + 120°\right)]$ (10.5.5)

$\lambda_c = L_1 i_c + M_1 \left(i_a + i_b\right) + M[i_A \cos\left(\theta_2 + 120°\right) + i_B \cos\left(\theta_2 - 120°\right) + i_C \cos\theta_2]$ (10.5.6)

Rotor: $\lambda_A = L_2 i_A + M_2 \left(i_B + i_C\right) + M[i_a \cos\theta_2 + i_b \cos\left(\theta_2 - 120°\right) + i_c \cos\left(\theta_2 + 120°\right)]$ (10.5.7)

$\lambda_B = L_2 i_B + M_2 \left(i_A + i_C\right) + M[i_a \cos\left(\theta_2 + 120°\right) + i_b \cos\theta_2 + i_c \cos\left(\theta_2 - 120°\right)]$ (10.5.8)

$\lambda_C = L_2 i_C + M_2 \left(i_A + i_B\right) + M[i_a \cos\left(\theta_2 - 120°\right) + i_b \cos\left(\theta_2 + 120°\right) + i_c \cos\theta_2]$ (10.5.9)

In general we can assume

$$i_a + i_b + i_c = 0 \tag{10.5.10}$$

and

$$i_A + i_B + i_C = 0 \tag{10.5.11}$$

because the great majority of induction machines have no neutral connection, or the external circuitry is such that there are no zero-sequence currents.

Let the self-inductance of the equivalent two-phase stator coils and the corresponding quantity for the rotor coils be introduced as followed:

$$L_{11} = L_1 - M_1 \tag{10.5.12}$$

and

$$L_{22} = L_2 - M_2 \tag{10.5.13}$$

By making use of Equations 10.5.10 through 10.5.13, we can simplify the flux-linkage equations as given here:

Stator:

$$\lambda_a = L_{11}i_a + M\left[i_A \cos\theta_2 + i_B \cos\left(\theta_2 + 120°\right) + i_C \cos\left(\theta_2 - 120°\right)\right] \qquad (10.5.14)$$

$$\lambda_b = L_{11}i_b + M[i_A \cos\left(\theta_2 - 120°\right) + i_B \cos\theta_2 + i_C \cos\left(\theta_2 + 120°\right)] \qquad (10.5.15)$$

$$\lambda_c = L_{11}i_c + M[i_A \cos\left(\theta_2 + 120°\right) + i_B \cos\left(\theta_2 - 120°\right) + i_C \cos\theta_2] \qquad (10.5.16)$$

Rotor:

$$\lambda_A = L_{22}i_A + M[i_a \cos\theta_2 + i_b \cos\left(\theta_2 - 120°\right) + i_c \cos\left(\theta_2 + 120°\right)] \qquad (10.5.17)$$

$$\lambda_B = L_{22}i_B + M\left[i_a \cos\left(\theta_2 + 120°\right) + i_b \cos\theta_2 + i_c \cos\left(\theta_2 - 120°\right)\right] \qquad (10.5.18)$$

$$\lambda_C = L_{22}i_C + M[i_a \cos\left(\theta_2 - 120°\right) + i_b \cos\left(\theta_2 + 120°\right) + i_c \cos\theta_2] \qquad (10.5.19)$$

If we now substitute the flux-linkage expressions in the voltage relations of Equations 10.5.2 and 10.5.3, we obtain an algebraically complicated set of nonlinear differential equations; the nonlinearity is introduced by the trigonometric terms of Equations 10.5.14 through 10.5.19 because θ_2 is a function of time. The algebra can be greatly simplified, however, by an appropriate transformation and the resulting equations made linear for all constant-speed cases. We can try by splitting the air-gap mmfs into components along the two perpendicular axes (the direct and quadrature axes) as is done for synchronous machines, even though now no obvious geometric feature of the machine, such as salient poles, dictates the specific choice for the location of the axes.

Let us choose the d-axis to coincide with the phase-a axis at $t = 0$, so that its displacement from the phase-a axis at any time t is given by ωt. And let us choose *synchronously rotating axes,*[6] rotating at synchronous speed as determined by the electrical angular velocity ω of the impressed stator voltages, with the q-axis rotating 90 electrical degrees ahead of the d-axis in the direction of rotation. The new stator-current variables, i_{1d} and i_{1q}, are now defined as follows:

$$i_{1d} = k_d\left[i_a \cos\omega t + i_b \cos\left(\omega t - 120°\right) + i_c \cos\left(\omega t + 120°\right)\right] \qquad (10.5.20)$$

$$i_{1q} = -k_q\left[i_a \sin\omega t + i_b \sin\left(\omega t - 120°\right) + i_c \sin\left(\omega t + 120°\right)\right] \qquad (10.5.21)$$

where each of the constants k_d and k_q is taken to be 2/3. Expressed in matrix form, we have

$$\begin{bmatrix} i_{1d} \\ i_{1q} \\ i_{10} \end{bmatrix} = \frac{2}{3} \begin{bmatrix} \cos\omega t & \cos\left(\omega t - 120°\right) & \cos\left(\omega t + 120°\right) \\ -\sin\omega t & -\sin\left(\omega t - 120°\right) & -\sin\left(\omega t + 120°\right) \\ \frac{1}{2} & \frac{1}{2} & \frac{1}{2} \end{bmatrix} \begin{bmatrix} i_a \\ i_b \\ i_c \end{bmatrix} \qquad (10.5.22)$$

in which the zero-sequence component i_{10} is zero because of Equation 10.5.10 and the factor 1/3 is chosen for the zero-sequence component. The phase components can then be expressed in terms of the d, q, 0 components as follows:

$$\begin{bmatrix} i_a \\ i_b \\ i_c \end{bmatrix} = \begin{bmatrix} \cos\omega t & -\sin\omega t & 1 \\ \cos\left(\omega t - 120°\right) & -\sin\left(\omega t - 120°\right) & 1 \\ \cos\left(\omega t + 120°\right) & -\sin\left(\omega t + 120°\right) & 1 \end{bmatrix} \begin{bmatrix} i_{1d} \\ i_{1q} \\ i_{10} \end{bmatrix} \qquad (10.5.23)$$

where i_{10} is zero, as stated already.

---

[6] Other choices of axes are covered in the problems at the end of the chapter.

The component currents i_{1d} and i_{1q} produce the same magnetic fields as the actual phase currents. The flux-linkage and voltage transformations are performed using the same transformation matrix as for the currents. Equations 10.5.22 and 10.5.23 can accordingly be written twice more, once with λ replacing i, and once with v replacing i.

Corresponding transformations for rotor quantities need now be made in relation to the same synchronously rotating dq-axes. Let θ_s be the angle from the rotor phase-A axis to the d-axis. If the rotor is rotating at slip S, the d-axis is advancing continuously with respect to a point on the rotor at the rate

$$\frac{d\theta_s}{dt} = p\theta = S\omega \tag{10.5.24}$$

The rotor dq-component currents are then defined as

$$\begin{bmatrix} i_{2d} \\ i_{2q} \\ i_{20} \end{bmatrix} = \frac{2}{3} \begin{bmatrix} \cos\theta_s & \cos(\theta_s - 120°) & \cos(\theta_s + 120°) \\ -\sin\theta_s & -\sin(\theta_s - 120°) & -\sin(\theta_s + 120°) \\ \frac{1}{2} & \frac{1}{2} & \frac{1}{2} \end{bmatrix} \begin{bmatrix} i_A \\ i_B \\ i_C \end{bmatrix} \tag{10.5.25}$$

where i_{20} is zero because of Equation 10.5.11. The phase components are then expressed in terms of the dq-components as

$$\begin{bmatrix} i_A \\ i_B \\ i_C \end{bmatrix} = \begin{bmatrix} \cos\theta_s & -\sin\theta_s & 1 \\ \cos(\theta_s - 120°) & -\sin(\theta_s - 120°) & 1 \\ \cos(\theta_s + 120°) & -\sin(\theta_s + 120°) & 1 \end{bmatrix} \begin{bmatrix} i_{2d} \\ i_{2q} \\ i_{20} \end{bmatrix} \tag{10.5.26}$$

where i_{20} is zero, as stated already. Once more, exactly the same transformations hold good for the rotor flux-linkages λ_A, λ_B, and λ_C in terms of λ_{2d} and λ_{2q}, and for the rotor voltages v_A, v_B, and v_C in terms of v_{2d} and v_{2q}.

Recognizing that the angle θ_2 of Equation 10.5.1 can be replaced by

$$\theta_2 = \omega t - \theta_s \tag{10.5.27}$$

the flux-linkage equations are obtained as follows after simplification by use of trigonometric reduction formulas:

$$\textbf{Stator:} \quad \lambda_{1d} = L_{11}i_{1d} + L_{12}i_{2d} \tag{10.5.28}$$
$$\lambda_{1q} = L_{11}i_{1q} + L_{12}i_{2q} \tag{10.5.29}$$

$$\textbf{Rotor:} \quad \lambda_{2d} = L_{22}i_{2d} + L_{12}i_{1d} \tag{10.5.30}$$
$$\lambda_{2q} = L_{22}i_{2q} + L_{12}i_{1q} \tag{10.5.31}$$

where $L_{12} = \frac{3}{2}M$.

Next, by making use of the transformation matrix of Equation 10.5.22 and voltage relations of Equation 10.5.2, the expressions for the dq-component stator voltages v_{1d} and v_{1q} can be obtained in terms of λ_{1d} and λ_{1q}. A similar process is used for the rotor voltages v_{2d} and v_{2q}. The final results follow:

$$\textbf{Stator:} \quad v_{1d} = r_1 i_{1d} + p\lambda_{1d} - \omega\lambda_{1q} \tag{10.5.32}$$
$$v_{1q} = r_1 i_{1q} + p\lambda_{1q} + \omega\lambda_{1d} \tag{10.5.33}$$

$$\textbf{Rotor:}\quad v_{2d} = r_2 i_{2d} + p\lambda_{2d} - \lambda_{2q}\left(p\theta_s\right) \tag{10.5.34}$$

$$v_{2q} = r_2 i_{2q} + p\lambda_{2q} + \lambda_{2d}\left(p\theta_s\right) \tag{10.5.35}$$

Equations 10.5.28 through 10.5.35 constitute the basic idealized induction machine relations for analysis in dq-variables. The quantity $\left(p\theta_s\right)$ in the speed-voltage terms of Equations 10.5.34 and 10.5.35 is the slip angular velocity given by Equation 10.5.24, which is the relative angular velocity of the synchronously rotating dq-axes with respect to the rotor. A positive value of the slip S corresponds to motor action, when $\left(p\theta_s\right)$ is positive; negative S corresponds to generator action, when $\left(p\theta_s\right)$ is negative. If the rotor speed is constant, $p\theta_s$ is constant and Equations 10.5.32 through 10.5.35 become linear differential equations with constant coefficients.

The $p\lambda_{1d}$ and $p\lambda_{1q}$ terms in Equations 10.5.32 and 10.5.33 are often neglected because they are usually small compared with the terms $\omega\lambda_{1q}$ and $\omega\lambda_{1d}$. This approximation, for a synchronous machine, corresponds to ignoring the dc-component of the stator short-circuit current following a transient. The r_1 terms are frequently neglected in transient analysis, while they are usually included in steady-state analysis.

The instantaneous power input to the three-phase stator is given by

$$p_1 = v_a i_a + v_b i_b + v_c i_c \tag{10.5.36}$$

or, in terms of the d-variables, it can be shown to be

$$p_1 = \frac{3}{2}\left(v_{1d}i_{1d} + v_{1q}i_{1q}\right) \tag{10.5.37}$$

The electromagnetic torque can be obtained as the power associated with the speed voltages divided by the corresponding speed in mechanical radians per second. The power associated with the speed-voltage terms of Equations 10.5.34 and 10.5.35 is given by

$$\frac{3}{2}\left[-\lambda_{2q}i_{2d}\left(p\theta_s\right) + \lambda_{2d}i_{2q}\left(p\theta_s\right)\right]$$

The term $p\theta_s$ is positive for motor action and the rotor goes backward with respect to the synchronously rotating dq-axes; the corresponding speed is therefore $[-p\theta_s]2/poles]$, expressed in mechanical radians per second. The electromagnetic torque, positive for motor action, is then given by

$$T = \frac{3}{2} \cdot \frac{\text{poles}}{2}\left(\lambda_{2q}i_{2d} - \lambda_{2d}i_{2q}\right) \tag{10.5.38}$$

This result is compatible with the concept of torque production from interacting magnetic fields: the d-axis magnetic field interacting with the q-axis mmf, and the q-axis magnetic field with the d-axis mmf. The sine of the space-displacement angle has a magnitude of unity for both interactions. This torque is available to furnish the rotational losses, drive the load, and provide acceleration. The inertia torque required to accelerate the rotating mass is given by

$$T_{inertia} = J\frac{d\omega_0}{dt} = J\frac{d^2\theta_0}{dt^2} \tag{10.5.39}$$

where J is the moment of inertia of the rotor and the mechanical equipment coupled to it, θ_0 is the shaft-position angle, and ω_0 is the shaft angular velocity, the angular measurement being in mechanical radians.

■|||||||||||||

10.6 Induction-Machine Electrical Transients

Let us consider an induction machine operating as a motor or a generator, and let a *three-phase short-circuit* occur at its terminals. Because of the "trapped" flux linkages with the rotor circuits, in either case, the machine will feed current into the fault; this fault current will, of course, decay to zero in time. Let us determine its initial magnitude, while neglecting the stator-phase resistance r_1 as well as the $p\lambda_{1d}$ and $p\lambda_{1q}$ terms in the stator-voltage equations 10.5.32 and 10.5.33. That is to say, we will consider the fundamental component alone, ignoring the dc component in the short-circuit current. Further, let us consider the machine to be operating at small values of slip so that the speed-voltage terms $\lambda_{2q}(p\theta_s)$ and $\lambda_{2d}(p\theta_s)$ in the rotor-voltage equations 10.5.34 and 10.5.35 can be ignored. Such an assumption is justified because an induction machine usually operates at small values of slip; also, the short-circuit current decays so rapidly that the speed is not changed appreciably.

The voltage equations 10.5.32 through 10.5.35 are then simplified as follows:

$$\textbf{Stator:} \quad v_{1d} = -\omega\lambda_{1q} \tag{10.6.1}$$

$$v_{1q} = \omega\lambda_{1d} \tag{10.6.2}$$

$$\textbf{Rotor:} \quad v_{2d} = r_2 i_{2d} + p\lambda_{2d} = 0 \tag{10.6.3}$$

$$v_{2q} = r_2 i_{2q} + p\lambda_{2q} = 0 \tag{10.6.4}$$

Note that the rotor voltages in Equations 10.6.3 and 10.6.4 are set equal to zero to indicate either a squirrel-cage rotor or a wound rotor with short-circuited slip rings. The rotor external impedance, if any, for the wound rotor can be included along with the rotor parameters.

Solving for i_{2d} from Equation 10.5.30, we obtain

$$i_{2d} = \frac{\lambda_{2d} - L_{12} i_{1d}}{L_{22}} \tag{10.6.5}$$

and, upon substitution into Equation 10.5.28,

$$\lambda_{1d} = \left(L_{11} - \frac{L_{12}^2}{L_{22}}\right) i_{1d} + \frac{L_{12}}{L_{22}}\lambda_{2d} \tag{10.6.6}$$

which may be written as

$$\lambda_{1d} = L' i_{1d} + \frac{L_{12}}{L_{22}}\lambda_{2d} \tag{10.6.7}$$

where

$$L' = L_{11} - \frac{L_{12}^2}{L_{22}} \tag{10.6.8}$$

Similarly, from Equations 10.5.29 and 10.5.31, we obtain

$$\lambda_{1q} = L' i_{1q} + \frac{L_{12}}{L_{22}}\lambda_{2q} \tag{10.6.9}$$

The quantity L' is known as the *transient inductance* of an induction machine, which is analogous to the direct-axis transient inductance L'_d of a synchronous machine.

The voltages v_{1d} and v_{1q} go to zero when a three-phase short circuit occurs at the terminals; it follows from Equations 10.6.1 and 10.6.2 that the stator-flux linkages λ_{1d} and λ_{1q} must also go to zero. Note that the dc component of the stator current, which physically prevents λ_{1d} and λ_{1q} from changing suddenly, is ignored here by neglecting the $p\lambda_{1d}$ and $p\lambda_{1q}$ terms. With λ_{1d} and λ_{1q} each equal to zero under these circumstances, in order to maintain the rotor-flux linkages constant at their initial prefault values λ_{2d0} and λ_{2q0}, the stator component currents must be, from Equations 10.6.7 and 10.6.9,

$$i_{1d} = -\frac{(L_{12}/L_{22})}{L'}\lambda_{2d0} \tag{10.6.10}$$

$$i_{1q} = -\frac{(L_{12}/L_{22})}{L'}\lambda_{2q0} \tag{10.6.11}$$

The equations we just derived yield *initial* values immediately after the short circuit takes place, i.e., at $t = 0_+$; these values would continue to exist if no rotor resistance caused their decay. The instantaneous phase currents can be obtained from the transformation equation 10.5.23. The corresponding rms stator current is given by

$$\bar{I}_1 = I_{1d} + jI_{1q} = \frac{i_{1d}}{\sqrt{2}} + j\frac{i_{1q}}{\sqrt{2}} = -\frac{1}{X'}\omega\frac{L_{12}}{L_{22}}\left(\frac{\lambda_{2d0}}{\sqrt{2}} + j\frac{\lambda_{2q0}}{\sqrt{2}}\right) \tag{10.6.12}$$

where $X' = \omega L'$ is known as the *transient reactance*. The rms magnitude of $\bar{I}_1$ is given by

$$I_1 = \frac{1}{X'}\omega\frac{L_{12}}{L_{22}}\frac{1}{\sqrt{2}}\sqrt{\lambda_{2d0}^2 + \lambda_{2q0}^2} \tag{10.6.13}$$

which is known as the *initial symmetrical short-circuit current*.

Next we shall examine how this short-circuit current decays, and how we can evaluate it from prefault conditions.

Decay of the Short-Circuit Current

Because of the rotor resistance r_2, the short-circuit current will decay to zero over time. The decrement can be found by examining the decay of rotor currents i_{2d} and i_{2q}. Substituting the values of λ_{2d} from Equation 10.5.30 into Equation 10.6.3, we have

$$r_2 i_{2d} + L_{22}pi_{2d} + L_{12}pi_{1d} = 0 \tag{10.6.14}$$

With $\lambda_{1d} = 0$ from Equation 10.5.28,

$$i_{1d} = -\frac{L_{12}}{L_{11}}i_{2d} \tag{10.6.15}$$

Using Equation 10.6.15 in Equation 10.6.14, we obtain

$$r_2 i_{2d} + \frac{1}{L_{11}}\left(L_{11} - \frac{L_{12}^2}{L_{22}}\right)L_{22}pi_{2d} = 0$$

or

$$i_{2d} + \frac{L'}{L_{11}}\frac{L_{22}}{r_2}\frac{di_{2d}}{dt} = 0 \tag{10.6.16}$$

in which the quantity $(L'/L_{11})(L_{22}/r_2)$ is known as the *short-circuit time constant* T'. The ratio (L_{22}/r_2) is the *time constant of the rotor circuit alone*, T'_0, of the induction machine that is responsible for the decay of rotor transients when the stator is open-circuited. Because of the coupling between the rotor and stator windings, with the stator short-circuited, the apparent inductance is lower. T' can be expressed in terms of T'_0 as

$$T' = \frac{L'}{L_{11}}\frac{L_{22}}{r_2} = T'_0 \frac{L'}{L_{11}} = T'_0 \frac{X'}{X_{11}} \tag{10.6.17}$$

Identical results are obtained by considering the decay of the rotor component i_{2q}. The exponential decay of the transient current whose initial rms magnitude is given by Equation 10.6.13 is characterized by the short-circuit time constant T', which is similar to the direct-axis short-circuit transient time constant T'_d of a synchronous machine. The results are about the same as for a synchronous machine except that the short-circuit current in an induction machine decays to zero. Note again that the dc component of the short-circuit current has been ignored in this discussion.

Transient Equivalent Circuit

We will now evaluate the flux linkages λ_{2d0} and λ_{2q0} from the prefault operating conditions of the machine and, in the process, develop a transient circuit of the induction machine.

Let the rms value of each of the stator-phase currents prior to the short circuit be I_{10}, with its dq-components I_{1d0} and I_{1q0} given by $i_{1d0}/\sqrt{2}$ and $i_{1q0}/\sqrt{2}$, respectively. Rewriting Equation 10.6.7 for prefault conditions and multiplying by ω, we obtain

$$\omega \frac{L_{12}}{L_{22}}\lambda_{2d0} = \omega\lambda_{1d0} - X'i_{1d0} \tag{10.6.18}$$

Replacing $\omega\lambda_{1d0}$ by v_{1q0} per Equation 10.6.2 and expressing Equation 10.6.18 in phasor form, we get

$$\omega \frac{L_{12}}{L_{22}}\frac{\bar{\lambda}_{2d0}}{\sqrt{2}} = \bar{V}_{1q0} - X'\bar{I}_{1d0} \tag{10.6.19}$$

where V_{1q0} is given by $v_{1q0}/\sqrt{2}$. Similarly, Equation 10.6.9 yields

$$\omega \frac{L_{12}}{L_{22}}\frac{\bar{\lambda}_{2q0}}{\sqrt{2}} = -\bar{V}_{1d0} - X'\bar{I}_{1q0} \tag{10.6.20}$$

Equations 10.6.19 and 10.6.20 can be combined to give

$$\omega \frac{L_{12}}{L_{22}}\left(\frac{\bar{\lambda}_{2d0}}{\sqrt{2}} + j\frac{\bar{\lambda}_{2q0}}{\sqrt{2}}\right) = \frac{1}{j}[(\bar{V}_{1do} + j\bar{V}_{1q0}) - jX'(\bar{I}_{1d0} + j\bar{I}_{1q0})] = \frac{1}{j}(\bar{V}_{10} - jX'\bar{I}_{10}) = \frac{1}{j}\bar{E}'_1 \tag{10.6.21}$$

in which $\bar{V}_{10}$ is the prefault rms terminal voltage, and $\bar{E}'_1$ is given by

$$\bar{E}'_1 = \bar{V}_{10} - jX'\bar{I}_{10} \tag{10.6.22}$$

$\bar{E}'_1$ is known as the *initial voltage behind the transient reactance* of an induction machine; it is proportional to the rotor-flux linkages. The rms magnitude of the initial symmetrical short-circuit current expressed earlier by Equation 10.6.13 is now given by

$$I_1 = E'_1/X' \tag{10.6.23}$$

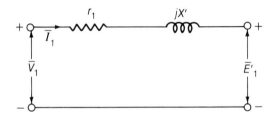

Figure 10.6.1 Simple transient equivalent circuit of an induction machine.

The rms magnitude of the symmetrical short-circuit current at any time t after the fault can now be expressed as

$$I'_1 = \frac{E'_1}{X'} e^{-t/T'} \tag{10.6.24}$$

in which the decrement characterized by the short-circuit time constant T' of Equation 10.6.17 is applied. In case an external reactance exists between the machine terminals and the fault, the time constant can be modified by adding that reactance to both the numerator and the denominator of the fraction in Equation 10.6.6.

From this discussion, it follows that an induction machine can be represented by a simple transient equivalent circuit of Figure 10.6.1, in which X' is the transient reactance, and r_1, the stator-phase resistance, can be added as shown for somewhat greater precision. The results are once again very similar to those obtained for a synchronous machine. In the case of an induction machine, however, while the initial short-circuit current is relatively high compared to its normal current, the electrical transient usually subsides quite rapidly compared to the duration of the motional transient. Hence the electrical transients in an induction machine are ignored in many problems. We will illustrate this statement with Example 10.6.1. (Note that large 3,600-rpm induction motors are among the principal exceptions to this statement.)

▌▌▌▌▌▌▌▌▌▌▌▌██████████████████████

EXAMPLE 10.6.1

A 500-hp, 440-volt, 60-Hz, three-phase, star-connected, six-pole squirrel-cage induction motor has a full-load efficiency of 0.95 and a lagging power factor of 0.9. Determine the motor rms short circuit current if a three-phase short circuit occurs on its supply lines at the motor terminals while the motor is operating in the steady state under rated full-load conditions.

The motor parameters in ohms per phase referred to the stator are given as follows:

$$X_{l1} = 0.075; \qquad X_{l2} = 0.075; \qquad X_m = 3.00; \qquad r_1 = 0.009; \qquad r_2 = 0.008$$

Solution

Since $X_{l1} = \omega(L_{11} - L_{12})$, $X_{l2} = \omega(L_{22} - L_{12})$, $X_m = \omega L_{12}$, and $X' = \omega L' = \omega[L_{11} - (L_{12}^2/L_{22})]$, the motor transient reactance can be worked out as

$$X' = X_{l1} + X_m - \frac{X_m^2}{X_{l2} + X_m}$$

or

$$X' = 0.075 + 3.00 - \frac{(3.00)^2}{0.075 + 3.00} = 0.148 \ \Omega/\text{phase}$$

The prefault stator current with the motor operating in the steady state under rated full-load conditions is given by

$$I_1 = \frac{500 \times 746}{0.90 \times 0.95 \times 440 \times \sqrt{3}} = 572.46 \text{ A}$$

Choosing the terminal voltage as the reference phasor, the prefault voltage behind transient reactance is then

$$\bar{E}'_1 = \frac{440}{\sqrt{3}} - (0.009 + j0.148)(572.46\angle - \cos^{-1} 0.90)$$

or

$$\bar{E}'_1 = 225\angle - 19.2^\circ \ \text{V}$$

The initial rms short-circuit current, from Equation 10.6.23, is

$$\frac{225}{0.148} = 1,520 \text{ A}$$

The open-circuit time constant T'_0 given by L_{22}/r_2 is

$$T'_0 = \frac{3.00 + 0.075}{2\pi \times 60 \times 0.008} = 1.02 \text{ s}$$

and the short-circuit time constant T', given by Equation 10.6.17, is

$$T' = 1.02 \times \frac{0.148}{3.075} = 0.05 \text{ s}$$

The rms short-circuit current is then given by Equation 10.6.24 as

$$I_1 = 1,520e^{-t/0.05} \text{ A}$$

The short-circuit time constant of 0.05 second corresponds to three cycles on a 60-Hz base; that is to say, the short-circuit current decreases to 36.8% of its initial value in three cycles. It will substantially disappear in about 10 cycles. Thus we can see that the electrical transient usually decays rapidly, even though the initial short-circuit current of an induction machine might be relatively high compared to its normal current.

10.7 Induction-Machine Dynamics

For the study of electrical transients in ac machines, the researcher must compromise between the detailed analysis necessary when conditions within a single machine are the only concern and the more approximate models that must be adopted for simplified representation of the machine as a single element within a large dynamic system. From the systems point of view, then, we must decide on reasonable approximations allowable from the engineering perspective and derive simple equivalents for the analysis of dynamic problems.

In many power-system transient-stability studies, the system loads are represented by shunt impedances, which are treated as static impedances. Not all loads behave as static impedances, however; one notable exception is the case of large induction-motor loads. The load inertia must be considered to account for mechanical transients, that might have significant effect on generator swings. Several representations can be used for induction-motor loads, depending upon the degree of refinement desired:

1. Simple impedance
2. Steady-state equivalent circuit
3. Steady-state behavior considering only the motor's mechanical transients
4. Transient equivalent circuit considering both mechanical and rotor electrical transients
5. Transient behavior considering the motor's mechanical, rotor electrical, and stator electrical transients

Let us further consider the model given by case 4 for the analysis of induction-machine dynamics. The mechanical transients are identified by the differential equation for the slip:

$$\frac{d\mathbf{S}}{dt} = \frac{1}{2H}(T_m - T_e) \tag{10.7.1}$$

where H is the inertia constant in kWs/kVA, T_m is the mechanical load torque in per-unit notation, T_e is the electromagnetic torque in per-unit notation, and $\mathbf{S}$ is the motor slip in per-unit notation. The differential equation describing the rate of change of voltage E_1' behind the transient reactance X' of Figure 10.6.1 is thus given by[7]

$$\frac{d\bar{E}_1'}{dt} = -j\omega\mathbf{S}\bar{E}' - \frac{1}{T_0'}[\bar{E}_1' - j(X - X')\bar{I}_1] \tag{10.7.2}$$

where T_0' is the rotor open-circuit time constant in seconds, X is the reactance in per-unit notation with open-circuit rotor, X' is the per unit reactance with blocked rotor, and $\bar{I}_1$ in per-unit notation (from Figure 10.6.1, page 496) is given by

$$\bar{I}_1 = \frac{\bar{V}_1 - \bar{E}_1'}{r_1 + jX'} \tag{10.7.3}$$

Equation 10.7.2, along with the swing equation 10.7.1 representing the rotor acceleration plus the equations for rotor electrical and mechanical torques, are solved to account for the transient behavior of an induction machine including both mechanical and rotor electrical transients. A step-by-step method of solution follows:

1. Let the disturbance be initiated on the system at a time t_n. The electrical torque of the induction motor is computed corresponding to t_n.
2. The change in motor speed in a certain time interval Δt is calculated from Equation 10.7.1, based on the difference between the electrical and mechanical torques of the motor. The motor speed at $(t_n + \Delta t)$ is then found.

---

[7] D. S. Brereton, D. G. Lewis, and C. C. Young, "Representation of Induction Motor Loads During Power System Stability Studies," *AIEE Transactions,* vol. 76. pt. III, pp. 45–61. (August 1957).

3. The change in $\bar{E}_1'$, proportional to the change in rotor flux linkage, during Δt is computed from Equation 10.7.2 based on the conditions at t_n.

4. The procedure is then repeated at each time interval.

Besides power-system stability studies involving large concentrations of induction-motor loads, common induction-machine dynamic problems are associated with starting and stopping and with the ability of the motor to continue its operation during serious disturbances of the supply system. A typical industrial problem might be to analyze the starting situation of a large motor, the associated voltage reduction caused by the heavy inrush of the starting current, and the effect of voltage reduction on the operation of other parallel motors in the system. The period of heavy current inrush during starting of an induction motor usually lasts for about the time required to reach the slip $\mathbf{S}_{\text{max T}}$ corresponding to maximum torque, after which it comes down to the normal running value.

Piecewise linear representations of the torque-speed characteristic are often used in analytical solutions. Since the curve is practically linear in normal operating region of small slips, the relations

$$T_e = k\mathbf{S} \tag{10.7.4}$$

where k is a constant, can be used. Alternatively, when a wider region of the torque-speed curve is involved, a number of straight lines might be needed; or sometimes a reasonably good representation, as given by Equation 7.4.12, is used. When the terminal voltage is varied, to include the effect that the elctromagnetic torque varies as the square of the voltage, Equation 10.7.4 is modified as

$$T_e = k'V_1^2\mathbf{S} \tag{10.7.5}$$

where k' is a constant and V_1 is the applied voltage. Typical analytical expressions are used to represent the mechanical load torque as a function of slip, such as

$$T_m = [A(1 - \mathbf{S})^2 + B(1 - \mathbf{S}) + C]P \tag{10.7.6}$$

where P is the per-unit net power transferred to the rotor through the air gap, and A, B, and C are constants of the mechanical load speed-torque characteristic.

■■■■■■■■■■ IIII
BIBLIOGRAPHY

Adkins, B. *The General Theory of Electrical Machines.* New York: John Wiley & Sons, 1957.

Alger, P. L. *Induction Machines—Their Behavior and Uses.* 2d ed. New York: Gordon and Breach, 1970.

Concordia, C. *Synchronous Machines.* New York: John Wiley & Sons, 1951.

Fitzgerald, A. E., and C. Kingsley Jr. *Electric Machinery.* 2d ed. New York: McGraw-Hill, 1961.

Kimbark, E. W. *Power System Stability: Synchronous Machines.* New York: Dover, 1968.

Knowlton, A. E., ed. *Standard Handbook for Electrical Engineers.* 8th ed. New York: McGraw-Hill, 1949.

Meisel, J. *Principles of Electromechanical Energy Conversion.* New York: McGraw-Hill, 1966.

Sarma, M. S. *Synchronous Machines (Their Theory, Stability and Excitation Systems).* New York: Gordon and Breach, 1979.

Westinghouse Electric Corporation Central Station Engineers. *Electrical Transmission and Distribution Reference Book.* 4th ed. East Pittsburgh, Pa.: Westinghouse Electric Corporation, 1964.

White, D. C., and H. H. Woodson. *Electromechanical Energy Conversion.* New York: John Wiley & Sons, 1959.

Woodson, H. H., and J. R. Melcher. *Electromechanical Dynamics—Part I: Discrete Systems.* New York: John Wiley & Sons, 1968.

▮▮▮▮▮▮▮▮▮▮▮

PROBLEMS

10–1.   Following the procedure indicated in the text, obtain the flux-linkage relations given by Equations 10.1.41 through 10.1.43 starting from Equations 10.1.38 through 10.1.40

10–2.   Develop the armature voltage relations given by Equations 10.1.47 through 10.1.49 starting from Equations 10.1.1 through 10.1.3, indicating clearly the transformations you carry out.

10–3.   A three-phase, 13,800-V (rms line-to-line voltage), 60-Hz, two-pole, wye-connected synchronous machine operates at its rated speed. The maximum value of the mutual inductance between the field winding and any one of the armature phase windings is 0.04 H. Calculate the required field current for the machine to develop normal rated voltage on open circuit.

10–4.   Suppose the negative-sequence stator currents from a generator have the following form:

$$i_a = (t + \alpha); \qquad i_b = 1.0\cos(t + \alpha + 120°); \qquad i_c = 1.0\cos(t + \alpha - 120°)$$

Let the machine be operating at synchronous speed under steady-state conditions. Evaluate the corresponding i_d, i_q, and i_0.

10–5.   Consider a machine operating at synchronous speed under steady-state conditions, and let dc stator currents flow from the generator having the following magnitudes:

$$i_a = 1.0; \qquad i_b = i_c = -1/2$$

Find the corresponding i_d, i_q, and i_0.

10–6.   Consider the flux-linkage and armature-voltage relations of a synchronous machine given by Equations 10.1.35 through 10.1.37, 10.1.41 through 10.1.43, and 10.1.47 through 10.1.49.

    a. For the case of the machine operating at steady-state synchronous speed, rewrite these equations making all possible simplifications

    b. Repeat part a with the additional constraint that the armature is open-circuited. If the rated frequency is 60 Hz and $L_{afd} = 0.05$ H, calculate the line-to-line voltage of the machine corresponding to a field current of 1,500 A.

    c. Repeat part a with the additional constraint that the armature is short-circuited. Neglect armature resistance. If the rated frequency is 60 Hz, $L_{afd} = 0.05$ H, and $L_d = 0.025$ H, calculate the line current for a wye-connected generator corresponding to a field current of 1,500 A.

10–7.   A three-phase, 60-Hz, 12.1-kV, 20-MVA, four-pole synchronous generator has the following inductances and resistances:

$$L_1 = 0.00214 \text{ H}; \qquad r_{fd} = 0.0208 \text{ } \Omega$$
$$L_{ad} = 0.02145 \text{ H}; \qquad L_{a1d} = 0.017 \text{ H}; \qquad L_{a1q} = 0.007 \text{ H}$$
$$L_{aq} = 0.0191 \text{ H}; \qquad L_{1d1d} = 0.0226 \text{ H}; \qquad r_{1q} = 0.0611 \text{ } \Omega$$
$$L_{afd} = 0.045 \text{ H}; \qquad L_{fd1d} = 0.0541 \text{ H}; \qquad r_a = 0.0147 \text{ } \Omega$$
$$L_{fdfd} = 0.1538 \text{ H}; \qquad r_{1d} = 0.0315 \text{ } \Omega; \qquad L_{1q1q} = 0.00446 \text{ H}$$

Convert all of the values to per-unit quantities.

10–8. A three-phase, 60-Hz, 13-kV, 25-MVA, wye-connected synchronous generator has $X_d = 1.2$ per unit and a time constant of 1,000 radians. Find X_d in ohms and the time constant in seconds.

10–9. Starting from the fundamental flux-linkage equations, justify the equivalent circuits shown in Figures 10.1.4 and 10.1.5.

10–10. A four-pole, three-phase, 10-MVA, 11.8kV, 60-Hz turbogenerator is subjected to a sudden three-phase short circuit from an unloaded condition when the open-circuit voltage is 5.9 kV. An oscillogram of one of the armature currents is taken and the measurements made on the ordinates of the envelope of the oscillogram are given in the following table. The scale factor of the oscillogram based on the instantaneous value is 1,930 amperes per centimeter. Compute the direct-axis reactance X_d, X_d'', X_d' in per-unit notation, and the time constants T_d', T_d'', and T_a in seconds.

Time (cycles)	Ordinates of Envelopes (cm)		Time (cycles)	Ordinates of Envelopes (cm)	
	Upper	Lower		Upper	Lower
0	2.63	−0.63	15	0.86	−0.62
1	2.24	−0.50	20	0.75	−0.63
2	1.96	−0.44	25	0.67	−0.61
3	1.73	−0.41	30	0.63	−0.59
4	1.56	−0.42	40	0.54	−0.54
5	1.44	−0.44	50	0.48	−0.48
6	1.33	−0.45	60	0.43	−0.43
7	1.24	−0.48	90	0.32	−0.32
8	1.16	−0.50	120	0.25	−0.25
10	1.05	−0.55	∞	0.15	−0.15

10–11. Consider a three-phase short circuit of a synchronous machine occuring at the armature terminals of the machine initially unloaded and normally excited. Which of the following are affected by the time at which the short-circuiting switch is closed?

a. The dc component in phase a

b. The dc component in direct axis

c. The dc component in field

d. The torque

10–12. If $i_d(s)$ is given to be

$$i_d(s) = \frac{1}{X_d'} \frac{s + 1/T_{d0}'}{s(s + 1/T_d')(s^2 + 2as + 1)}$$

where a is a constant, find the sustained (steady-state) value of $i_d(t)$.

10–13. A synchronous machine is initially operating at unity power factor, rated kVA, and rated terminal voltage. The machine parameters are given as follows:

$$X_d = X_q = 1.0; \quad X_d' = 0.15; \quad X_d'' = 0.10; \quad X_q'' = 0.12; \quad r = 0$$

Corresponding to a three-phase terminal short circuit, determine the following values:

a. The initial subtransient fundamental frequency fault current

b. The maximum dc offset that can occur in any phase current. Comment on the base of which this is expressed in per unit.

c. The initial transient fundamental frequency fault current (using the voltage behind transient reactance) and the dc field current corresponding to this value (using E'_q)

d. The steady-state fault current

10–14.   Consider a synchronous machine with no amortisseur windings and negligible armature resistance. Let a sudden three-phase short circuit occur at the armature terminals of the machine, initially unloaded, normally excited, and run at synchronous speed, that is to say, in per-unit notation, at $t = 0_-$ (just prior to short circuit),

$$e_d = 0; \qquad e_q = 1; \qquad e_{fd} = \frac{r_{fd}}{X_{ad}}$$

at $t = 0_+$ (just after short circuit),

$$e_d = 0; \qquad e_q = 0; \qquad e_{fd} = \frac{r_{fd}}{X_{ad}}$$

Apply the performance equations of the synchronous machine and show that

$$i_d(t) = \frac{1}{X_d} + \left(\frac{1}{X'_d} - \frac{1}{X_d}\right)e^{-t/T'_d} - \frac{1}{X'_d}\cos t\, e^{-t/T_a}$$

where T_a is the armature time constant given by (1/a), in which

$$a = \frac{r}{2}\frac{X'_d + X_q}{X'_d X_q}$$

10–15.   Let

$$i_d(t) = \left(\frac{1}{X''_d} - \frac{1}{X'_d}\right)e^{-t/T''_d} + \left(\frac{1}{X'_d} - \frac{1}{X_d}\right)e^{-t/T'_d} + \frac{1}{X_d} - \frac{1}{X''_d}\cos t\, e^{-t/T_a}$$

$$i_q(t) = \frac{1}{X''_q}\sin t\, e^{-t/T_a}$$

and

$$i_0(t) = 0$$

Show that

$$i_a(t) = \left[\left(\frac{1}{X''_d} - \frac{1}{X'_d}\right)e^{-t/T''_d} + \left(\frac{1}{X'_d} - \frac{1}{X_d}\right)e^{-t/T'_d} + \frac{1}{X_d}\right]\cos(t + \alpha)$$

$$- \left[\frac{1}{2}\left(\frac{1}{X''_d} + \frac{1}{X''_q}\right)\cos\alpha + \frac{1}{2}\left(\frac{1}{X''_d} - \frac{1}{X''_q}\right)\cos(2t + \alpha)\right]e^{-t/T_a}$$

10–16.   Consider the conditions of Problem 10–14 and show that

$$i_{fd}(t) = \frac{1}{X_{ad}} + \frac{X_{ad}}{X_{fdfd}X'_d}e^{-t/T'_d} - \frac{X_{ad}}{X_{fdfd}X'_d}\cos t\, e^{-t/T_a}$$

10–17. Consider the conditions of Problem 10–14 in order to calculate the short-circuit torque as in Equation 10.1.35. Neglect damping for the first half-cycle and show that

$$T = \frac{1}{X'_d} \sin t + \left(\frac{1}{2X_q} - \frac{1}{2X'_d} \right) \sin 2t$$

10–18. The short-circuit torque of a typical synchronous machine is given by $T = 3.44 \sin \phi - 1.095 \sin 2\phi$ for the first half-cycle.

    a. Sketch the alternating components as well as the total torque as functions of the angle ϕ for the first half-cycle.

    b. Find the angle ϕ corresponding to the maximum torque and calculate the maximum torque in terms of the full-load torque.

10–19. Equation 10.4.10 gives the transient power-angle equation for a salient-pole synchronous machine with negligible armature resistance. Show that the condition for maximum transient power is given by

$$\cos \delta = \frac{X_q E'_q}{4(X_q - X'_d)V_t} - \sqrt{\left[\frac{X_q E'_q}{4(X_q - X'_d)V_t} \right]^2 + \frac{1}{2}}$$

10–20. A waterwheel generator with negligible armature resistance has the following per-unit parameters:

$$X_d = 1.0; \qquad X_q = 0.6; \qquad X'_d = 0.3$$

Let the machine operate as a generator connected to an infinite bus having a voltage of 1.0 per unit. Plot the steady-state and transient power-angle characteristics for the following cases:

    a. The no-load excitation and initial field flux

    b. The full-load excitation with unit armature current at 0.8 lagging power factor and initial field flux

10–21. The power-angle characteristic of a synchronous machine is given by $P = (630 \sin \delta)$ kW. Compute the maximum shaft load in hp that can be suddenly applied to the motor operating initially unloaded. Neglect damping.

10–22.   a. Consider a round-rotor synchronous generator with a synchronous impedance Z_s per phase and a terminal voltage V_t per phase, operating in parallel with an infinite bus whose voltage remains constant regardless of load fluctuations. The phasor diagram is shown in Figure P10–22. Because of some

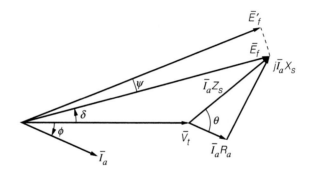

Figure P10–22

disturbance, let the power angle change by ψ as shown with $E_f' = E_f$; it causes the machine to develop an additional power, thereby keeping in synchronism. This additional power is known as the *synchronizing power*. Derive a general expression for the synchronizing power per phase and discuss the special cases when

    (i) ψ is very small

    (ii) ψ is very small and $R_a \ll X_s$

  b. A 1000-kVA, 3300-V, 60-Hz, four-pole, three-phase, wye-connected, round-rotor synchronous generator has a synchronous reactance of $1\,\Omega$ per phase; the armature resistance is negligible. The generator is operating at full load and 0.8 lagging power factor on an infinite bus. Calculate the synchronizing power per phase and synchronizing torque per phase if a disturbance causes the power angle to swing by one mechanical degree.

10-23. By carrying out the manipulations needed, show that Equations 10.5.28 through 10.5.31 are correct.

10-24. Carry out the manipulations described in the text to arrive at the Equations 10.5.32 through 10.5.35.

10-25. The steady-state analysis of a polyphase induction machine (presented in Chapter 7) can also be accomplished from the mathematical model set up for the machine in Section 10.5. Let the three stator-phase currents be given by

$$i_a = \sqrt{2}I_1 \cos(\omega t + \alpha)$$

$$i_b = \sqrt{2}I_1 \cos(\omega t + \alpha - 120°)$$

$$i_c = \sqrt{2}I_1 \cos(\omega t + \alpha + 120°)$$

where α is an arbitrary phase angle.

  a. Show that the currents, viewed from the synchronously rotating dq-axes, appear as steady dc quantities.

  b. In steady-state analysis, the researcher is usually interested in relations involving rms currents and voltages rather than instantaneous values. So

    (i)   express the stator-phase current and voltage as well as the rotor-phase current in phasor form;

    (ii) obtain the voltage equations for the stator and rotor by making use of the flux-linkage equations; and

    (iii) show that they can be arranged in the following form:

$$\bar{V}_1 = r_1\bar{I}_1 + jX_{l1}\bar{I}_1 + jX_m(\bar{I}_1 + \bar{I}_2); \qquad 0 = \frac{r_2}{S}\bar{I}_2 + jX_{l2}\bar{I}_2 + jX_m(\bar{I}_1 + \bar{I}_2)$$

where X_{l1} (the stator leakage reactance per phase) is $\omega(L_{11} - L_{12})$; X_{l2} (the rotor leakage reactance per phase) is $\omega(L_{22} - L_{12})$; and X_m (the magnetizing reactance per phase) is ωL_{12}.

  c. With all quantities referred to the stator, obtain the per-phase equivalent circuit of a polyphase induction machine satisfying the volt-ampere equations of part b, and show that it is the same as Figure 7.2.5c developed from a generalized transformer point of view.

  d. Through the application of Equation 10.5.38, show that the expression for the internal electromagnetic torque is the same as Equation 7.4.1.

10-26. The dq-transformation presented in the text makes use of synchronously rotating axes. As an alternative, some induction machine analysts have used axes that are fixed with respect to the stator structure and are labeled as α-axis and β-axis. The α-axis coincides with that of phase a and the β-axis is located 90° ahead of the α-axis in the direction of rotation. Let the axis of each rotor phase be displaced by an angle θ from the axis of the correspondingly labeled stator phase.

  a. By replacing ωt by zero in Equations 10.5.22 and 10.5.23 and simplifying, obtain the transformation equations relating abc-phase currents and $\alpha\beta0$-currents for the stator.

  b. By replacing θ_s by $(-\theta)$ in Equations 10.5.25 and 10.5.26, obtain the transformation equations for the rotor quantities.

c. Show that the component flux-linkage equations are the same as Equations 10.5.28 through 10.5.31 if we replace d and q by α and β, respectively.

d. Show that the voltage equations are then

$$\textbf{Stator:}\quad e_{1\alpha} = r_1 i_{1\alpha} + p\lambda_{1\alpha}; \qquad e_{1\beta} = r_1 i_{1\beta} + p\lambda_{1\beta}$$

$$\textbf{Rotor:}\quad e_{2\alpha} = r_2 i_{2\alpha} + p\lambda_{2\alpha} + \lambda_{2\beta}(p\theta); \qquad e_{2\beta} = r_2 i_{2\beta} + p\lambda_{2\beta} - \lambda_{2\alpha}(p\theta)$$

e. Show that, in the steady state, the component currents and voltages in the $\alpha\beta$ system are stator-frequency quantities.

10–27. As an alternative to the use of synchronous rotating dq-axes for the induction-machine analysis, some analysts have used dq-axes fixed with respect to the rotor. Let the axis of each rotor phase be displaced by an angle θ from the axis of the correspondingly labeled stator phase.

a. By replacing ωt by θ in Equations 10.5.22 and 10.5.23, obtain the transformation equations relating abc-phase currents and the new $dq0$-currents for the stator.

b. By replacing θ_s by zero in Equations 10.5.25 and 10.5.26, obtain the transformation equations for the rotor quantities.

c. Show that the component flux-linkage equations are the same as Equations 10.5.28 through 10.5.31.

d. Show that the voltage equations are then

$$\textbf{Stator:}\quad e_{1d} = r_1 i_{1d} + p\lambda_{1d} - \lambda_{1q}(p\theta); \qquad e_{1q} = r_1 i_{1q} + p\lambda_{1q} + \lambda_{1d}(p\theta)$$

$$\textbf{Rotor:}\quad e_{2d} = r_2 i_{2d} + p\lambda_{2d}; \qquad e_{2q} = r_2 i_{2q} + p\lambda_{2q}$$

e. Show that, in the steady state, the component currents and voltages in the new dq system are slip-frequency quantities.

10–28. This problem is concerned with the analysis of a two-phase induction machine instead of the three-phase machine presented in the text, with the use of synchronously rotating dq-axes. Let the magnetic axis of phase b be 90° ahead of the phase a axis in the direction of rotation. Let there be two corresponding windings, A and B, on the rotor with their magnetic axes displaced by 90°. At any instant of time, let the axis of the rotor phase be displaced by an angle θ_2 from the axis of the correspondingly labeled stator phase. Following the general procedure adopted in the text,

a. develop the flux-linkage equations corresponding to Equations 10.5.14 through 10.5.19.

b. Show that the appropriate dq-transmission of variables is typically given by the following:

$$\textbf{Stator:}\quad i_{1d} = i_a \cos \omega t + i_b \sin \omega t; \qquad i_{1q} = -i_a \sin \omega t + i_b \cos \omega t$$

$$\textbf{Rotor:}\quad i_{2d} = i_A \cos \theta_s + i_B \sin \theta_s; \qquad i_{2q} = -i_A \sin \theta_s + i_B \cos \theta_s$$

c. Obtain the relations in terms of dq variables for phase variables.

d. Show that the flux-linkage equations in terms of dq variables are still given by Equations 10.5.28 through 10.5.31; find L_{11}, L_{22}, and L_{12} in terms of L_1, L_2, and M.

e. Show that the component voltage Equations 10.5.32 through 10.5.35 hold good in this case.

f. Show that the instantaneous power input to the two-phase stator is given by

$$p_1 = v_{1d} i_{1d} + v_{1q} i_{1q}$$

g. Show that the motor torque is given by

$$T = \frac{poles}{2}(\lambda_{2q} i_{2d} - \lambda_{2d} i_{2q})$$

10–29. When unbalanced voltages are applied, the method of three-phase symmetrical components is often used for steady-state analysis. The *abc*-phasors are related to the symmetrical-component phasors by the following relationship:

$$\bar{I}_a = \bar{I}_a^+ + \bar{I}_a^- + \bar{I}_a^0$$

$$\bar{I}_b = \bar{I}_a^+ e^{-j120°} + \bar{I}_a^- e^{-j120°} + \bar{I}_a^0$$

$$\bar{I}_c = \bar{I}_a^+ e^{j120°} + \bar{I}_a^- e^{-j120°} + \bar{I}_a^0$$

where the $\bar{I}_a^+$ set is balanced positive-sequence set of components with an *abc*-phase sequence, and the $\bar{I}_a^-$ set is a balanced negative-sequence set of components with an *acb*-phase sequence.

  a. Obtain the transformation relating the symmetrical-component phasors and the *abc* phasors.

  b. Assuming that zero-sequence-component currents and voltages do not exist for purposes of this problem, and assuming that the principle of superposition holds good, outline a method for computing the torque-slip characteristic of an induction motor with unbalanced applied voltages.

10–30. A 400-hp, 440-V, 60-Hz, three-phase, wye-connected, four-pole wound-rotor induction motor has the following parameters in ohms per phase referred to the stator:

$$X_{l1} = X_{l2} = 0.05; \qquad X_m = 2.5; \qquad r_1 = 0.005; \qquad r_2 = 0.006$$

The motor is supplied at the rated terminal voltage through a step-down transformer that can be represented by a series reactance of 0.03 ohm per phase, The induction motor is operating at its full load with an efficiency and power factor of 90% each, with its slip rings short-circuited. If a three-phase short circuit occurs at the high-voltage terminals of the transformer bank, determine the initial symmetrical short-circuit current in the motor and show that it is decremented.

10–31. For an induction motor started at full voltage, develop an expression for the time t required to reach the speed corresponding to $S_{max\,T}$. The motor is unloaded at starting, and rotational losses can be neglected. Let J be the combined inertia of the rotor and the connected mechanical equipment, and let the torque-slip curve be given by the relation

$$\frac{T}{T_{max}} = \frac{2}{(S/S_{max\,T}) + (S_{max\,T}/S)}$$

where T_{max} is the maximum torque and $S_{max\,T}$ is the slip at maximum torque.

11

Direct-Current Machine Dynamics

11.1 Dynamic Models
11.2 Dynamic Analysis
11.3 Metadynes and Amplidynes

The steady-state operation and performance of direct-current machines are discussed in Chapters 5 and 9. This chapter studies the dynamic behavior of direct-current machines to examine the response to sudden changes. The effects of inductance and inertia, which are negligible during the steady-state operation, come into play during a rapid transition from one operating condition to another and must therefore be considered when analyzing the dynamic behavior of a dc machine. As loads change and shift, the dynamic characteristics of the individual machines taken together determine whether the system can react in a stable manner to such disturbances. Because of the versatility of dc machines and the ease with which they can be controlled, dc motors with solid-state controls are often used in applications requiring a wide range of motor speeds or precise control of motor output. The solid-state control of dc motors is presented in some detail in Section 12.2 as a part of the chapter on power semiconductor-controlled drives. Our primary objective in this chapter is to develop mathematical models of dc machines and analyze their dynamic characteristics.

11.1 Dynamic Models

Because of the complexity of dynamic-system problems, the usual simplifying assumptions made in many problems involving the behavior of a dc machine as a system component are given here:

a. The air-gap flux distribution produced by field windings is symmetrical about the center line of the field poles, which is known as the field axis or the direct axis.

b. The axis of the armature-mmf wave is fixed in space along the quadrature-axis; the brushes are narrow, and commutation is linear.

c. Because the armature-mmf axis is perpendicular to the field axis, the armature-mmf has no effect on the total direct-axis flux; that is to say, the demagnetizing effect of armature reaction is ignored.

d. The effects of magnetic saturation will be neglected, at least for the time being, thereby allowing superposition and considering inductances to be independent of the currents.

e. A two-pole machine is considered for the model; the results can be applied to a P-pole machine.

The schematic representation of the model of a dc machine is shown in Figure 11.1.1 with motor and generator conventions. The arrow representing the reference direction for a current also represents that of its associated magnetic field. Based on Equations 5.3.24 and 5.3.26, one can write

$$e_a = K\phi_d\omega_m = ki_f\omega_m \tag{11.1.1}$$

$$T_e = K\phi_d i_a = ki_f i_a \tag{11.1.2}$$

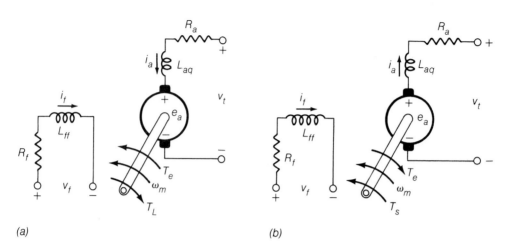

Figure 11.1.1 Schematic representation of the model of a dc machine. (a) Motor conventions. (b) Generator conventions.

where K and k are constants, ϕ_d is the direct-axis air-gap flux that is linearly proportional to the field current i_f, ω_m is the angular velocity corresponding to the speed of rotation, i_a is the armature current, e_a is the generated speed voltage, and T_e is the electromagnetic torque developed by the machine. The system performance is described by these equations together with the differential equation of motion of the mechanical system, the volt-ampere equations for the armature and field circuits, as well as the magnetization curve, which is taken here to be linear.

The voltage equation for the field circuit is given by

$$v_f = L_{ff} \frac{di_f}{dt} + R_f i_f = L_{ff} p i_f + R_f i_f \qquad (11.1.3)$$

where p is the derivative operator d/dt, and v_f, i_f, R_f, and L_{ff} are the terminal voltage, current, resistance, and self-inductance of the field circuit, respectively.

Note that the axis of the field mmf lies along the direct axis, and the mutual inductance between the field and armature circuits is zero because their axes are perpendicular to each other. The voltage equation for the armature circuit of the motor in Figure 11.1.1a is given by

$$v_t = e_a + L_{aq} p i_a + R_a i_a = k i_f \omega_m + L_{aq} p i_a + R_a i_a \qquad (11.1.4)$$

where v_t, R_a, and L_{aq} are the terminal voltage, resistance, and self-inductance of the armature circuit, respectively. The subscript q is used with the inductance L_{aq} to indicate that the axis of the armature-mmf is along the quadrature axis. The inductance L_{aq} includes the effect of any quadrature-axis stator windings in series with the armature, like the interpoles and compensating windings discussed in Section 9.5. The dynamic equation for the mechanical system of a motor is given by

$$T_e = k i_f i_a = J p \omega_m + B \omega_m + T_L \qquad (11.1.5)$$

or

$$T_e - T_L = J p \omega_m + B \omega_m \qquad (11.1.6)$$

where J is the combined polar moment of inertia of the load and the rotor of the motor, B is the equivalent viscous friction constant of the load and the motor (for the motor, it is used for approximating the rotational losses), and T_L is the mechanical load torque (delivered to the load) opposing rotation.

For the case of a dc generator shown in Figure 11.1.1b, the armature voltage and torque equations are given by

$$v_t = e_a - L_{aq}pi_a - R_a i_a = ki_f \omega_m - L_{aq}pi_a - R_a i_a \tag{11.1.7}$$

$$T_S - T_e = Jp\omega_m + B\omega_m \tag{11.1.8}$$

where

$$T_e = ki_f i_a \tag{11.1.9}$$

and T_S is the mechanical driving torque applied to the shaft in the direction of rotation.

The field and armature currents and the speed can be considered to be *state variables* because the state of a physical system can be described in terms of its stored energy. The energy storage is associated with the magnetic fields produced by the field and armature currents and with the kinetic energy of the rotating parts. Equations 11.1.3 to 11.1.9 are first-order differential equations containing product nonlinearities, such as $i_f i_a$ and $i_f \omega_m$, of these state variables. These equations, together with the torque-speed characteristics of the mechanical system connected to the shaft and the Kirchhoff-law equations for the circuits connected to the armature and field terminals, describe the system performance.

▇▇▇▇▇▇▇▇▇▇▇‖‖‖‖‖
11.2 Dynamic Analysis

The *transfer function* is a means by which the dynamic characteristics of electromechanical devices are described. The transfer function is a mathematical formulation that relates the output variable of a device to the input variable. For linear devices, the transfer function is independent of the input quantity and solely dependent on the parameters of the device together with any operations of time, like differentiation and integration, that it may possess. To obtain the transfer function, one usually goes through three steps: (i) determining the governing equation for the device expressed in terms of the output and input variables; (ii) Laplace transforming the governing equation, assuming all initial conditions to be zero; and (iii) rearranging the equation to yield the ratio of the output to input variable. The properties of transfer functions are summarized as follows:

- A transfer function is defined only for a linear time-invariant system.
- The transfer function between an input variable and an output variable of a system is defined as the Laplace transform of the output to the Laplace transform of the input, or as the Laplace transform of the impulse response. The impulse response of a linear system is defined as the output response of the system when the input is a unit impulse function.
- All initial conditions of the system are assumed to be zero.
- The transfer function is independent of the input.

The *block diagram* is a pictorial representation of the equations of the system. Each block represents a mathematical operation, and the blocks are interconnected to satisfy the governing equations of the system. The block diagram thus provides a chart of the procedure to be followed in combining the simultaneous equations, from which useful information can often be obtained without finding a

Figure 11.2.1 Basic building block of a block diagram.

complete analytical solution. The block-diagram technique has been highly developed in connection with studies of feedback control systems,[1] often leading to programming a problem for solution on an analog computer.

The simple configuration shown in Figure 11.2.1 is actually the basic building block of a complex block diagram. The arrows on the diagram imply that the block diagram has a unilateral property; in other words, signal can pass only in the direction of the arrows. A box is the symbol for multiplication; the input quantity is multiplied by the function in the box to obtain the output. With circles indicating summing points (in an algebraic sense), and with boxes or blocks denoting multiplication, any linear mathematical expression can be represented by block-diagram notation, as in Figure 11.2.2 for an *elementary feedback control system.*

The block diagrams of complex feedback control systems usually contain several feedback loops, which might have to be simplified in order to evaluate an overall transfer function for the system. A few of the block diagram reduction manipulations are given in Table 11.2.1; no attempt is made to cover all the possibilities.

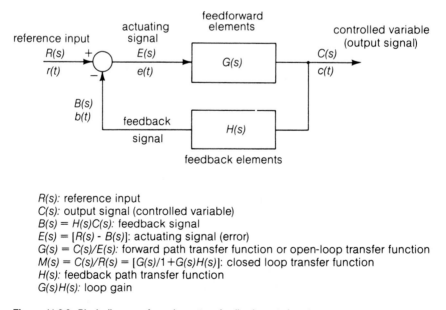

$R(s)$: reference input
$C(s)$: output signal (controlled variable)
$B(s) = H(s)C(s)$: feedback signal
$E(s) = [R(s) - B(s)]$: actuating signal (error)
$G(s) = C(s)/E(s)$: forward path transfer function or open-loop transfer function
$M(s) = C(s)/R(s) = [G(s)/1+G(s)H(s)]$: closed loop transfer function
$H(s)$: feedback path transfer function
$G(s)H(s)$: loop gain

Figure 11.2.2 Block diagram of an elementary feedback control system.

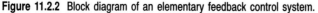

[1] B. C. Kuo, *Automatic Control Systems,* 4th ed. (Englewood Cliffs, N.J.: Prentice-Hall, 1982).

Table 11.2.1 Some of the Block Diagram Reduction Manipulations

original block diagram	manipulation	modified block diagram
$R \rightarrow \boxed{G_1} \rightarrow \boxed{G_2} \rightarrow C$	cascaded elements	$R \rightarrow \boxed{G_1 G_2} \rightarrow C$
$R \rightarrow \boxed{G_1},\ \boxed{G_2} \rightarrow C$ (+/+ summing)	addition or subtraction (eliminating the auxiliary forward path)	$R \rightarrow \boxed{G_1 \pm G_2} \rightarrow C$
$R \rightarrow \boxed{G} \rightarrow C$ (pick-off)	shifting of the pick-off point ahead of the block	$R \rightarrow \boxed{G} \rightarrow C$, with $\boxed{G}$ in feedback
$R \rightarrow \boxed{G} \rightarrow C$ (pick-off behind)	shifting of the pick-off point behind the block	$R \rightarrow \boxed{G} \rightarrow C$, with $\boxed{1/G}$ in feedback
$R \rightarrow \boxed{G} \rightarrow + \bigcirc - \rightarrow E$, C	shifting the summing point ahead of the block	$R \rightarrow + \bigcirc - \rightarrow \boxed{G} \rightarrow E$, $\boxed{1/G} \leftarrow C$
$R \rightarrow + \bigcirc - \rightarrow E \rightarrow \boxed{G} \rightarrow$, C	shifting the summing point behind the block	$R \rightarrow \boxed{G} \rightarrow + \bigcirc - \rightarrow E$, $\boxed{G} \leftarrow C$
$R \rightarrow + \bigcirc - \rightarrow \boxed{G} \rightarrow C$, $\boxed{H}$	removing H from feedback path	$R \rightarrow \boxed{1/H} \rightarrow + \bigcirc - \rightarrow \boxed{H} \rightarrow \boxed{G} \rightarrow C$
$R \rightarrow + \bigcirc - \rightarrow \boxed{G} \rightarrow C$, $\boxed{H}$	eliminating the feedback path	$R \rightarrow \boxed{\dfrac{G}{1 + GH}} \rightarrow C$

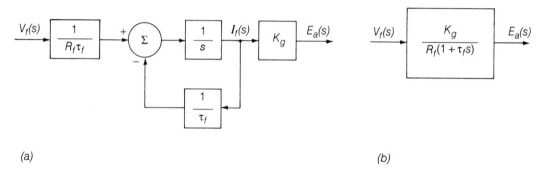

(a)   (b)

Figure 11.2.3 Block diagram representing Equations 11.2.3 and 11.2.4.

Separately Excited DC Generator

Let us consider the electrical transients in a separately excited dc generator resulting from changes in excitation. The analysis will be made on a linear basis, while neglecting the effects of saturation. Further, let the generator speed be constant so that the dynamics of the mechanical drive do not enter into the problem. For the dc generator of Figure 11.1.1b, the voltage equation for the field circuit is given by

$$v_f = R_f i_f + L_{ff} p i_f = R_f (1 + \tau_f p) i_f \tag{11.2.1}$$

where $\tau_f = L_{ff}/R_f$ is the time constant of the field circuit. The Laplace transformation of Equation 11.2.1 with zero initial conditions yields

$$V_f(s) = I_f(s)R_f + L_{ff}sI_f(s) = R_f(1 + \tau_f s)I_f(s) \tag{11.2.2}$$

Equation 11.2.2 can be rearranged as follows:

$$sI_f(s) = \frac{1}{\tau_f}\left[\frac{V_f(s)}{R_f} - I_f(s)\right] \tag{11.2.3}$$

With an integrator $1/s$ in the forward path and $I_f(s)$ as the state variable, the block diagram corresponding to Equation 11.2.3 is given in Figure 11.2.3a. Multiplication of the output $I_f(s)$ by a constant K_g gives the generated emf $E_a(s)$ at a speed ω_m. K_g is the slope of the air-gap line of the magnetization curve (taken at the speed ω_m) representing the relationship between e_a and i_f. This operation is also shown in the block diagram of Figure 11.2.3a. The transfer function relating the armature induced voltage to the field winding voltage is then given by

$$\frac{E_a(s)}{V_f(s)} = \frac{K_g I_f(s)}{V_f(s)} = \frac{K_g}{R_f(1 + \tau_f s)} \tag{11.2.4}$$

which is represented in Figure 11.2.3b.

Let us next consider the generator dynamics including the effect of the load. Let R_L and L_L be the load resistance and load inductance, and R_a and L_{aq} be the armature winding resistance and leakage inductance, respectively. Then one can write

$$e_a = i_a R_a + L_{aq} p i_a + i_a R_L + L_L p i_a$$

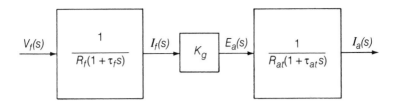

Figure 11.2.4 Block diagram of Equation 11.2.8 representing the generator dynamics including the effect of the load.

or

$$e_a = i_a(R_a + R_L) + (L_{aq} + L_L)pi_a \tag{11.2.5}$$

Laplace transformation gives

$$E_a(s) = I_a(s)(R_a + R_L)(1 + s\tau_{at})$$

where τ_{at} is the armature circuit time constant given by

$$\tau_{at} = \frac{L_{at}}{R_{at}} = \frac{L_{aq} + L_L}{R_a + R_L} \tag{11.2.6}$$

The transfer function then becomes

$$\frac{I_a(s)}{E_a(s)} = \frac{1}{R_{at}(1 + s\tau_{at})} \tag{11. 2.7}$$

The total transfer function relating the armature current to the field voltage is given by

$$\frac{I_a(s)}{V_f(s)} = \frac{E_a(s)}{V_f(s)}\frac{I_a(s)}{E_a(s)} = \frac{K_g}{R_f(1 + \tau_f s)}\frac{1}{R_{at}(1 + \tau_{at}s)} \tag{11.2.8}$$

which is represented in the block diagram of Figure 11.2.4 with two state variables $I_f(s)$ and $I_a(s)$ for the second-order system. After finding the armature current i_a, the electromagnetic torque T_e can then be determined from Equation 11.1.2 or by

$$T_e = \frac{e_a}{\omega_m}i_a \tag{11.2.9}$$

The block diagram can thus be extended piece by piece to take into account as many additions as required.

‖‖‖‖‖‖‖‖‖‖
EXAMPLE 11.2.1

A separately excited dc generator has the following parameters:

$$R_f = 100\ \Omega; \qquad L_{ff} = 20\ H; \qquad R_a = 0.1\ \Omega; \qquad L_{aq} = 0.1\ H$$

$$K_g = 100\ V \text{ per field ampere at rated speed}$$

The load connected to the generator has a resistance $R_L = 4.5$ ohms and an inductance $L_L = 2.2$ H. Assume that the prime mover is rotating at rated speed, the load switch is closed, and the generator is initially unexcited. Determine the armature current as a function of time when a 230-V dc supply is suddenly applied to the field winding, assuming the generator speed to be essentially constant.

Solution

The transfer function relating the armature current to the field voltage is given by Equation 11.2.8:

$$\frac{I_a(s)}{V_f(s)} = \frac{K_g}{R_f(1 + \tau_f s)} \frac{1}{R_{at}(1 + \tau_{at}s)}$$

With $R_{at} = R_a + R_L = 0.1 + 4.5 = 4.6 \ \Omega$; $\tau_f = L_{ff}/R_f = 20/100 = 0.2$; and $\tau_{at} = L_{at}/R_{at} = (L_{aq} + L_L)/(R_a + R_L) = (0.1 + 2.2)/(0.1 + 4.5) = 0.5$,

$$\frac{I_a(s)}{V_f(s)} = \frac{100}{100(1 + 0.2s)} \frac{1}{4.6(1 + 0.5s)}$$

The Laplace transform of the field voltage for a step change of 230 V is

$$V_f(s) = 230/s$$

Hence

$$I_a(s) = \frac{230}{s} \frac{100}{100(1 + 0.2s)} \frac{1}{4.6(1 + 0.5s)}$$

or

$$I_a(s) = \frac{500}{s(s + 5)(s + 2)} = \frac{K_1}{s} + \frac{K_2}{s + 5} + \frac{K_3}{s + 2}$$

where

$$K_1 = \frac{500}{(s + 5)(s + 2)}\bigg|_{s=0} = 50$$

$$K_2 = \frac{500}{s(s + 2)}\bigg|_{s=-5} = 33.3$$

$$K_3 = \frac{500}{s(s + 5)}\bigg|_{s=-2} = -83.3$$

Taking the inverse Laplace transform, we get the current buildup of

$$i_a(t) = 50 + 33.3e^{-5t} - 83.3e^{-2t}$$

The effect of the smaller of the two time constraints, τ_f, on the current buildup can be ignored when its value is less than about one-quarter of the longer one.

Separately Excited DC Motor

Let us next consider the case of a separately excited dc motor with constant field excitation; we will investigate how the speed of the motor responds to changes in the voltage applied to the armature

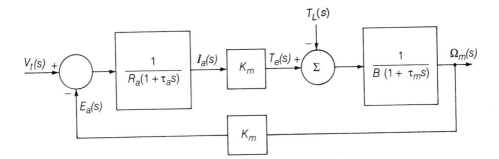

Figure 11.2.5 Block diagram representing Equations 11.2.15 and 11.2.16.

terminals. The analysis involves electrical transients in the armature circuit and the dynamics of the mechanical load driven by the motor. At a constant motor-field current I_f, the electromagnetic torque and the generated emf are given by

$$T_e = K_m i_a \tag{11.2.10}$$

$$e_a = K_m \omega_m \tag{11.2.11}$$

where $K_m = k I_f$ is a constant, which is also the ratio e_a/ω_m; in terms of the magnetization curve, e_a is the generated emf corresponding to the field current I_f at the speed ω_m. Let us now try to find the transfer function that relates $\Omega_m(s)$ to $V_t(s)$.

The differential equation for the motor armature current i_a is given by Equation 11.1.4, which can be rearranged as

$$R_a(1 + \tau_a p)i_a = v_t - e_a \tag{11.2.12}$$

where v_t is the terminal voltage applied to the motor, e_a is the back emf given by Equation 11.2.11, R_a and L_a include the series resistance and inductance of the armature circuit and electrical source put together, and $\tau_a = L_a/R_a$ is the *electrical time constant* of the armature circuit. The electromagnetic torque is given by Equation 11.2.10, and from the dynamic Equation 11.1.6 for the mechanical system, the acceleration is given by

$$(B + Jp)\omega_m = T_e - T_L \tag{11.2.13}$$

or

$$B(1 + \tau_m p)\omega_m = T_e - T_L \tag{11.2.14}$$

where $\tau_m = J/B$ is the *mechanical time constant,* and the load torque T_L, in general, is a function of speed.

Laplace transforms of Equations 11.2.12 and 11.2.14 lead to the following:

$$I_a(s) = \frac{V_t(s) - E_a(s)}{R_a(1 + \tau_a s)} = \frac{V_t(s) - K_m \Omega_m(s)}{R_a(1 + \tau_a s)} \tag{11.2.15}$$

$$\Omega_m(s) = [T_e(s) - T_L(s)]\frac{1}{B}\frac{1}{(1 + \tau_m s)} \tag{11.2.16}$$

The corresponding block diagram representing these operations is given in Figure 11.2.5 in terms of the state variables $I_a(s)$ and $\Omega_m(s)$, with $V_t(s)$ as input.

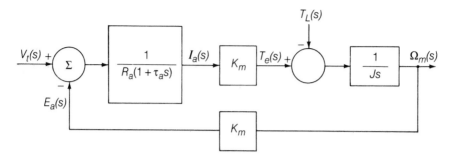

Figure 11.2.6 Block diagram representing Equation 11.2.18

The application of the closed loop transfer function $M(s)$ shown in Figures 11.2.2 to 11.2.5 yields the following transfer function relating $\Omega_m(s)$ and $V_t(s)$, with $T_L = 0$:

$$\frac{\Omega_m(s)}{V_t(s)} = \frac{K_m/[R_a(1 + \tau_m s))B(1 + \tau_m s)]}{1 + [K_m^2/R_a(1 + \tau_a s)B(1 + \tau_m s)]} \tag{11.2.17}$$

With mechanical damping B neglected, Equation 11.2.17 reduces to

$$\frac{\Omega_m(s)}{V_t(s)} = \frac{1}{K_m[\tau_i s(\tau_a s + 1) + 1]} \tag{11.2.18}$$

where $\tau_i = JR_a/K_m^2$ is the *inertial time constant*, and the corresponding block diagram is shown in Figure 11.2.6. The transfer function relating speed to load torque with $V_t = 0$ can be obtained from Figure 11.2.6 by eliminating the feedback path as follows:

$$\frac{\Omega_m(s)}{T_L(s)} = -\frac{(1/Js)}{1 + (1/Js)[K_m^2/R_a(1 + \tau_a s)]} = -\frac{(\tau_a s + 1)}{Js(\tau_a s + 1) + (K_m^2/R_a)} \tag{11.2.19}$$

Expressing the torque equation for the mechanical system as

$$T_e = K_m i_a = Jp\omega_m + B\omega_m + T_L \tag{11.2.20}$$

then dividing by K_m, and substituting $\omega_m = e_a/K_m$, we obtain

$$i_a = \frac{J}{K_m^2}\frac{de_a}{dt} + \frac{B}{K_m^2}e_a + \frac{T_L}{K_m} \tag{11.2.21}$$

Equation 11.2.21 can be identified to be the node equation for a parallel C_{eq}–G_{eq}–Z_L circuit with

$$C_{eq} = J/K_m^2; \quad G_{eq} = B/K_m^2; \quad \text{and} \quad Z_L = K_m e_a/T_L \tag{11.2.22}$$

and a common voltage of e_a. An analog electrical circuit can then be drawn as in Figure 11.2.7 for a separately excited dc motor, in which the inertia is represented by a capacitance, the damping by a shunt conductance, and the load-torque component of current is shown flowing through an equivalent

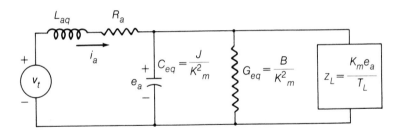

Figure 11.2.7 Analog electrical circuit for a separately excited dc motor.

impedance Z_L. The time constants, τ_i associated with inertia and τ_m associated with the damping-type load torque that is proportional to speed (in terms of the analog-circuit notations), are given by

$$\tau_i = R_a C_{eq} \quad \text{and} \quad \tau_m = C_{eq}/G_{eq} \tag{11.2.23}$$

Note in the above analysis that K_m is proportional to the constant motor field current I_f.

The self-inductance of the armature can often be neglected except for a motor driving a load that has rapid torque pulsations of appreciable magnitude.

▌▌▌▌▌▌▌▌▌▌▌

EXAMPLE 11.2.2

A 5-hp, 220-V, separately excited dc motor has the following parameters: $R_a = 0.5 \ \Omega$; $k = 2$ H; $R_f = 220 \ \Omega$; $L_{ff} = 110$ H. The armature winding inductance is negligible. The torque required by the load is proportional to the speed, and the combined constants of the motor armature and the load are $J = 3$ kg·m^2 and $B = 0.3$ kg·m^2/s.

Consider the armature-controlled dc motor whose speed is made to respond to variations in the applied motor armature voltage v_t. Let the field current be maintained constant at 1 A.

a. Develop a block diagram relating the motor speed and the motor applied voltage, and find the corresponding transfer function.

b. Compute the steady-state speed corresponding to a step-applied armature voltage of 220 V.

c. How long does the motor take to reach 0.95 of the steady-state speed of part b?

d. Determine the value of the total effective viscous damping coefficient of the motor-load configuration.

Solution

a. Modifying Figure 11.2.5 for the negligible armature winding inductance, one has the following:

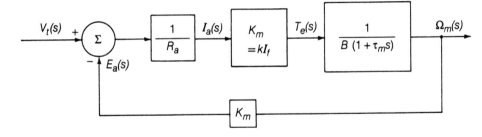

Figure 11.2.8

$$\frac{\Omega_m(s)}{V_t(s)} = \frac{K_m/[R_aB(1 + \tau_m s)]}{1 + K_m^2/[R_aB(1 + \tau_m s)]} = \frac{K_m}{K_m^2 + R_aB} \frac{1}{1 + \tau_m's}$$

where

$$\tau_m' = \tau_m \frac{R_aB}{R_aB + K_m^2} = \frac{JR_a}{R_aB + K_m^2}$$

b. $\omega_{ss} = \lim_{s \to 0} s\Omega_m(s)$

Noting that $V_t(s) = \frac{220}{s}$, we obtain

$$\omega_{ss} = 220 \frac{K_m}{R_aB + K_m^2} = (220)(2)/(0.5 \times 0.3 + 4) = 440/4.15 = 106 \text{ rad/s}$$

c. $\tau_m' = JR_a/(R_aB + K_m^2) = 3 \times 0.5/4.15 = 0.36$

For the motor to reach 0.95 of ω_{ss},

$$e^{-t/0.36} = 0.05$$

or

$$t = 0.36 \times 3 = 1.08 \text{ seconds.}$$

d. Total effective viscous damping $= B + (K_m^2/R_a) = 0.3 + (4/0.5) = 8.3 \text{ kg} \cdot m^2/s$

Effect of Saturation

The dynamic analysis of dc machines has so far been studied on a linear basis with magnetic saturation neglected. Such a treatment is justified if the transient response takes place in substantially the linear region. The response to small disturbances, however, can be treated on an incrementally linear basis. For more comprehensive studies, in general, transient investigations might require a nonlinear analysis because of saturation.

The linearized block diagram of a separately excited dc generator can be modified rather easily, as shown in Figure 11.2.9, to take care of magnetic saturation by feeding back the output e_a through the magnetization curve to obtain i_f. The equation to be satisfied is

$$v_f = R_f i_f + L_{ff} p i_f \tag{11.2.24}$$

or

$$v_f - R_f i_f = L_{ff} p i_f = \frac{K_{ff}}{K_g} p e_a \tag{11.2.25}$$

or

$$p e_a = \frac{K_g}{\tau_f} \left(\frac{v_f}{R_f} - i_f \right) \tag{11.2.26}$$

where K_g is the slope of the air-gap line in generated volts per field ampere at the speed ω_m, and $\tau_f = L_{ff}/R_f$ is the unsaturated value of the field-circuit time constant. Note that, when saturation is considered,

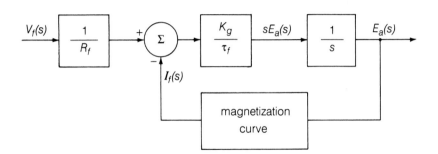

Figure 11.2.9 Block diagram of a separately excited dc generator including saturation.

the inductance of the field winding is no longer a constant. Laplace transformation of Equation 11.2.26 gives

$$sE_a(s) = \frac{K_g}{\tau_f}\left[\frac{V_f(s)}{R_f} - I_f(s)\right]$$

(11.2.27)

as represented in the block diagram of Figure 11.2.9.

Dynamic Analysis with Time-Domain Techniques

Now that the dynamic analysis with the transfer-function approach has been presented, let us take up the dynamic analysis with time-domain techniques as a direct-solution approach, in which we seek a closed-form solution of the describing equations in the time domain. Rather than pursuing an analytical solution of the equations, we can take advantage of the simulation-diagram techniques, which are not presented here in detail in view of the scope of this text. The reader should however be aware of the prepackaged digital-computer programs that furnish all the desired information with minimum effort.

The voltage equation for the field circuit of a separately-excited dc machine is given by

$$v_f = R_f i_f + L_{ff}\frac{di_f}{dt}$$

or

$$L_{ff}\frac{d}{dt}[i_f(t)] = -R_f i_f(t) + v_f(t)$$

(11.2.28)

The voltage equation for the armature circuit of a motor of Figure 11.1.1a is

$$v_t = e_a + L_{aq}\frac{di_a}{dt} + R_a i_a = ki_f\omega_m + L_{aq}\frac{di_a}{dt} + R_a i_a$$

or

$$L_{aq}\frac{d}{dt}[i_a(t)] = -R_a i_a(t) - ki_f(t)\omega_m(t) + v_t(t)$$

(11.2.29)

Equations 11.2.28 and 11.2.29 can be put in matrix form as

$$\begin{bmatrix} L_{ff} & 0 \\ 0 & L_{aq} \end{bmatrix}\frac{d}{dt}\begin{bmatrix} i_f(t) \\ i_a(t) \end{bmatrix} = \begin{bmatrix} -R_f & 0 \\ -k\omega_m(t) & -R_a \end{bmatrix}\begin{bmatrix} i_f(t) \\ i_a(t) \end{bmatrix} + \begin{bmatrix} v_f(t) \\ v_t(t) \end{bmatrix}$$

(11.2.30)

Premultiplying the equation through by the inverse of the inductance-coefficient matrix of the left-hand side,

$$\frac{d}{dt}\begin{bmatrix} i_f(t) \\ i_a(t) \end{bmatrix} = \begin{bmatrix} -R_f/L_{ff} & 0 \\ -(k/L_{aq})\omega_m(t) & -R_a/L_{aq} \end{bmatrix}\begin{bmatrix} i_f(t) \\ i_a(t) \end{bmatrix} + \begin{bmatrix} 1/L_{ff} & 0 \\ 0 & 1/L_{aq} \end{bmatrix}\begin{bmatrix} v_f(t) \\ v_t(t) \end{bmatrix}$$

(11.2.31)

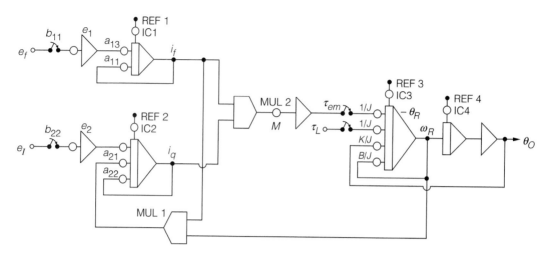

Figure 11.2.10 Analog-computer simulation diagram.

In a compact matrix form, we have

$$\dot{\mathbf{X}}(t) = \mathbf{A}[\omega_m(t)]\mathbf{X}(t) + \mathbf{B}\mathbf{u}(t) \tag{11.2.32}$$

Defining $\omega_m = d\theta_1(t)/dt = \theta_2(t)$, the dynamic equation for the mechanical system of a motor is given by

$$\frac{d}{dt}[\theta_2(t)] = -\frac{K}{J}\theta_1(t) - \frac{B}{J}\theta_2(t) - \frac{1}{J}(T_L - T_e) \tag{11.2.33}$$

where K is the rotational stiffness (analogous to a spring constant) in N · m/rad. Expressed in matrix form, we get

$$\frac{d}{dt}\begin{bmatrix} \theta_1(t) \\ \theta_2(t) \end{bmatrix} = \begin{bmatrix} 0 & 1 \\ -K/J & -B/J \end{bmatrix}\begin{bmatrix} \theta_1(t) \\ \theta_2(t) \end{bmatrix} + \begin{bmatrix} 0 \\ 1/J \end{bmatrix}[(T_e - T_L)] \tag{11.2.34}$$

where

$$T_e = \begin{bmatrix} i_f(t) & i_a(t) \end{bmatrix}\begin{bmatrix} 0 & 0 \\ K & 0 \end{bmatrix}\begin{bmatrix} i_f(t) \\ i_a(t) \end{bmatrix} \tag{11.2.35}$$

The above equations may be written as

$$\dot{\boldsymbol{\theta}}(t) = \mathbf{C}\boldsymbol{\theta}(t) + \mathbf{D}\mathbf{v}(t) \tag{11.2.36}$$

and

$$T_e = \mathbf{X}^T(t)\mathbf{G}\mathbf{X}(t) \tag{11.2.37}$$

Equations 11.2.32, 11.2.36, and 11.2.37 form the *state equations* of the separately excited dc motor. These constitute a system of nonlinear differential equations with time-varying coefficients (and with cross-coupling), the analytical solution of which is indeed cumbersome. Figure 11.2.10 (drawn here with no developmental details[2]) shows an analog-computer simulation diagram for the solution of the state variables.

---

[2] M. G. Rekoff Jr., *Analog Computer Programming*, Merrill, Columbus, Ohio, 1967.

Instead of going through the development of simulation techniques, one can directly go to CSMP (Continuous System Modeling Program)[3], A specialized fortran-language program developed by IBM for analog-computer simulation to be performed on a digital computer. Upon execution, the computer printout provides numerical and graphical solutions as a function of the electrical and mechanical state variables, as well as the electromagnetic torque.[4]

11.3 Metadynes and Amplidynes

We have so far considered the dc machines with brushes located only in the quadrature axis. The separately excited generator can be looked upon as a single-stage rotating power amplifier having a gain and one or more time constants. By introducing a shunt or series field, we can obtain performance control through feedback. The shunt generator can then be seen to be similar to a feedback amplifier. With the object of increasing the control possibilities like sensitivity or gain, decreasing the effective time constants or response time, and decreasing the sensitivity of the system to uncontrolled external disturbances, *cross-field machines* have been developed for control applications. Cross-field machines with more than two brush sets per pair of poles, called *metadynes,* can be used in a wide variety of applications requiring high-speed response, high power amplification, or some other special built-in characteristic. The purpose of this section is to examine the effects of additional brushes located in the direct axis; we shall concern ourselves with metadyne generators, with emphasis on the most commonly used form, the *amplidyne.*

The basic two-pole metadyne has a field winding and two pairs of commutator brushes, one pair in line with the field winding in the direct *d*-axis and the other pair at right angles to the field winding in the quadrature *q*-axis, as shown in Figure 11.3.1. With the generator driven at a constant speed and with magnetic saturation neglected, the voltage generated in the armature between the quadrature-axis brushes is proportional to the field current i_f:

$$e_{aq} = k_{qf}i_f \tag{11.3.1}$$

where k_{qf} is a constant. If the quadrature-axis brushes are short-circuited, a weak control-field current will produce a relatively much larger quadrature-axis armature current because the impedance of the short-circuited armature is small. A corresponding flux-density wave will be centered on the quadrature

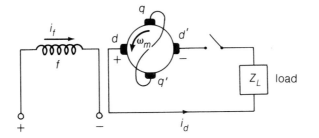

Figure 11.3.1 Cross-field machine (basic two-pole metadyne).

[3] *System/360 Continuous System Modeling Program (360 A-CX-16X), User's manual,* 3rd ed., IBM. Corp., White Plains, NY, 1928

[4] Van E. Mablekos, *Electric Machine Theory for Power Engineers.* New York: Harper & Row, 1980, Chapter 2, pp. 133-144

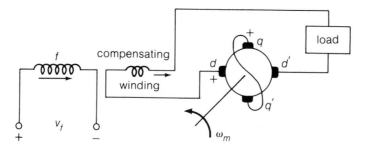

Figure 11.3.2 Basic two-pole amplidyne.

axis and will produce an effect similar to that of a fictitious stator winding on the quadrature axis. With a pair of brushes dd' placed on the commutator in the d-axis, the speed voltage generated in the armature by its rotation in the q-axis flux will appear across these brushes. Assuming constant speed and negligible saturation, this voltage will be proportional to the q-axis armature current i_q:

$$e_{ad} = k_{dq}i_q \tag{11.3.2}$$

where k_{dq} is a constant. Now, if a load Z_L is connected to the d-axis brushes as shown in Figure 11.3.1, the d-axis armature current i_d will result, which in turn will produce an mmf *opposing* the control field mmf. Under the assumed conditions of constant speed and negligible saturation, the q-axis generated emf is given by

$$e_{aq} = k_{qf}i_f - k_{qd}i_d \tag{11.3.3}$$

where k_{qd} is a constant. The metadyne generator is thus a two-stage power amplifier with rapid response and with strong negative current feedback from the final output stage to the input. Because of the effect of the negative feedback, the power amplification will be reduced. For a given value of the field current, the metadyne maintains nearly constant output current i_d over a wide range of load impedance, giving approximately a *constant-current* characteristic.

  The most commonly used form of the metadyne is the amplidyne, which consists of the basic metadyne generator along with a *cumulative compensating winding* on the d-axis connected in series with the d-axis load current, as shown in Figure 11.3.2. If the compensating winding is designed to produce a flux as nearly as possible equal and opposite to the flux produced by the d-axis armature current, the negative feedback effect of the load current is canceled and the control-field winding has almost complete control over the d-axis flux. Very little control-field power input is then required to produce a large current in the short-circuited q-axis of the armature. Power amplification of the order of 20,000:1 can easily be obtained. The fully compensated amplidyne gives approximately a *constant-voltage* characteristic. Any degree of compensation can be used, however, by varying the number of compensating turns.

  Assuming perfect compensation, constant speed, and negligible saturation, the transfer function relating the d-axis generated emf E_{ad} to the control-field applied voltage V_f can be obtained as

$$\frac{E_{ad}}{V_f} = \frac{k_{qf}/R_f}{\tau_f s + 1} \frac{k_{dq}/R_{aq}}{\tau_{aq}s + 1} \tag{11.3.4}$$

where R_f and τ_f are the resistance and time constant of the control field, and R_{aq} and τ_{aq} are the resistance and time constant of the q-axis armature-circuit. The d-axis armature circuit resistance is

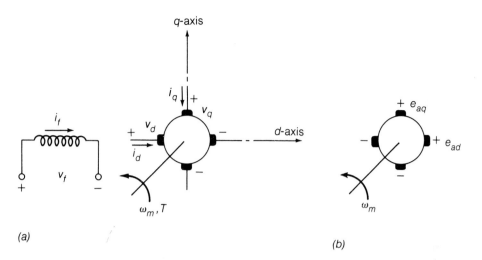

Figure 11.3.3 Metadyne as a generalized machine: (a) schematic diagram of a two-pole metadyne with voltages impressed on both axes, (b) reference directions for speed voltages.

considered along with the load, while the d-axis armature inductance is usually neglected. The time lag is esentially due to the q-axis time constant τ_{aq}. Various auxiliary or control-field windings can be added to either axis of an amplidyne to improve performance characteristics or to obtain characteristics to meet performance specifications.

Metadyne as a Generalized Machine

The two-axis metadyne can be regarded as a basic prototype or generalized machine. Let us consider the schematic diagram of a simple two-pole metadyne shown in Figure 11.3.3 with voltages v_d and v_q applied to the d-axis and q-axis brushes, respectively, and v_f applied to the d-axis stator field circuit. Let i_d, i_q, and i_f be the currents supplied to the respective circuits, as shown in Figure 11.3.3. For the *reference directions* shown in Figure 11.3.3, the flux linkages λ_f, λ_d, and λ_q associated with the field winding, the d-axis armature winding, and the q-axis armature winding can be expressed in a matrix form as follows:

$$\begin{bmatrix} \lambda_f \\ \lambda_d \\ \lambda_q \end{bmatrix} = \begin{bmatrix} L_{ff} & L_{fd} & 0 \\ L_{fd} & L_{ad} & 0 \\ 0 & 0 & L_{aq} \end{bmatrix} \begin{bmatrix} i_f \\ i_d \\ i_q \end{bmatrix} \tag{11.3.5}$$

where L_{ff}, L_{ad}, and L_{aq} are the self-inductances of the field winding, the d-axis armature winding, and the q-axis armature winding, respectively. L_{fd} is the mutual inductance between the field winding and the d-axis armature winding; it is a positive quantity for the chosen reference directions because the mmfs of the field current and of the d-axis armature current are in the same direction.

Noting that the reference directions for speed voltages e_{aq} and e_{ad} as shown in Figure 11.3.3b, the voltage equations can now be written as follows with the chosen reference directions:

$$v_f = R_f i_f + p\lambda_f \tag{11.3.6}$$
$$v_d = R_{ad} i_d + p\lambda_d - e_{ad} \tag{11.3.7}$$
$$v_q = R_{aq} i_q + p\lambda_q + e_{aq} \tag{11.3.8}$$

where R_f, R_{ad}, and R_{aq} are the resistances of the respective windings. For a two-pole machine with sinusoidally distributed windings, with saturation neglected, the speed voltages can be expressed as follows:

$$e_{ad} = \omega_m L_{aq} i_q \tag{11.3.9}$$

$$e_{aq} = \omega_m(L_{af} i_f + L_{ad} i_d) \tag{11.3.10}$$

where ω_m is the angular velocity corresponding to the speed of rotation, and L_{af} is the coefficient relating the no-load speed voltage generated in the q-axis to the field current and the speed. The voltage Equations 11.3.6 to 11.3.8 can be expressed in matrix form as follows, after making use of Equations 11.3.5, 11.3.9, and 11.3.10:

$$\begin{bmatrix} v_f \\ v_d \\ v_q \end{bmatrix} = \begin{bmatrix} R_f + L_{ff}p & L_{fd}p & 0 \\ L_{fd}p & R_{ad} + L_{ad}p & -\omega_m L_{aq} \\ \omega_m L_{af} & \omega_m L_{ad} & R_{aq} + L_{aq}p \end{bmatrix} \begin{bmatrix} i_f \\ i_d \\ i_q \end{bmatrix} \tag{11.3.11}$$

The magnetic torque in the positive direction of rotation as shown in Figure 11.3.3a is given by the power input corresponding to the speed voltages divided by ω_m:

$$T = \frac{e_{aq} i_q - e_{ad} i_d}{\omega_m} = L_{af} i_f i_q + (L_{ad} - L_{aq}) i_d i_q \tag{11.3.12}$$

▰▰▰▰▰ ▌▌▌▌
BIBLIOGRAPHY

Fitzgerald, A. E., and C. Kingsley Jr. *Electric Machinery*. 2d ed. New York: McGraw-Hill, 1961.

Fitzgerald, A. E.; C. Kingsley Jr.; and A. Kusko. *Electric Machinery*. 3d ed. New York: McGraw-Hill, 1971.

Kusko, A. *Solid-State DC Motor Drives*. Cambridge, Mass.: M.I.T. Press, 1969.

Mablekos, Van E., *Electric Machine Theory for Power Engineers*. New York: Harper & Row, 1980.

Matsch, L. W. *Electromagnetic and Electromechanical Machines*. New York: Intext, 1972.

Morgan, A. T. *General Theory of Electrical Machines*. London: Heyden & Son, 1979.

Pearman, R. A. *Power Electronics—Solid State Motor Control*. Reston, Va.: Reston, 1980.

Say, M. G. and E. O. Taylor. *Direct Current Machines*. New York: John Wiley & Sons, Halsted Press, 1980.

Slemon, G. R., and A. Straughen. *Electric Machines*. Reading, Mass.: Addison-Wesley, 1980.

White, D. C., and H. H. Woodson. *Electromechanical Energy Conversion*. New York: John Wiley & Sons, 1959.

Woodson, H. H., and J. R. Melcher. *Electromechanical Dynamics—Part I: Discrete Systems*. New York: John Wiley & Sons, 1968.

▰▰▰▰▰▰▰ ▌▌▌▌▌▌▌▌▌▌
PROBLEMS

11–1.   A separately excited dc generator has the following parameters:

Field winding resistance $R_f = 60\ \Omega$

Field winding inductance $L_{ff} = 60$ H

Armature resistace $= 1\ \Omega$

Armature inductance $L_{aq} = 0.4$ H

Generated emf constant $K_g = 120$ V per field ampere at rated speed

The armature terminals of the generator are connected to a low-pass filter with a series inductance of $L = 1.6$ H and a shunt resistance of $R = 1$ Ω. Determine the transfer function relating the output voltage $V_t(s)$ across the shunt resistance R and the input voltage $V_f(s)$ applied to the field winding.

11–2. A separately excited dc generator, running at a constant speed, supplies a load having a 1-Ω resistance in series with a 1-H inductance. The armature resistance is 0.1 Ω and its inductance is negligible. The field, having a resistance of 50 Ω and an inductance of 5 H, is suddenly connected to a 100-V source.

Determine the armature current build-up as a function of time, if the generator voltage constant K_g is 50 V per field ampere at rated speed.

11–3. The following test data are taken on a 20-hp, 250-V, 600-rpm dc shunt motor:

$$R_f = 150 \ \Omega; \qquad \tau_f = 0.5 \ \text{s}; \qquad R_a = 0.15 \ \Omega; \qquad \tau_a = 0.05 \ \text{s}$$

When the motor is driven at rated speed as a generator with no load, a field current of 2 A produces an armature emf of 250 V.

Determine the following: (a) L_{ff}, the self-inductance of the field circuit; (b) L_{aq}, the self-inductance of the armature circuit; (c) the coefficient k relating the speed voltage to the field current; and (d) the friction coefficient B_L of the load at rated load and rated speed, assuming that the torque required by the load T_L is proportional to the speed.

11–4. A separately excited dc generator can be treated as a power amplifier when driven at constant angular velocity ω_m. If the armature circuit is connected to a load having a resistance R_L, obtain an expression for the voltage gain $V_L(s)/V_f(s)$.

With $R_a = 0.1$ Ω, $R_f = 10$ Ω, $R_L = 1$ Ω and $K_g = 100$ V/field-amp at rated speed, determine the voltage gain and the power gain if the generator is operating under steady state with 25 V applied across the field.

11–5. Consider the motor of Problem 11–3 to be initially running at constant speed with an impressed armature voltage of 250 V, with the field separately excited by a constant field current of 2 A. Let the motor be driving a pure-inertia load with the combined polar moment of inertia of the armature and load of 3 Kg · m². The rotational losses of the motor can be neglected.

a. Determine the speed.

b. Neglecting the self-inductance of the armature, obtain the expressions for the armature current and the speed as functions of time if the applied armature voltage is suddenly increased from 250 V to 260 V.

c. Repeat part b, including the effect of the armature self-inductance.

11–6. A separately excited dc motor carries a load of $(330 \ \dot{\omega}_m + \omega_m)$N · m. The armature resistance is 1 Ω and its inductance is negligible. If 100 V is suddenly applied across the armature while the field current is constant, obtain an expression for the motor speed buildup as a function of time, given the motor torque constant K_m to be 10 N · m/A.

11–7. Consider the motor of Example 11.2.2 in the text to be operated as a separately excited dc generator at a constant speed of 900 rpm, with a constant field current of 1.5 A. Let the load current be initially zero.

Determine the armature current and the armature terminal voltage as functions of time for a suddenly applied load impedance consisting of a resistance of 11.5 Ω and an inductance of 0.1 H.

11–8. Consider a motor supplied by a generator, each with a separate and constant field excitation. Assume the internal voltage E of the generator to be a constant, and neglect the armature reaction of both machines. With the motor running with no external load and the system being in steady state, let a lead torque be suddenly increased from zero to T. The machine parameters are given below:

Armature inductance of motor + generator, $L = 0.008$ H

Armature resistance of motor + generator, $R = 0.04$ Ω

Internal voltage of the generator, $E = 400$ V

Moment of inertia of motor armature and load, $J = 42$ kg $\cdot$ m^2

Motor constant, $K_m = 4.25$ N $\cdot$ m/A

No-load armature current, $i_0 = 35$ A

Suddenly applied torque, $T = 2{,}000$ N$\cdot$m

Determine the following:

a. The undamped angular frequency of the transient speed oscillations, and the damping ratio of the system

b. The initial speed, final speed, and the speed drop in rpm

11–9.   Consider a separately excited dc motor having a constant field current and constant applied armature voltage. It is accelerating a pure inertia load from rest. Neglecting the armature inductance and the rotational losses, show that by the time the motor reaches its final speed the energy dissipated in the armature resistance is equal to the energy stored in the rotating parts.

11–10.   Consider the shunt motor of Problem 9–16 to be coupled to a pure-inertia load. Let the polar moment of inertia of the load and the motor combined be 2 kg$\cdot$m^2. Neglect the rotational losses of the motor and the self-inductance of the motor armature.

a. Assuming the magnetic circuit of the motor to be linear, calculate k of Equation 11.1.1 from the full-load data given in Problem 9–16.

b. While the motor accelerates on step 3 of the starting box, determine the armature current and motor speed as functions of time. (Note that this problem deals with the transient response of a dc shunt motor when the initial values are other than zero.)

11–11.   This problem deals with an alternative derivation of the torque and voltage equations for a dc machine in a dynamic state from the coupled-circuit viewpoint. Consider a simple two-pole dc machine comprising a salient-pole stator with a field winding f and a cylindrical rotor with a finely distributed armature winding a. Neglect magnetic saturation. Let

$$L_{aa} = \frac{L_{ad} + L_{aq}}{2} + \frac{L_{ad} - L_{aq}}{2} \cos 2\theta$$

$$L_{af} = -L_{af} \cos \theta$$

$$L_{ff} = L_{ff} = \text{constant}$$

Where θ is the angle between the field and armature axes, measured from the direct axis or the centerline of the stator poles.

a. First consider what would happen if connections were made to the armature winding at fixed points with respect to the rotor through slip rings. Then consider the effect of a commutator, noting that the brushes always make contact with the coils on the armature winding which are passing through the quadrature axis. By evaluating the electromechanical-coupling term $dL_{af}/d\theta$ at $\theta = \pi/2$ in terms of armature conductors Z_a, the number of parallel paths a, and the direct-axis flux ϕ_d, show that the equations for e_a, v_t, and T_e are the same as obtained in the text.

b. Considering the field circuit, with the assumptions of linear commutation and a narrow commutating zone, show that the rotational voltage induced in the field winding by the active part of the armature winding is canceled by the transformer voltage induced by the coils undergoing commutation. Then the field-voltage equation will be the same as Equation 11.1.3 as if the armature winding were a stationary coil in the quadrature axis.

11–12.   Determine the parameters of the analog capacitive circuit (Figure 11.2.7 in the text) for the motor in Example 11.2.2 and its connected load. With the aid of the equivalent circuit, obtain the expression for the armature current under the conditions. Include also a 3.5-Ω starting resistance in series with the armature to limit the starting current.

11-13. Neglecting the self-inductance of the armature circuit, show that the time constant of the equivalent capacitive circuit for a separately excited dc motor with no load is $R_a R_{eq} C_{eq}/(R_a + R_{eq})$, where $R_{eq} = 1/G_{eq}$ in Figure 11.2.7 of the text.

11-14. Consider the dc motor of Example 11.2.2 to be operating at rated voltage under steady state with a field current of 1 A and with the starting resistance in series with the armature reduced to zero.

a. Obtain the equivalent capacitive circuit neglecting the armature self-inductance and calculate the steady armature current.

b. If the field current is suddenly reduced to 0.8 A while the armature applied voltage is constant at 220 V, compute the initial armature current $i_a(0)$ on the basis that the kinetic energy stored in the rotating parts cannot change instantaneously.

c. Determine the final armature current $i_a(\infty)$ for the condition of part b.

d. Obtain the time constant τ'_{am} of the armature current for the condition of part b and express the armature current as a function of the time on the basis that

$$i_a = i_a(\infty) + [i_a(0) - i_a(\infty)]e^{-t/\tau'_{am}}$$

11-15. A separately excited dc motor, having a constant field current, accelerates a pure inertia load from rest. If the system is represented by an electrical equivalent circuit, with symbols as shown in Figure P11–15, express R, L, and C in terms of the motor parameters.

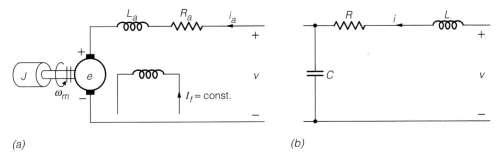

(a)   (b)

Figure P11–15

11-16. Figure P11–16 represents the Ward-Leonard system for controlling the speed of the motor M. With the generator field voltage v_{fg} as the input and the motor speed ω_m as the output, obtain an expression for

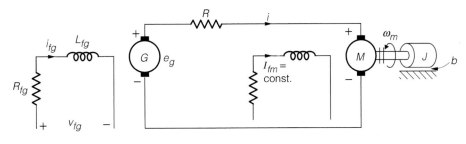

Figure P11–16

the transfer function for the system, assuming idealized machines. Let the load on the motor be given by $J\dot{\omega}_m + b\omega_m$. The generator runs at constant angular velocity ω_g.

11–17 A separately excited dc generator has the following parameters:

$$R_f = 100 \ \Omega; \qquad L_{ff} = 25 \ \text{H};$$
$$R_a = 0.25 \ \Omega; \qquad L_{aq} = 0.02 \ \text{H};$$
$$K_g = 100 \ \text{V per field ampere at rated speed}$$

a. The generator is driven at rated speed, and a field circuit voltage $V_f = 200$ V is suddenly applied to the field winding.
   (i)   Find the armature generated voltage as a function of time.
   (ii) Calculate the steady-state armature voltage.
   (iii) How much time is required for the armature voltage to rise to 90% of its steady-state value?

b. The generator is driven at rated speed, and a load consisting of $R_L = 1 \ \Omega$ and $L_L = 0.15$ H in series is connected to the armature terminals. A field circuit voltage $V_f = 200$ V is suddenly applied to the field winding. Determine the armature current as a function of time.

11–18 For the position control system shown in Figure P11–18, let the potentiometer transducers give a voltage of 1 V per radian of position. The transfer function of the servo amplifier is $G(s) = 10(1 + 0.01571s)/(7 + s)$. Let the initial angular position of the radar be zero, and the transfer function between the motor control phase voltage V_a and radar position θ be $M(S) = 2.733/s(1 + 0.01571s)$

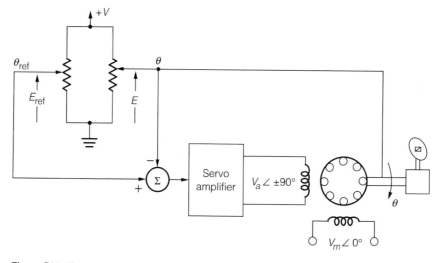

Figure P11–18

a. Obtain the transfer function of the system.

b. For a step change in the command angle of 180° ($= \pi$ radians), find the time response of the angular position of the antenna.

11–19 A separately excited dc motor has the following parameters:

$$R_a = 0.5 \ \Omega; \qquad L_{aq} \simeq 0; \qquad B \simeq 0$$

The machine generates an open-circuit armature voltage of 220 V at 2000 rpm and with a field current of 1.0 ampere.

The motor drives a constant load torque $T_L = 25$ N·m. The combined inertia of motor and load is $J = 2.5$ kg·m^2. With field current $I_f = 1.0$ A, if the armature terminals are connected to a 220-V dc source,

a. Obtain expressions for speed (ω_m) and armature current (i_a) as a function of time.

b. Find the steady-state values of the speed and armature current.

11-20 The process of plugging a motor involves reversing the polarity of the supply to the armature of the machine. *Plugging* corresponds to applying a step voltage of $-Vu(t)$ to the armature of the machine, where V is the rated terminal voltage.

A separately excited 200-V dc motor operates at rated voltage with constant excitation on zero load. The torque constant of the motor is 2 N·m/A; its armature resistance is 0.5 Ω; and the total moment of inertia of the rotating parts is 4 kg · m^2. Neglect the rotational losses and the armature inductance. Obtain an expression for the speed of the machine after plugging as a function of time, and calculate the time taken for the machine to stop.

11-21 The Ward-Leonard system discussed in Chapter 9 is a highly flexible arrangement for effecting position and speed control of a separately excited dc motor. The schematic diagram of such a system is shown in Figure P11–21, including a separately excited dc generator, the armature of which is connected directly to the armature of the separately excited dc motor driving a mechanical load. Let J be the combined polar moment of inertia of the load and motor, and B be the combined viscous friction constant of the load and motor. Assuming that the mechanical angular speed of the generator ω_{mG} is a constant, develop the block diagram for the system and obtain an expression for the transfer function $\Omega_{mM}(s)/V_{fG}(s)$.

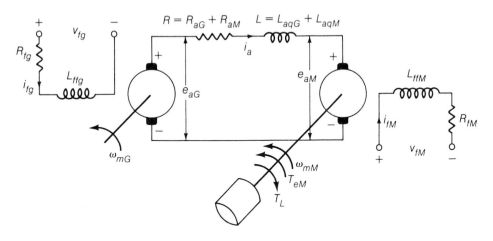

Figure P11–21

11-22 Consider the elementary motor-speed regulator scheme shown in Figure P11–22 for a separately excited dc motor, whose armature is supplied from a solid-state controlled rectifier. The motor speed is measured by means of a dc tachometer generator, and its voltage e_t is compared with a reference voltage E_R. The error voltage ($E_R - e_t$) is amplified and made to control the output voltage of the power-conversion equipment, so as to maintain substantially constant speed at the value set by the reference voltage.

Let the armature-circuit parameters be R_a and L_a, and the speed-voltage constant of the motor be K_m with units V · s/rad. Assume that the combination of A and P is equivalent to a linear controlled voltage source $v_s = K_A$ (error voltage), with negligible time lag and gain K_A. Assume also that the load torque T_L is independent of the speed, with zero damping. Neglect no-load rotational losses.

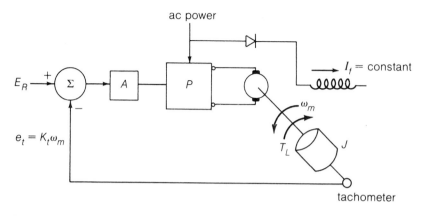

Figure P11–22

a. Develop the block diagram for the feedback speed-control system with E_R/K_t, the steady-state no-load speed setting, as input and Ω_m as output. K_t is the tachometer speed-voltage constant in volts/rpm.

b. With $T_L = 0$, evaluate the transfer function Ω_m/E_R.

c. With $E_R = 0$, obtain the transfer function Ω_m/T_L.

d. Find the expressions for the underdamped natural frequency ω_n, the damping factor α, and the damping ratio $\xi = \alpha/\omega_n$.

e. For a step input ΔE_R, obtain the final steady-state response $\Delta\omega_m(\infty)$; i.e., evaluate $\Delta\omega_m(\infty)/\Delta E_R$.

f. Evaluate $\Delta\omega_m(\infty)/\Delta T_L$, for a step input ΔT_L of a load torque.

11–23 Consider the motor of Example 11.2.2 in the text to be used as a field-controlled dc machine. Let the armature be energized from a constant current source of 15 A. Assume no saturation.

a. Develop a block diagram relating the motor speed and the applied field voltage.

b. Determine the steady-state speed for a step-applied field voltage of 220 V.

c. How long does the motor take to reach 0.95 of the steady-state speed of part b?

11–24 The output voltage of a 10-kW, 240-V dc generator is regulated by means of the closed-loop system shown in Figure P11–24. The generator parameters are given below:

$$R_f = 150\ \Omega; \qquad L_{ff} = 75\ \text{H}; \qquad R_a = 0.5\ \Omega; \qquad K_E = 150\ \text{V/field-amp at 1,200 rpm}$$

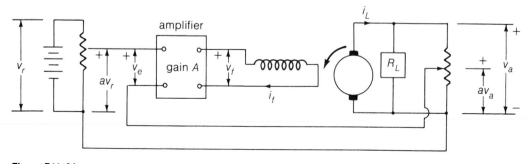

Figure P11–24

The self-inductance of the armature is negligible. The amplifier has an amplification factor of $A = 10$, and the potentiometers are set such that a is unity and the reference voltage v_r is 250 V. The generator is driven by an induction motor the speed of which is almost 1,200 rpm when the generator output is zero and 1,140 rpm when the generator delivers an armature current of 42 A.

a. Compute the steady-state armature terminal voltage at no-load and when the generator is delivering 42 A.

b. Calculate the time constant for part a.

c. With the value of the field current as in part a, for an armature current of 42 A, calculate the steady-state armature voltage.

11–25 Perturbations in the voltage applied to the field or to the armature or in the torque of a dc motor can be regarded as small-signal inputs. The response of the motor to such inputs can be determined to a good degree of approximation by means of linearization techniques by considering the perturbations to be superimposed on quiescent operating points. Let the following equations with an additional subscript 0 represent the quiescent operating conditions for a dc motor:

$$T_{e0} = B\,\Omega_{m0} + T_{L0} = kI_{f0}I_{a0}; \qquad V_{f0} = R_f I_{f0}; \qquad V_{t0} = R_{a0}I_{a0} + k\,\Omega_{m0}I_{f0}$$

Let a small-signal input produce quantities with additional subscripts 1, superimposed on the quiescent points, as shown below:

$$T_e = T_{e0} + T_{e1}; \qquad \omega_m = \Omega_{m0} + \omega_{m1}; \qquad i_f = I_{f0} + i_{f1}$$
$$v_t = V_{t0} + v_{t1}; \qquad i_a = I_{a0} + i_{a1}$$

When the above equations are substituted in the basic motor equations and the quiescent quantities are subtracted, show that the following equations result when the products $(i_{f1}\,i_{a1})$ and $(i_{f1}\,\omega_{m1})$ are neglected:

$$T_{e1} = (Jp + B)\omega_{m1} + T_{L1} = k(I_{f0}I_{a1} + I_{a0}i_{f1})$$
$$v_{f1} = (L_{ff}p + R_f)i_{f1}$$
$$v_{t1} = (L_{aq}p + R_a)i_{a1} + k(I_{f0}\omega_{m1} + \Omega_{m0}i_{f1})$$

Develop a linearized block diagram based on the above equations for a dc shunt motor.

11–26 Pulsating loads produce oscillations in the torque that are superimposed on the average value of torque and are accompanied by oscillations in the speed and armature current. If the torque oscillations are assumed to vary sinusoidally at an angular frequency of r rad/s, the torque and voltage can be expressed as phasors by substituting the imaginary quantity jr for p in the equations developed in Problem 11–25.

$$(B + jrJ)\bar{\Omega}_{m1} + \bar{T}_{L1} = k(I_{f0}\bar{I}_{a1} + I_{a0}\bar{I}_{f1})$$
$$\bar{V}_{f1} = (R_f + jrL_{ff})\bar{I}_{f1}$$
$$\bar{V}_{t1} = (R_a + jrL_{aq})\bar{I}_{a1} + k(I_{f0}\bar{\Omega}_{m1} + \Omega_{m0}\bar{I}_{f1})$$

where $\bar{\Omega}_{m1}, \bar{T}_{L1}, \bar{I}_{a1}, \bar{V}_{t1}, \bar{I}_{f1}$, and $\bar{V}_{f1}$ are phasors. Note that, if the voltage applied to the field is constant, $\bar{V}_{f1} = 0$ and $\bar{I}_{f1} = 0$ regardless of variations in the armature current, because there is no inductive coupling between the armature and the field. Apply the above analysis with phasor relationships for small oscillations in solving the following problem.

A 200-hp, 250-V, dc shunt motor has the following parameters:

$$R_a = 0.025 \ \Omega; \qquad L_{aq} = 0.001 \ \text{H}; \qquad R_f = 25 \ \Omega; \qquad L_{ff} = 10 \ \text{H}; \qquad k = 0.5 \ \text{H};$$

The polar moment of inertia of the motor armature and the load together is 40 kg·m^2;

The load torque in N·m is $T_L = 2{,}500 + 600\sin 20t$

The no-load speed is 480 rpm. The rotational losses of the motor are negligible. If the motor is supplied from a 250-V dc source of negligible impedance, find the motor-armature current and the speed as functions of time.

11–27   Consider the armature of the motor in Problem 11–26 to be energized at a constant terminal voltage of 250 V. Let the voltage applied to the field be $v_f = 150 + 60 \sin 3t$, and let the load be purely inertial with the combined polar moment of inertia of the motor armature and the load of 40 kg·m^2.

a.   Determine the quiescent values of the field current, armature current, and speed.

b.   Express the corresponding variable components in part a as phasors and as functions of time.

c.   Obtain the expressions for the corresponding total quantities in part a as functions of time.

11–28   Repeat Problem 11–27 but let the voltage applied to the field be constant at 250 V and the voltage applied to the armature be

$$v_t = 150 + 60 \sin 3t$$

11–29   a.   For the case of a separately excited dc generator, consider Equations 11.1.3 and 11.1.7 through 11.1.9, and develop the state equations similar to those of a motor.

b.   Suppose ω_m provided by the prime mover is a constant; then comment on the nature of the system of equations.

c.   Develop a corresponding analog-computer simulation diagram.

11–30   Develop the block diagram of a simple metadyne generator, shown in Figure 11.3.1 of the text, with v_f as the input and i_d as the output, while operating at rated speed.

Also obtain the block diagram of the basic amplidyne corresponding to Equation 11.3.4.

11–31   A 5-kW, 240-V amplidyne driven at its rated speed of 1,800 rpm has the following parameters:

$$R_f = 1{,}000\ \Omega; \qquad L_{ff} = 200\ \text{H}; \qquad L_{af} = 1.5\ \text{H}; \qquad R_{aq} = 1\ \Omega; \qquad L_{aq} = 0.3\ \text{H}$$

Assuming the machine to be completely compensated, express the open-circuit output voltage e_{ad} as a function of time after a constant voltage $v_f = 20$ V is applied to the field winding.

11–32   Consider the steady-state operation of the basic metadyne generator and the amplidyne with all currents and voltages constant. Obtain the three voltage equations for the field, d-axis, and q-axis circuits. Express the output voltage in terms of the field current and the output current.

11–33   A cross-field generator has the following parameters:

Inductance of d-axis or q-axis armature windings = 0.05 H

Resistance of d-axis or q-axis armature windings = 1 Ω

Resistance of compensating winding = 1 Ω

Mutual inductance between field winding and the d-axis armature winding = 1 H

If the machine rotates at a constant speed of 200 rad/s, plot curves of output current versus output voltage

a.   When operating as a fully-compensated amplidyne with a field current of (i) 0.1 A, (ii) 0.05 A.

b.   When operating as a metadyne with a field current of (i) 1 A, (ii) 0.5 A.

11–34   A two-pole, fully compensated amplidyne has the following parameters:

Field inductance = 25 H;   Field resistance = 100 Ω;

Armature inductance = 0.05 H;   Armature resistance = 1 Ω;

Compensating winding resistance = 1 Ω;   Field/armature mutual inductance = 1 H

If a step-function voltage of 100 V is applied to the field, determine the output current as a function of time if the machine is connected to a load of 25-Ω resistance and 5-H inductance. Assume the speed to remain constant at 200 rad/s.

12

Power Semiconductor-Controlled Drives

Power electronics deals with the applications of solid-state electronics for the control and conversion of electric power. Conversion techniques require switching power semiconductor devices on and off. The development of solid-state motor drive packages has progressed to the point at which they can solve practically any power-control problem. This chapter describes fundamentals common to all electric drives: dc drives fed by controlled rectifiers and choppers; squirrel-cage induction motor drives controlled by ac voltage controllers, inverters, and cycloconverters; slip-power controlled wound-rotor induction motor drives; and inverter-controlled and cycloconverter-controlled synchronous motor drives, including brushless dc and ac motor drives. Even though the detailed study of such power electronic circuits and components would require a book in itself, some familiarity becomes important to an understanding of modern motor applications. This chapter is only a modest introduction.

12.1 Introduction to Power Electronic Drives

The essential components of an electric drive controlled by a power semiconductor converter are shown in the block diagram of Figure 12.1.1.

The converter regulates the flow of power from the source to the motor in such a way that the motor speed-torque and speed-current characteristics become compatible with the load requirements. The low-voltage control unit, which may consist of integrated transistorized circuits or a microprocessor, is electrically isolated from the converter-motor circuit and controls the converter. The sensing unit required for closed-loop operation or protection, or both, is used to sense the power circuit's electrical parameters, such as converter current, voltage, and motor speed. The command signal forms an input to the control unit adjusting the operating point of the drive. The complete electric drive system shown in Figure 12.1.1 must be treated as an integrated system.

A motor operates in two modes—motoring and braking. Supporting its motion, it converts electrical energy to mechanical energy while motoring. In braking, while opposing the motion, it works as a generator converting mechanical energy to electrical energy, which is consumed in some part of the circuit. The motor can provide motoring and braking operations in both forward and reverse

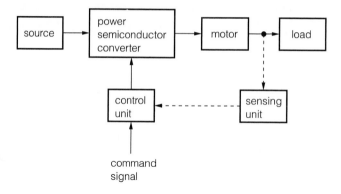

Figure 12.1.1 Essential components of an electric drive.

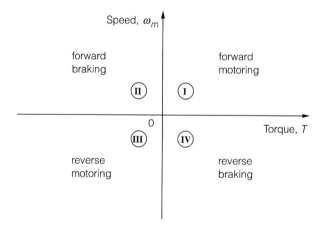

Figure 12.1.2 Four-quadrant operation of drives.

directions. Figure 12.1.2 illustrates the four-quadrant operation of drives. The continuous as well as transient torque and power limitations of a drive in the four quadrants of operation are shown in Figure 12.1.3 for speeds below and above *base speed* ω_{mb}, which is the highest drive speed at the rated flux.

Motors commonly used in variable speed drives are induction motors, dc motors, and synchronous motors. For the control of the motors, various types of converters are needed, as exemplified in Table 12.1.1. A variable speed drive can use a single converter or more than one. All converters have harmonics in their input and output. Some converters suffer from a poor power factor, particularly at low output voltages. The main advantages of converters are high efficiency, fast response, flexibility of control, easy maintenance, reliability, low weight and volume, less noise, and long life. The power semiconductor converters have virtually replaced the conventional power controllers such as mercury-arc rectifiers and magnetic amplifiers.

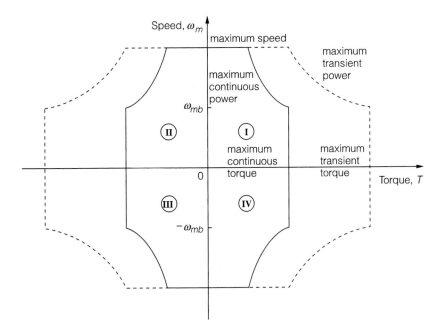

Figure 12.1.3 Continuous as well as transient torque and power limitations of a drive.

Table 12.1.1 Converters and their functions for the control of motors

Converter	Conversion Function	Applications
Controlled rectifiers	AC to variable dc	Control of dc motors, induction motors, and synchronous motors
Choppers	Fixed voltage dc to variable voltage dc	Control of dc motors and induction motors
AC voltage controllers	Fixed voltage ac to variable voltage ac at same frequency	Control of induction motors
Inverters (voltage source or current source)	DC to fixed or variable voltage and frequency ac, voltage or current sources	Control of induction motors and synchronous motors
Cycloconverters	Fixed voltage and frequency ac to variable voltage and frequency ac	Induction motors and synchronous motors

Most of the drive specifications are governed by the load requirements, which in turn depend on normal running needs, transient operational needs, and needs related to location and environment. Other specifications are governed by the available source and its capacity, as well as other aspects like harmonics, power factor, reactive power, regenerated power, and peak current.

Power Semiconductor Devices

Since the advent of the first thyristor or silicon-controlled rectifier (SCR) in 1957, tremendous advances have been made in the power semiconductor devices during the past two decades. The devices can broadly be divided into four types:

- Power diodes
- Thyristors
- Power bipolar junction transistors (BJTs)
- Power MOSFETs (metal-oxide semiconductor field-effect transistors)

Thyristors can be subdivided into seven categories:

- Forced-commutated thyristor
- Line-commutated thyristor
- Gate-turn-off thyristor (GTO)
- Reverse-conducting thyristor (RCT)
- Static-induction thyristor (SITH)
- Gate-assisted turn-off thyristor (GATT)
- Light-activated silicon-controlled rectifier (LASCR)

Taken from the literature, ratings of these devices are given in Table 12.1.2.

Before we can examine the power electronic circuits, we must understand what each device does.

The Diode.   A diode is a semiconductor device that conducts current in one direction only: from its anode to its cathode. Figure 12.1.4 shows the symbol for this device, while Figure 12.1.5 gives us the voltage-current characteristic. When a voltage is applied to the diode in the forward direction, a large current flow results. When a voltage is applied to the diode in the reverse direction, the current flow is limited to a very small value; however, if a large enough reverse voltage is applied, the diode will eventually breakdown and allow current to flow in the reverse direction. The maximum reverse voltage of a diode is known as its peak inverse voltage (*PIV*). Fast-recovery diodes are essential for high-frequency switching of power converters.

The Thyristor.   Figure 12.1.6 shows the symbol of a three-wire thyristor or SCR (silicon-controlled rectifier), while Figure 12.1.7 gives its voltage-current characteristic. A thyristor (with three terminals, called an anode, a cathode, and a gate) is a controlled rectifier or diode; its voltage-current characteristic with the gate lead open is the same as that of a PNPN diode, or a two-wire thyristor. Observe the following facts about a thyristor or SCR from Figure 12.1.7:

- It turns on when the voltage v_D applied to it exceeds a break-over voltage V_{BO}.
- Its breakover-voltage level can be controlled by its gate current i_G.
- It turns off when the current i_D passing through it drops below the holding current i_H.
- It blocks all current flow in the reverse direction until the maximum reverse voltage is exceeded.

Table 12.1.2 Typical Ratings of Power Semiconductor Devices

Type		Voltage/current rating	Switching time (μs)	On voltage/current*
Diodes	General purpose	3 kV/3.5 kA		1.6 V/10 kA
	High speed	3 kV/1 kA	2–5	3 V/3 kA
	Schottky	40 V/60 A	0.23	0.58 V/60 A
Forced-turned-off thyristors	Reverse blocking	3 kV/1 kA	400	2.5 V/10 kA
	High speed	1.2 kV/1.5 kA	20	2.1 V/4.5 kA
	Reverse blocking	2.5 kV/400 A	40	2.7 V/1.25 kA
	Reverse conducting	2.5 kV/1 kA/R400 A	40	2.1 V/1 kA
	GATT	1.2 kV/400 A	8	2.8 V/1.25 kA
	Light triggered	6 kV/1.5 kA	200–400	2.4 V/4.5 kA
TRIACs		1.2 kV/300 A		1.5 V/420 A
Self-turned-off thyristors	GTO	3.6 kV/600 A	25	2.5 V/1 kA
	SITH	4 kV/2.2 kA	6.5	2.3 V/400 A
Power transistors	Single	400 V/250 A	9	1 V/250 A
		400 V/40 A	6	1.5 V/49 A
		630 V/50 A	1.7	0.3 V/20 A
	Darlington	900 V/200 A	40	2 V
SITs Power MOSFETS		1.2 kV/10 A	0.55	1.2 Ω
		500 V/8.6 A	0.7	0.6 Ω
		1 kV/4.7 A	0.9	2 Ω
		500 V/10 A	0.6	0.4 Ω

Source: F. Harashima, "State of the Art on Power Electronics and Electrical Drives in Japan," 3rd IFAC Symposium on Control in Power Electronics and Electrical Drives, Lausanne, Switzerland, 1983, Tutorial Session and Survey Papers, pp. 23–33.

*Note: On-voltage is the on-state voltage drop of the device at the specified current.

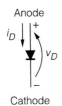

Figure 12.1.4 Typical symbol of a diode.

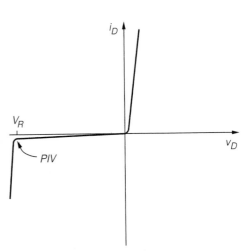

Figure 12.1.5 Diode voltage-current characteristic.

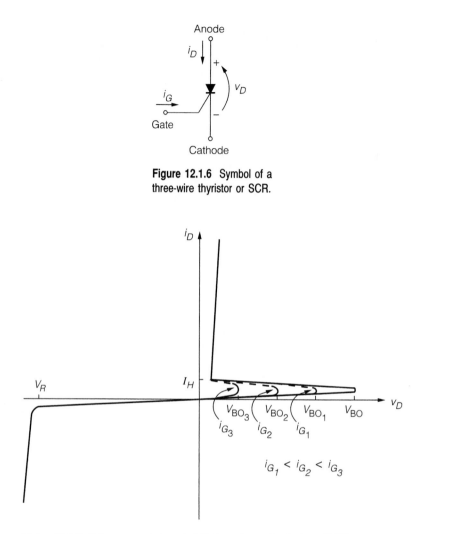

Figure 12.1.6 Symbol of a three-wire thyristor or SCR.

Figure 12.1.7 Voltage-current characteristic for a three-wire thyristor (SCR).

Once the thyristor is in a conduction mode, the gate circuit has no control and the thyristor continues to conduct. A conducting thyristor can be turned off by making the anode potential equal to or less than the cathode potential. The line-commutated thyristors are turned off due to the sinusoidal nature of the input voltage, while the forced-commutated thyristors are turned off by an extra circuit called commutation circuitry.

A TRIAC is a device that behaves like two thyristors connected in inverse parallel (back to back) and having only one gate terminal. Its symbol and current-voltage characteristic are shown in Figures 12.1.8 and 12.1.9, respectively. It can conduct in either direction once its breakover voltage is exceeded. It is widely used for low-power ac applications.

GTOs and SITHs are self-turned-off thyristors, which do not require any commutation circuit. They are turned on by applying a short positive pulse to the gates and turned off by applying a short negative

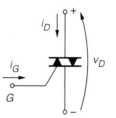

Figure 12.1.8 Symbol of a TRIAC.

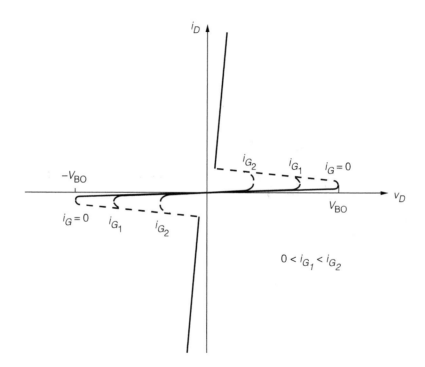

Figure 12.1.9 TRIAC voltage-current characteristic.

pulse to the gates. An RCT can be considered a thyristor with an inverse-parallel diode. RCTs and GATTs are widely used for high-speed switching, especially in traction applications. LASCRs are suitable for high-voltage power systems, especially in HVDC.

The Power Transistor (Power BJT). A bipolar transistor has three terminals: base, emitter, and collector. Its symbol is shown in Figure 12.1.10, while its voltage-current characteristic is given in Figure 12.1.11. Note that its collector current i_C is directly proportional to its base current i_B over a very wide range of collector-to-emitter voltages, v_{CE}. If the base-drive voltage is withdrawn, the transistor remains in the nonconduction or off mode.

By connecting two transistors in cascade to form a *Darlington power transistor,* a substantial improvement in gain is obtained, with a consequent reduction in the base drive.

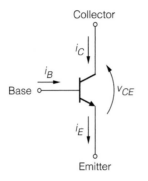

Figure 12.1.10 Symbol of a power transistor.

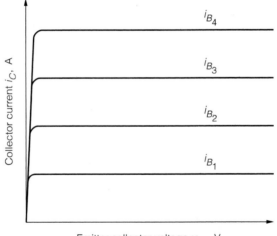

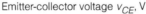

Figure 12.1.11 Power transistor voltage-current characteristic.

The Insulated Gate Bipolar Transistor (IGBT).   The IGBT is controlled by the voltage applied to a gate rather than by current flowing into the base as in the power transistor. With very high impedance of the control gate, the device is equivalent to the combination of a MOSFET and a power transistor. Its symbol is shown in Figure 12.1.12. Because it can switch much more rapidly than a conventional power transistor, it is used in high-power, high-frequency applications.

Figure 12.1.13 illustrates a comparison of the relative power-handling capabilities and speeds of SCRs, GTO thyristors, and power transistors (PTRs). Notice that power transistors can handle less power than either type of thyristor, but power transistors can switch more than ten times faster.

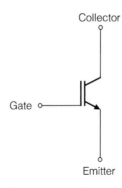

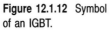

Figure 12.1.12 Symbol of an IGBT.

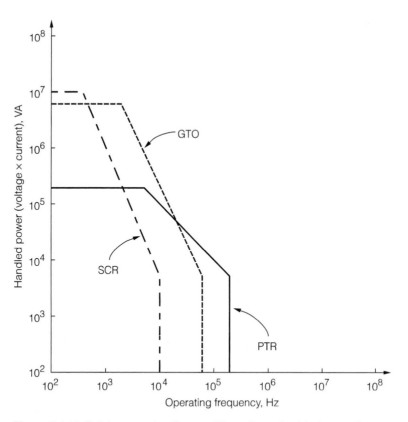

Figure 12.1.13 Relative power-handling capability and speeds of thyristors and power transistors.

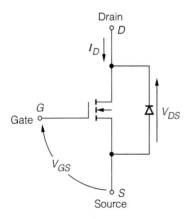

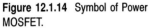

Figure 12.1.14 Symbol of Power
MOSFET.

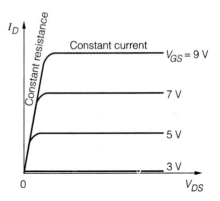

Figure 12.1.15 Drain current vs. drain-to-
source voltage characteristics of a power
MOSFET.

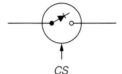

CS

Figure 12.1.16 Symbol
of a self-commutated
semiconductor switch.

The Power MOSFET.   The circuit symbol is shown in Figure 12.1.14 for an *n*-channel MOSFET. Unlike the power transistor, which is a current-controlled device, the MOSFET is a voltage-controlled device. The device has an integral antiparallel diode, which allows reverse current of the same magnitude as that of the main device. The I_D (drain current) versus V_{DS} (drain-to-source voltage) characteristics are shown in Figure 12.1.15, with V_{GS} (gate-to-source voltage) as a parameter. Observe the two distinct regions: the constant-resistance region and the constant-current region. As seen from Figure 12.1.15, the threshold voltage (above which appreciable drain current flows) is about 3 V.

The Self-Commutated Semiconductor Switch.   A common symbol for the self-commutated semi-conductor switch is shown in Figure 12.1.16. The control signal (either voltage or current) is denoted by CS, and the diode gives the direction in which the switch can conduct current. GTOs, power transistors, and MOSFETS are classified as self-commutated semi-conductor devices because they can be turned off by their respective control signals: a GTO by a gate pulse, a power transistor by a base drive, and a MOSFET by a gate-to-source voltage. A thyristor, on the other hand, is a naturally commutated device that cannot be turned off by its gate signal. A thyristor combined with a forced commutation circuit behaves like a self-commutated semiconductor device, however. The self-commutation capability makes its turn-off independent of the polarity of the source voltage, the load voltage, or the nature of load. Self-commutated semiconductor switches are suitable for applications in converters fed from a dc source, such as inverters and choppers.

In summary, Table 12.1.3 gives the symbols and the *v–i* characteristics of the commonly used power semiconductor devices.

Table 12.1.3 Symbols and v–i Characteristics of Some Power Semiconductor Devices

Device	Symbols	Characteristics
Diode		
Thyristor		
GTO		
TRIAC		
LASCR		
NPN BJT		
PNP BJT		
n-Channel MOSFET		
p-Channel MOSFET		

Control Characteristics of Power Semiconductor Switching Devices

Figure 12.1.17 illustrates the output voltages and control characteristics of some commonly used power switching devices. The switching devices can be classified as follows:

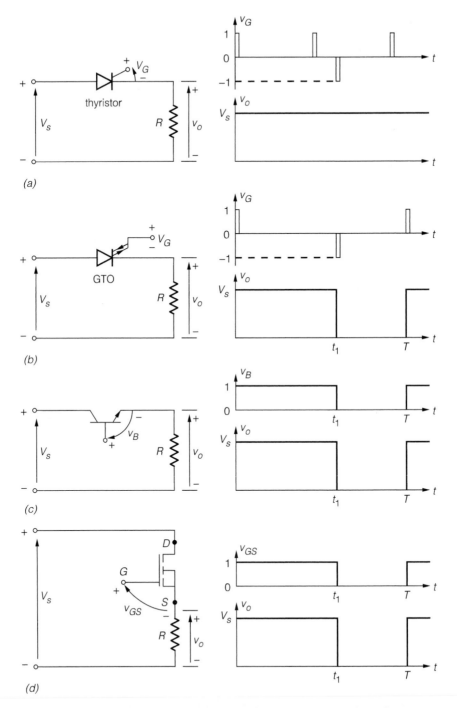

Figure 12.1.17 Output voltages and control characteristics of some commonly used power switching devices. (a) Thyristor switch. (b) GTO switch. (c) Transistor switch. (d) MOSFET switch.

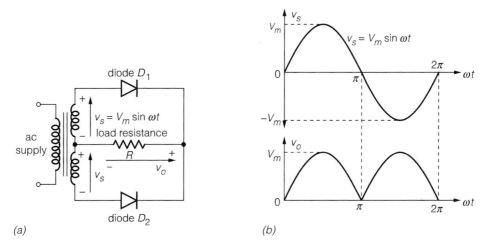

Figure 12.1.18 Single-phase rectifier circuit with diodes. (a) Circuit diagram. (b) Voltage waveforms.

- Uncontrolled turn on and off (diode)
- Controlled turn on and uncontrolled turn off (SCR)
- Controlled turn on and off (BJT, GTO, MOSFET)
- Continuous gate signal requirement (BJT, MOSFET)
- Pulse gate requirement (GTO, SCR)
- Bipolar voltage capability (SCR)
- Unipolar voltage capability (BJT, GTO, MOSFET)
- Bidirectional current capability (RCT, TRIAC)
- Unidirectional current capability (BJT, diode, GTO, MOSFET, SCR)

Power Electronic Circuits

These circuits can be classified as follows:

- Diode rectifiers
- ac-dc converters (controlled rectifiers)
- ac-ac converters (ac voltage controllers)
- dc-dc converters (dc choppers)
- dc-ac converters (inverters)
- Static switches (contactors), supplied by either ac or dc

Figure 12.1.18 shows a single-phase rectifier circuit converting ac voltage into a fixed dc voltage. It can be extended to three-phase supply. Figure 12.1.19 shows a single-phase ac-dc converter with two natural line-commutated thyristors. The average value of the output voltage is controlled by varying the conduction time of the thyristors. Three-phase input can also be converted.

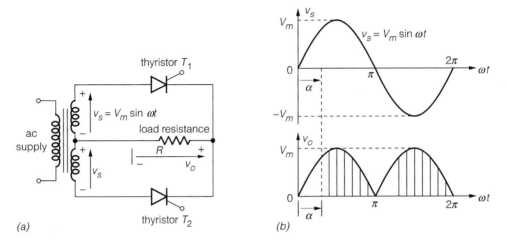

Figure 12.1.19 Single-phase ac-dc converter with two natural line-commutated thyristors. (a) Circuit diagrams. (b) Voltage waveforms.

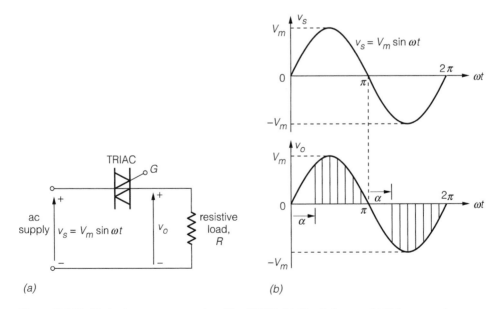

Figure 12.1.20 Single-phase ac-ac converter with a TRIAC. (a) Circuit diagram. (b) Voltage waveforms.

Figure 12.1.20 shows a single-phase ac-ac converter with a TRIAC to obtain a variable ac output voltage from a fixed ac source. The output voltage is controlled by changing the conduction time of the TRIAC.

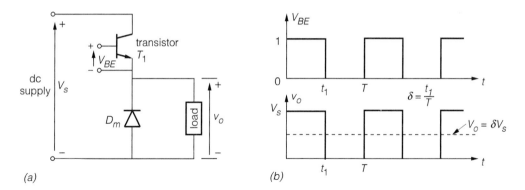

Figure 12.1.21 A dc-dc converter. (a) Circuit diagram. (b) Voltage waveforms.

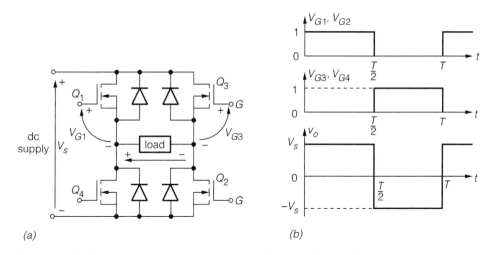

Figure 12.1.22 Single-phase dc-ac converter (inverter). (a) Circuit diagram. (b) Voltage waveforms.

Figure 12.1.21 shows a dc-dc converter in which the average output voltage is controlled by changing the conduction time t_1 of the transistor. The chopping period is T, the duty cycle of the chopper is δ, and the conduction time t_1 is given by δT.

Figure 12.1.22 shows a single-phase dc-ac converter, known as an inverter, in which the output voltage is controlled by varying the conduction time of transistors. Note that the voltage is of alternating form when transistors Q_1 and Q_2 conduct for one-half period and Q_3 and Q_4 conduct for the other half.

▌▌▌▌▌▌▌▌▌▌▌

EXAMPLE 12.1.1

Consider a diode circuit with an RLC load as shown in Figure 12.1.23, and analyze it for $i(t)$ when the switch S is closed at $t = 0$. Treat the diode to be ideal so that the reverse recovery time and the forward voltage drop are negligible. Allow for general initial conditions at $t = 0$ to have nonzero current and a capacitor voltage $v_C = V_0$.

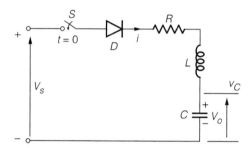

Figure 12.1.23 A diode circuit with RLC load.

Solution

The KVL equation for the load current is given by

$$L \frac{di}{dt} + Ri + \frac{1}{C} \int i \, dt + v_C \text{ (at } t = 0) = V_S$$

Differentiating and then dividing both sides by L, we obtain

$$\frac{d^2i}{dt^2} + \frac{R}{L} \frac{di}{dt} + \frac{i}{LC} = 0$$

Note that the capacitor will be charged to the source voltage V_s under steady-state conditions, when the current will be zero. While the forced component of the current is zero in the solution of the previous second-order homogeneous differential equation, we can solve for the natural component. The characteristic equation in the frequency domain is

$$s^2 + \frac{R}{L} s + \frac{1}{LC} = 0$$

whose roots are given by

$$s_{1,2} = -\frac{R}{2L} \pm \sqrt{\left(\frac{R}{2L}\right)^2 - \frac{1}{LC}}$$

For the second-order circuit, the damping factor α and the resonant frequency ω_0 are given by

$$\alpha = \frac{R}{2L}; \qquad \omega_0 = \frac{1}{\sqrt{LC}}$$

(Note: the ratio of α/ω_0 is known as *damping ratio, δ.*) Substituting, we get

$$s_{1,2} = -\alpha \pm \sqrt{\alpha^2 - \omega_0^2}$$

Three possible cases arise for the solution of the current that will depend on the values of α and ω_0:

Case 1: $\alpha = \omega_0$;   the roots are than equal, $s_1 = s_2$. The current is said to be *critically damped.* The solution of the current is of the form

$$i(t) = (A_1 + A_2t)e^{s_1t}$$

Case 2: $\alpha > \omega_0$;   the roots are unequal and real. The circuit is said to be *overdamped.* The solution is then

$$i(t) = A_1e^{s_1t} + A_2e^{s_2t}$$

Case 3: $\alpha < \omega_0$;   the roots are complex. The circuit is *underdamped.* Letting $s_{1,2} = -\alpha \pm j\omega_r$, where ω_r is known as the damped resonant frequency or *ringing frequency* given by $\sqrt{\omega_0^2 - \alpha^2}$, the solution takes the form

$$i(t) = e^{-\alpha t}(A_1 \cos \omega_r t + A_2 \sin \omega_r t)$$

Observe that the current consists of a damped or decaying sinusoidal. The constants A_1 and A_2 are determined from the initial conditions of the circuit.

The current waveform can be sketched taking the conduction time of the diode into account.

EXAMPLE 12.1.2

A single-phase half-wave rectifier circuit is shown in Figure 12.1.24 with a purely resistive load R.

a. As a function of ωt, plot v_s, v_L, i, and v_D for the range of 0 to 2π.
b. Determine the following values: (i) efficiency; (ii) form factor; (iii) ripple factor; (iv) transformer utilization factor; and (v) peak inverse voltage (PIV) of the diode.

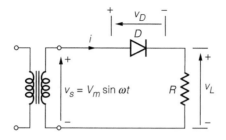

Figure 12.1.24

Solution

a. During the positive half cycle of the input voltage, the diode D conducts and the input voltage appears across the load. During the negative half cycle of the input voltage, the output voltage is

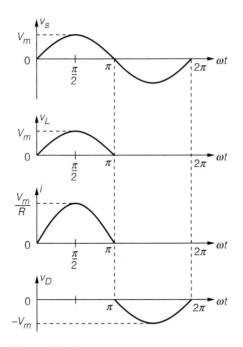

Figure 12.1.25

zero when the diode is said to be in a *blocking condition*. The waveforms of the input voltage are shown in Figure 12.1.25.

b. The average voltage V_{dc} is given by

$$V_{dc} = \frac{1}{T} \int_0^T v_L(t)\, dt$$

because $v_L(t) = 0$ for $T/2 \le t \le T$. In our case,

$$V_{dc} = \frac{1}{T} \int_0^{T/2} V_m \sin \omega t\, dt = -\frac{V_m}{\omega T}\left(\cos\frac{\omega T}{2} - 1\right)$$

Using the relationships $f = 1/T$ and $\omega = 2\pi f$, we obtain

$$V_{dc} = \frac{V_m}{\pi} = 0.318 V_m \qquad \text{and} \qquad I_{dc} = \frac{V_{dc}}{R} = \frac{0.318 V_m}{R}$$

The RMS value of a periodic wave form is given by

$$V_{rms} = \left[\frac{1}{T} \int_0^T v_L^2(t)\, dt\right]^{1/2}$$

For a sinusoidal voltage of $v_L(t) = V_m \sin \omega t$ for $0 \le t \le T/2$, the rms value of the output voltage is

$$V_{rms} = \left[\frac{1}{T}\int_0^{T/2} (V_m \sin \omega t)^2 \, dt\right]^{1/2} = \frac{V_m}{2} = 0.5V_m \quad \text{and} \quad I_{rms} = \frac{V_{rms}}{R} = \frac{0.5V_m}{R}$$

The output dc power is given by

$$P_{dc} = V_{dc}I_{dc} = \frac{(0.318V_m)^2}{R}$$

The output average power P_{ac} is given by

$$P_{ac} = V_{rms}I_{rms} = \frac{(0.5V_m)^2}{R}$$

(i)   The efficiency or the rectification ratio of the rectifier, which is a figure of merit used for comparison, is then

$$\eta = P_{dc}/P_{ac} = \frac{(0.318V_m)^2}{(0.5V_m)^2} = 0.404 \quad \text{or} \quad 40.4\%$$

(ii)   The form factor is $FF = V_{rms}/V_{dc}$, which gives a measure of the shape of output voltage. In our case,

$$FF = \frac{0.5V_m}{0.318V_m} = 1.57 \quad \text{or} \quad 157\%$$

(iii)   The ripple factor is $RF = V_{ac}/V_{dc}$ which gives a measure of the ripple content. In our case,

$$RF = \sqrt{(V_{rms}/V_{dc})^2 - 1} = \sqrt{FF^2 - 1} = \sqrt{1.57^2 - 1} = 1.21 \quad \text{or} \quad 121\%$$

(iv)   The transformer utilization factor is given by

$$TUF = \frac{P_{dc}}{V_s I_s}$$

where V_s and I_s are the rms voltage and rms current of the transformer secondary, respectively. In our case, the rms voltage of the transformer secondary is

$$V_s = \left[\frac{1}{T}\int_0^{T} (V_m \sin \omega t)^2 \, dt\right]^{1/2} = \frac{V_m}{\sqrt{2}} = 0.707 \, V_m$$

The rms value of the transformer secondary current is the same as that of the load:

$$I_s = \frac{0.5V_m}{R}$$

Hence,

$$TUF = \frac{0.318^2}{0.707 \times 0.5} = 0.286$$

(v) $PIV = V_m$, which is the peak reverse blocking voltage.

The *displacement factor* is given by $DF = \cos \phi$, where ϕ is the angle between the fundamental components of the input current and voltage.

The *harmonic factor* of the input current is given by

$$HF = \left(\frac{I_s^2 - I_1^2}{I_1^2} \right)^{1/2} = \left[\left(\frac{I_s}{I_1} \right)^2 - 1 \right]^{1/2}$$

where I_1 is the fundamental rms component of the input current.

The input *power factor* is given by

$$PF = \frac{I_1}{I_s} \cos \phi$$

For an ideal rectifier,

$$\eta = 1.00; \qquad V_{ac} = 0; \qquad FF = 1.0; \qquad RF = 0; \qquad TUF = 1.0; \qquad HF = 0; \qquad PF = 1.0$$

where V_{ac} is the effective value of the ac component of output voltage given by

$$V_{ac} = \sqrt{V_{rms}^2 - V_{dc}^2}$$

▬▬▬▬▬▬▬▬▬▬▬▬▬▬▬ ‖‖‖‖‖‖‖‖‖‖
12.2 Solid-State Control of DC Motors

Direct-current motors, which are easily controllable, have dominated the adjustable-speed drive field. The torque-speed characteristics of a dc motor can be controlled by adjusting the armature voltage or the field current, or by inserting resistance into the armature circuit. Solid-state motor controls are designed to use each of these modes for particular purposes. The control resistors in which much energy is wasted are being eliminated through the development of power semiconductor devices and the evolution of flexible and efficient converters. Thus, the inherently good controllability of a dc machine has been significantly increased in recent years. In this section, we will consider *rectifier control, chopper control,* and *closed-loop control* of dc motors. When a dc source of suitable and constant voltage is already available, designers can employ dc-to-dc converters or choppers. When only an ac source is available, phase-controlled rectifiers are used. When the steady-state accuracy requirement cannot be satisfied in an open-loop configuration, the drive is operated as a closed-loop system. Closed-loop rectifier drives are more widely used than chopper drives.

Rectifier Control of DC Motors

Controlled rectifier circuits are classified as fully controlled and half-controlled rectifiers, which are fed from either one-phase or three-phase supply. Figure 12.2.1 shows a fully controlled, rectifier-fed, separately excited dc motor drive and its characteristics. A transformer might be required if the motor voltage rating is not compatible with the ac source voltage. To reduce ripple in the motor current, a filter inductor can be connected in series between the rectifier and the motor armature. The field can be supplied from the same ac source supplying the armature, through a transformer and a diode bridge or a controlled rectifier. While single-phase controlled rectifiers are used up to a rating of 10 kW, and in special cases even up to 50 kW, three-phase controlled rectifiers are used for higher power ratings.

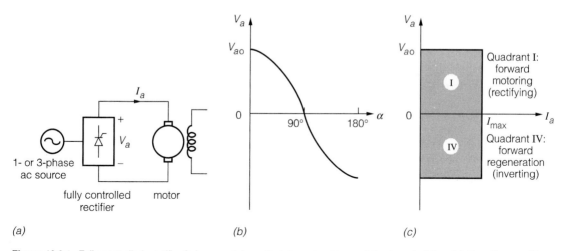

Figure 12.2.1 Fully controlled, rectifier-fed, separately excited dc motor drive and its characteristics. (a) Line diagram. (b) Output voltage versus firing angle curve. (c) Quadrants of operation.

V_a and I_a in Figure 12.2.1 denote the average values of the converter output voltage and current, respectively. Assuming continuous conduction when the armature currents flows continuously without becoming zero for a finite time interval, the variation of V_a with the firing angle α is shown in Figure 12.2.1b. Providing operation in the first and fourth quadrants of the V_a-I_a plane, as shown in Figure 12.2.1c, the fully controlled rectifiers are two-quadrant converters. I_{max} is the rated rectifier current. In quadrant 4, the rectifier works like a line-commutated inverter with a negative output voltage, and the power flows from the load to the ac source.

Let us now consider the single-phase, fully controlled, rectifier-fed, separately excited dc motor shown in Figure 12.2.2a. Note that the armature has been replaced by its equivalent circuit, in which R_a and L_a respectively represent the armature-circuit resistance and inductance (including the effect of a filter, if connected), and E is the back emf.

Figure 12.2.2b shows the source voltage and thyristor firing pulses. The pair T_1 and T_3 receives firing pulses from α to π, and the pair T_2 and T_4 receives firing pulses from $(\pi + \alpha)$ to 2π.

Only the *continuous conduction* mode of operation of the drive for motoring and regenerative braking will be considered here. The angle α can be greater or less than γ, which is the angle at which the source voltage v_s is equal to the back emf E.

$$\gamma = \sin^{-1}(E/V_m) \tag{12.2.1}$$

where V_m is the peak value of the supply voltage. For the case of $\alpha < \gamma$, waveforms are shown in Figure 12.2.3a for the motoring operation. It is possible to turn on thyristors T_1 and T_3 because $i_a > 0$, even though $v_s < E$. The same is true for thyristors T_2 and T_4.

When T_1 and T_3 conduct during the interval $\alpha \le \omega t \le (\pi + \alpha)$, the following volt-ampere equation holds:

$$v_s = E + i_a R_a + L_a \frac{di_a}{dt} \tag{12.2.2}$$

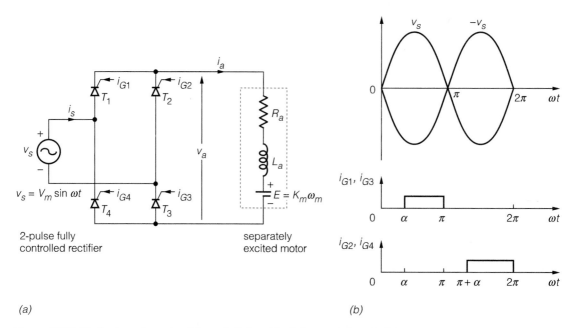

Figure 12.2.2 Single-phase, two-pulse, fully controlled, rectifier-fed, separately excited dc motor.

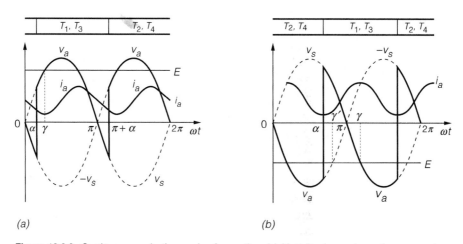

Figure 12.2.3 Continuous conduction mode of operation. (a) Motoring (case shown for $\alpha < \gamma$.) (b) Regenerative braking (case shown for $\alpha < \gamma'$.)

Multiplying both sides by $i_a \Delta t$, where Δt is a small time interval, we obtain

$$v_s i_a \Delta t = E i_a \Delta t + i_a^2 R_a \Delta t + L_a i_a \left(\frac{di_a}{dt} \right) \Delta t \tag{12.2.3}$$

in which the terms can be identified as the energy supplied or consumed by the respective elements. Figure 12.2.3b corresponds to the regenerative braking operation, in which

$$\gamma' = \pi - \gamma = \pi - \sin^{-1}(|E| / V_m) \tag{12.2.4}$$

and α can be greater or less than γ'.

During the *duty interval*, when T_1 and T_3 conduct, Equation 12.2.2 describes the motor operation with $E = K_m \omega_m$; thus

$$v_a = L_a \frac{di_a}{dt} + R_a i_a + K_m \omega_m = V_m \sin \omega t \tag{12.2.5}$$

Similarly, when T_2 and T_4 conduct,

$$v_a = L_a \frac{di_a}{dt} + R_a i_a + K_m \omega_m = -V_m \sin \omega t \tag{12.2.6}$$

where ω is the supply frequency, ω_m is the motor speed, and K_m is the motor back emf constant.

From α to $(\pi + \alpha)$ in the output voltage waveform of Figure 12.2.3a, Equation 12.2.5 holds and its solution can be found as

$$i_a(\omega t) = \frac{V_m}{z} \sin(\omega t - \psi) - \frac{K_m \omega_m}{R_a} + K_1 e^{-t/\tau_a}, \quad \text{for } \alpha \leq \omega t \leq (\pi + \alpha) \tag{12.2.7}$$

where

$$z = \left[R_a^2 + (\omega L_a)^2 \right]^{1/2} \tag{12.2.8}$$

$$\tau_a = L_a / R_a \tag{12.2.9}$$

$$\psi = \tan^{-1}(\omega L_a / R_a) \tag{12.2.10}$$

and K_1 is a constant.

The first term on the right-hand side of Equation 12.2.7 is due to the ac source; the second term is due to the back emf; and the third represents the combined transient component of the ac source and back emf. In the steady state, however,

$$i_a(\alpha) = i_a(\pi + \alpha) \tag{12.2.11}$$

Subject to the constraint in Equation 12.2.11, the steady-state expression of current can be obtained. Flux being a constant, recall that the average motor torque depends only on the average value or the dc component of armature current, while the ac components produce only pulsating torques with zero-average value. Thus the motor torque is given by

$$T_a = K_m I_a \tag{12.2.12}$$

To obtain the average value of I_a under steady state, we can use the following equation:

$$\text{Average motor voltage } V_a = \text{average voltage drop across } R_a$$
$$+ \text{ average voltage drop across } L_a \tag{12.2.13}$$
$$+ \text{ back emf}$$

in which

$$V_a = \frac{1}{\pi} \int_\alpha^{\pi+\alpha} V_m \sin(\omega t)\, d(\omega t) = \frac{2V_m}{\pi} \cos \alpha \tag{12.2.14}$$

The rated motor voltage will be equal to the maximum average terminal voltage, $2V_m/\pi$.

$$\text{Average drop across } R_a = \frac{1}{\pi} \int_\alpha^{\pi+\alpha} R_a i_a(\omega t)\, d(\omega t) = R_a I_a \tag{12.2.15}$$

$$\text{Average drop across } L_a = \frac{1}{\pi} \int_\alpha^{\pi+\alpha} L_a \left(\frac{di_a}{dt}\right) d(\omega t) = \frac{\omega}{\pi} \int_{i_a(\alpha)}^{i_a(\pi+\alpha)} L_a\, di_a$$

$$= \frac{\omega L_a}{\pi} \left[i_a(\pi + \alpha) - i_a(\alpha)\right] = 0 \tag{12.2.16}$$

Substituting, we obtain

$$V_a = I_a R_a + K_m \omega_m \tag{12.2.17}$$

for the steady-state operation of a dc motor fed by any converter. From Equations 12.2.14 and 12.2.17, it follows that

$$I_a = \frac{(2V_m/\pi) \cos \alpha - K_m \omega_m}{R_a} \tag{12.2.18}$$

Substituting in Equation 12.2.12, we get the following equation after rearranging the terms:

$$\omega_m = \frac{2V_m}{\pi K_m} \cos \alpha - \frac{R_a}{K_m^2} T_a \tag{12.2.19}$$

which is the relationship between the speed and torque under steady state. The ideal no-load speed ω_{m0} is obtained when I_a is equal to zero.

$$\omega_{m0} = V_m/K_m, \quad 0 \le \alpha \le \pi/2 \tag{12.2.20}$$

$$\omega_{m0} = V_m \sin \alpha/K_m, \quad \pi/2 \le \alpha \le \pi \tag{12.2.21}$$

For torques less than the rated value, a low-power drive operates predominantly in the *discontinuous conduction* mode, for which a zero-armature-current interval exists besides the duty interval. With continuous conduction, as seen from Equation 12.2.19, the speed-torque characteristics are parallel straight lines whose slope depends on R_a, the armature circuit resistance. A filter inductor is sometimes included to reduce the discontinuous conduction zone, although such an addition will lead to an increase of losses, armature circuit time constant, noise, cost, weight, and volume of the drive.

‖‖‖‖‖‖‖‖‖ ▬▬▬▬▬▬▬▬▬

EXAMPLE 12.2.1

Consider a 3-hp, 220-V, 1800-rpm separately excited dc motor controlled by a single-phase fully-controlled rectifier with an ac source voltage of 230 V at 60 Hz. Assume that the full-load efficiency of the motor is 88%, and enough filter inductance is added to ensure continuous conduction for any torque greater than 25% of rated torque. The armature circuit resistance is 1.5 ohms.

a. Determine the value of the firing angle to obtain rated torque at 1,200 rpm.
b. Compute the firing angle for the rated braking torque at −1,800 rpm.
c. With armature circuit inductance of 30 mH, calculate the motor torque for $\alpha = 60°$ and 500 rpm, assuming that the motor operates in the continuous conduction mode.
d. Find the firing angle corresponding to a torque of 35 N·m and speed of 480 rpm, assuming continuous conduction.

Solution

a. $V_m = \sqrt{2}(230) = 325.27$ V

$$I_a = \frac{3 \times 746}{0.88(220)} = 11.56 \text{ A}$$

$E = 220 - (11.56 \times 1.5) = 202.66$ V at rated speed of 1,800 rpm

$$\omega_m = \frac{1,800 \times 2\pi}{60} = 188.57 \text{ rad/s}$$

$$K_m = E/\omega_m = \frac{202.66}{188.57} = 1.075$$

For continuous conduction, Equation 12.2.18 holds:

$$\frac{2V_m}{\pi} \cos \alpha = I_a R_a + K_m \omega_m = I_a R_a + E$$

At rated torque,

$$I_a = 11.56 \text{ A}$$

$$\text{Back emf at 1,200 rpm} = E_1 = \frac{1,200}{1,800} \times 202.66 = 135.11$$

Substituting these values, one gets

$$\frac{2 \times 325.27}{\pi} \cos \alpha = (11.56)(1.5) + 135.11 = 152.45$$

or

$$\cos \alpha = \frac{152.45 \times \pi}{2 \times 325.27} = 0.7365 \quad \text{or} \quad \alpha = 42.6°$$

b. At −1,800 rpm,

$$E = -202.66 \text{ V}$$

So it follows that

$$\frac{2 \times 325.27}{\pi} \cos \alpha = (11.56)(1.5) - 202.66 = -185.32$$

or

$$\cos \alpha = -\frac{185.32 \times \pi}{2 \times 325.27} = -0.8953 \quad \text{or} \quad \alpha = 153.5°$$

c. From part a, $K_m = 1.075$. Corresponding to a speed of 500 rpm,

$$\omega_m = \frac{500 \times 2\pi}{60} = 52.38 \text{ rad/s}$$

From Equation 12.2.9, we have

$$52.38 = \frac{2 \times 325.27}{\pi \times 1.075} \cos 60° - \frac{1.5}{(1.075)^2}T_a \quad \text{or} \quad T_a = 33.82 \text{ N} \cdot \text{m}$$

d. From part a, $K_m = 1.075$. From Equation 12.2.19, we obtain

$$\frac{480 \times 2\pi}{60} = \frac{2\pi \times 325.27}{\pi \times 1.075} \cos \alpha - \frac{1.5}{(1.075)^2} \times 35$$

or

$$\cos \alpha = 0.497 \quad \text{or} \quad \alpha = 60.2°$$

The most widely used dc drive is the three-phase, fully-controlled, six-pulse, bridge-rectifier-fed, separately excited dc motor drive shown in Figure 12.2.4. With a phase difference of 60°, the firing of thyristors occurs in the same sequence as they are numbered. The line commutation of an even-numbered thyristor takes place with the turning on of the next even-numbered thyristor, and similarly for odd-numbered thyristors. Thus, each thyristor conducts for 120° and only two thyristors (one odd-numbered and one even-numbered) conduct at a time.

Considering the continuous conduction mode for the motoring operation, as shown in Figure 12.2.5, for the converter output voltage cycle form $\omega t = \alpha + \frac{\pi}{3}$ to $\omega t = \alpha + \frac{2\pi}{3}$,

$$V_a = \frac{3}{\pi}\int_{\alpha+\pi/3}^{\alpha+2\pi/3} V_m \sin \omega t \, d(\omega t) = \frac{3}{\pi}V_m \cos \alpha = V_{ao} \cos \alpha \tag{12.2.22}$$

where the line voltage $v_{AB} = V_m \sin \omega t$ is taken as the reference. From Equations 12.2.12, 12.2.17, and 12.2.22, we get

$$\omega_m = \frac{3V_m}{\pi K_m}\cos \alpha - \frac{R_a}{K_m^2}T_a \tag{12.2.23}$$

For normalization, taking the base voltage V_B as the maximum-average converter output voltage $V_{ao} = 3V_m/\pi$, and the base current as the average motor current (that will flow when $\omega_m = 0$ and $V_a = V_B$) $I_B = V_B/R_a = 3V_m/\pi R_a$, the normalized speed and torque are given by

$$\text{Speed} = \omega_{mn} = \frac{E}{V_B} = \frac{\pi E}{3V_m} \tag{12.2.24}$$

$$\text{Torque} = T_{an} = I_{an} = \frac{\pi R_a}{3V_m}(I_a) \tag{12.2.25}$$

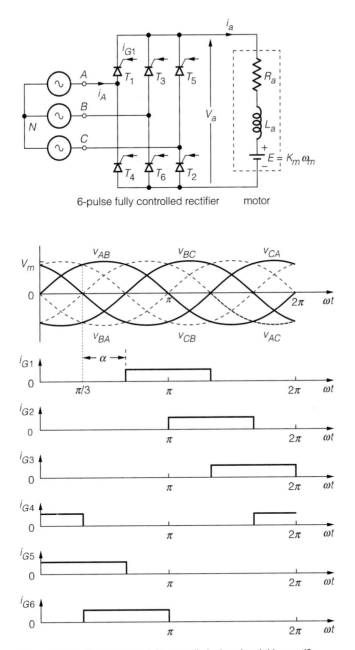

Figure 12.2.4 Three-phase, fully controlled, six-pulse, bridge-rectifier-fed, separately excited dc motor.

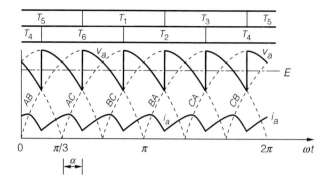

Figure 12.2.5 Continuous conduction mode for the motoring operation.

EXAMPLE 12.2.2

Consider the 220-V, 1,800-rpm dc motor of Example 12.2.1, controlled by a three-phase fully controlled rectifier from an ac source of 60 Hz. The armature-circuit resistance and inductance are 1.5 Ω and 30 mH, respectively.

a. When the motor is operating in continuous conduction, find the ac source voltage required to get rated voltage across the motor terminals.

b. With the ac source voltage obtained in part a, compute the motor speed corresponding to $\alpha = 60°$ and $T_a = 25$ N·m assuming continuous conduction.

c. Let the motor drive a load whose torque is constant and independent of speed. The minimum value of the load torque is 1.2 N·m. Calculate the inductance that must be added to the armature circuit to get continuous conduction for all operating points, given that, for $\psi = \tan^{-1}(\omega L_a/R_a) = 1.5$ radians, and $T_{an} > 0.006$, all points on the $T_{an}-\omega_{mn}$ plane lie on the right of the boundary between continuous and discontinuous conductions.

Solution

a. From Equation 12.2.22 with $\alpha = 0$, we get

$$220 = \frac{3\sqrt{2}V}{\pi} \quad \text{or} \quad V = 163 \text{ V line-to-line}; \quad V_m = 230.5 \text{ V}$$

b. From Equation 12.2.23, we obtain

$$\omega_m = \frac{3 \times 230.5 \times \cos 60°}{\pi \times 1.075} - \frac{1.5}{(1.075)^2}25 = 69.9 \text{ rad/s} = 667 \text{ rpm}$$

c. From Equation 12.2.25, the normalized torque corresponding to 1.2 N·m

$$T_{an} = \frac{\pi R_a}{3V_m}\left(\frac{T_a}{K_m}\right) = \frac{\pi \times 1.5}{3 \times 230.5}\left(\frac{1.2}{1.075}\right) = 0.0076$$

The straight line $T_{an} = 0.0076$ is on the right of the boundary for

$$\psi = 1.5 \text{ rad} = \tan^{-1}\frac{\omega L}{R_a}$$

Therefore,

$$L_a = \frac{R_a}{\omega} \tan \psi = \frac{1.5}{2\pi \times 60} \tan 85.9° = 55.5 \text{ mH}$$

External inductance needed = 55.5 − 30 = 25.5 mH

With the availability of self-commutated semiconductor switches, pulse-width (either equal or sinusoidal) modulation techniques are being employed for controlled rectifiers. The dual converter with simultaneous control is shown in Figure 12.2.6, using armature current reversal for multiquadrant operation. The current control forms an integral part of such drives in order to limit the current within safe values. Schemes employing field current reversal have also been developed for multiquadrant operation. Controlled flywheeling[1] is implemented for improving the power factor, reducing the armature-current ripple and discontinuous-conduction region.

Chopper Control of DC Motors

For the control of dc motors, choppers offer advantages such as high efficiency, control flexibility, light weight, small size, quick response, and regeneration down to very low speeds. Let us consider the steady-state analysis and performance of a separately excited dc motor fed by a one-quadrant step-down chopper. Let it be class A, which can vary the output voltage from V to zero according to $V_a = \delta V$, where δ is the *duty ratio*. Figure 12.2.7 shows the drive scheme in which an L–C filter connected between the source and the chopper reduces fluctuations in the source current and voltage.

Idealized steady-state output voltage and armature current waveforms are shown in Figure 12.2.8, for the chopper operating in continuous conduction mode. Let us now consider steady-state analysis for continuous conduction.

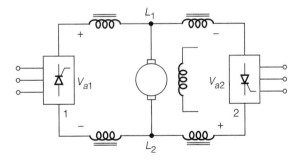

Figure 12.2.6 Dual converter with simultaneous control using armature current reversal.

---

[1] W. Drurry, *et al.,* "Performance of thyristor bridge converters employing flywheeling," *Proc. IEEE,* 127B, no.4 (July 1980): 268–76; P. S. Bhat and G. K. Dubey, "Performance and analysis of dc motor fed by 1-phase converter with controlled flywheeling," *IEEE Journal* 65, (April 1985): 167–73.

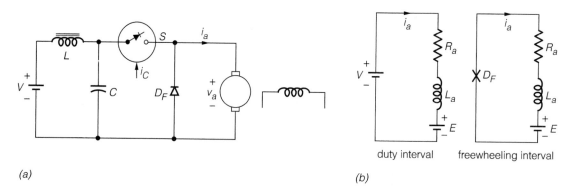

(a)

(b)

Figure 12.2.7 Class-A chopper-controlled, separately excited dc motor. (a) Chopper drive. (b) Equivalent circuits.

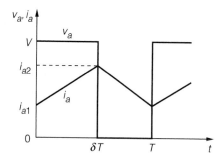

Figure 12.2.8 Voltage and current waveforms.

Time-Ratio Control (TRC). For the interval $0 \leq t \leq \delta T$, from Figure 12.2.7b,

$$R_a i_a + L_a \frac{di_a}{dt} + E = V \tag{12.2.26}$$

With initial condition $i_a(0) = i_{a1}$, Equation 12.2.26 yields

$$i_a = \frac{V - E}{R_a} \left(1 - e^{-t/\tau_a}\right) + i_{a1} e^{-t/\tau_a} \tag{12.2.27}$$

where $\tau_a = L_a/R_a$. The current i_{a2} at the end of the duty interval is given by

$$i_{a2} = \frac{V - E}{R_a} \left(1 - e^{-\delta T/\tau_a}\right) + i_{a1} e^{-\delta T/\tau_a} \tag{12.2.28}$$

For the freewheeling interval $\delta T \leq t \leq T$, from Figure 12.2.7b,

$$R_a i_a + L_a \frac{di_a}{dt'} + E = 0 \tag{12.2.29}$$

where $t' = t - \delta T$. With initial current of i_{a2} at $t' = 0$, Equation 12.2.29 yields

$$i_a = -\frac{E}{R_a}\left(1 - e^{-t'/\tau_a}\right) + i_{a2}e^{-t'/\tau_a} \tag{12.2.30}$$

For $t' = (1 - \delta)T$, $i_a = i_{a1}$ in the steady state. Thus we obtain

$$i_{a1} = -\frac{E}{R_a}\left(1 - e^{-(1-\delta)T/\tau_a} + i_{a2}e^{-(1-\delta)T/\tau_a}\right) \tag{12.2.31}$$

Solving Equations 12.2.28 and 12.2.31, we get

$$i_{a1} = \frac{V}{R_a}\left(\frac{e^{\delta T/\tau_a} - 1}{e^{T/\tau_a} - 1}\right) - \frac{E}{R_a} \tag{12.2.32}$$

$$i_{a1} = \frac{V}{R_a}\left(\frac{1 - e^{-\delta T/\tau_a}}{1 - e^{-T/\tau_a}}\right) - \frac{E}{R_a} \tag{12.2.33}$$

The current ripple Δi_a is the given by

$$\Delta i_a = \frac{i_{a2} - i_{a1}}{2} = \frac{V}{2R_a}\left[\frac{1 + e^{T/\tau_a} - e^{\delta T/\tau_a} - e^{(1-\delta)T/\tau_a}}{e^{T/\tau_a} - 1}\right] \tag{12.2.34}$$

Taking the steady-state average voltage drop across the inductance to be zero,

$$V_a = E + R_a I_a \tag{12.2.35}$$

where V_a and I_a are the average values of armature terminal voltage and current, respectively, given by

$$V_a = \frac{1}{T}\int_0^T v_a\,dt = \frac{1}{T}\int_0^{\delta T} V\,dt = \delta V = E + I_a R_a \tag{12.2.36}$$

or $I_a = (\delta V - E)/R_a$. Therefore, the average motor torque T_a is given by

$$T_a = K_m I_a \tag{12.2.37}$$

and the speed is

$$\omega_m = \frac{\delta V}{K_m} - \frac{R_a}{K_m^2}T_a \tag{12.2.38}$$

by using $E = K_m \omega_m$.

EXAMPLE 12.2.3

Consider a 230-V, separately excited dc motor with an armature resistance of 1.5 Ω, controlled by a chopper circuit at a frequency of 400 Hz and an input voltage of 230 V. If the armature takes 20 A when driving a load at 600 rpm with constant torque, find the duty ratio to reduce the speed to 400 rpm while maintaining constant load torque.

Solution

At 600 rpm, with $\delta = 1$, $E = V_a - I_a R_a = 230 - 20(1.5) = 200$ V.

At 400 rpm, $E_1 = 200 \times 400/600 = 133$ V.

The average chopper output voltage $V_{a1} = E_1 + I_a R_a = 133 + 20(1.5) = 163$ V. Therefore,

$$\delta = V_{a1}/V = 163/230 = 0.71$$

Current-Limit Control (CLC).   The values of δ and T are adjusted so that the current varies between prescribed limits i_{a1} and i_{a2}. Equation 12.2.28 yields

$$\delta T = \tau_a \ln \left(\frac{V - E - R_a i_{a1}}{V - E - R_a i_{a2}} \right) \tag{12.2.39}$$

Equation 12.2.31 yields

$$(1 - \delta)T = -\tau_a \ln \left(\frac{E + R_a i_{a1}}{E + R_a i_{a2}} \right) \tag{12.2.40}$$

Adding the last two equations, we get

$$T = \tau_a \ln \left[\left(\frac{V - E - R_a i_{a1}}{V - E - R_a i_{a2}} \right) \left(\frac{E + R_a i_{a2}}{E + R_a i_{a1}} \right) \right] \tag{12.2.41}$$

With T derived, δ can be determined from Equation 12.2.39. Then I_a and T_a can be obtained from Equations 12.2.36 and 12.2.37, respectively. The CLC yields constant-torque characteristics with speed-torque curves parallel to the speed axis.

Multiquadrant Control of Chopper-Fed DC Motors.   Figure 12.2.9 shows a schematic diagram of a Class E chopper circuit for the four-quadrant operation. With S_2 kept closed continuously, controlling S_1 and S_4, we obtain a two-quadrant chopper operating in quadrants 1 and 2 with a variable positive terminal voltage and the armature current in either direction. With S_3 kept closed continuously, controlling S_1 and S_4, a two-quadrant chopper is obtained operating in quadrants 3 and 4 with a variable negative terminal

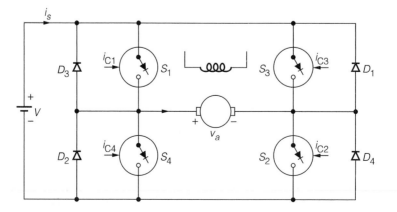

Figure 12.2.9 Schematic diagram of a Class E chopper circuit for four-quadrant operation.

voltage and the armature current in either direction. Other and better methods of control exist, but they are not within the scope of this book.

Closed-Loop Control of DC Drives.   Closed-loop rectifier, variable-speed drives are more widely used than chopper drives. They are adopted when open-loop configurations cannot meet steady-state accuracy needs of the drive. Single quadrant and multiquadrant drives have been developed, but they are to be outside the scope of this text.

12.3 Solid-State Control of Induction Motors

For our next discussions, you might find it helpful to review Section 7.5. The speed-control methods employed in power semiconductor-controlled induction motor drives are listed here:

- Variable terminal voltage control (for either squirrel-cage or wound-rotor motors)
- Variable frequency control (for either squirrel-cage or wound-rotor motors)
- Rotor resistance control (for wound-rotor motors only)
- Injecting voltage into the rotor circuit (for wound-rotor motors only)

AC Voltage Controllers

Common applications for these controllers are found in fan, pump, and crane drives. Figure 12.3.1 shows three-phase symmetrical ac voltage-controlled circuits for wye-connected and delta-connected

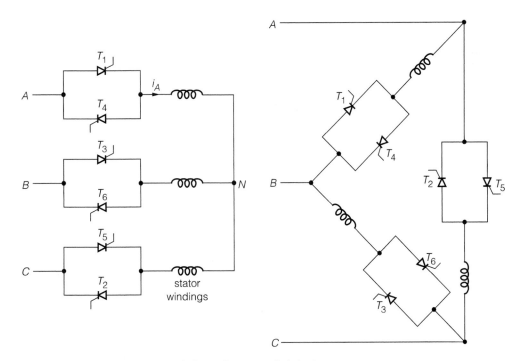

Figure 12.3.1 Three-phase symmetrical ac voltage-controlled circuits.

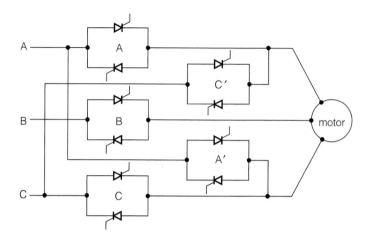

Figure 12.3.2 Four-quadrant ac voltage controller.

stators, in which the thyristors are fired in the sequence that they are numbered, with a phase difference of 60°. The four-quadrant operation with plugging is obtained by the use of a typical circuit shown in Figure 12.3.2. Closed-loop speed-control systems have also been developed for single-quadrant and multiquadrant operation. Induction motor starters that realize energy savings are one of the ac voltage controller applications.

Frequency-Controlled Induction-Motor Drives

The converters employed for variable-frequency drives can be classified as

- Voltage-source inverter
- Current-source inverters
- Cycloconverters, which allow a variable-frequency ac supply with voltage-source or current-source characteristics obtained from a fixed-frequency voltage source

Figure 12.3.3 shows a three-phase voltage-source inverter circuit along with the corresponding voltage and current waveforms. The motor connected to terminals A, B, and C can have wye or delta connection. Operating as a six-step inverter, the inverter generates a cycle of line or phase voltage in six steps. The following Fourier-series expressions describe the voltages v_{AB} and v_{AN}:

$$v_{AB} = \frac{2\sqrt{3}}{\pi}V_d\left[\sin\left(\omega t + \frac{\pi}{6}\right) + \frac{1}{5}\sin\left(5\omega t - \frac{\pi}{6}\right) + \frac{1}{7}\sin\left(7\omega t + \frac{\pi}{6}\right) + \cdots\right] \qquad (12.3.1)$$

$$v_{AN} = \frac{2}{\pi}V_d\left[\sin\omega t + \frac{1}{5}\sin 5\omega t + \frac{1}{7}\sin 7\omega t + \cdots\right] \qquad (12.3.2)$$

The rms value of the fundamental component of the phase voltage v_{AN} is given by

$$V_1 = \frac{\sqrt{2}}{\pi}V_d \qquad (12.3.3)$$

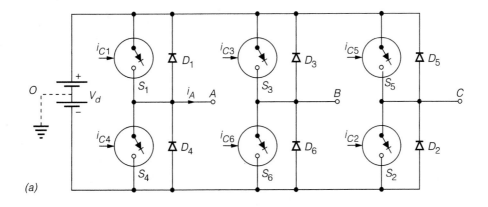

(a)

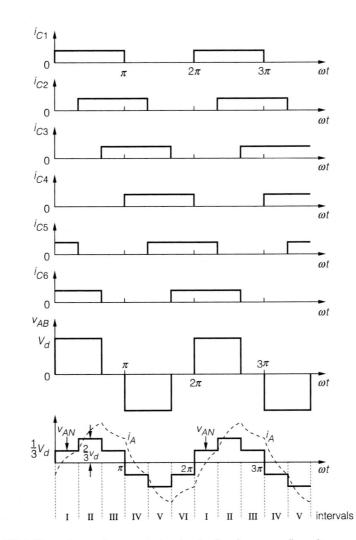

(b)

Figure 12.3.3 Three-phase voltage-source inverter circuit and corresponding voltage, and current-waveforms.

From Figure 12.3.3b, the rms value of the phase voltage is given by

$$V = \left[\frac{1}{\pi} \left\{ \int_0^{\pi/3} \left(\frac{1}{3} V_d \right)^2 d(\omega t) + \int_{\pi/3}^{2\pi/3} \left(\frac{2}{3} V_d \right)^2 d(\omega t) + \int_{2\pi/3}^{\pi} \left(\frac{1}{3} V_d \right)^2 d(\omega t) \right\} \right]^{1/2}$$

$$= \frac{\sqrt{2}}{3} V_d \tag{12.3.4}$$

For a wye-connected stator, the waveform of the phase current is shown in Figure 12.3.3, which is also the output current i_A of the inverter. The output voltage of a six-step inverter can be controlled by controlling either the dc input voltage or the ac output voltage with multiple inverters.

EXAMPLE 12.3.1

A 440-V, 60-Hz, 6-pole, wye-connected squirrel-cage induction motor with a full-load speed of 1,170 rpm has the following parameters per phase referred to the stator:

$$R_1 = 0.2 \ \Omega; \qquad R_2' = 0.1 \ \Omega; \qquad X_{l1} = 0.75 \ \Omega; \qquad X_{l2}' = 0.7 \ \Omega; \qquad X_M = 20 \ \Omega$$

(See Chapter 7, Figure 7.4.1, page 310, for the notation.)

Consider the motor to be fed by a six-step inverter, which in turn is fed by a six-pulse, fully controlled rectifier.

a. Let the rectifier be fed by an ac source of 440 V and 60 Hz. Find the rectifier firing angle that will obtain rated fundamental voltage across the motor.

b. Calculate the inverter frequency at 570 rpm and rated torque when the motor is operated at a constant flux.

c. Now let the drive be operated at a constant V/f ratio. Compute the inverter frequency at 570 rpm and half the rated torque. Neglect the derating due to harmonics and use the approximate equivalent circuit of Figure 7.4.1, with jX_M shifted to the supply terminals.

Solution

a. From Equation 12.3.1, the fundamental rms line voltage of a six-step inverter is

$$V_L = \frac{\sqrt{6}}{\pi} V_d$$

For a six-pulse rectifier, from Equation 12.2.22, $V_d = (3/\pi) V_m \cos \alpha$, where V_m is the peak ac source line voltage. Thus,

$$V_L = \frac{3\sqrt{6}}{\pi^2} V_m \cos \alpha \quad \text{or} \quad \cos \alpha = \frac{V_L}{V_m} \frac{\pi^2}{3\sqrt{6}}$$

With $V_L = 440$ V and $V_m = 440\sqrt{2}$ V,

$$\cos \alpha = \frac{\pi^2}{3\sqrt{6}\sqrt{2}} = 0.95 \text{ or } \alpha = 18.26°$$

b. For a given torque, the motor operates at a fixed slip speed for all frequencies as long as the flux is maintained constant. At rated torque, the slip speed $N_{Sl} = 1{,}200 - 1{,}170 = 30$ rpm. Hence, synchronous speed at 570 rpm is $N_S = N + N_{Sl} = 570 + 30 = 600$ rpm. Therefore the inverter frequency $= (600/1{,}200)60 = 30$ Hz.

c. Based on the equivalent circuit, it can be shown that the torque for a constant (V/f) ratio is

$$T = \frac{3}{\omega_S}\left[\frac{V_{rated}^2(R_2'/S)}{(R_1 + R_2'/S)^2 + (X_{l1} + X_{l2}')^2}\right]$$

With $a = f/f_{rated}$

$$T = \frac{3}{\omega_S}\left[\frac{V_{rated}^2(R_2'/aS)}{\left(\frac{R_1}{a} + \frac{R_2'}{aS}\right)^2 + (X_{l1} + X_{l2}')^2}\right], \quad a < 1$$

Here

$$\omega_S = \frac{120 \times 60}{6} \times \frac{2\pi}{60} = 40\pi = 125.66 \text{ rad/s}; \qquad V_{rated} = \frac{440}{\sqrt{3}} = 254 \text{ V}$$

$$S = \frac{30}{1{,}200} = 0.025$$

With $a = 1$,

$$T_{rated} = \frac{3}{125.66}\left[\frac{(254)^2(0.1/0.025)}{\left(0.2 + \frac{0.1}{0.025}\right)^2 + (0.75 + 0.7)^2}\right] = 312.11 \text{ N}\cdot\text{m}$$

At half the the rated torque,

$$0.5 \times 312.11 = \frac{3}{125.66}\left[\frac{(254)^2(0.1/aS)}{\left(\frac{0.2}{a} + \frac{0.1}{aS}\right)^2 + (1.45)^2}\right]$$

or

$$\left(\frac{0.2}{a} + \frac{0.1}{aS}\right)^2 + 2.1 = \frac{0.987}{aS}$$

Also,

$$S = \frac{a\omega_S - \omega_m}{a\omega_S}$$

or

$$a = \frac{\omega_m}{(1 - S)\omega_S} = \frac{59.7}{(1 - S)125.66} = \frac{0.457}{(1 - S)}$$

(Note: $\omega_m = 570 \times 2\pi/60 = 59.7$ rad/s.)

Based on the last two equations, we can solve for a and S using an iterative procedure:

$$a = 0.4864; \qquad S = 0.0235$$

Therefore, frequency $= 0.4864 \times 60 = 29.2$ Hz.

The inverter can also be operated as a pulse-width-modulated (PWM) inverter. Figure 12.3.4 shows the schemes: part a when the supply is dc, and part b when the supply is ac.

Figure 12.3.5 shows a three-phase current-source inverter circuit along with the corresponding voltage and current waveforms. With two switches conducting at a time (one from the group of S_1, S_3, and S_5, and other from the group of S_2, S_4, and S_6), the waveform of the line current i_A is a six-step square wave.

The capacitor bank helps to reduce the rate of change of machine-phase currents and consequent voltage spikes. The following Fourier series represents the current i_A:

$$i_A = \frac{2\sqrt{3}}{\pi} I_d \left[\sin\left(\omega t + \frac{\pi}{6}\right) + \frac{1}{5}\sin\left(5\omega t - \frac{\pi}{6}\right) + \frac{1}{7}\sin\left(7\omega t + \frac{\pi}{6}\right) + \cdots \right] \tag{12.3.5}$$

The fundamental rms current is given by

$$I_S = \frac{\sqrt{6}}{\pi} I_d \tag{12.4.6}$$

and the rms total current is

$$I_{rms} = \left[\frac{1}{\pi} \int_0^{2\pi/3} I_d^2 \, d(\omega t) \right]^{1/2} = \left(\sqrt{2/3}\right) I_d \tag{12.3.7}$$

Current-controlled PWM inverters have also been developed.

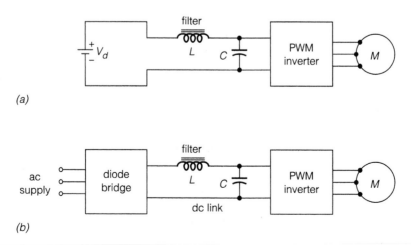

(a)

(b)

Figure 12.3.4 PWM inverter drives.

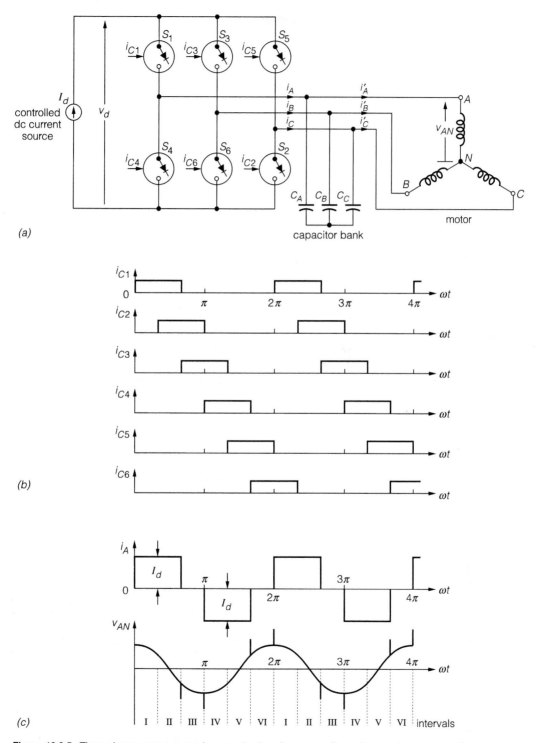

(a)

(b)

(c)

Figure 12.3.5 Three-phase current-source inverter circuit and corresponding voltage and current waveforms.

||||||||||||

EXAMPLE 12.3.2

Let the motor of Example 12.3.1 be supplied from a current source inverter with constant flux maintained at rated value. Using the equivalent circuit of Figure 7.4.1 (page 310), find the stator current when the machine operates at rated torque and frequency.

Solution

For the rated operation,

$$\text{Synchronous speed} = N_S = \frac{120 \times 60}{6} = 1{,}200 \text{ rpm or } 125.7 \text{ rad/s}$$

$$\text{Rated full-load slip} = \frac{1{,}200 - 1{,}170}{1{,}200} = 0.025$$

$$\text{Rotor impedance } Z_r' = \frac{0.1}{0.025} + j0.7 = 4 + j0.7 = 4.06\angle 9.93° \text{ } \Omega$$

Impedance viewed from input terminals is

$$(0.2 + j0.75) + \frac{(4.06\angle 9.93°)(j20)}{4 + j20.7} = 3.8 + j2.12 = 4.35\angle 29.16° \text{ } \Omega$$

$$\text{Stator current } \bar{I}_S = \frac{440/\sqrt{3}}{4.35\angle 29.16°} = 58.4\angle -29.16°$$

From Equation 12.3.6, $I_d = (\pi/\sqrt{6})58.4 = 74.9$ A. The rms total stator current, from Equation 12.3.7, is

$$I_{\text{rms}} = \sqrt{2/3}\,(74.9) = 61.1 \text{ A}$$

Cycloconverter-controlled induction-motor drives are employed in high-power drives that have a low speed-range, like rolling mills and mine winders. Figure 12.3.6 shows a three-phase cycloconverter with three dual converters, with their reference waves shifted in phase by 120° with respect to each other. Three separate secondary windings are used to feed the dual converters to avoid interaction between dual converters. Several other cycloconverter configurations exist, which are not discussed here.

Slip-Power Controlled Wound-Rotor Induction-Motor Drives

Slip power is that portion of the air-gap power that is not converted into mechanical power. The methods involving rotor-resistance control and voltage injection into the rotor circuit belong to this class of slip-power control. In order to implement these techniques using power semiconductor devices, these schemes are considered: static rotor resistance control; the static Scherbius drive; and the static Kramer drive.

The rotor circuit resistance can be smoothly varied statically (instead of mechanically) by using the principles of a chopper. Figure 12.3.7 shows a scheme in which the slip-frequency ac rotor voltages are converted into dc by a three-phase diode bridge and applied to an external resistance R; the self-commutated semiconductor switch S, operating periodically with a period T, remains on for an interval t_{on} in each period, with a duty ratio δ defined by (t_{on}/T).

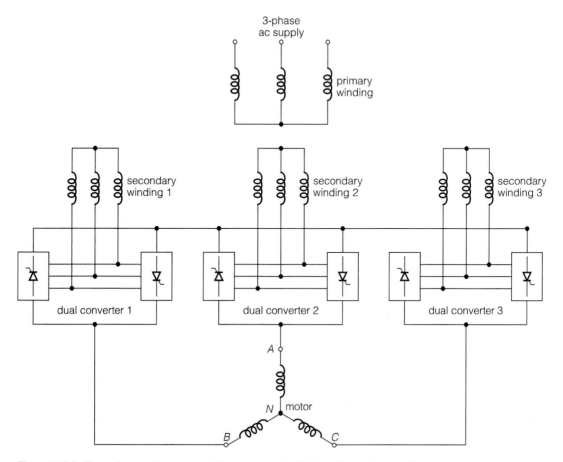

Figure 12.3.6 Three-phase cycloconverter with dual converters fed by a three-phase transformer.

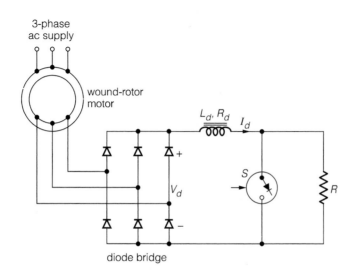

Figure 12.3.7 Static rotor resistance control of a wound-rotor induction motor.

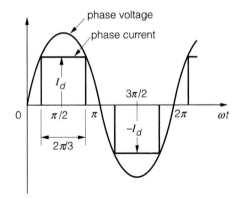

Figure 12.3.8 Rotor voltage and current waveforms.

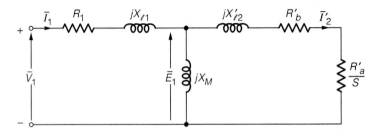

Figure 12.3.9 Per-phase fundamental equivalent circuit of the drive with static rotor resistance control.

The filter inductor helps to reduce the ripple in current I_d and eliminate discontinuous conduction. Figure 12.3.8 shows input phase voltage waveforms of the diode bridge and a six-step rotor phase current waveform, assuming ripple-free I_d. For a period T of the switch operation, the energy absorbed by resistance R is given by $I_d^2 R(T - t_{on})$, while the average power consumed by R is

$$\frac{1}{T}[I_d^2 R(T - t_{on})] \quad \text{or} \quad I_d^2 R(1 - \delta)$$

The effective value of resistance is then given by

$$R^* = (1 - \delta)R \tag{12.3.8}$$

The total resistance across the diode bridge is

$$R_t = R^* + R_d = R_d + (1 - \delta)R \tag{12.3.9}$$

The per-phase power absorbed by R_t is $\left(\frac{1}{3}I_d^2 R_t\right)$, where I_d is related to the rms value of the rotor phase current, $\sqrt{3/2}\,I_{rms}$. Thus, the effective per-phase resistance R_{eff} is given by

$$R_{eff} = 0.5R_t \tag{12.3.10}$$

From the Fourier analysis of the rotor phase current with quarter-wave symmetry, the fundamental rotor current is $[(\sqrt{6}/\pi)I_d]$ or $[(3/\pi)I_{rms}]$. It can be shown (as in Problem 12–20) that the per-phase fundamental equivalent circuit of the drive referred to the stator is given by Figure 12.3.9, where

$$R_a = R_2 + R_{eff} \tag{12.3.11}$$

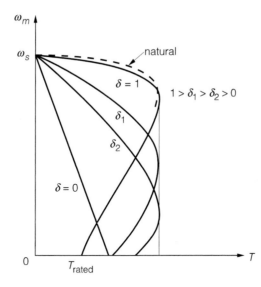

Figure 12.3.10 Effect of static rotor resistance control on speed-torque curves.

and

$$R_b = \left(\frac{\pi^2}{9-1}\right) R_a \tag{12.3.12}$$

As in Chapter 7, the primed notation denotes referral to the stator, and R_2 is the per-phase resistance of the rotor. The resistance R_a'/S accounts for the mechanical power developed and the fundamental rotor copper loss, whereas the resistance R_b' represents the effect of the rotor harmonic copper loss.

With jX_M shifted to the stator terminals in Figure 12.3.9 as an approximation, we get

$$\bar{I}_2' = \frac{\bar{V}_1}{[R_1 + R_b' + (R_a'/S)] + j(X_{l1} + X_{l2}')} \tag{12.3.14}$$

and

$$T = \frac{3}{\omega_s}(I_2')^2(R_a'/S) \tag{12.3.15}$$

Figure 12.3.10 shows the nature of speed-torque characteristics for different values of the duty ratio δ.

▮▮▮▮▮▮▮▮▮▮▮
EXAMPLE 12.3.3

A 440-V, 60-Hz, six-pole, wye-connected wound-rotor induction motor with a full-load speed of 1,170 rpm has the following per-phase parameters referred to the stator:

$$R_1 = R_2' = 0.5\ \Omega; \qquad X_{l1} = X_{l2}' = 2\ \Omega; \qquad X_M = 40\ \Omega$$

and a stator-to-rotor turns ratio of 2.5.

The scheme of Figure 12.3.7 is employed for speed control with $R_d = 0.02\ \Omega$ and $R = 1\ \Omega$. For a speed of 1,000 rpm at 1.5 times the rated torque, find the duty ratio δ, neglecting friction and windage, and using the equivalent circuit with jX_M moved over to stator terminals.

Solution
Full-load torque without rotor resistance control is

$$T = \frac{3}{\omega_S}\left[\frac{V_1^2(R_2'/S)}{\left(R_1 + \frac{R_2'}{S}\right)^2 + \left(X_{l1} + X_{l2}'\right)^2}\right]$$

With $V_1 = 440/\sqrt{3} = 254$ V; $\omega_S = 125.7$ rad/s; full-load slip = $(1{,}200 - 1{,}170)/1{,}200 = 0.025$, then

$$T = \frac{3}{125.7}\left[\frac{(254)^2(0.5/0.025)}{\left(0.5 + \frac{0.5}{0.025}\right)^2 + (4)^2}\right] = 70.6\ \text{N}\cdot\text{m}$$

With rotor resistance control,

$$T = \frac{3}{\omega_S}\left[\frac{V_1^2(R_a'/S)}{\left[R_1 + R_b' + (R_a'/S)\right]^2 + \left(X_{l1} + X_{l2}'\right)^2}\right]$$

With $R_b' = \left(\frac{\pi^2}{9} - 1\right)R_a' = 0.0966R_a'$, and slip = $(1{,}200 - 1{,}000)/1{,}200 = 0.167$, we obtain

$$1.5 \times 70.6 = \frac{3}{125.7}\left[\frac{(254)^2\left(R_a'/0.167\right)}{\left[0.5 + 0.0966R_a' + (R_a'/0.167)\right]^2 + (4)^2}\right]$$

Thus, $R_a'^2 - 2.18R_a' + 0.44 = 0$, or $R_a' = 0.225$ or 1.955 ohms. The value of 0.225 being less than R_2' is not feasible; therefore, $R_a' = 1.955$ ohms.

From Equation 12.3.11, $R_{eff} = (R_a' - R_2')/(\text{turns ratio})^2 = (1.955 - 0.5)/(25)^2 = 0.233$ ohm. From Equations 12.3.10 and 12.3.9, we calculate

$$(1 - \delta) = \frac{2R_{eff} - R_d}{R} = \frac{(2 \times 0.233) - 0.02}{1} = 0.446 \quad \text{or} \quad \delta = 0.554$$

Instead of wasting the slip power in the rotor circuit resistance, as suggested by Scherbius, we can feed it back to the ac mains by using the scheme of Figure 12.3.11, known as a *static Scherbius drive*. The equivalent circuit of Figure 12.3.9 gets modified by a counter emf (V_2'/S) in series with the rotor parameters (referred to the stator), accounting for the power fed back. The injected voltage is always in phase with the rotor current I_2' and is proportional to $\cos\alpha$, where α is the inverter firing angle. Figure 12.3.12 shows the nature of speed-torque characteristics of a static Scherbius drive with

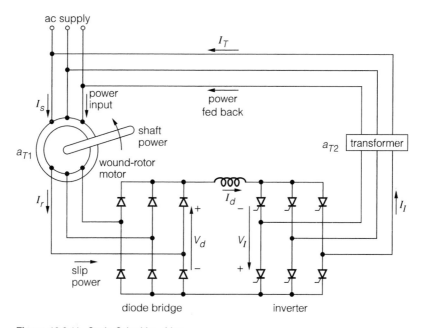

Figure 12.3.11 Static Scherbius drive.

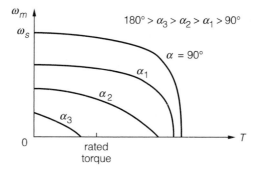

Figure 12.3.12 Speed-torque characteristics of a subsynchronous static Scherbius drive.

α as a parameter. Since speed control is possible only in the subsynchronous speed range, this type of slip-power recovery scheme is called a subsynchronous converter cascade.

The slip power can be converted to mechanical power (with the aid of an auxiliary motor mounted on the induction-motor shaft), which supplements the main motor power, thereby delivering the same power to the load at different speeds. Based on Kramer's suggestion, Figure 12.3.13 shows such a scheme with a commutatorless dc motor, which is a synchronous motor fed by a load-commutated inverter.

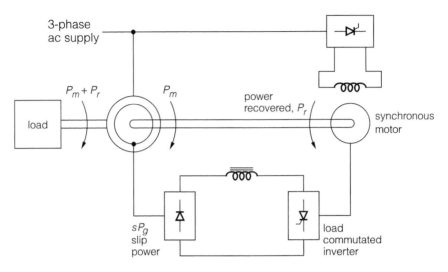

Figure 12.3.13 Scheme of a commutatorless Kramer drive.

12.4 Solid-State Control of Synchronous Motors

The speed of a synchronous motor can be controlled by changing its supply frequency. With variable frequency control, two modes of operation are possible: true synchronous mode (employed with voltage source inverters), in which the supply frequency is controlled from an independent oscillator; and self-controlled mode, in which the armature supply frequency is changed proportionately so that the armature field always rotates at the same speed as the rotor. The true synchronous mode is used only in multiple synchronous-reluctance and permanent-magnet motor drives, in applications such as paper mills, textile mills, and fiber-spinning mills, because of the problems associated with hunting and stability. Variable-speed synchronous-motor drives, commonly operated in the self-controlled mode, are superior to or competitive with induction-motor or dc-motor variable-speed drives.

Drives fed from a load-commutated current-source inverter or a cycloconverter find applications in high-speed/high-power drives such as compressors, conveyers, traction, steel mills, and ship propulsion. The drives fed from a line-commutated cycloconverter are used in low-speed gearless drives for mine hoists and ball mills in cement production. Self-controlled permanent-magnet synchronous-motor drives are replacing the dc-motor drives in servo applications.

Self-control can be applied to all variable-frequency converters, whether they are voltage-source inverters, current-source inverters, current-controlled PWM inverters, or cycloconverters. Rotor position sensors (rotor position encoders with optical or magnetic sensors, or armature terminal voltage sensors) are used for speed tracking. In the optical rotor position encoder shown in Figure 12.4.1 for a four-pole synchronous machine, the semiconductor switches are fired at a frequency proportional to the motor speed. A circular disk, with two slots (S' and S'') on an inner radius and a large number of slots on the outer periphery, is mounted on the rotor shaft. Four stationary optical sensors P_1 to P_4 (with corresponding light-emitting diodes and photo transistors) are placed as shown in Figure 12.4.1. Whenever the sensor faces a slot, an output results. Waveforms caused by the sensors are also shown in Figure 12.4.1.

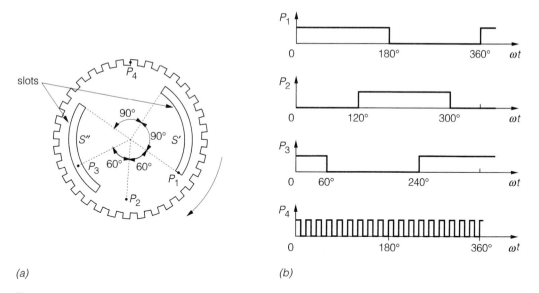

(a)   (b)

Figure 12.4.1 An optical rotor position encoder and its output waveforms.

Detailed discussion of this kind of microprocessor control of current-fed synchronous-motor drives is outside the scope of this text.

||||||||||||

EXAMPLE 12.4.1

A 1,500-hp, 6,600-V, six-pole, 60-Hz, three-phase, wye connected synchronous motor, with a synchronous reactance of 36 ohms, negligible armature resistance, and unity power factor at rated power, is controlled from a variable-frequency source with constant (V/f) ratio. Determine the armature current, torque angle, and power factor at full-load torque, one-half of the rated speed, and rated field current. (Neglect friction, windage, and core loss.)

Solution
The simplified per-phase equivalent circuits of the motor are shown in Figure 12.4.2.

$$\text{Phase voltage } V = 6{,}600/\sqrt{3} = 3{,}810.6 \text{ V}$$

$$\bar{I}'_m = \frac{V\angle 0°}{jX_s} = \frac{3{,}810.6}{36\angle 90°} = 105.9\angle -90° \text{ A}$$

With unity power factor at rated power, we obtain

$$1{,}500 \times 746 = \sqrt{3} \times 6{,}600 \times I_s; \qquad \text{Rated armature current } I_s = 97.9 \text{ A}$$

$$\bar{I}'_f = \bar{I}'_m - \bar{I}'_s = 105.9\angle -90° -97.9\angle 0° = -97.9 - j105.9 = 144.2\angle -132.8° \text{ A}$$

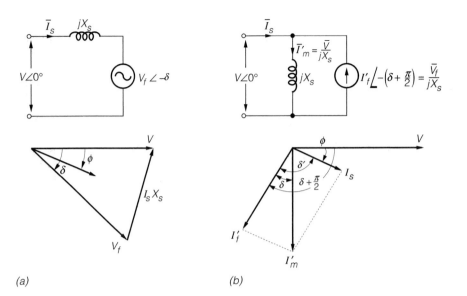

Figure 12.4.2

Synchronous speed is $(120 \times 60)/6 = 1{,}200$ rpm or $\omega_s = 125.7$ rad/s.

$$\text{Rated torque} = \frac{1{,}500 \times 746}{125.7} = 8902 \text{ N} \cdot \text{m}$$

$$\sin \delta = \frac{1{,}500 \times 746}{3 \times (6{,}600/\sqrt{3}) \times 144.2} = 0.6788$$

or torque angle $\delta = 42.75°$ electrical or $14.25°$ mechanical.

One-half of rated speed $= 600$ rpm. With a constant (V/f) ratio, I'_m is constant. Since torque $T = (3/\omega_s)VI'_f \sin \delta = (3X_s/\omega_s)I'_m I'_f \sin \delta$, δ is unchanged.

Noting that

$$\bar{I}_s = \bar{I}'_m - \bar{I}'_f; \qquad I_s \sin \phi = -I'_m + I'_f \cos \delta; \qquad I_s \cos \phi = I'_f \sin \delta;$$

then, for constant values of δ and I'_m, I_s and power factor remain the same, which is true for any speed. Therefore,

$$I_s = 97.9 \text{ A}; \qquad \delta = 42.75\,° \text{ electrical}; \qquad \text{power factor} = 1.0$$

Brushless (and Commutatorless) DC and AC Motors

Figure 12.4.3 shows brushless dc and ac motor schemes. The inverter, controlled from a dc source, can be a current-source inverter, current-controlled PWM inverter, or voltage-source inverter. The cycloconverter, fed from an ac supply, can be controlled to produce either a voltage or a current source. The rotor-position sensor and inverter in Figure 12.4.3a cause the same effect as the brushes and commutator in a dc motor. However, the angle δ' between the armature current $\bar{I}_s$ and field current $\bar{I}'_f$ can be suitably chosen, instead of $90°$ as in a dc motor. (See Figure 12.4.2; $\delta' = \frac{\pi}{2} + \delta - \phi$.)

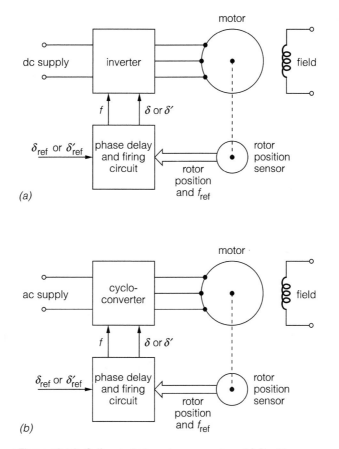

Figure 12.4.3 Self-controlled synchronous motors. (a) Brushless dc motor. (b) Brushless ac motor.

A brushless dc motor system has a linear speed-torque characteristic of the conventional dc permanent-magnet motor with constant armature voltage. Having versatile control characteristics of a dc machine, it does not suffer commutator-related limitations.

Current-Source Inverter and Load Commutation

Figure 12.4.4a shows a synchronous motor fed by a current-source inverter. Load commutation, caused by voltages v_{AN}, v_{BN}, and v_{CN} induced in the load, is possible because a synchronous motor can operate at a leading power factor. Synchronous-motor drives with load-commutated inverters are quite popular in high-power applications. The commutating inductance L_c per phase in Figure 12.4.4 is taken to be the subtransient inductance of the synchronous motor due to the very short duration of commutation transients. Waveforms of line voltages are shown in Figure 12.4.4b. Note that the current transfer from thyristor T_5 to thyristor T_1 can occur when the line voltage v_{AC} is positive; the firing angle α for thyristor T_1 is measured from the instant v_{AC} is zero and increasing. The duration of each firing pulse is 120°. Thyristors are fired in the sequence of their numbers, with a 60° interval.

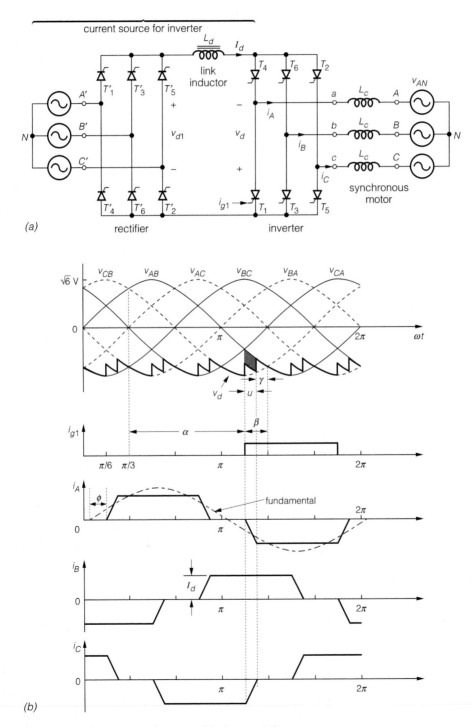

Figure 12.4.4 Current-source inverter and load commutation.

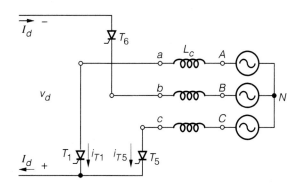

Figure 12.4.5 Equivalent circuit for the commutation interval from thyristor T_5 to thyristor T_1 in the inverter of Figure 12.4.4.

For $\alpha > 90°$ as shown in Figure 12.4.4b, assuming T_5 and T_6 to be conducting initially, with the dc-link voltage $v_d = v_{CB}$, the equivalent circuit for the commutation interval is given by Figure 12.4.5, while T_1 is fired at angle α. With constant dc-link current I_d during this interval, notice that the positive line voltage v_{AC} reduces i_{T5} to zero while increasing i_{T1} to I_d in the period u, the *commutation-overlap angle*. With T_5 now turned off, T_1 and T_6 conduct, making $v_{AB} = v_d$. Waveforms of the dc-link voltage v_d and motor-phase currents can be developed by considering various conduction and commutation intervals. With negative average value of v_d and positive I_d, the power flows from the dc link to the machine, which then operates as a motor.

The synchronous machine shown in Figure 12.4.4a, with induced-voltage sensor working as a rotor-position sensor, always operates in the self-control mode because the frequency of the inverter operation is the same as the frequency of induced voltages and the firing is synchronized with the induced voltages.

From Figure 12.4.5, it follows that

$$i_{T1} + I_{T5} = I_d \tag{12.4.1}$$

$$L_c \frac{di_{T1}}{dt} - v_{AC} - L_c \frac{di_{T5}}{dt} = 0 \tag{12.4.2}$$

With constant I_d, Equation 12.4.1 yields

$$\frac{di_{T5}}{dt} = -\frac{di_{T1}}{dt} \tag{12.4.3}$$

Using Equation 12.4.3, we get, from Equation 12.4.2,

$$\frac{di_{T1}}{dt} = \frac{v_{AC}}{2L_c} \tag{12.4.4}$$

With rms value V of phase voltage,

$$v_{AB} = \sqrt{3}(\sqrt{2}V)\sin \omega t \tag{12.4.5}$$

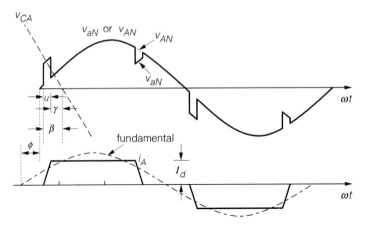

Figure 12.4.6 Waveforms of motor terminal voltage and current.

(see Figure 12.4.4b) and

$$v_{AC} = \sqrt{3}(\sqrt{2}V)\sin(\omega t - 60°) \tag{12.4.6}$$

The dc-link voltage is then found to be

$$v_d = -L_c \frac{di_{T1}}{dt} + v_{AB} = v_{AB} - 0.5v_{AC} \tag{12.4.7}$$

At $\omega t = \alpha + u$, $i_{T5} = 0$ and $i_{T1} = I_d$ when the commutation period ends. It follows from Equation 12.4.4 that

$$I_d = \frac{1}{2\omega L_c} \int_{\alpha+(\pi/3)}^{\alpha+(\pi/3)+u} v_{AC}d(\omega t) \tag{12.4.8}$$

Using Equation 12.4.6, we obtain

$$\cos\alpha - \cos(\alpha + u) = \frac{2\omega L_c I_d}{\sqrt{6}V} \tag{12.4.9}$$

The angle β (see Figure 12.4.3b), which is the angle of lead with respect to the instant when v_{AC} ceases to be positive, is known as the *commutation lead angle,* given by

$$\beta = 180° - \alpha \tag{12.4.10}$$

Substituting in Equation 12.4.9, we get

$$\cos(\beta - u) - \cos\beta = \frac{2\omega L_c}{\sqrt{6}V}I_d \tag{12.4.11}$$

Let us now define the *margin angle* to be

$$\gamma = \beta - u \tag{12.4.12}$$

which is the duration when T_5 remains reverse biased after turn-off until v_{AC} is zero. It can be shown that waveforms of the motor terminal voltage and current are as in Figure 12.4.6, with the machine power-factor angle ϕ given by $(\beta - 0.5u)$ leading. The machine will always operate with a leading power

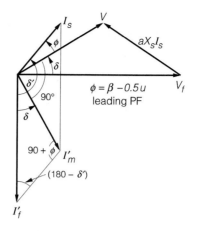

Figure 12.4.7 Phasor diagram for motoring
operation of a synchronous machine fed by a
load-commutated current-source inverter.

factor with load commutation. For $\alpha > 90°$ or $\beta < 90°$, the phasor diagram for motoring operation with load commutation is shown in Figure 12.4.7.

Neglecting the commutation overlap, for a six-step current waveform, the fundamental rms current and total rms current are given by

$$I_s = \frac{\sqrt{6}}{\pi}I_d \tag{12.4.13}$$

$$I_{rms} = \sqrt{2/3}\,I_d \tag{12.4.14}$$

The effect of overlap causes a phase shift of I_s by $0.5u$, while keeping approximately the same magnitude of I_s. With zero commutation overlap, the average value of v_d is

$$V_d = \frac{3}{\pi}\int_{\alpha+(\pi/3)}^{\alpha+(2\pi/3)} v_{AB}d(\omega t) = \frac{3\sqrt{6}}{\pi}V\cos\alpha \tag{12.4.15}$$

The effect of commutation angle u is to change the dc-link voltage from v_{AB} to $(v_{AB} - 0.5v_{AC})$, as per Equation 12.4.7, and hence change the average dc-link voltage:

$$V_d = \frac{3\sqrt{6}V}{\pi}\cos\alpha - \frac{3}{\pi}\omega L_c I_d \tag{12.4.16}$$

(See Problem 12–24.)

Load commutation is not feasible at very low speeds (less than 10% of base speed) because the induced voltages are too small to provide satisfactory performance. A pulse mode of operation (not discussed here) is then employed. Load commutation is also used for cycloconverters.

▌▐▌▐▌▐▌▐▌▐▌▐▌▐███████████████

EXAMPLE 12.4.2

A 6-MW, 6,600-V, six-pole, 60-Hz, three-phase, wye-connected synchronous motor, with a synchronous reactance of 10 ohms, subtransient reactance of 2 ohms, negligible armature resistance, and unity power factor at rated power, of a self-controlled drive is fed from a load-commutated current source inverter. The field is controlled to maintain a constant flux below base speed. The machine is operated at a constant commutation-lead angle of 55°. Neglecting friction, windage, and core loss, determine the margin angle, power factor, power developed, and torque when the machine is operating at the rated armature current and speed.

Solution

Per-phase voltage $V = 6{,}600/\sqrt{3} = 3{,}810.6$ V

$$\text{Rated speed} = \frac{1{,}200 \times 60}{6} = 1{,}200 \text{ rpm} \quad \text{or} \quad \omega_s = 125.7 \text{ rad/s}$$

$$\text{Rated rms armature current} = \frac{6 \times 10^6}{\sqrt{3} \times 6{,}600} = 524.9 \text{ A}$$

From Equations 12.4.13 and 12.4.14,

$$I_d = \sqrt{3/2}I_{rms} = \sqrt{3/2} \times 524.9 = 642.9 \text{ A}; \qquad I_s = \frac{\sqrt{6}}{\pi}I_d = \frac{\sqrt{6}}{\pi} \times 642.9 = 501.3 \text{ A}$$

The terminal voltage at rated speed will be the rated value, since the machine operates as a constant flux. From Equations 12.4.11 and 12.4.12,

$$\cos\gamma - \cos 55° = \frac{2 \times 2}{\sqrt{6} \times 3{,}810.6} \times 642.9 = 0.2755$$

$$\cos\gamma = 0.8491 \quad \text{or} \quad \gamma = 31.9°$$

Therefore, $u = \beta - \gamma = 55 - 31.9 = 23.1°$.

$$\phi = \beta = 0.5u = 55 - 11.5 = 43.45°$$

Therefore,

Power factor $= 0.726$ leading

Power developed $= \sqrt{3} \times 6{,}600 \times 501.3 \times 0.726 = 4.16$ MW

$$\text{Torque} = \frac{4.16 \times 10^6}{125.7} = 33.1 \times 10^3 \text{ N} \cdot \text{m}$$

12.5 Applications for the Electronic Control of Motors

A wide range of variable-speed drive systems are now available, each having particular advantages and disadvantages. The potential user needs to evaluate each application on its own merit, with consideration given to both technical limitations of particular systems and long term economics. An indication of possible applications of the electronic control of motors is given in Table 12.5.1.

Table 12.5.1 Applications for the electronic control of motors

Converter	Motor	Typical output range, kW	Practical speed range for a one- or two-quadrant drive	Quadrants of operation	Supply	Variables governing speed	Applications
DC drives							
Single-phase rectifier with center-tapped transformer	Separately excited or permanent magnet dc	0–5	1:50	Quad 1 driving Quad 2 dynamic or mechanical braking Reversal or armature connections or dual converter for four-quadrant operation	120-V single-phase ac	Armature terminal pd Field current	Small variable-speed drives in general
Single-phase fully controlled bridge rectifier	Separately excited or permanent magnet dc	0–10	1:50	Quad 1 driving Quad 2 dynamic braking Reversal of armature connections or dual converter for four-quadrant operation	240-V single-phase ac	Armature terminal pd Field current	Processing machinery, machine tools
Single-phase half-controlled bridge rectifier	Separately excited or permanent magnet dc	0–10	1:50	Quad 1 driving Quad 2 dynamic braking Reversal of armature connections for four-quadrant operation	240-V single-phase ac	Armature terminal pd Field current	Processing machinery, machine tools
Three-phase bridge rectifier	Separately excited or permanent magnet dc	10–200 (10–50 for PM motor)	1:50	Quad 1 driving Quad 2 dynamic braking Dual converter for regenerative braking with four-quadrant operation	240-V to 480-V three-phase ac	Armature terminal pd Field current	Hoists, machine tools, centrifuges, calendar rollers
Three-phase bridge rectifier	Separately excited dc	200–1,000	1:50	Quad 1 driving Quad 2 dynamic braking Dual converter for regenerative braking with four-quadrant operation	460-V to 600-V three-phase ac	Armature terminal pd Field current	Winders, rolling mills
Chopper (dc-to-dc converter)	Separately excited dc		1:50	Alternatives are quad 1 only; quads 1 and 2; quads 1 and 4; all four	600-V to 1,000-V dc	Armature terminal pd Field current	Electric trains, rapid transit streetcars, trolley, buses, cranes
AC drives							
Ac power controller	Class D squirrel-cage induction motor	0–25	1:2	Quad 1 driving Quad 4 plugging with source phase-sequence reversal Quad 3 driving Quad 2 plugging	240-V three-phase ac	Stator terminal pd	Centrifugal pumps and fans
Slip-energy recovery system	Wound-rotor induction motor	Up to 20,000	1:2	Quad 1 driving, with source phase-sequence reversal Quad 3 driving	Up to 5,000-V three-phase ac	Rotor terminal pd	Centrifugal pumps and fans
Voltage-source inverter and additional converters	Class B or C squirrel-cage induction motors or synchronous motors	15–250	1:10	Quad 1 driving Quad 2 regenerative braking, with appropriate additional converters Quad 3 driving Quad 4 regenerative breaking	240-V to 600-V three-phase ac	Stator terminal pd and frequency	Group drives in textile machinery and run-out tables
Current-source inverter and additional converters	Class B or C squirrel cage induction motors	15–500	1:10	Quad 1 driving Quad 2 regenerative braking, with inverter control signal sequence reversal Quad 3 driving Quad 4 regenerative braking	240-V to 600-V three-phase ac	Stator current and frequency	Single-motor drives, centrifuges, mixers, conveyors, etc.
Current-source inverter and additional converters	Three-phase synchronous motors	Up to 15,000	1:50	Quad 1 driving Quad 2 regenerative breaking, with inverter control signal sequence reversal Quad 3 driving Quad 4 regenerative braking	Up to 5,000-V three-phase ac	Stator current and frequency	Single-motor drives, processing machinery of all kinds

From S. A. Nasar, ed., *Handbook of Electric Machines.* (New York: McGraw-Hill, 1987): Table 8.1, pp. 8–68, 69. Reproduced by permission of McGraw-Hill, Inc.

▋▋▋▋▋IIII
BIBLIOGRAPHY

Campbell, S. J. *Solid-State AC Motor Controls.* New York: Marcel Dekker, Inc., 1987.

Dewan, S.B., G. R. Slemon, and A. Straughen. *Power Semiconductor Drives.* New York: Wiley Interscience, 1984.

Dubey, G. K. *Power Semiconductor Controlled Drives.* Englewood Cliffs, NJ: Prentice-Hall, 1989.

Finney, D. *Variable Frequency AC Motor Drive Systems.* London: Peter Peregrinus, Ltd., 1988.

Griffith, D. C. *Uninterruptible Power Supplies.* New York: Marcel Dekker, Inc., 1989.

IEEE Conference Publication No. 234. *Power Electronics and Variable-Speed Drives.* London: IEEE, 1984.

IEEE Press Publication. *Adjustable Speed AC Drive Systems.* B. K. Rose, ed. New York: IEEE, 1981.

IEEE Press Publication. *Modern Power Electronics.* B. K. Rose, ed. New York: IEEE, 1992.

Kenjo, T., and S. Nagamori. *Permanent-Magnet and Brushless DC Motors.* Oxford: Clarendon Press, 1985.

Nasar, S. A., *Handbook of Electric Machines.* New York: McGraw-Hill Book Company, 1987.

Rajagopalan, V. *Computer-Aided Analysis of Power Electronic Systems.* New York: Marcel Dekker, Inc., 1987.

Rashid, M. H. *Power Electronics,* Englewood Cliffs, NJ.: Prentice-Hall, 1988.

Shepherd, W., and L. N. Hulley. *Power Electronics and Motor Control.* Cambridge: Cambridge University Press, 1987.

Slemon, G. R. *Electric Machines and Drives.* Reading, MA: Addison-Wesley Publishing Co., Inc., 1992.

Sokira, T. J., and W. Jaffe. *Brushless DC Motors.* Blue Ridge Summit, PA: Tab Books, Inc., 1990.

▋▋▋▋▋▋▋IIIIIIIIIII
PROBLEMS

12-1 a. Consider a diode circuit with RC load as shown in Figure P12–1. With the switch closed at $t = 0$ and with the initial condition at $t = 0$ that $v_c = 0$, obtain the functional forms of $i(t)$ and $v_c(t)$, and plot them.

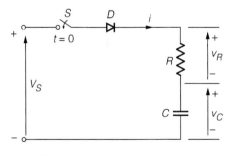

Figure P12–1

b. Then consider a diode circuit with an RL load with the initial condition at $t = 0$ that $i = 0$. With the switch closed at $t = 0$, obtain the functional forms of $i(t)$ and $v_L(t)$.

c. In part b, if $t \gg (L/R)$, describe what happens. If an attempt is then made to open switch S, comment on what is likely to happen.

d. Next, consider a diode circuit with an LC load with the initial condition at $t = 0$ that $i = 0$ and $v_c = 0$. With the switch closed at $t = 0$, obtain the waveforms of $i(t)$ and $v_c(t)$.

12-2   a.   Figure P12–2a contains a *freewheeling diode* D_m, commonly connected across an inductive load to provide a path for the current in the inductive load when the switch S is opened after time t (during which the switch was closed). Consider the circuit operation in two modes, with mode 1 beginning when the switch is closed at $t = 0$, and mode 2 starting when the switch is opened after the current i has reached its steady state in mode 1. Obtain the waveforms of the currents $i(t)$ and $i_f(t)$.

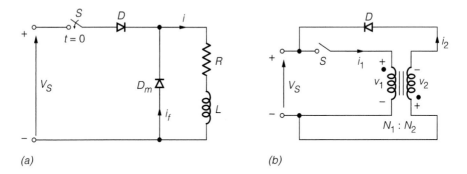

(a)   (b)

Figure P12–2   (a) Diode circuit with a freewheeling diode. (b) Energy-recovery diode circuit with a feedback winding.

   b.   Consider the energy-recovery diode circuit shown in Figure P12–2b along with a *feedback winding*. Assume the transformer to have a magnetizing inductance of L_m and an ideal turns ration of $a = N_2/N_1$. Let mode 1 begin when the switch S is closed at $t = 0$, and mode 2 start at $t = t_1$, when the switch is opened.

Let t_1 and t_2 be the durations of modes 1 and 2, respectively. Develop the equivalent circuits for the two modes of operation, and obtain the various waveforms for the currents and voltages.

12-3   a.   Consider a full-wave rectifier circuit with a center-tapped transformer, as shown in Figure P12–3, with a purely resistive load of R. Let $v_s = V_m \sin \omega t$. Determine (i) efficiency, (ii) form factor; (iii) ripple factor; (iv) *TUF;* and (v) *PIV* of diode D_1. Compare the performance with that of a half-wave rectifier. (See Example 12.1.2.)

   b.   Let the rectifier in part a have an RL load. Obtain expressions for output voltage $v_L(t)$ and load current $i_L(t)$ by using the Fourier series.

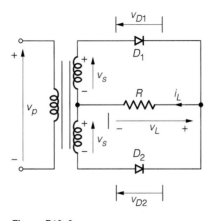

Figure P12–3

12-4   Consider a full-wave single-phase bridge rectifier circuit with dc motor load, as shown in Figure P12–4a. Let the transformer turns-ratio be unity. Let the load be such that the motor draws a ripple-free armature current of I_a. Given the waveforms for the input current and input voltage of the rectifier, as in Figure P12–4b, Determine (i) *HF* of input current and (ii) input *PF* of the rectifier.

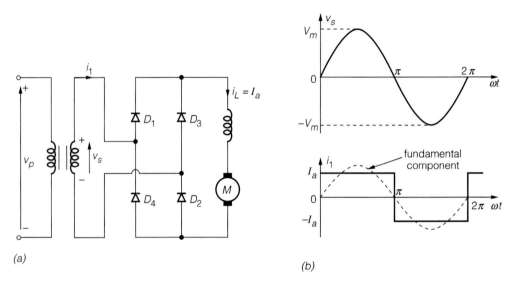

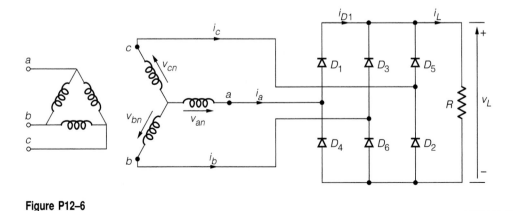

Figure P12–4   (a) Circuits diagram. (b) Waveforms.

12-5   a.   Consider a three-phase star or wye half-wave rectifier with a purely resistive load of R. Determine (i) efficiency; (ii) form factor; (iii) ripple factor; (iv) *TUF*; and (v) *PIV* of each diode.

   b.   Express the output voltage of a three-phase rectifier in Fourier series.

12-6   Consider a three-phase, full-wave bridge rectifier, shown in Figure P12–6, with a purely resistive load of R. For each diode, determine (i) efficiency; (ii) form factor; (iii) ripple factor; (iv) *TUF*; (v) *PIV*.

Figure P12–6

12-7   Consider the discontinuous-conduction motoring modes of operation of a single-phase, fully controlled, rectifier-fed, separately excited dc motor.

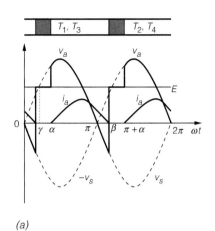

(a)

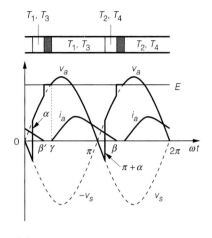

(b)

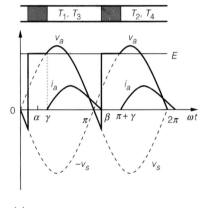

(c)

Figure P12–7 (a) Motoring mode ($\alpha > \gamma$). (b) Motoring mode ($\alpha < \gamma$). (c) Motoring mode ($\alpha < \gamma$).

a. Refer to Figure P12–7a, for $\alpha > \gamma$. Let i_a flow from α to β and stay zero from β to ($\pi + \alpha$). Obtain the steady-state motor speed-torque relationship.

b. Refer to Figure P12–7b, for $\alpha < \gamma$. Each cycle of the output voltage consists of two duty intervals and a zero-current interval. Obtain the steady-state motor performance equations for speed as a function of torque.

c. Refer to Figure P12–7c, for $\alpha < \gamma$. In this mode, current begins to flow at γ instead of at α as in part a. Obtain the steady-state motor speed-torque equation.

12–8 a. Consider the operating mode of Problem 12–7a. Find the expression for ω_{mc}, the critical speed, which is the speed on the boundary between continuous and discontinuous conductions.

b. Consider the operating mode of Problem 12–7b. Obtain the expression for ω_{mc} for motoring with $\alpha < \gamma$.

12-9   Consider Example 12.2.1 in the text.

a. Calculate the firing angle corresponding to a torque of 35 N·m and speed of $(-1,350)$ rpm, assuming continuous conduction. What is the quadrant of operation in the torque-speed relationship?

b. Find the motor speed at the rated torque and $\alpha = 160°$ for the regenerative braking in the second quadrant.

c. With armature circuit inductance of 30 mH, compute motor torques for

    (i) $\alpha = 60°$ and speed $= 1,200$ rpm.   (ii) $\alpha = 130°$ and speed $= -1,920$ rpm.

    Comment on the quadrant of operation and the operating mode in each case.

d. Obtain the firing angle corresponding to a torque of 22 N·m and speed of 880 rpm, in the discontinuous conduction mode of Figure P12–7a of Problem 12–7a.

12-10   Consider the mode of operation, shown in Figure P12–10, of a three-phase, fully controlled, rectifier-fed, separately excited dc motor.

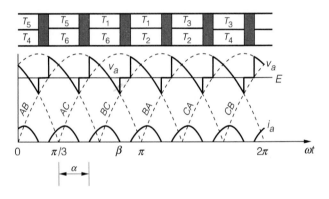

Figure P12-10

a. Obtain the speed-torque relation for this case with discontinuous conduction.

b. Obtain an expression for the critical speed ω_{mc}.

12-11   For a three-phase, fully controlled, rectifier-fed, separately excited dc motor, corresponding to ideal no-load operation, find the expression for the no-load speeds. Comment on whether no-load speeds could be negative. Compared to one-phase case, would you expect a considerable reduction in the zone of discontinuous conduction?

12-12   Consider the motor of Example 12.2.2 in the text. Calculate the motor speed for

a. $\alpha = 120$; $T_a = 25$ N·m   b. $\alpha = 60°$; $T_a = 5$ N·m

Assume continuous conduction in both cases.

12-13   Consider Example 12.2.3 in the text. Compute the minimum value of the armature inductance in order to limit the maximum armature-current ripple to 10% of the rated current.

12-14   Consider the CLC (current limit control) analysis in Section 12.2 of the motoring operation of a separately excited dc motor. Show that the average current and torque are nearly constant for all speeds for given values of i_{a1} and i_{a2}.

12-15   Consider Example 12.3.1 of the text. For part c, compute the corresponding rms fundamental stator current.

12-16   A 60-Hz, six-pole, wye-connected, three-phase induction motor, with the following parameters, is controlled by variable-frequency control with constant (V/f) ratio.

$$R_1 = R_2' = 0.025 \ \Omega; \qquad X_{l1} = X_{l2}' = 0.125 \ \Omega$$

Use the approximate equivalent circuit of Figure 7.4.1 (page 310) with jX_M moved over to the supply terminals. For an operating frequency of 15 Hz, find

    a. The maximum motoring torque as a ratio of its value at the rated frequency.

    b. The starting torque and rotor current in terms of their values at the rated frequency.

12-17   Given that the motor of Problem 12–16 has a full-load slip of 0.05, compute the motor speed corresponding to rated torque and frequency of 30 Hz.

12-18   Consider the motor of Example 12.3.1 in the text. Let the motor be controlled by variable frequency at a constant flux of rated value. By using the equivalent circuit of Figure 7.4.1 (page 310),

    a. Determine the motor speed and the stator current at one-half the rated torque and 30 Hz.

    b. Redo part a assuming the speed-torque curves to be straight lines for $S < S_{max}$.

12-19   For the motor of Example 12.3.2 in the text, calculate

    a. The inverter frequency for a speed of 600 rpm and rated torque.

    b. The motor speed and stator current for one-half of rated torque and inverter frequency of 30 Hz.

12-20   For the static rotor resistance control considered in Section 12.3 for an induction-motor drive, develop the fundamental frequency equivalent circuit of Figure 12.3.9.

12-21   Consider the induction motor drive of Example 12.3.3 in the text. Compute the motor speed corresponding to $\delta = 0.65$ and 1.5 times the rated full-load torque.

12-22   Let the simplified per-phase equivalent circuit of an under excited cylindrical-rotor synchronous machine be given by a source $V_f \angle -\delta$ in series with an impedance jX_s; Let $V \angle 0°$ and $I_s \angle -\phi$ be the applied voltage at the terminals and the current drawn by the motor, respectively.

    a. With $\bar{V}$ as reference, draw a phasor diagram for the motor.

    b. Develop an equivalent circuit with a current source and draw the corresponding phasor diagram with $\bar{V}$ as reference.

    c. Obtain expressions for power and torque for parts a and b.

    d. Now consider the variable-frequency operation of the synchronous motor, with the per-unit frequency a to be (f/f_{rated}). Taking $\bar{V}_f$ as reference in the quadrature axis, draw the phasor diagram and comment on the effect of variable frequency.

12-23   Consider the motor of Example 12.4.1 in the text.

    a. Determine the armature current and power factor at one-half the rated speed, one-half the rated torque, and rated field current.

    b. Find the torque and field current corresponding to rated armature current, 1.25 times the rated speed, and unity power factor.

12-24   Justify Equation 12.4.16.

12-25   Consider the motoring operation of a synchronous machine fed by a load-commutated current-source inverter as presented in Section 12.4 of the text. Comment on and justify the following statements:

    a. Speed control below base speed is obtained by the control of the rectifier-output voltage.

    b. For a wound-field motor, speed control above base speed is obtained by reducing the field current.

    c. For a permanent-magnet motor, speed control above base speed is obtained by increasing the commutation-lead angle.

12-26   For the motor of Example of 12.4.2 in the text,

    a. Repeat the calculations for a speed of 1,800 rpm, if the machine is operated at the rated terminal voltage above rated speed.

    b. Find the dc-link voltage for the conditions of the example. Justify the negative sign associated with the answer.

A

Units, Constants, and Conversion Factors for the SI System

Table A.1 Physical Quantities

Physical Quantity	Si Unit	Symbol
length	meter	m
mass	kilogram	kg
time	second	s
current	ampere	A
admittance	siemen (A/V)	S
angle	radian	rad
angular acceleration	radian per second squared	rad/s^2
angular velocity	radian per second	rad/s
apparent power	voltampere (V·A)	VA
area	square meter	m^2
capacitance	farad (C/V)	F
charge	coulomb (A·s)	C
conductance	siemen (A/V)	S
electric field intensity	volt/meter	V/m
electric flux	coulomb (A·s)	C
electric flux density	coulomb/square meter	C/m^2
energy	joule (N·m)	J
force	newton (kg·m/s^2)	N
frequency	hertz (1/s)	Hz
impedance	ohm (V/A)	Ω
inductance	henry (Wb/A)	H
linear acceleration	meter per second squared	m/s^2
linear velocity	meter per second	m/s
magnetic field intensity	ampere/meter	A/m
magnetic flux	weber (V·s)	Wb
magnetic flux density	tesla (Wb/m^2)	T
magnetomotive force	ampere or ampere-turn	A or At
moment of inertia	kilogram-meter squared	$kg·m^2$
power	watt (J/s)	W
pressure	pascal (N/m^2)	Pa
reactance	ohm (V/A)	Ω
reactive power	voltampere reactive	var
resistance	ohm (V/A)	Ω
resistivity	ohm·meter	$\Omega·m$
susceptance	siemen (A/V)	S
torque	newton·meter	N·m
voltage	volt (W/A)	V
volume	cubic meter	m^3

Table A.2 Prefixes

Prefix	Symbol	Meaning
exa	E	10^{18}
peta	P	10^{15}
tera	T	10^{12}
giga	G	10^{9}
mega	M	10^{6}
kilo	k	10^{3}
hecto	h	10^{2}
deka	da	10^{1}
deci	d	10^{-1}
centi	c	10^{-2}
milli	m	10^{-3}
micro	μ	10^{-6}
nano	n	10^{-9}
pico	p	10^{-12}
femto	f	10^{-15}
alto	a	10^{-18}

Table A.3 Physical Constants

Quantity	Symbol	Value	Unit
permeability constant	μ_0	1.257×10^{-6}	H/m
permittivity constant	ε_0	8.854×10^{-12}	F/m
gravitational acceleration constant	g_0	9.807	m/s^2
speed of light in a vacuum	c	0.2998×10^{9}	m/s

Table A.4 Conversion Factors

Physical Quantity	SI Unit	Equivalents
length	1 meter (m)	3.281 feet (ft) 39.37 inches (in)
angle	1 radian (rad)	57.30 degrees
mass	1 kilogram (kg)	0.0685 slugs 2.205 pounds (lb) 35.27 ounces (oz)
force	1 newton (N)	0.2248 pounds (lbf) 7.233 poundals 0.1×10^6 dynes 102 grams
torque	1 newton-meter (N·m)	0.738 pound-feet (lbf-ft) 141.7 oz-in 10×10^6 dyne-centimeter 10.2×10^3 gram-centimeter^2
moment of inertia	1 kilogram-meter^2 (kg·m^2)	0.738 slug-feet^2 23.7 pound-feet^2 (lb-ft^2) 54.6×10^3 ounce-inches^2 10×10^6 gram-centimeter^2 (g·cm^2)
energy	1 joule (J)	1 watt-second 0.7376 foot-pounds (ft-lb) 0.2778×10^{-6} kilowatt-hours (kWh) 0.2388 calorie (cal) 0.948×10^{-3} British Thermal Units (BTU) 10×10^6ergs
power	1 watt (W)	0.7376 foot-pounds/second 1.341×10^{-3} horsepower (hp)
resistivity	1 ohm-meter ($\Omega \cdot$ m)	0.6015×10^9 ohm-circular mil/foot 0.1×10^9 micro-ohm-centimeter
magnetic flux	1 weber (Wb)	0.1×10^9 maxwells or lines 0.1×10^6 kilolines
magnetic flux density	1 tesla (T)	10×10^3 gauss 64.52 kilolines/in.^{2}
magnetomotive force	1 ampere (A) or ampere-turn (At)	1.257 gilberts
magnetic field intensity	1 ampere/meter (A/m)	25.4×10^{-3} ampere/in. 12.57×10^{-3} oersted

Reference: Wilde, T. *Units and Conversion Charts,* IEEE Press, Newark, 1991.

B

Appendix B: Special Machines

We have discussed the basic operating principles of electrical machines in the text and presented the performance analyses of the more common and conventional types of machines. Certain special machines have features that distinguish them from the more conventional types. There is such a wide variety of these special machines that it is not practical to attempt to familiarize the reader with all of them. As an introduction to special machines, this appendix is devoted to some that are used to an appreciable extent in practice.

B.1 Reluctance Motors

Reluctance motors depend for their operation on the difference between the reluctances in the direct and quadrature axes; in effect, they are synchronous motors that operate without dc field excitation. Fractional horsepower motors are usually single-phase and find their application in such drives as electric clocks and other timing devices that require exact synchronous speed. In spite of their relatively large size and low power factor, polyphase reluctance motors in ratings up to 150 hp have been built because of their constructional simplicity and practically maintenance-free operation, owing to the lack of slip-rings, brushes, and dc field winding.

Reluctance motors are usually started as induction motors by making use of a squirrel-cage rotor from which some teeth have been removed so as to produce the desired number of salient poles, as, for example, the four-pole rotor shown in Figure B.1.1. In such rotors the absent teeth leave all the rotor bars and the entire end-rings intact. Any of the single-phase starting arrangements discussed in this book can be used for single-phase reluctance motors.

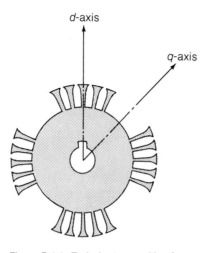

Figure B.1.1 Typical rotor punching for a four-pole reluctance synchronous motor.

B.2 Hysteresis Motors

The hysteresis loss in the rotor can be greatly magnified by constructing the rotor-core of an induction motor out of hardened magnet steel instead of from the usual annealed low-loss silicon-steel laminations. Because of hysteresis, the magnetization of the rotor lags behind the inducing mmf wave; the synchronous-motor action is produced by an angular shift between the axis of rotating primary mmf and the axis of secondary magnetization. Such hysteresis motors, having smooth rotor surfaces without secondary teeth or windings, yield substantially the same torque from standstill all the way up to synchronous speed, and they are practically noiseless. Because of the inherently small torque obtainable from hysteresis losses, hysteresis motors are limited to small sizes, from generally only a few watts to as high as $\frac{1}{7}$ hp. These motors have found applications in clocks and phonographs.

B.3 Permanent-Magnet Machines

A small synchronous machine (motor or generator) can have its excitation provided by a permanent magnet. A permanent-magnet synchronous motor is constructed by fitting the magnet inside the rotor cage, which is necessary for induction starting. A permanent-magnet generator, however, does not require a starting cage on its rotor. A major problem with such machines is to avoid demagnetization under service conditions; a generator might have its flux reduced to zero if a short-circuit occurs, and a motor is subjected to a reversal of flux every time it is started from standstill.

Permanent-magnet (PM) dc motors are built with power ratings ranging from a few watts to 100 kW or more. The PM motor is usually considered a linear device over its entire operating range of torque and speed, with a practically linear speed-torque characteristic. In the construction of small PM motors

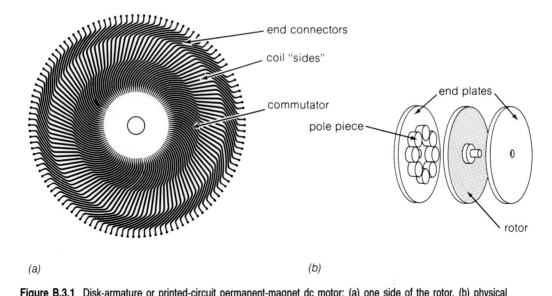

(a)   *(b)*

Figure B.3.1 Disk-armature or printed-circuit permanent-magnet dc motor: (a) one side of the rotor, (b) physical arrangement of the stator and rotor.

up to ratings of a few kilowatts, grain-oriented ferrite magnets are normally employed and magnetized during manufacture, before they are fitted into a stator. Typically, a 30% reduction in machine weight can be achieved by using permanent magnets instead of wound-field poles.

In large PM dc motors, it is not possible to assemble the large magnets, ferrite or alloy (such as Alnico), into the stator after they have been magnetized. It is therefore necessary to equip the pole pieces with magnetizing windings after the stator has been assembled with unmagnetized pole pieces. The presence of these windings, as well as the interpoles and compensating windings that can be provided to improve commutation, result in a machine structure somewhat similar to that of a dc shunt motor.

A motor configuration that is radically different from that of a conventional machine is the *disk-armature* or *printed-circuit motor,* which has been widely used with permanent-magnet materials. The rotor is made of a disk of nonconducting and nonmagnetic material; on both sides of the disk, printed in copper, is the entire armature winding and commutator with appropriate connections made from side to side, as illustrated in Figure B.3.1a. The stator consists essentially of two ferromagnetic end plates, each carrying the required number of permanent-magnet pole pieces, as shown in Figure B.3.1b. The low rotor inertia of such a configuration and consequent fast response of the motor make it suitable for control-system applications.

The rare-earth and cobalt (such as samarium-cobalt) permanent magnets seem to permit the production of PM motors with improved performance and reduced size at a reasonable cost. Such motors of about a 1-kW rating are already available, while higher ratings should be realizable in the near future. To use as little magnetic material as possible, besides the printed-circuit motor configuration, researchers developed the *stationary-armature* or the *"inside-out" configuration,* in which the armature winding and commutator are placed on the stator and the field poles are on the rotor. Two possible arrangements of the field poles on the rotor of the inside-out machines are shown in Figure B.3.2 with radial and circumferential magnetization.

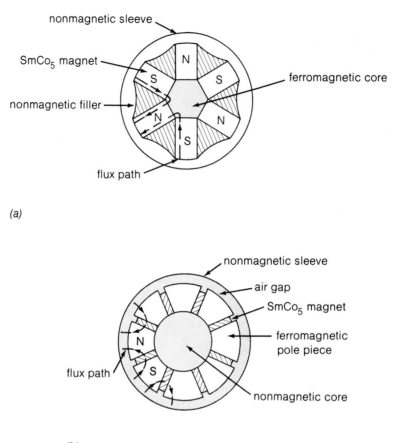

nonmagnetic sleeve

SmCo$_5$ magnet

ferromagnetic core

nonmagnetic filler

flux path

(a)

nonmagnetic sleeve

air gap

SmCo$_5$ magnet

ferromagnetic pole piece

nonmagnetic core

flux path

(b)

Figure B.3.2 Rotor configurations for stationary-armature or inside-out PM machines: (a) radial magnetization, (b) circumferential magnetization.

▌▌IIIIII
B.4 Stepping Motors

The stepping or stepper motor is a form of synchronous motor designed to rotate a specific number of degrees for each electrical pulse received by its control unit. Such a motor is used in digital control systems, where the motor receives open-loop commands as a train of pulses to turn a shaft or move a plate by a specific distance. A typical application is to position both the cutting tool and the workpiece very accurately during machine-tool operations in accordance with instructions on tape. With typical step angles of 15°, 7.5°, 5°, or 2° per pulse (depending on the angular resolution required for the particular application), stepper motors are built to follow signals as rapid as 1,200 pulses per second and to have equivalent power ratings up to several horsepower.

Stepper motors are usually designed with a multipole, multiphase stator winding; typically, three-phase and four-phase windings are used, with the number of poles determined by the required angular

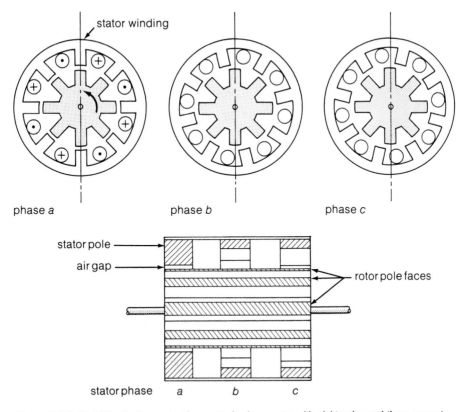

Figure B.4.1 Variable-reluctance stepping motor having a rotor with eight poles and three separate eight-pole stators arranged along the rotor.

change per input pulse. The rotors are either of the *variable-reluctance* or *permanent-magnet* type. Stepper motors are operated with an external drive logic circuit. With a train of pulses applied to the input of the drive circuit, the circuit delivers appropriate currents to the motor-stator windings to make the axis of the air-gap field step around in coincidence with the input pulses. By virtue of the reluctance torque or the permanent-magnet torque, or both, the rotor follows the axis of the air-gap magnetic field, depending on the pulse rate and the load torque, including inertia effects.

The *variable-reluctance stepping motor* shown in Figure B.4.1 has a rotor with eight poles and three separate eight-pole stators arranged along the rotor. When the poles of stator-phase a are energized with alternate polarity by a set of series-connected coils carrying current i_a, the rotor poles tend to align with the poles of stator-phase a as shown in Figure B.4.1. The stator-phase b is identical with the phase a stator, except that its poles are displaced by 15° in a counterclockwise direction. If the current i_a is now set equal to zero and the current i_b is established, the motor will develop a torque rotating the rotor counterclockwise by 15°. Stator-phase c has its poles displaced a further 15° anticlockwise rotation of the rotor. Finally, interrupting the current i_c and reestablishing the current i_a results in the completion of a 45° rotation of the rotor. Further current pulses in the *a-b-c* sequence produce further anticlockwise stepping motions. Reversing the current-pulse sequence to *a-b-c* sequence produces reversed rotation. It follows that the step angle of a motor is determined by the number of poles.

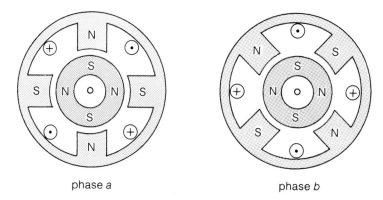

phase *a*   phase *b*

Figure B.4.2 Stepping motor with a permanent-magnet rotor.

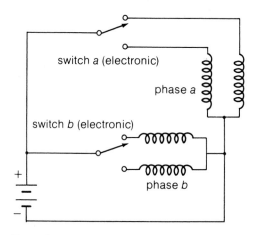

Figure B.4.3 Circuit with double coils in each phase
for the motor of Figure B.4.2.

The torque developed by a stepping motor of a given size can be increased if a *permanent-magnet rotor* is used. Figure B.4.2 illustrates the stator arrangement for a four-pole, two-phase motor with a permanent-magnet rotor, typically made of ferrite. Energizing the phase *a* winding only with the polarity shown holds the rotor in the position shown. The additional energizing of phase *b* winding with the polarity shown results in a movement of 22.5° in an anticlockwise direction. De-energizing the phase *a* winding results in a further 22.5° movement, and so on. To simplify the switching operation required to reverse stator polarities, double coils are usually employed in each phase, as shown in Figure B.4.3 for the motor of Figure B.4.2.

Typical variable-reluctance stepper motors operate at small steps, 15° or less, and at maximum position response rates up to 1,200 pps. Typical permanent-magnet types operate at larger steps, up to 90°, and at maximum response rates of 300 pps. Applications include table positioning for machine tools, tape drives, recorder-pen drives, and X-Y plotters.

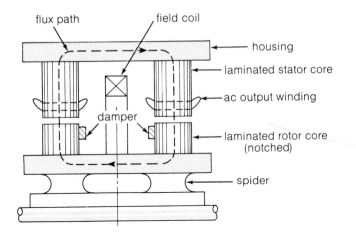

Figure B.5.1 Homopolar inductor alternator.

B.5 Inductor Alternators

The inductor alternator is a synchronous generator with stationary field and armature windings. It depends for its operation on a periodic variation in the reluctance of the air gap. Such machines can operate at high speeds to generate correspondingly high frequencies, usually from several hundred hertz up to 100,000 Hz, for use in such applications as steel and nonferrous melting by induction with high-frequency currents, for supplying power at radio frequency, and supplying military power demands, including lighting, electric motors, and sophisticated electronic equipment requiring precise voltage and frequency regulation and control. They are suitable for direct connection to high-speed prime movers such as gas turbines.

The flux through an armature coil of an inductor alternator is unidirectional; instead of periodically reversing its direction, it fluctuates between a maximum and a minimum value, with the design of the air gap being such that the induced voltage is practically sinusoidal. Inductor generators can generally be classified as *homopolar* or *heteropolar* types.

The *homopolar inductor alternator*, shown in Figure B.5.1, has two laminated stator cores and two laminated rotor cores. The ac output winding is arranged in the slots of the stator cores. The field winding, consisting of a coil that is wound concentric with the axis of the machine, produces the unidirectional flux. The rotor cores are notched with open slots so as to produce the desired waveform of the flux-density distribution. One cycle of voltage is generated as the rotor advances through a distance equal to that between the centers of two adjacent rotor teeth. The frequency is therefore given by

$$f = \frac{\text{(Number of rotor teeth)(Synchronous speed in rpm)}}{60} \tag{B.5.1}$$

Figure B.5.2 shows a cross-sectional view of a four-section homopolar inductor alternator, which can be viewed as the equivalent of a combination of two two-section machines, such as the one shown in Figure B.5.1. Approximate flux paths of such an alternator are indicated in the figure.

The *heteropolar inductor alternator,* illustrated in Figure B.5.3, has a stator housing and stator cores, as well as a rotor, all similar to those of the squirrel-cage induction motor. The stator punchings are

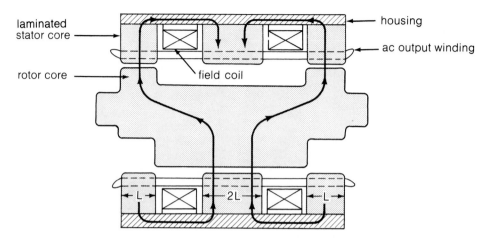

Figure B.5.2 Cross-sectional view of a four-section homopolar inductor alternator. Courtesy of U.S. Department of the Army.

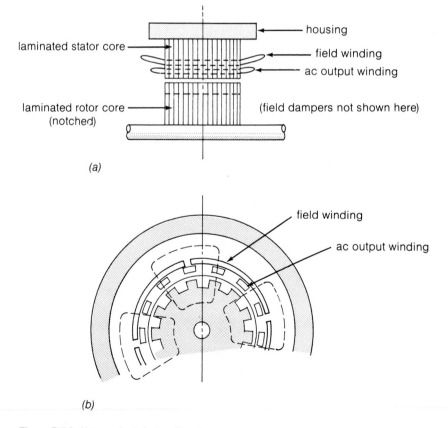

(a)

(b)

Figure B.5.3 Heteropolar inductor alternator.

notched to carry the ac output winding. The field winding, wound in several slots on the stator, produces a multipolar field, as indicated by the broken lines in Figure B.5.3b, in contrast with the unipolar field produced by the field winding in the homopolar type. As in the homopolar type, however, the field fluctuates in the heteropolar machine due to variations in the air-gap length. The voltage response to changes in the field current is more rapid in the heteropolar machine than in the homopolar machine.

B.6 AC Commutator Motors

AC commutator motors have armatures that are similar to those in dc machines; their field structures, however, are laminated to reduce eddy currents. The two distinct advantages of the ac commutator motors over induction motors are a wide speed range and a high starting torque. Also, series motors can be operated at several times the induction-motor synchronous speed. AC commutator motors can be grouped into two classes: *series motors*, which have characteristics similar to those of series-wound dc motors and whose speed at any given load can be varied by changing the applied voltage or, in some cases, by shifting the brushes; and *shunt motors,* whose speed is approximately constant for operation from a source of constant voltage and whose speed can be changed (independent of the load) by changing the voltage of the terminals of the motor, either by brush shifting or by providing suitably disposed and connected auxiliary coils.

The most common ac commutator machine is the single-phase series motor, although some three-phase series motors and three-phase shunt motors do exist. Polyphase commutator motors have inherently better commutating ability. Single-phase motors are generally limited to sizes below 10 hp, except for railway application. With the advent of solid-state diodes and thyristors, the ac commutator motor is being displaced by the combination of rectifiers and dc motors, at less cost and superior performance.

A dc series motor with a laminated field structure can be made to operate on ac supply; in fact, small series motors up to about $\frac{1}{2}$-hp rating, known as *universal motors,* are designed to operate on either dc or ac supply. Figure B.6.1 shows the schematic diagrams for the *straight series motor* and the *compensated series motor* with a compensating winding to overcome the armature mmf. Universal motors are generally used for small devices like vacuum cleaners, food mixers, and portable tools operating at speeds in the range of 3,000 to 11,000 rpm.

If, instead of the field and armature being connected in series, the supply voltage is connected directly across the field and the armature brushes are short-circuited, the simple *repulsion motor* is

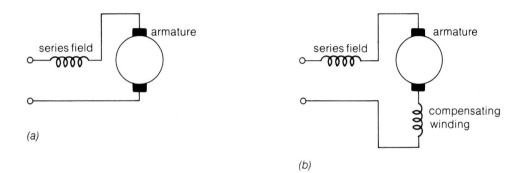

(a)

(b)

Figure B.6.1 AC single-phase series motor. (a) straight series motor. (b) compensated series motor.

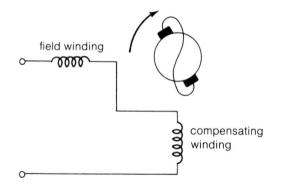

Figure B.6.2 Schematic diagram of a repulsion motor.

obtained, as shown in Figure B.6.2. Actually, the field and compensating functions are performed by a single stator winding, and the brush axis is placed at an angle α with respect to the axis of this winding. If the brushes are in line with the stator-winding axis (i.e., $\alpha = 0$), the motor becomes a short-circuited transformer, producing zero torque. With brushes located in a position corresponding to $\alpha = 90°$, the induced armature current is zero, and hence the torque is also zero. The brushes must therefore occupy some intermediate position so that the currents induced in the rotor by transformer action react on the stator flux and produce torque. When the brushes are shifted through a position of zero torque, the direction of rotation reverses.

The commutation of the repulsion motor is superior to that of the series motor up to synchronous speed and inferior at higher speeds because of the large short-circuit currents in the coils undergoing commutation. The armature, being isolated from the line, can be designed for any convenient low voltage. The repulsion motor has a torque-speed characteristic similar to that of the series motor, but it has the additional advantage that its speed can be adjusted by shifting the brushes. Prior to the widespread use of the capacitor-start single-phase induction motor, *repulsion-start induction-run motors* (see Section 7.9) were used in applications requiring high starting torque and practically constant running speed. Consider also the *repulsion-induction motor,* which has a squirrel cage in the bottom of the rotor slots, with the commutated winding occupying the top of the rotor slots; both repulsion and induction torques are combined in such a machine.

The *brush-shifting polyphase shunt motor,* also known as the *Schrage motor,* is essentially an inside-out induction motor with its secondary winding on the stator and its primary winding on the rotor connected to the supply line through slip rings, as illustrated in Figure B.6.3. An adjusting winding is embedded in the same rotor slots and connected to a commutator. Line-frequency voltages are induced in the adjusting winding by transformer action from the primary, and slip-frequency voltages appear between brushes on the commutator. The two sets of brushes, indicated by a and a' in the figure, can be adjusted as to angular position and relative spacing between them by means of a handwheel. The magnitude of the slip-frequency voltages inserted in series with the secondary windings depends on the spacing between the two sets of brushes, and the phase depends on their angular position. Thus, both the speed and power factor can be controlled. Machines of this type are mainly used in sizes up to 50 hp and in speed ranges of 6 to 1 or below, for a wide range of applications such as driving pumps, fans, paper mills and conveyors.

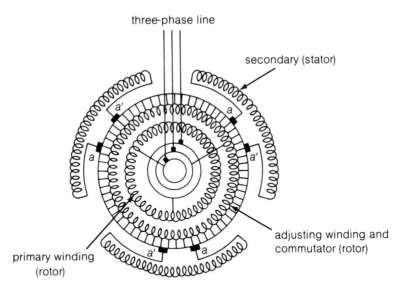

Figure B.6.3 Schematic diagram of a three-phase adjustable-speed brush-shifting Schrage-type motor.

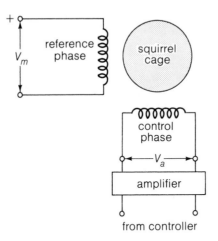

Figure B.7.1 Schematic diagram of a two-phase control motor.

B.7 Servo Components

Two-phase control motors are two-phase squirrel-cage induction motors used in some control systems. Their output ranges from a fraction of a watt to several hundred watts. As shown in the schematic diagram of Figure B.7.1, the two stator windings, phase *m* (or the *reference phase*) and phase *a* (or the

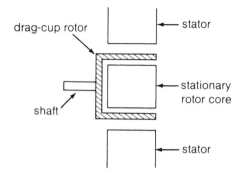

Figure B.7.2 Illustration of drag-cup rotor construction.

control phase), are identical and are displaced from each other by 90 electrical degrees. The reference phase is connected to a constant-voltage, constant-frequency source of V_m volts; the control phase is supplied with the voltage V_a from the controller, usually with an amplifier intervening. The error voltage V_a, which is the same frequency as V_m and phase-displaced from V_m by 90 electrical degrees, is proportional to the required amount of correction. *High-resistance rotors* are used for such control motors to assure high torque near zero speed and to prevent the motor from running as a single-phase motor when the error signal is zero. When the maximum power output is below a few watts, a *drag-cup rotor* construction is used, as illustrated in Figure B.7.2, in which the rotating member is like a thin metallic can with one end removed, thereby minimizing inertia. A stationary iron core, like a plug inside the cup, completes the magnetic circuit.

A small two-phase machine, often with drag-cup rotor construction, is used as an *ac tachometer* for measuring the angular velocity of a shaft. While the main winding, often referred to as the *fixed field* or reference field, is energized from a suitable alternating voltage of constant magnitude and frequency, a voltage of the same frequency is then generated in the auxiliary winding or *control field*. This voltage is applied to the high-impedance input circuit of an amplifier, so the auxiliary winding can be considered as open-circuited. Ideally, the electrical requirements are that the magnitude of the signal voltage generated in the auxiliary winding is linearly proportional to the speed and that the phase of this voltage is fixed with respect to the voltage applied to the main winding.

Self-synchronous devices, also known as *selsyns, synchros,* and *autosyns*, are used in many systems to synchronize the angular positions of two shafts at different locations when mechanical interconnection of the shafts is rather impractical. Three-phase selsyns are used in systems where high torque transmission is required, while single-phase selsyns are used for low torque transmission. Single-phase selsyns often find their applications for transmitting shaft-position information, for remote indication of the position of a device such as an elevator or a hoist, and for operating servomechanisms to control the position or motion of larger equipment.

The three-phase selsyn consists of two three-phase wound-rotor induction motors. As illustrated in Figure B.7.3, the stator windings are connected in parallel and energized from a three-phase source, while the rotors connected to their respective shafts are also electrically in parallel with each other. The single-phase selsyn, shown in Figure B.7.4, has a *selsyn-transmitter* or *generator* and a *selsyn-receiver* or *motor*, both of which have a single-phase winding (usually on the rotor) connected to a common ac voltage source; the other member of each (usually the stator) has a three-phase wye-connected winding,

three-phase line

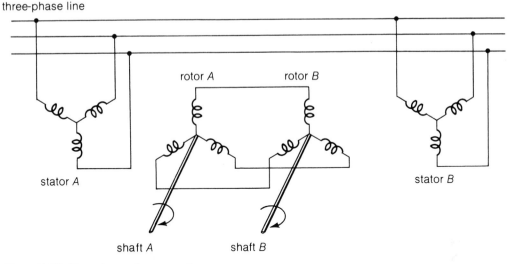

Figure B.7.3 Three-phase selsyn connections.

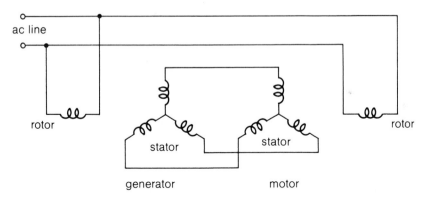

Figure B.7.4 Single-phase synchrotransmitter-motor system

and the two windings are connected in parallel. The rotors are of salient-pole construction, which makes a simple winding arrangement possible. When the single-phase rotor windings are excited, voltages are induced by transformer action in the wye-connected stator windings. If the two rotors are in the same space position relative to their stator windings, the transmitter and receiver stator-winding voltages are equal, no current circulates in these windings, and hence no torque is transmitted. If, however, the two rotor space positions do not correspond, the stator-winding voltages are unequal; currents circulating in the stator windings, in conjunction with the air-gap magnetic fields, produce torques tending to align the two rotors. The receiver is generally provided with mechanical damping to reduce oscillations. The motor-torque at standstill or for slow rotation can be shown to depend closely on the sine of the relative angular difference in position of the transmitter and receiver shafts.

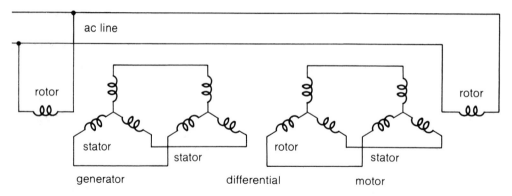

Figure B.7.5 Synchro generator-motor system with differential.

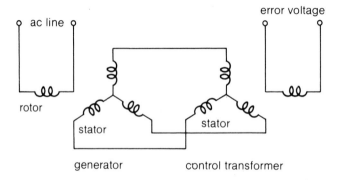

Figure B.7.6 Synchro generator-transformer system.

The addition of a third selsyn, known as a *differential selsyn,* as shown in Figure B.7.5, produces the rotation of its shaft as a function of the sum or difference of the rotation of two other shafts. The differential selsyn has a cylindrical rotor with a three-phase winding on the rotor as well as on the stator.

A *synchro generator-transformer system* with a *control transformer,* as illustrated in Figure B.7.6, is used to supply an error voltage, resulting from angular displacement between two shafts, to a corrective device like a two-phase control motor, which has a greater torque capability than the selsyn system.

B.8 Linear Induction Motors

Linear propulsion relates to a novel type of electric drive for ground-transportation systems. In such a scheme, the motor is split into two parts, one of which is carried by the vehicle and the other lies straight along the track. The force of interaction between these two structures is used directly as tractive effort, without the need for intermediate transmission or gear. The basic difference between rotating and linear motors is that, in the former, the air gap and the magnetic structures are *endless,* whereas, in the

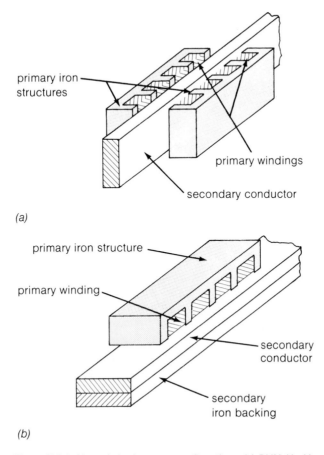

primary iron
structures

primary windings

secondary conductor

(a)

primary iron structure

primary winding

secondary
conductor

secondary
iron backing

(b)

Figure B.8.1 Linear induction motor configurations. (a) DLIM (double-sided). (b) SLIM (single-sided).

latter, the air gap and one of the magnetic structures (short side) are *open-ended.* All types of existing electric motors can be realized in the linear form. Further, the peculiar requirements of linear propulsion are giving rise to the development of novel configurations. As a result, a great variety of schemes is being considered, and only the test of time will decide their practicality.

The *linear induction motor* (LIM) has reached the most advanced stage of development. In its simplest form, it consists of a ferromagnetic structure, located on the underside of the vehicle and carrying a winding energized by a polyphase power supply, and a passive *reaction rail,* in the form of a bar of conducting material, usually aluminum, embedded in the guideway. In order to reduce the reluctance of the magnetic-flux path, the reaction rail is either sandwiched between two energized structures (double-sided LIM or DLIM) or backed by a passive ferromagnetic structure (single-sided LIM or SLIM), as illustrated in Figures B.8.1a and B.8.1b, respectively. The SLIM might require a more expensive rail system, but it alleviates dynamic stability problems encountered with the DLIM.

In its basic operation, the LIM is similar to its rotating counterpart. The currents flowing in the *primary* energized winding set up an electromagnetic wave that travels backward with respect to the vehicle at synchronous speed. When the vehicle speed differs from the synchronous speed, the *secondary* reaction rail slips with respect to this traveling wave. The motionally induced voltage then generates a system of ac currents that interact with the impressed electromagnetic wave to produce a net thrust. A repulsive force also develops which is, however, opposed by a force of attraction, whenever the secondary conductor is backed by ferromagnetic material. In addition to propulsion, the LIM can provide *levitation* and *guidance*. The translational or linear motion produced by a linear motor can proceed indefinitely, limited only by the length of its track.

The characteristic curve of rotational speed versus torque for a rotating induction motor is, for the case of a LIM, replaced by a curve of velocity versus thrust. The synchronous velocity of a LIM is given by

$$v_s = 2\tau f \text{ m/s} \tag{B.8.1}$$

where τ is the pole pitch and f is the source frequency. The slip of the LIM is then given by

$$S = \frac{v_s - v}{v_s} \tag{B.8.2}$$

In contrast with a rotating machine, the synchronous velocity of a LIM for a given frequency can be made to have any desired value, and it is not governed by the number of poles. Further, the number of poles on the primary need not be even, or indeed integral. The LIM operates at high slip with corresponding low efficiency, and its typical velocity-thrust curve has the same general shape as the speed-torque curve of an induction motor with a rotor of reasonably high resistance. Also, the large gap between the ferromagnetic surfaces of the primary and secondary members in a SLIM or between the two primary members in a DLIM strongly influences the operating characteristics of a LIM.

A phenomenon present in the LIM but that has no parallel in the rotating induction motor is known as the *end effect*. The LIM primary has an entry edge, at which the new secondary conductor continuously comes under the influence of the magnetic field, and an exit edge, at which the secondary conductor continuously leaves. By Lenz's law, the secondary current in the region of entry acts to prevent the build-up of gap-flux. As a result, when the motor is moving, the average gap-flux density for the first pair of poles near the entry edge tends to be significantly lower than that under later poles in the motor. Also by Lenz's law, a current that tends to maintain the flux persists in the secondary conductor after it has left the exit edge. This current produces resistive loss without any corresponding development of thrust. While the general direction of the current beneath the primary is transverse, substantial regions exist where the orthogonal relationship between current, flux, and motion is not satisfied; in such regions, little or no thrust is developed in the direction of motion; The end effect, which is naturally most pronounced at high speed, reduces the maximum thrust that the motor can produce.

Large LIMs usually have three-phase primary windings and are employed for transportation, materials handling, extrusion presses, and the pumping of liquid metal. Small LIMs are used as curtain pullers, sliding-door closers, and the like.

B.9 Linear Synchronous Motors

The *linear synchronous machine* (LSM), although not widely used at present, has such potential, particularly in the field of transportation, that a brief description of two principal types is in order. The LSM develops a thrust when it operates at the synchronous velocity given by Equation B.8.1. The required system of interacting currents must be excited by a separate dc supply. The power involved, is small,

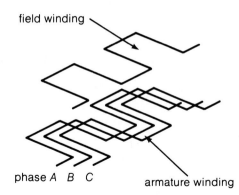

Figure B.9.1 Schematic arrangement of windings in an air-cored linear synchronous motor.

phase *A B C*   armature winding   field winding

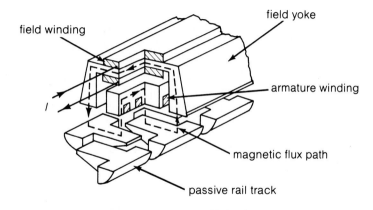

Figure B.9.2 Sketch of a claw-pole motor (Nadyne).

field winding   field yoke   armature winding   magnetic flux path   passive rail track   *I*

however, because it needs to cover only the ohmic loss; it can be drastically reduced by cryogenic cooling or totally eliminated by employing superconducting or permanent magnets. Even with no excitation, a small amount of thrust can be obtained from the variable-reluctance effect of the nonenergized structure having salient poles. By varying the frequency of the supply through a power conditioner, such as a thyristor inverter, operation at variable speed is achieved.

In its simplest form, an air-cored LSM consists of two arrays of conductors, as illustrated in Figure B.9.1. The armature winding carries the polyphase system of currents to set up the traveling wave, and the field winding carries the dc excitation current. Such a scheme is practical only when the field is superconducting and located on the vehicle.

The thrust obtained from a given system of conductor-currents can be increased by at least one order of magnitude by embedding both field and armature windings in iron structures. This configuration leads to *iron-cored LSMs*. By placing the field and armature windings on board the vehicle, significant savings in the cost of the track can be achieved. Of particular interest are those machine types in which the track does not carry an alternating flux and hence need not be thinly laminated: namely, the *claw-pole* or *Nadyne* motor (Figure B.9.2) and the *homopolar inductor* (Figure B.9.3).

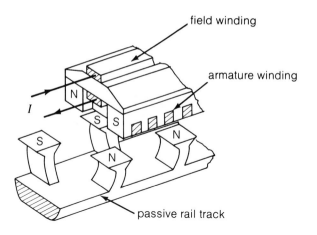

Figure B.9.3 Sketch of a homopolar linear synchronous motor.

In the intermediate speed range (150–350 km/h), iron-cored LSMs appear to have the edge over the LIMs because LSMs are less affected by end effects and can operate with larger clearances (3 cm), leading power factor, and better efficiency. Also, their strong attractive forces can be controlled independently of the thrust. At speeds of 350 km/h and above, the air-cored LSMs seem to have no competitors; they can operate with clearances up to 30 cm, and they require no power collectors. The practical realization, however, depends on major technological advances in superconducting magnets and power electronics.

B.10 Cryogenerators

The present development of technological bases for the production of future *cryogenerators* (also known as *superconductive generators*) indicates that large turboalternators with liquid-helium-cooled, superconducting field windings are feasible and that the projected advantages should be realizable through further intensive research and development. Only the future will show at what value of the machine rating the economic transition from the present conventional fluid cooling to cryo-cooling can be made, however. Superconductivity does permit generators to considerably exceed the present flux density limit of 2 to 2.5 T in the rotor iron, but one must forgo the ferromagnetic properties of steel. Current loadings and power density could be increased significantly with practically no change in subtransient and transient reactances and no deterioration in electrical stability.

Figure B.10.1 shows a typical basic layout of a superconducting generator.[1] The rotor is a rotating Dewar vessel containing a superconducting field winding of niobium-titanium (NbTi) alloy wire that is kept supercondutive by liquid helium at 4 to 5 K. The cold rotor interior is thermally insulated in the radial direction by a vacuum space and a radiation shield, and in the axial direction by mechanical connections to the shaft with a high thermal resistance. The rotor outer cylinder is an electrical damping

[1] Abegg, K., "The Growth of Turbogenerators," *Philosophical Transactions of the Royal Society of London* 275, 1248 (August 1973): 51–67.

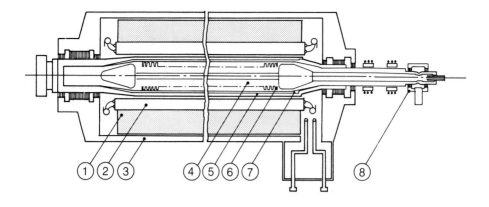

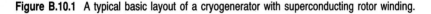

1. laminated iron shield
2. armature winding and supporting structure
3. stator housing
4. inner rotor
5. outer rotor shield (damper cylinder)
6. superconducting winding
7. rotor thermal insulation (vacuum)
8. rotor vacuum seals

Figure B.10.1 A typical basic layout of a cryogenerator with superconducting rotor winding.

screen and a shield protecting the rotor winding from transient influences of the armature. The stator winding is directly water cooled; its support structure must be of a nonmagnetic material and must resist high short-circuit forces and steady-state fatigue loads. Immediate surroundings must be shielded from the strong magnetic fields; the best rating per unit volume is achieved by using a shield of laminated magnetic iron, while minimum mass per unit volume results by using a conducting copper screen with a rating per unit volume reduced to about two-thirds.

Superconductive generation appears to have the following potential advantages as compared to the conventional fluid-cooling technology: reduced size and weight, higher efficiency, higher voltage and lower current levels, improved system performance, higher rotor critical speeds, and reduced cost.

Serious design challenges are presented in these key problem areas: the electromagnetic rotor shield, overspeed control requirements, the optimized superconducting rotor, and the optimized stator winding and its structure. In the final analysis, relative reliability, availability, and cost become overriding considerations in a practical sense before the new class of cryogenerators can be put into commercial operation on a large scale.

Answers to Odd–Numbered Problems

Chapter 1

1–1 **a.**

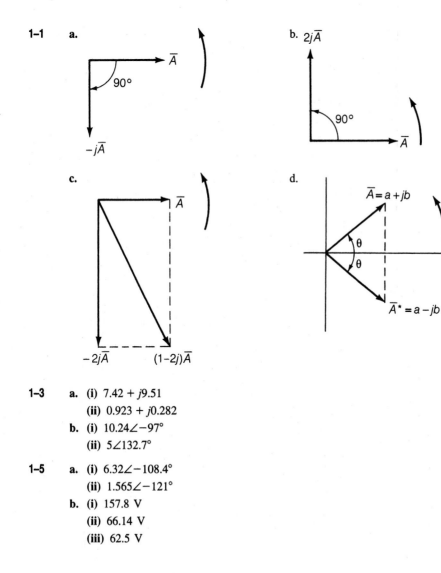

b. $2j\overline{A}$

c.

d.

1–3 **a.** (i) $7.42 + j9.51$
     (ii) $0.923 + j0.282$
  b. (i) $10.24\angle -97°$
     (ii) $5\angle 132.7°$

1–5 **a.** (i) $6.32\angle -108.4°$
     (ii) $1.565\angle -121°$
  b. (i) 157.8 V
     (ii) 66.14 V
     (iii) 62.5 V

1–7   **a.**

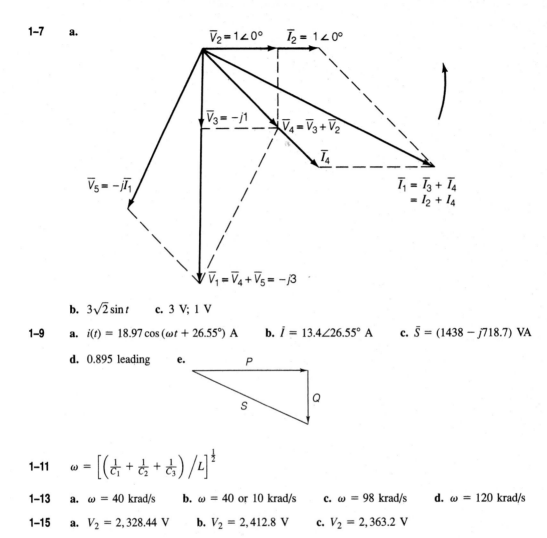

$$\overline{V}_2 = 1\angle 0° \qquad \overline{I}_2 = 1\angle 0°$$

$$\overline{V}_3 = -j1$$

$$\overline{V}_4 = \overline{V}_3 + \overline{V}_2$$

$$\overline{I}_4$$

$$\overline{V}_5 = -j\overline{I}_1$$

$$\overline{I}_1 = \overline{I}_3 + \overline{I}_4$$
$$= \overline{I}_2 + \overline{I}_4$$

$$\overline{V}_1 = \overline{V}_4 + \overline{V}_5 = -j3$$

b. $3\sqrt{2}\sin t$ **c.** 3 V; 1 V

1–9   **a.** $i(t) = 18.97\cos(\omega t + 26.55°)$ A   **b.** $\overline{I} = 13.4\angle 26.55°$ A   **c.** $\overline{S} = (1438 - j718.7)$ VA

d. 0.895 leading   **e.**

$$P$$
$$S \qquad Q$$

1–11 $\omega = \left[\left(\frac{1}{C_1} + \frac{1}{C_2} + \frac{1}{C_3}\right)\Big/L\right]^{\frac{1}{2}}$

1–13   **a.** $\omega = 40$ krad/s   **b.** $\omega = 40$ or 10 krad/s   **c.** $\omega = 98$ krad/s   **d.** $\omega = 120$ krad/s

1–15   **a.** $V_2 = 2,328.44$ V   **b.** $V_2 = 2,412.8$ V   **c.** $V_2 = 2,363.2$ V

Chapter 2

2–1   **a.** $\overline{V}_{AN} = \frac{100}{\sqrt{3}}\angle -30°$; $\overline{V}_{BN} = \frac{100}{\sqrt{3}}\angle -150°$; $\overline{V}_{CN} = \frac{100}{\sqrt{3}}\angle 90°$

b. $\overline{V}_{AN} = \frac{100}{\sqrt{3}}\angle 30°$; $\overline{V}_{BN} = \frac{100}{\sqrt{3}}\angle 150°$; $\overline{V}_{CN} = \frac{100}{\sqrt{3}}\angle -90°$

2–3   **a.** $I_L = 32.8$ A $= I_{ph}$ **b.** $I_{ph} = 18.94$ A

2–5   **a.** 20 A; 0.8 lagging; 6 kVA; kW $= 4.8$; kVAR $= 3.6$

b. Fig. 2.2.2b of the text applies except that the diagram as a whole must be rotated such that $\overline{V}_{AB}$ becomes horizontal (i.e. a reference) with 0° angle.

c. zero

2-7   **a.** 21.2 kW   **b.** 45.56 kW

2-9   **a.** 360.54 kW   **b.** 94.08 kW   **c.** 21.6 kW

2-11   162 V

2-13   **a.** $P_{\text{TOTAL}} = 24$ kW   **b.** $\text{kVA}_{\text{TOTAL}} = 25$ **c.** overall PF = 0.96 leading

2-15 -124.3 kVAR; 134 kVA; 0.373 leading PF

2-17   **a.** 12.25 MVA; 7.35 MVAR   **b.** 2.53 MVAR; 0.9 lagging   **c.** 18.5 μF

2-19   751.34 V 784.2 V; 16.325 kW 12.824 kVAR

2-21 $274.2\angle\pm40.89°$ Ω/phase; no

2-23   **a.** $13.5\angle75.6°$ Ω/phase   **b.** $13.5\angle-75.6°$ Ω/phase

Chapter 3

3-1 $\nabla \cdot \bar{D} = \rho;\ \nabla \cdot \bar{B} = 0;\ \nabla \times \bar{E} = -\dfrac{\partial \bar{B}}{\partial t};\ \nabla \times \bar{H} = \bar{J} + \dfrac{\partial \bar{D}}{\partial t};$ In low-frequency problems, the dielectric effects or displacement currents can be neglected. The material regions can be considered as void of the volume-charge density. $\nabla \cdot \bar{J} = -\dfrac{\partial \rho}{\partial t}$

3-3 $\bar{F}_{21} = -\dfrac{\mu_0}{2\pi}\dfrac{I_1 I_2}{r}\bar{a}_r$ N/m, where $\bar{a}_r$ is the unit vector from $d\bar{\ell}_1$ to $d\bar{\ell}_2$. If I_2 and I_1 are in the same direction, the force pulls the wires together; when the wires carry current in opposite directions, the wires are repelled by the magnetic force.

3-5   **a.** UBL; 1 is positive with respect to 2

       b. $I = \dfrac{UBL}{R}$; $P = (UBL)^2/R$; direction of current is from 1 to 2 through resistor R

       c, d. Energy is transferred from the mech. source to the resistor. $\bar{F} = -BLI\hat{U}$;

$$P = \frac{(BLU)^2}{R}$$ which is same as the elec. power dissipated in the resistor.

3-7 $B_g = 90$ mT

3-9   0.0006 Wb

3-11   1,508 At

3-13   0.7 A

3-15   **a.** $B_2 = 0.8$ T   **b.** $B_o = 1.0$ T   **c.** $I = 11.94$ A

3-17 $B_1 = 0.387$ T $= B_2 = B_3$

3-19 $L = \dfrac{\mu a(r_2 - r_1)N^2}{\pi(r_2 + r_1)}$; 0.32 H

3-21   10 W; 1,273 turns

3-23   Energy at point a = 2 J; total energy at point b = 3.5 J; coenergy at point a = 2 J; total coenergy at point b = 6.5 J

3-25   **a.** v_1 = 0; v_2 = 0; W_m = 5 μJ   **b.** v_1 = 10 cos 100t mV; v_2 = 0.1 cos 100t V; W_m = 0 at t = 0
   c. v_1 = {−10 sin t + 3 cos (t + 30°)} mV; v_2 = {− sin t + 30 cos (t + 30°)} mV; W_m = 1.775 μJ at
   t = 0

3-27

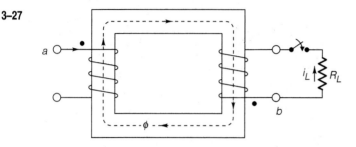

3-29   **a.** $\phi = \mu_0 I \ell/(4\pi)$ Wb   **b.** $w_m = \dfrac{\mu_0 I^2 \rho^2}{8\pi^2 R^4}$ J/m^3 **c.** $W_m = \dfrac{\mu_0 I^2 \ell}{16\pi}$ J   **d.** $L = \dfrac{\mu_0 \ell}{8\pi}$ H

3-31   0.5 T

3-33   **a.** 12.5 cm^2 **b.** 31.23 cm^2 **c.** 1.044 T   **d.** 60.2 cm^2

Chapter 4

4-1   **a.** e = 75.4 cos 377t V   **b.** 53.32 V

4-3   **a.** 4.13 mWb   **b.** ϕ = −2.05 cos 377t − 0.068 cos 1,131t mWb; ϕ_{rms} = 1.45 mWb

4-5   0.632

4-7   **a.** 212 A; 2,500 V; 0.77   **b.** 0.978   **c.** (0.135 + j0.6) Ω; (0.0054 + j0.024) Ω
   d. 11.79∠39.61° Ω

4-9   **a.** 2.3%   **b.** 98.2%   **c.**

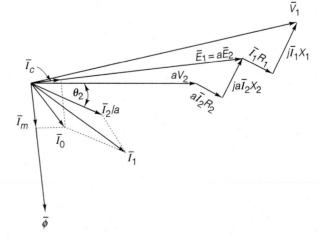

4–11   387.9 A, which is 0.62 full-load

4–13   **a.** 0.375 A   **b.** 2.47 A   **c.** 0.15

4–15   0.964

4–17   **a.**

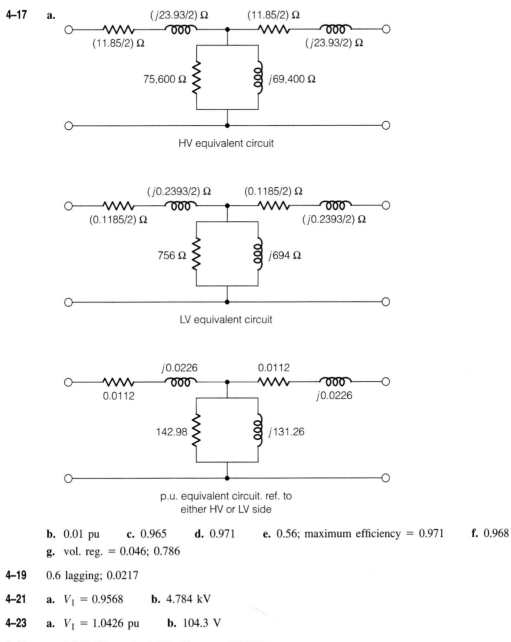

$(j23.93/2)\ \Omega$ $(11.85/2)\ \Omega$

$(11.85/2)\ \Omega$ $(j23.93/2)\ \Omega$

$75{,}600\ \Omega$ $j69{,}400\ \Omega$

HV equivalent circuit

$(j0.2393/2)\ \Omega$ $(0.1185/2)\ \Omega$

$(0.1185/2)\ \Omega$ $(j0.2393/2)\ \Omega$

$756\ \Omega$ $j694\ \Omega$

LV equivalent circuit

$j0.0226$ 0.0112

0.0112 $j0.0226$

142.98 $j131.26$

p.u. equivalent circuit. ref. to
either HV or LV side

b. 0.01 pu   **c.** 0.965   **d.** 0.971   **e.** 0.56; maximum efficiency = 0.971   **f.** 0.968
g. vol. reg. = 0.046; 0.786

4–19   0.6 lagging; 0.0217

4–21   **a.** $V_1 = 0.9568$ **b.** 4.784 kV

4–23   **a.** $V_1 = 1.0426$ pu   **b.** 104.3 V

4–25   **a.** 1,012 W   **b.** 2,973 W   **c.** 99.02%

4–27   **a.** 0.076   **b.** 0.95   **c.** 0.948

4–29 **a.**

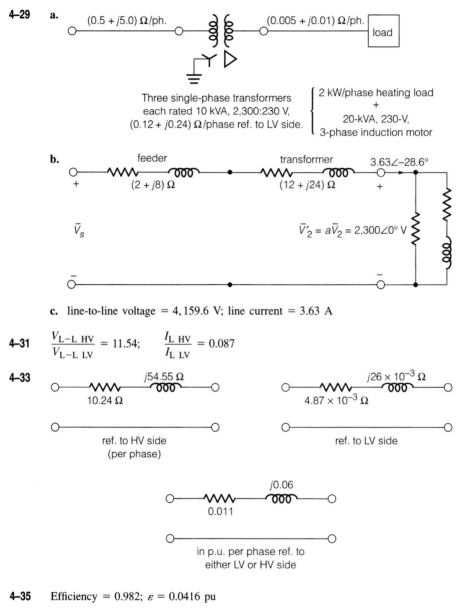

Three single-phase transformers each rated 10 kVA, 2,300:230 V, $(0.12 + j0.24)$ Ω/phase ref. to LV side.

2 kW/phase heating load
+
20-kVA, 230-V, 3-phase induction motor

b.

c. line-to-line voltage = 4, 159.6 V; line current = 3.63 A

4–31 $\dfrac{V_{\text{L–L HV}}}{V_{\text{L–L LV}}} = 11.54;$ $\dfrac{I_{\text{L HV}}}{I_{\text{L LV}}} = 0.087$

4–33

j54.55 Ω

10.24 Ω

ref. to HV side
(per phase)

j26 × 10⁻³ Ω

4.87 × 10⁻³ Ω

ref. to LV side

j0.06

0.011

in p.u. per phase ref. to
either LV or HV side

4–35 Efficiency = 0.982; ε = 0.0416 pu

4–37 $0.954\angle-3.85°$ pu; 0.08 pu

4–39 **a.** I_H = 1.506 kA; I_X = 8.0 kA   **b.** I_H = 1.506 kA; I_X = 13.86 kA

4–41

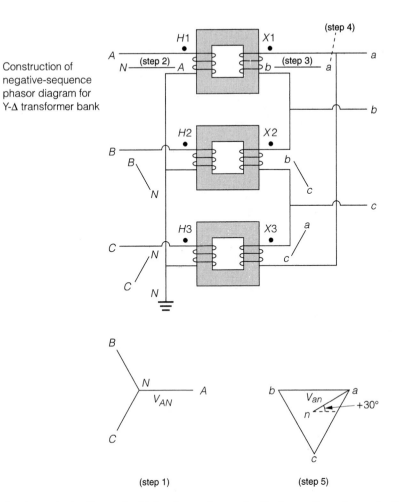

(step 4)

Construction of
negative-sequence
phasor diagram for
Y-Δ transformer bank

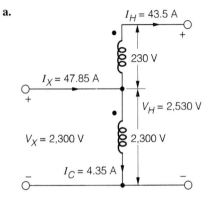

(step 1)   (step 5)

As shown in the figure, the high-voltage phasors *lag* the low-voltage phasors by 30°. Thus the negative-sequence phase shift is the reverse of the positive-sequence phase shift.

4–43   a.

$I_H = 43.5$ A

230 V

$I_X = 47.85$ A

$V_H = 2,530$ V

$V_X = 2,300$ V   2,300 V

$I_C = 4.35$ A

b. Autotransformer kVA rating = 110; kVA transformed by em induction = 10

c. $\eta = 0.997$

4-45   **a.** $Z_1 = j0.026$; $Z_2 = j0.026$; $Z_3 = j0.14$ **b.** $V_2 = 0.843$ pu

4-47   **a.** $Z_1 = j0.023$; $Z_2 = j0.058$; $Z_3 = j0.078$

b.

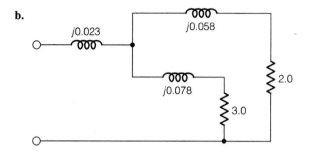

4-49   **a.** $E_s = 83.6$ V   **b.** $f_\ell = 15.3$ Hz; $f_h = 19,116$ Hz

4-51   **a.** $S = 2.22\, f B_m J A_i k_w A_w$ **b.** $S = 300$ kVA

Chapter 5

5-1   With independent variables q and x: $\quad F_e = -\dfrac{\partial W_e}{\partial x}(q,x);\qquad F_e = +\dfrac{\partial W'_e}{\partial x}(q,x) - q\dfrac{\partial v}{\partial x}(q,x)$

With independent variables v and x: $\quad F_e = -\dfrac{\partial W_e}{\partial x}(v,x) + v\dfrac{\partial q}{\partial x}(v,x);\qquad F_e = +\dfrac{\partial W'_e}{\partial x}(v,x)$

5-3 $F_e = -\dfrac{1}{2} i^2 \dfrac{N^2 A \mu_0}{(x+t)^2}$

5-5 $F_e = 333.3$ N

5-7   **a.** $F_e = -\dfrac{I^2 k_1 k_2}{2(k_2 g + k_3)^2}$ **b.** 600 N

5-9   **a.** $F_e = -\dfrac{\mu_0 a^2 N^2 I^2}{(b+g)^2}$ **b.** 256.4 N

5-11   **a.** ± 377 rad/s; ± 188.5 rad/s   **b.** 188.5 W; 141.4 W

5-13   **a.** 314 rad/s   **b.** 62.8 W   **c.** 1.2 A

5-15 $F_e = -\dfrac{V_m^2 \cos^2 \omega t}{(k_1 + k_2)\omega^2}$; F_e is independent of x.

5-17   **a.** $T_e = -i_a i_f L \sin \theta + i_b i_f L \cos \theta$ **b.** $T_e = -I_a I_f L \sin \delta$

   c. $v_{ta} = L_{aa} \dfrac{di_a}{dt} + e_{af}$ where $e_{af} = -\omega L I_f \sin(\omega t + \delta)$; $v_{tb} = L_{aa} \dfrac{di_b}{dt} + e_{bf}$; $e_{bf} = +\omega L I_f \cos(\omega t + \delta)$

5-19   **a.** $T_e = -I_a^2 L_2 \sin 2\delta - I_a I_f L \sin \delta$
   b. Additional term proportional to $\sin 2\delta$ that does not depend on I_f
   c. With a negative value of δ and a positive value of T_e, the machine will run as a synchronous motor; it will run as a generator, if driven mechanically, with a positive value of δ and a negative value of T_e.
   d. With $I_f = 0$, $T_e = -I_a^2 L_2 \sin 2\delta$; The machine will run as a 2-phase reluctance motor or generator.

5-21 $T_e = 0.98 \cos^2 314 t \left[\dfrac{\sin \theta + 4 \sin 2\theta}{(6 + \cos \theta + 2 \cos 2\theta)^2} \right]$ N·m

5-23   No reluctance torque is produced: parts c, d, e, f.   Reluctance torque is produced: parts a, b, g, h.

5-25   5,100 or 2,100 rpm in either direction

5-27 $v = 2N\ell r B_m \omega \cos 2\omega t$

5-29   **a.** rms phase voltage = 160 V
   b. rms line-to-line voltage = 277 V; with $\omega = 377$, $e_{ab} = 277\sqrt{2} \sin(\omega t + 30°)$;
       $e_{bc} = 277\sqrt{2} \sin(\omega t - 90°)$; $e_{ca} = 277\sqrt{2} \sin(\omega t + 150°)$

5-31   1,187.3 N·m

5-33   300, 150, 100 rpm

5-35   **a.** 655.2 V   **b.** 1,310.4 V   **c.** 65.52 kW; 347.6 N·m
   d. Maximum developed power or torque remains unchanged.

5-37 $B = 1.16$ T

5-39   rpm = 283.8

5-41   **a.** 50 A   **b. (i)** 56 A;   **(ii)** 1,200 rpm;   **(iii)** 178 N·m   **c.** 1,320 rpm; $I_a = 0$

5-43   **a.** $P = 8$ **b.** $3\frac{1}{3}$ Hz   **c.** 50 rpm with respect to rotor; 750 rpm with respect to stator

5-45   **a.** 1,746 rpm   **b.** 1.8 Hz   **c.** 1,800 rpm   **d.** zero

5-47   **a.** 1.97   **b.** 118.33 Hz

5-49   180 Hz

5-51 $C = 48.23$ μF with $f = 60$ Hz

5-53   **a.** 0.85   **b.** 0.82

5–55

$L/\tau \ (= k)$	D (meters)	L (meters)	$\tau \ (= \pi D/32)$ (meters)
1.0	3.06	0.3	0.3
1.5	2.67	0.393	0.262
1.75	2.54	0.435	0.25
2.0	2.43	0.475	0.24
2.25	2.34	0.516	0.23

Chapter 6

6–1

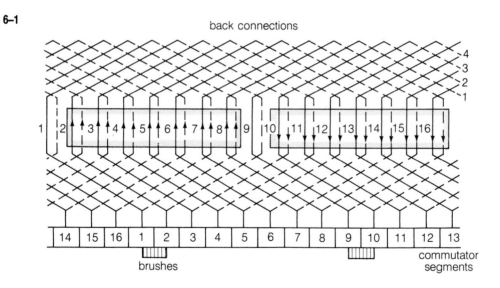

6–3   radial form

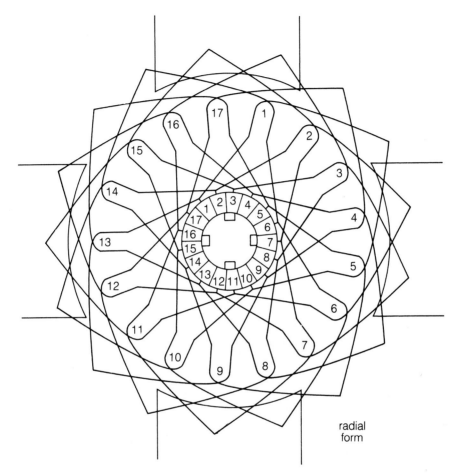

radial
form

6–5   **a.** 100 A   **b.** 50 A

6–7

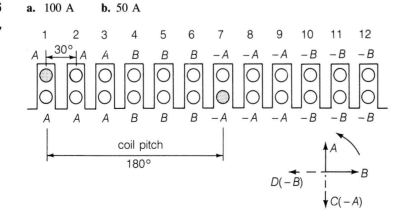

6-9   **a.** 60 electrical degrees phase spreading; full-pitch coils; $k_{d1} = 0.966$; $k_{p1} = 1$

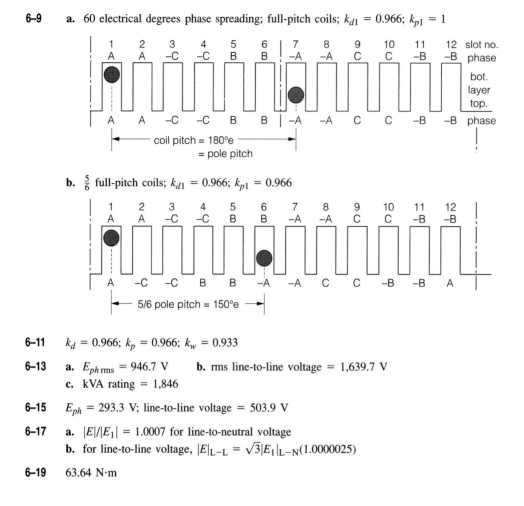

b. $\frac{5}{6}$ full-pitch coils; $k_{d1} = 0.966$; $k_{p1} = 0.966$

6-11 $k_d = 0.966$; $k_p = 0.966$; $k_w = 0.933$

6-13   **a.** $E_{ph\,\text{rms}} = 946.7$ V   **b.** rms line-to-line voltage = 1,639.7 V
c. kVA rating = 1,846

6-15 $E_{ph} = 293.3$ V; line-to-line voltage = 503.9 V

6-17   **a.** $|E|/|E_1| = 1.0007$ for line-to-neutral voltage
b. for line-to-line voltage, $|E|_{\text{L-L}} = \sqrt{3}|E_1|_{\text{L-N}}(1.0000025)$

6-19   63.64 N·m

Chapter 7

7-1   **a.** 3,600 rpm   **b.** 180 rpm   **c.** 3,600 rpm   **d.** 0   **e.** 3 Hz   **f.** 50 V; 2.5 V

7-3 $R_1 = 0.1\ \Omega$; $X_{\ell 1} = X'_{\ell 2} = 0.141\ \Omega$; $X_M = 6.159\ \Omega$; $R'_2 = 0.105\ \Omega$; 880 W

7-5   0.083; 62.5 N·m

7-7   **a.**
$$\frac{T}{T_{\text{max}}} = \frac{1 + \sqrt{1 + Q^2}}{1 + \frac{1}{2}\left(\dfrac{S}{S_{\text{maxT}}} + \dfrac{S_{\text{maxT}}}{S}\right)\sqrt{1 + Q^2}}$$

with $R''_1 = 0$, $Q = \infty$, $\dfrac{T}{T_{\text{max}}} = \dfrac{2}{\dfrac{S}{S_{\text{maxT}}} + \dfrac{S_{\text{maxT}}}{S}}$

b. $\dfrac{I_2'}{I_{2maxT}'} = \sqrt{\dfrac{(1 + \sqrt{1 + Q^2})^2 + Q^2}{[1 + \frac{S_{maxT}}{S}\sqrt{1 + Q^2}]^2 + Q^2}}$

As $Q \to \infty$, $\dfrac{I_2'}{I_{2maxT}'} = \dfrac{\sqrt{2}(S/S_{maxT})}{\sqrt{(S/S_{maxT})^2 + 1}}$

7–9   0.5; 0.134

7–11   **a.** 0.1   **b.** 38,060 W   **c.** 459.2 hp   **d.** 0.3   **e.** $3I_{2FL}'$ **f.** 1.2 T_{FL}
　　　　g. $4I_{2FL}'$

7–13   18.8 A; 0.844 lagging; 1,176 rpm; 7 hp; 42.5 N·m; 86.3%

7–15 $I_2' \simeq KS$; $T \simeq K_1 S$; $P_m \simeq K_2 S$

7–17   **a.** $X_{\ell 1} = X_{\ell 2}' = 0.41\ \Omega$; $X_M = 35.63\ \Omega$, $R_1 = 0.5\ \Omega$; $R_2' = 0.583\ \Omega$
　　　　b. 3.94 hp; 16.42 N·m; 0.755   **c.** 0.61; 86.8 N·m

7–19 $R_C = 136.8\ \Omega$; $X_M = 12.5\ \Omega$; $R_1 = R_2' = 0.115\ \Omega$; $X_{\ell 1} = X_{\ell 2}' = 0.17\ \Omega$

7–21   70.7%

7–23   **a.** 1.52; 2,705 N·m   **b.** 231 A; 2,083 N·m

7–25   243.36 A; $P = 867$ kW; $Q = -465$ kVAR

7–27   **Part 1 a.** one-half   **b.** 4 times   **c.** same
　　　　Part 2 a. one-half;   **b.** 2 times;   **c.** doubled

7–29   3.57 A; 0.645 lagging; 0.21 hp; 1,710 rpm; 0.87 N·m; 0.616

7–31   174.3 μF

Chapter 8

8–1   **a.** 0.256   **b.** −0.053   **c.** 0.894 leading

8–3   0.969

8–5   0.987 Ω/phase; 0.92 pu; 0.836 Ω/phase; 0.775 pu; 1.29

8–7   **a.** 59.35×10^3 N·m; 87.05×10^3 N·m   **b.** 978 A

8–9   **a.** $\delta = -10.25°$; $I_f = 9.1$ A   **b.** 16.15 A; 6.1 A   **c.** 0.83 leading; 14,350 W
　　　　d. 72.9 N·m

8–11   **a.** $\bar{E}_{fm} = 0.825\angle -51°$; $\bar{E}_{fg} = 1.753\angle 26°$ **b.** 0.8 pu

8–13   **a.** 3,179 V; −17.76°   **b.** 230.5 A; 0.985 lagging

8–15   179.26 kVA

8–17 **a.** 0.993 leading **b.** 19 kVAR

8–19 **a.** 0.333 pu **b.** -1.333 pu; 0.242 leading

8–21 **a.** 32.6°; 1.73 pu **b.** 1.5 pu

8–23

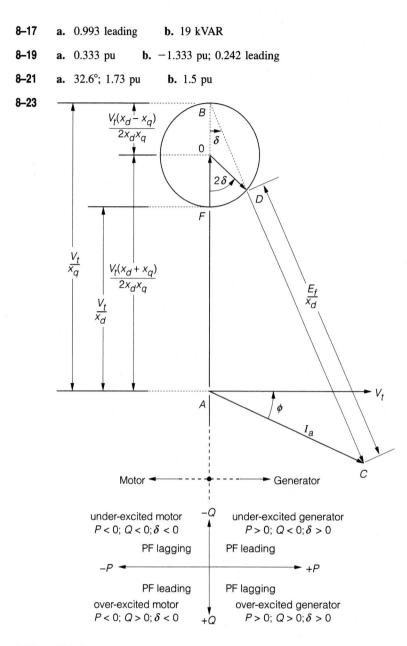

8–25 881 A

8–27 **a.** 0.84 lagging; 138.1 A; 0.76 lagging; 2.45°; 2.7° **b.** 0.701 lagging; 149.75 A

8–29 **a.** **(i)** 60 MW **(ii)** 45 MVAR **(iii)** 3,138 A **(iv)** 0.8 lagging **(v)** 23.93 kV; 21.67 kV **(vi)** 25.63°; 22.5°

b. **(i)** 60 MW   **(ii)** 75.2 MVAR;   14.8 MVAR   **(iii)** 4,016 A; 2,386 A   **(iv)** 0.625 lagging; 0.971 lagging   **(v)** 28.7 kV; 17.94 kV   **(vi)** 21.14°; 27.5°

c. **(i)** 72 MW; 48 MW   **(ii)** 70 MVAR; 20 MVAR   **(iii)** 4,200 A; 2,174 A   **(iv)** 0.717 lagging; 0.924 lagging   **(v)** 28.7 kV; 17.8 kV   **(vi)** 25.64°; 21.8°

Chapter 9

9–1   **a.** 270.2 V   **b.** 260.2 V

9–3   1,170.6 rpm

9–5 $RPM_{FL} = 1,199$

9–7   **a.** 254.5 V   **b.** 253.9 V

9–9   272.2 V

9–11   963 rpm

9–13   **a.** 66.67 Ω   **b.** 46 Ω

9–15

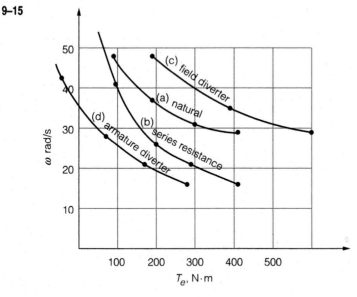

9–17   1,400 A; 1,600 A; 130 V

9–19   **a.** 240 V; 15 A; 25 A   **b.** 248 V; 12 A; 22 A

9–21   **a.** 0.9 A; 522 rpm   **b.** 544 rpm   **c.** **(i)** 159 Ω   **(ii)** 238.65 Ω   **d.** 4.7 Ω

9–23   1,289 rpm   **a.** 1,096 rpm   **b.** 161 hp   **c.** 1,045.5 N·m

9–25   0.0075 Ω

9-27   **a.** $77 \, \Omega$ **b.** 470 V at 1,000 rpm   **c.** 460 V

9-29   2.6 turns/pole; 6.5 times that of the series-field winding

9-31   **a.** 89%   **b.** 150.6 hp

9-33   **a.** 600 N·m   **b.** 908 rpm   **c.** 76.52 hp   **d.** 87.5%

9-35   774.8 rpm

Chapter 10

10-3   747 A

10-5 $i_d = \cos t$; $i_q = -\sin t$; $i_0 = 0$

10-7   0.11; 1.1; 0.9845; 1.1; 1.1996; 0.00043; 1.1; 1.235; 1.1178; 0.0046; 0.9845; 0.0414; 0.002; 1.1436

10-11   Only part a

10-13   **a.** 10.05   **b.** 10.05 pu where 1 pu is the peak value of the rated armature current
c. 6.73; 6.53   **d.** 1.414

10-17 $T = \dfrac{1}{X_d'} \sin t + \left(\dfrac{1}{2X_q} - \dfrac{1}{2X_d'} \right) \sin 2t$

10-21   610 hp

10-25   **a.** $i_{1d} = \sqrt{2} I_1 \cos \alpha$; $i_{1q} = \sqrt{2} I_1 \sin \alpha$

b. $\bar{V}_1 = r_1 \bar{I}_1 + jX_{l1} \bar{I}_1 + jX_m(\bar{I}_1 + \bar{I}_2)$; $0 = \dfrac{r_2}{S} \bar{I}_2 + jX_{\ell 2} \bar{I}_2 + jX_m(\bar{I}_1 + \bar{I}_2)$

c.

d. $T = \dfrac{3}{\omega_s} I_2^2 \dfrac{r_2}{S}$

10-29   **a.** $\begin{bmatrix} \bar{I}_a \\ \bar{I}_b \\ \bar{I}_c \end{bmatrix} = \begin{bmatrix} 1 & 1 & 1 \\ 1 & a^2 & a \\ 1 & a & a^2 \end{bmatrix} \begin{bmatrix} \bar{I}_a^0 \\ \bar{I}_a^+ \\ \bar{I}_a^- \end{bmatrix}$ and $\begin{bmatrix} \bar{I}_a^0 \\ \bar{I}_a^+ \\ \bar{I}_a^- \end{bmatrix} = \frac{1}{3} \begin{bmatrix} 1 & 1 & 1 \\ 1 & a & a^2 \\ 1 & a^2 & a \end{bmatrix} \begin{bmatrix} \bar{I}_a \\ \bar{I}_b \\ \bar{I}_c \end{bmatrix}$

b. By superposition, the positive and negative sequence voltages can be applied separately to the motor, and the resulting torques, powers, and currents superimposed to find the resultants.

10-31 $t = \dfrac{J}{2T_{max}} \left(\dfrac{1 - S_{maxT}^2}{2S_{maxT}} + S_{maxT} \ln \dfrac{1}{S_{maxT}} \right)$

Chapter 11

11-1 $\dfrac{V_t(S)}{V_f(S)} = \dfrac{1}{(1 + S)^2}$

11-3 **a.** 75 h **b.** 0.0075 h **c.** 1.99 h **d.** 3.78

11-5 **a.** 62.8 rad/s **b.** $66.7e^{-35.2t}$ A; $65.3 - 2.5e^{-35.2t}$ rad/s
c. $54.2e^{-10t} \sin 24.6t$ A; $65.31 - 2.7e^{-10t} \cos(24.6t - 22.2°)$ rad/s

11-7 $23.57(1 - e^{-109.1t})$ A; $(271.04 - 13.9e^{-109.1t})$ V

11-9 $\frac{1}{2}J\omega_m^2$

11-15 $R \leftrightarrow R_a$; $L \leftrightarrow L_a$; $C \leftrightarrow J/(kI_f)^2$

11-17 **a.** **(i)** $200(1 - e^{-4t})$ V **(ii)** 200 V **(iii)** 0.575 sec **b.** $i_a(t) = 160 - 351e^{-4t} + 191e^{-7.35t}$

11-19 **a.** $\omega_m(t) = 198.2(1 - e^{-0.8824t})$; $i_a = 23.8 + 416.2e^{-0.8824t}$ **b.** 198.2 rad/s; 23.8 A

11-21 $G(S) = \dfrac{k_G \omega_{mG} k_M I_{fM}}{(R_{fG} + L_{ffG}S)[k_M^2 I_{fM}^2 + (R + LS)(B + JS)]}$

11-23 **a.**

b. 100 rad/s **c.** 30.47 S

11-25

11-27   **a.** 83.3   **b.** $1.54\angle -50.2°$; $-844\angle 20.7°$; $-21.1\angle -69.3°$

c. $i_a(t) = i_{a1}(t) = -844 \sin(3t + 20.7°)$; $\omega_m(t) = 83.3 - 21.1 \sin(3t - 69.3°)$;
$i_f(t) = 6 + 1.54 \sin(3t - 50.2°)$

11-29   **a.** $T_e = \begin{bmatrix} i_f(t) \mid i_a(t) \end{bmatrix} \begin{bmatrix} 0 & 0 \\ k & 0 \end{bmatrix} \begin{bmatrix} i_f(t) \\ i_a(t) \end{bmatrix}$

b. Equations are in the form of $\bar{X}(t) = AX(t) + Bu(t)$, $T_e = X^T(t)GX(t)$

c.

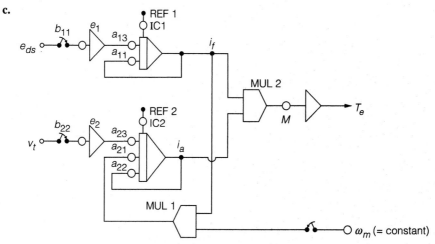

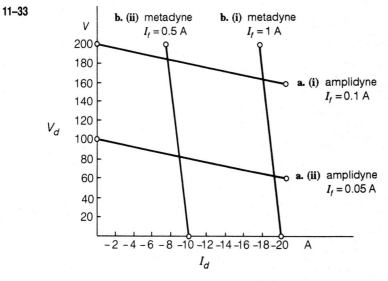

simulation diagram for the separately excited generator

11-31 $e_{ad}(t) = 320.4 + 638.9e^{-5t} - 959.3e^{-3.33t}$

11-33

Chapter 12

12-1   **a.** $i(t) = \dfrac{V_S}{R} e^{-t/RC}$; $v_C(t) = V_S(1 - e^{-t/RC})$ **b.** $i(t) = \dfrac{V_S}{R}(1 - e^{-tR/L})$; $v_L(t) = V_S e^{-tR/L}$

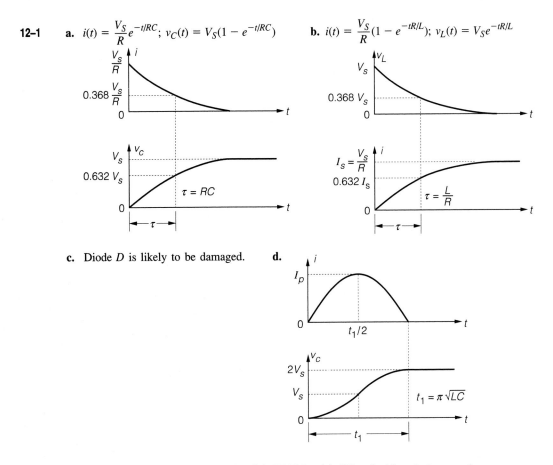

c. Diode D is likely to be damaged.   **d.**

12-3   **a.** (i) 81%   (ii) 1.11   (iii) 48.2%   (iv) 57.32%   (v) $2V_m$; significantly improved

b. $v_L(t) = \dfrac{2V_m}{\pi} - \dfrac{4V_m}{3\pi} \cos 2\omega t - \dfrac{4V_m}{15\pi} \cos 4\omega t - \dfrac{4V_m}{35\pi} \cos 6\omega t - \cdots$

$i_L(t) = I_{dC} - \dfrac{4V_m}{\pi \sqrt{R^2 + (n\omega L)^2}} \left[\dfrac{1}{3} \cos (2\omega t - \theta_2) - \dfrac{1}{15} \cos (4\omega t - \theta_4) - \cdots \right]$

12-5   **a.** (i) 96.77%   (ii) 101.65%   (iii) 18.24%   (iv) 0.6643   (v) $\sqrt{3}V_m$

b. $v_L(t) = 0.827 V_m (1 + \frac{2}{8} \cos 3\omega t - \frac{2}{35} \cos 6\omega t + \cdots)$

12-7   **a.** $\omega_m = \dfrac{V_m(\cos \alpha - \cos \beta)}{K(\beta - \alpha)} - \dfrac{\pi R_a}{K^2(\beta - \alpha)} T_a$

b. $\omega_m = \dfrac{V_m}{(\beta - \gamma)K}(2 \cos \alpha + \cos \gamma + \cos \beta) - \dfrac{\pi R_a}{K^2(\beta - \gamma)} T_a$

c. $\omega_m = \dfrac{V_m(\cos \gamma - \cos \beta)}{K(\beta - \gamma)} - \dfrac{\pi R_a}{K^2(\beta - \gamma)} T_a$

12–9   **a.** $\alpha = 119.9°$; Fourth-quandrant operation   **b.** 1,882 rpm

   c. **(i)** The motor operates in the discontinuous conduction mode. First-quadrant operation 10.8 N·m

   (ii) The motor operates in the fourth quadrant (regenerative braking). This is a continuous conduction mode; $T_a = 59.6$ N·m

   d. $\alpha \simeq 60°$; $\beta \simeq 228°$

12–11   No-load speeds can be negative for $\frac{2\pi}{3} < \alpha < \pi$. Yes, a considerable reduction in the zone of discontinuous conduction will result.

12–13   36 mH

12–15   32 A

12–17   529 rpm

12–19   **a.** 31.5 Hz   **b.** 584.4 rpm; 33.64 A

12–21   1,032 rpm

12–23   **a.** 57.25 A; 0.854 leading   **b.** 7,123 N·m; 129.5 A

12–25   **a.** I_S and the motor torque increase. The machine speeds up.

   b. I_d and the machine torque increase. The motor speed increases.

   c. Increasing torque. The motor speed increases. Terminal voltage increases with speed. Efficiency decreases.

Index

639